2018

# Contractor's Pricing Guide:
# Residential Repair & Remodeling Costs
## with RSMeans data

**Gordian RSMeans data**

Construction Publishers & Consultants
1099 Hingham Street, Suite 201
Rockland, MA 02370
United States of America
1-800-448-8182
www.RSMeans.com

Copyright 2017 by The Gordian Group Inc.
All rights reserved.
Cover photo © iStock.com/quackersnaps

Printed in the United States of America
ISSN 1542-9156
ISBN 978-1-946872-19-7

0348  $59.99  per copy (in United States)
Price is subject to change without prior notice.

# Contractor's Pricing Guide:

## Residential Repair & Remodeling Costs with RSMeans data

**GORDIAN®**

**2018**

**Thomas Lane,** Senior Editor

**Chief Data Officer**
Noam Reininger

**Engineering Director**
Bob Mewis, CCP

**Contributing Editors**
Christopher Babbitt
Sam Babbitt
Michelle Curran
Matthew Doheny (8)
Cheryl Elsmore
Linval Gentles
John Gomes (13, 41)
Derrick Hale, PE (2, 31, 32, 33, 34, 35, 44, 46)
Wafaa Hamitou (11, 12)

Joseph Kelble (14, 21, 22, 23, 25, 26, 27, 28, 48)
Charles Kibbee (1, 4)
Gerard Lafond, PE
Thomas Lane (6, 7)
Genevieve Medeiros
Elisa Mello
Ken Monty
Marilyn Phelan, AIA (9, 10)
Stephen C. Plotner (3, 5)
Callum Riley
Stephen Rosenberg
Jeff Sessions
Gabe Sirota

Matthew Sorrentino
Kevin Souza
Keegan Spraker
Tim Tonello
Jen Walsh
David Yazbek

**Product Manager**
Andrea Sillah

**Production Manager**
Debbie Panarelli

**Production**
Jonathan Forgit
Mary Lou Geary

Sharon Larsen
Sheryl Rose

**Technical Support**
Judy Abbruzzese
Gary L. Hoitt

**Cover Design**
Blaire Collins

**Data Analytics**
Tim Duggan
Todd Glowac
Matthew Kelliher-Gibson

*Numbers in italics are the divisional responsibilities for each editor. Please contact the designated editor directly with any questions.

# Table of Contents

# Foreword

## The Value of RSMeans data from Gordian

Since 1942, RSMeans data has been the industry-standard materials, labor, and equipment cost information database for contractors, facility owners and managers, architects, engineers, and anyone else that requires the latest localized construction cost information. Over 75 years later, the objective remains the same: to provide facility and construction professionals with the most current and comprehensive construction cost database possible.

With the constant influx of new construction methods and materials, in addition to ever-changing labor and material costs, last year's cost data is not reliable for today's designs, estimates, or budgets. The RSMeans data engineers invest over 22,000 hours in cost research annually and apply real-world construction experience to identify and quantify new building products and methodologies, adjust productivity rates, and adjust costs to local market conditions across the nation. This unparalleled construction cost expertise is why so many facility and construction professionals rely on RSMeans data year over year.

## About Gordian

Gordian originated in the spirit of innovation and a strong commitment to helping clients reach and exceed their construction goals. In 1982, Gordian's Chairman and Founder, Harry H. Mellon, created Job Order Contracting while serving as Chief Engineer at the Supreme Headquarters Allied Powers Europe. Job Order Contracting is a unique indefinite delivery/indefinite quantity (IDIQ) process, which enables facility owners to complete a substantial number of repair, maintenance, and construction projects with a single, competitively awarded contract. Realizing facility and infrastructure owners across various industries could greatly benefit from the time and cost saving advantages of this innovative construction procurement solution, he established Gordian in 1990.

Continuing the commitment to providing the most relevant and accurate facility and construction data, software, and expertise in the industry, Gordian enhanced the fortitude of its data with the acquisition of RSMeans in 2014. And in an effort to expand its facility management capabilities, Gordian acquired Sightlines, the leading provider of facilities benchmarking data and analysis, in 2015.

## Our Offerings

Gordian is the leader in facility and construction cost data, software, and expertise for all phases of the building life cycle. From planning to design, procurement, construction, and operations, Gordian's solutions help clients maximize efficiency, optimize cost savings, and increase building quality with its highly specialized data engineers, software, and unique proprietary data sets.

## Our Commitment

At Gordian, we do more than talk about the quality of our data and the usefulness of its application. We stand behind all of our RSMeans data—from historical cost indexes to construction materials and techniques—to craft current costs and predict future trends. If you have any questions about our products or services, please call us toll-free at 800-448-8182 or visit our website at www.gordian.com.

# How the Cost data Is Built: An Overview

## A Powerful Construction Tool

You have in your hands one of the most powerful construction tools available today. A successful project is built on the foundation of an accurate and dependable estimate. This data will enable you to construct such an estimate.

For the casual user the data is designed to be:

- quickly and easily understood so you can get right to your estimate.
- filled with valuable information so you can understand the necessary factors that go into the cost estimate.

For the regular user, the data is designed to be:

- a handy reference that can be quickly referred to for key costs.
- a comprehensive, fully reliable source of current construction costs, so you'll be prepared to estimate any project.
- a source for project costs, product selections, and alternate materials and methods.

To meet all of these requirements we have organized the data into the following clearly defined sections.

## Unit Price Section

The cost data has been divided into 23 sections representing the order of construction. Within each section is a listing of components and variations of those components that are applicable to that section. Costs are shown for the various actions that must be done to that component.

## Reference Section

This section includes information on Location Factors and a listing of abbreviations.

### Location Factors

You can adjust total project costs to over 900 locations throughout the U.S. and Canada by using the data in this section.

### Abbreviations

A listing of the abbreviations used throughout, along with the terms they represent, is included.

## Index

A comprehensive listing of all terms and subjects will help you quickly find what you need when you are not sure where it falls in the order of construction.

## The Scope of This data

This data is designed to be as comprehensive and easy to use as possible. To that end we have made certain assumptions and limited its scope in two key ways:

1. We have established material prices based on a national average.
2. We have computed labor costs based on residential wage rates.

## Project Size

This data is intended for use by those involved primarily in residential repair and remodeling construction costing between $10,000 and $100,000.

With reasonable exercise of judgment, the figures can be used for any building work. For other types of projects, such as new home construction or commercial buildings, consult the appropriate RSMeans data set for more information.

# How to Use the Cost data: The Details

## What's Behind the Numbers? The Development of Cost data

RSMeans data engineers continually monitor developments in the construction industry in order to ensure reliable, thorough, and up-to-date cost information. While overall construction costs may vary relative to general economic conditions, price fluctuations within the industry are dependent upon many factors. Individual price variations may, in fact, be opposite to overall economic trends. Therefore, costs are constantly tracked, and complete updates are performed yearly. Also, new items are frequently added in response to changes in materials and methods.

## Costs in U.S. Dollars

All costs represent U.S. national averages and are given in U.S. dollars. The City Cost Index (CCI) with RSMeans data or Location Factors can be used to adjust costs to a particular location. The CCI or Location Factors for Canada can be used to adjust U.S. national averages to local costs in Canadian dollars. No exchange rate conversion is necessary because it has already been factored in.

## Material Costs

RSMeans data engineers contact manufacturers, dealers, distributors, and contractors all across the U.S. and Canada to determine national average material costs. If you have access to current material costs for your specific location, you may wish to make adjustments to reflect differences from the national average. Included within material costs are fasteners for a normal installation. RSMeans data engineers use manufacturers' recommendations, written specifications, and/or standard construction practices for the sizing and spacing of fasteners. Adjustments to material costs may be required for your specific application or location. The manufacturer's warranty is assumed. Extended warranties are not included in the material costs. **Material costs do not include sales tax.**

## Labor Costs

Labor costs are based upon a mathematical average of trade-specific wages in 30 major U.S. cities. The type of wage (union, open shop, or residential) is identified on the inside back cover of printed publications or is selected by the estimator when using the electronic products. Markups for the wages can also be found on the inside back cover of printed publications and/or under the labor references found in the electronic products.

- If wage rates in your area vary from those used, or if rate increases are expected within a given year, labor costs should be adjusted accordingly.

Labor costs reflect productivity based on actual working conditions. In addition to actual installation, these figures include time spent during a normal weekday on tasks, such as material receiving and handling, mobilization at site, site movement, breaks, and cleanup.

Productivity data is developed over an extended period so as not to be influenced by abnormal variations and reflects a typical average.

## Equipment Costs

Equipment costs include not only rental but also operating costs for equipment under normal use. The operating costs include parts and labor for routine servicing, such as the repair and replacement of pumps, filters, and worn lines. Normal operating expendables, such as fuel, lubricants, tires, and electricity (where applicable), are also included. Extraordinary operating expendables with highly variable wear patterns, such as diamond bits and blades, are excluded. These costs are included under materials. Equipment rental rates are obtained from industry sources throughout North America—contractors, suppliers, dealers, manufacturers, and distributors.

Rental rates can also be treated as reimbursement costs for contractor-owned equipment. Owned equipment costs include depreciation, loan payments, interest, taxes, insurance, storage, and major repairs.

Equipment costs do not include operators' wages.

## Factors Affecting Costs

Costs can vary depending upon a number of variables. Here's a listing of some factors that affect costs and points to consider.

**Quality**—The prices for materials and the workmanship upon which productivity is based represent sound construction work. They are also in line with industry standards and manufacturer specifications and are frequently used by federal, state, and local governments.

**Overtime**—We have made no allowance for overtime. If you anticipate premium time or work beyond normal working hours, be sure to make an appropriate adjustment to your labor costs.

**Productivity**—The productivity, daily output, and labor-hour figures for each line item are based on an eight-hour work day in daylight hours in moderate temperatures, and up to a 14' working height unless otherwise indicated. For work that extends beyond normal work hours or is performed under adverse conditions, productivity may decrease.

**Size of Project**—The size, scope of work, and type of construction project will have a significant impact on cost. Economies of scale can reduce costs for large projects. Unit costs can often run higher for small projects.

**Location**—Material prices are for metropolitan areas. However, in dense urban areas, traffic and site storage limitations may increase costs. Beyond a 20-mile radius of metropolitan areas, extra trucking or transportation charges may also increase the material costs slightly. On the other hand, lower wage rates may be in effect. Be sure to consider both of these factors when preparing an estimate, particularly if the job site is located in a central city or remote rural location. In addition, highly specialized subcontract items may require travel and per-diem expenses for mechanics.

**Other Factors—**

- season of year
- contractor management
- weather conditions
- local union restrictions
- building code requirements
- availability of:
  - adequate energy
  - skilled labor
  - building materials
- owner's special requirements/restrictions
- safety requirements
- environmental considerations
- access

**Unpredictable Factors**—General business conditions influence "in-place" costs of all items. Substitute materials and construction methods may have to be employed. These may affect the installed cost and/or life cycle costs. Such factors may be difficult to evaluate and cannot necessarily be predicted on the basis of the job's location in a particular section of the country. Thus, where these factors apply, you may find significant but unavoidable cost variations for which you will have to apply a measure of judgment to your estimate.

## Final Checklist

Estimating can be a straightforward process provided you remember the basics. Here's a checklist of some of the steps you should remember to complete before finalizing your estimate.

Did you remember to:

- factor in the City Cost Index or Location Factors for your locale?
- take into consideration which items have been marked up and by how much?
- mark up the entire estimate sufficiently for your purposes?

- read the background information on techniques and technical matters that could impact your project time span and cost?
- include all components of your project in the final estimate?
- double check your figures for accuracy?
- call RSMeans data engineers if you have any questions about your estimate or the data you've used? Remember, Gordian stands behind all of our products, including our extensive RSMeans data solutions. If you have any questions about your estimate, about the costs you've used from our data, or even about the technical aspects of the job that may affect your estimate, feel free to call the Gordian RSMeans editors at 1-800-448-8182.

**Access Quarterly Data Updates**

rsmeans.com/2018books

# Unit Price Section

## Table of Contents

# Table of Contents (cont.)

# RSMeans data: Unit Prices—How They Work

The prices in this publication include all overhead and profit mark-ups for the installing contractor. If a general contractor is involved, an additional 10% should be added to the cost shown in the "Total" column.

The costs for material, labor, and any equipment are shown here. Material costs include delivery and a 10% mark-up. No sales taxes are included. Labor costs include all mark-ups as shown on the inside back cover of this data set. Equipment costs also include a 10% mark-up.

The cost in the Total column should be multiplied by the quantity of each item to determine the total cost for that item. The sum of all items' total costs becomes the project's total cost.

The data set is arranged into 23 categories that follow the order of construction for residential projects. The category is shown as white letters on a black background at the top of each page.

A brief specification of the cost item is shown here.

Graphics representing some of the cost items are found at the beginning of each category.

The cost data is arranged in a tree structure format. Major cost sections are left justified and are shown as bold headings within a grey background. Cost items and sub-cost items are found under the bold headings. The complete description should read as "bold heading, cost item, sub-cost item."

For example, the identified item's description is "Beam/Girder, Solid Wood Beam, 2" x 8"."

Each action performed to the cost item is conveniently listed to the right of the description. See the explanation below for the scope of work included as part of the action.

The unit of measure upon which all the costs are based is shown here. An abbreviations list is included within the Reference Section found at the back of this data set.

## Rough Frame / Structure

**Rough Framing**

| Beam / Girder | | Unit | Material | Labor | Equip. | Total | Specification |
|---|---|---|---|---|---|---|---|
| **Solid Wood Beam** | | | | | | | |
| 2" x 8" | | | | | | | |
| | Demolish | L.F. | | .96 | | .96 | Cost includes material and labor to |
| | Install | L.F. | .98 | 1.43 | | 2.41 | install 2" x 8" beam. |
| | Demolish and Install | L.F. | .98 | 2.39 | | 3.37 | |
| | Clean | L.F. | .04 | .32 | | .36 | |
| | Paint | L.F. | .14 | 2.09 | | 2.23 | |
| | Minimum Charge | Job | | 232 | | 232 | |
| 2" x 10" | | | | | | | |
| | Demolish | L.F. | | 1.19 | | 1.19 | Cost includes material and labor to |
| | Install | L.F. | 1.55 | 1.55 | | 3.10 | install 2" x 10" beam. |
| | Demolish and Install | L.F. | 1.55 | 2.74 | | 4.29 | |
| | Clean | L.F. | .06 | .38 | | .44 | |
| | Paint | L.F. | .19 | 1.29 | | 1.48 | |
| | Minimum Charge | Job | | 232 | | 232 | |
| 2" x 12" | | | | | | | |
| | Demolish | L.F. | | 1.44 | | 1.44 | Cost includes material and labor to |
| | Install | L.F. | 1.97 | 1.69 | | 3.66 | install 2" x 12" beam. |
| | Demolish and Install | L.F. | 1.97 | 3.13 | | 5.10 | |
| | Clean | L.F. | .07 | .44 | | .51 | |
| | Paint | L.F. | .22 | 3.09 | | 3.31 | |
| | Minimum Charge | Job | | 232 | | 232 | |
| 4" x 8" | | | | | | | |
| | Demolish | L.F. | | 2.94 | | 2.94 | Cost includes material and labor to |
| | Install | L.F. | 4.57 | 2.13 | .55 | 7.25 | install 4" x 8" beam. |
| | Demolish and Install | L.F. | 4.57 | 5.07 | .55 | 10.19 | |
| | Clean | L.F. | .06 | .38 | | .44 | |
| | Paint | L.F. | .19 | 1.29 | | 1.48 | |
| | Minimum Charge | Job | | 232 | | 232 | |
| 4" x 10" | | | | | | | |
| | Demolish | L.F. | | 3.68 | | 3.68 | Cost includes material and labor to |
| | Install | L.F. | 5.45 | 2.25 | .57 | 8.27 | install 4" x 10" beam. |
| | Demolish and Install | L.F. | 5.45 | 5.93 | .57 | 11.95 | |
| | Clean | L.F. | .07 | .44 | | .51 | |
| | Paint | L.F. | .22 | 1.50 | | 1.72 | |
| | Minimum Charge | Job | | 232 | | 232 | |
| 4" x 12" | | | | | | | |
| | Demolish | L.F. | | 4.41 | | 4.41 | Cost includes material and labor to |
| | Install | L.F. | 6.05 | 2.37 | .61 | 9.03 | install 4" x 12" beam. |
| | Demolish and Install | L.F. | 6.05 | 6.78 | .61 | 13.44 | |
| | Clean | L.F. | .08 | .51 | | .59 | |
| | Paint | L.F. | .22 | 3.09 | | 3.31 | |
| | Minimum Charge | Job | | 232 | | 232 | |

| Action | Scope |
|---|---|
| Demolish | Includes removal of the item and hauling the debris to a truck, dumpster, or storage area. |
| Install | Includes the installation of the item with normal placement and spacing of fasteners. |
| Demolish and Install | Includes the demolish action and the install action. |
| Reinstall | Includes the reinstallation of a removed item. |
| Clean | Includes cleaning of the item. |
| Paint | Includes up to two coats of finish on the item (unless otherwise indicated). |
| Minimum Charge | Includes the minimum labor or the minimum labor and equipment cost. Use this charge for small quantities of material. |

| Scaffolding | Temporary Electrical Service | Air Compressor |
|---|---|---|

| Construction Fees | | Unit | Material | Labor | Equip. | Total | Specification |
|---|---|---|---|---|---|---|---|
| Asbestos Test | Minimum Charge | Job | | 271 | | 271 | Minimum charge for testing. |
| Architectural | Minimum Charge | Day | 2075 | | | 2075 | Minimum labor charge for architect. |
| Engineering | Minimum Charge | Day | 1900 | | | 1900 | Minimum labor charge for field engineer. |

| Construction Permits | | Unit | Material | Labor | Equip. | Total | Specification |
|---|---|---|---|---|---|---|---|
| Job (less than 5K) | Minimum Charge | Job | 59 | | | 59 | Permit fees, job less than 5K. |
| Job (5K to 10K) | Minimum Charge | Job | 118 | | | 118 | Permit fees, job more than 5K, but less than 10K. |
| Job (10K to 25K) | Minimum Charge | Job | 239 | | | 239 | Permit fees, job more than 10K, but less than 25K. |
| Job (25K to 50K) | Minimum Charge | Job | 455 | | | 455 | Permit fees, job more than 25K, but less than 50K. |
| Job (50K to 100K) | Minimum Charge | Job | 835 | | | 835 | Permit fees, job more than 50K, but less than 100K. |
| Job (100K to 250K) | Minimum Charge | Job | 1700 | | | 1700 | Permit fees, job more than 100K, but less than 250K. |
| Job (250K to 500K) | Minimum Charge | Job | 3275 | | | 3275 | Permit fees, job more than 250K, but less than 500K. |
| Job (500K to 1M) | Minimum Charge | Job | 5850 | | | 5850 | Permit fees, job more than 500K, but less than 1 million. |

1

| Trade Labor | | Unit | Material | Labor | Equip. | Total | Specification |
|---|---|---|---|---|---|---|---|
| **Carpenter** | | | | | | | |
| Daily (8 hours) | | | | | | | |
| | Install | Day | | 465 | | 465 | Daily labor rate for one carpenter. |
| | Minimum Charge | Job | | 232 | | 232 | |
| Weekly (40 hours) | | | | | | | |
| | Install | Week | | 2325 | | 2325 | Weekly labor rate for one carpenter. |
| | Minimum Charge | Job | | 232 | | 232 | |
| **Drywaller** | | | | | | | |
| Daily (8 hours) | | | | | | | |
| | Install | Day | | 465 | | 465 | Daily labor rate for one carpenter. |
| | Minimum Charge | Job | | 232 | | 232 | |
| Weekly (40 hours) | | | | | | | |
| | Install | Week | | 2325 | | 2325 | Weekly labor rate for one carpenter. |
| | Minimum Charge | Job | | 232 | | 232 | |
| **Roofer** | | | | | | | |
| Daily (8 hours) | | | | | | | |
| | Install | Day | | 435 | | 435 | Daily labor rate for one skilled roofer. |
| | Minimum Charge | Job | | 218 | | 218 | |
| Weekly (40 hours) | | | | | | | |
| | Install | Week | | 2175 | | 2175 | Weekly labor rate for one skilled |
| | Minimum Charge | Job | | 218 | | 218 | roofer. |
| **Painter** | | | | | | | |
| Daily (8 hours) | | | | | | | |
| | Install | Day | | 385 | | 385 | Daily labor rate for one painter. |
| | Minimum Charge | Job | | 193 | | 193 | |
| Weekly (40 hours) | | | | | | | |
| | Install | Week | | 1925 | | 1925 | Weekly labor rate for one painter. |
| | Minimum Charge | Job | | 193 | | 193 | |
| **Mason** | | | | | | | |
| Daily (8 hours) | | | | | | | |
| | Install | Day | | 465 | | 465 | Daily labor rate for one skilled mason. |
| | Minimum Charge | Job | | 232 | | 232 | |
| Weekly (40 hours) | | | | | | | |
| | Install | Week | | 2325 | | 2325 | Weekly labor rate for one skilled |
| | Minimum Charge | Job | | 232 | | 232 | mason. |
| **Electrician** | | | | | | | |
| Daily (8 hours) | | | | | | | |
| | Install | Day | | 525 | | 525 | Daily labor rate for an electrician. |
| | Minimum Charge | Job | | 263 | | 263 | |
| Weekly (40 hours) | | | | | | | |
| | Install | Week | | 2625 | | 2625 | Weekly labor rate for an electrician. |
| | Minimum Charge | Job | | 263 | | 263 | |
| **Plumber** | | | | | | | |
| Daily (8 hours) | | | | | | | |
| | Install | Day | | 525 | | 525 | Daily labor rate for a plumber. |
| | Minimum Charge | Job | | 262 | | 262 | |
| Weekly (40 hours) | | | | | | | |
| | Install | Week | | 2625 | | 2625 | Weekly labor rate for a plumber. |
| | Minimum Charge | Job | | 262 | | 262 | |
| **Common Laborer** | | | | | | | |
| **Daily (8 hours)** | | | | | | | |
| | Install | Day | | 355 | | 355 | Daily labor rate for a common |
| | Minimum Charge | Job | | 178 | | 178 | laborer. |

# Job Costs

## Trade Labor

| Trade Labor | | Unit | Material | Labor | Equip. | Total | Specification |
|---|---|---|---|---|---|---|---|
| Weekly (40 hours) | | Week | | 1775 | | 1775 | Weekly labor rate for a common |
| | Install | Week | | 1775 | | 1775 | laborer. |
| | Minimum Charge | Job | | 178 | | 178 | |

## Temporary Utilities

| Temporary Utilities | | Unit | Material | Labor | Equip. | Total | Specification |
|---|---|---|---|---|---|---|---|
| **Power Pole** | | | | | | | |
| Job Site Electricity | | | | | | | |
| | Install | Month | 59.50 | | | 59.50 | Includes average cost of electricity per |
| | Minimum Charge | Job | | 263 | | 263 | month. |
| **Water Service** | | | | | | | |
| Job Site Water | | | | | | | |
| | Install | Month | 76 | | | 76 | Includes average costs for water for |
| | Minimum Charge | Month | 76 | | | 76 | one month. |
| **Telephone** | | | | | | | |
| | Install | Month | 94.50 | | | 94.50 | Includes average cost for local and |
| | Minimum Charge | Month | 94.50 | | | 94.50 | long distance telephone service for |
| | | | | | | | one month. |
| **Chemical Toilet** | | | | | | | |
| | Install | Month | | | 315 | 315 | Monthly rental of portable toilet. |
| | Minimum Charge | Day | | | 15.70 | 15.70 | |
| **Office Trailer** | | | | | | | |
| Place and Remove | | | | | | | |
| | Install | Job | 395 | | | 395 | Includes cost to a contractor to deliver, |
| | Minimum Charge | Job | | 212 | | 212 | place and level an office trailer and |
| | | | | | | | remove at end of job. |
| **Storage Container** | | | | | | | |
| Place and Remove | | | | | | | |
| | Install | Job | 222 | | | 222 | Includes cost to a contractor to deliver |
| | Minimum Charge | Job | | 212 | | 212 | and place a storage container and |
| | | | | | | | remove at end of job. |
| **Fencing** | | | | | | | |
| | Install | L.F. | 3.51 | | | 3.51 | Includes material costs for temporary |
| | Minimum Charge | Job | | 178 | | 178 | fencing. |
| **Railing** | | | | | | | |
| | Install | L.F. | 19.70 | 12.70 | .61 | 33.01 | Includes material and labor to install |
| | Minimum Charge | Job | | 178 | | 178 | primed steel pipe. |

## Equipment Rental

| Equipment Rental | | Unit | Material | Labor | Equip. | Total | Specification |
|---|---|---|---|---|---|---|---|
| **Flatbed Truck** | | | | | | | |
| Daily Rental | | | | | | | |
| | Install | Day | | | 207 | 207 | Daily rental of flatbed truck. |
| | Minimum Charge | Job | | | 207 | 207 | |
| Weekly Rental | | | | | | | |
| | Install | Week | | | 1025 | 1025 | Weekly rental of flatbed truck. |
| | Minimum Charge | Job | | | 207 | 207 | |
| Monthly Rental | | | | | | | |
| | Install | Month | | | 4150 | 4150 | Monthly rental of flatbed truck. |
| | Minimum Charge | Job | | | 207 | 207 | |

For customer support on your Contractor's Pricing Guide: Residential Repair & Remodeling, call 800.448.8182.

# Job Costs

| Equipment Rental | | Unit | Material | Labor | Equip. | Total | Specification |
|---|---|---|---|---|---|---|---|
| **Dump Truck** | | | | | | | |
| Daily Rental | | | | | | | |
| | Install | Day | | | 370 | 370 | Daily rental of dump truck. |
| | Minimum Charge | Day | | | 370 | 370 | |
| Weekly Rental | | | | | | | |
| | Install | Week | | | 1850 | 1850 | Weekly rental of dump truck. |
| | Minimum Charge | Day | | | 370 | 370 | |
| Monthly Rental | | | | | | | |
| | Install | Month | | | 7450 | 7450 | Monthly rental of dump truck. |
| | Minimum Charge | Day | | | 370 | 370 | |
| **Forklift** | | | | | | | |
| Daily Rental | | | | | | | |
| | Install | Day | | | 305 | 305 | Daily rental of forklift. |
| | Minimum Charge | Day | | | 305 | 305 | |
| Weekly Rental | | | | | | | |
| | Install | Week | | | 1525 | 1525 | Weekly rental of forklift. |
| | Minimum Charge | Day | | | 305 | 305 | |
| Monthly Rental | | | | | | | |
| | Install | Month | | | 6150 | 6150 | Monthly rental of forklift. |
| | Minimum Charge | Day | | | 305 | 305 | |
| **Bobcat** | | | | | | | |
| Daily Rental | | | | | | | |
| | Install | Day | | | 191 | 191 | Daily rental of wheeled, skid steer |
| | Minimum Charge | Day | | | 191 | 191 | loader. |
| Weekly Rental | | | | | | | |
| | Install | Week | | | 955 | 955 | Weekly rental of wheeled, skid steer |
| | Minimum Charge | Day | | | 191 | 191 | loader. |
| Monthly Rental | | | | | | | |
| | Install | Month | | | 3825 | 3825 | Monthly rental of wheeled, skid steer |
| | Minimum Charge | Day | | | 191 | 191 | loader. |
| **Air Compressor** | | | | | | | |
| Daily Rental | | | | | | | |
| | Install | Day | | | 116 | 116 | Daily rental of air compressor. |
| | Minimum Charge | Day | | | 116 | 116 | |
| Weekly Rental | | | | | | | |
| | Install | Week | | | 580 | 580 | Weekly rental of air compressor. |
| | Minimum Charge | Day | | | 116 | 116 | |
| Monthly Rental | | | | | | | |
| | Install | Month | | | 2325 | 2325 | Monthly rental of air compressor. |
| | Minimum Charge | Day | | | 116 | 116 | |
| **Jackhammer** | | | | | | | |
| Daily Rental | | | | | | | |
| | Install | Day | | | 11.70 | 11.70 | Daily rental of air tool, jackhammer. |
| | Minimum Charge | Day | | | 11.70 | 11.70 | |
| Weekly Rental | | | | | | | |
| | Install | Week | | | 58.50 | 58.50 | Weekly rental of air tool, jackhammer. |
| | Minimum Charge | Day | | | 11.70 | 11.70 | |
| Monthly Rental | | | | | | | |
| | Install | Month | | | 234 | 234 | Monthly rental of air tool, |
| | Minimum Charge | Day | | | 11.70 | 11.70 | jackhammer. |
| **Concrete Mixer** | | | | | | | |
| Daily Rental | | | | | | | |
| | Install | Day | | | 147 | 147 | Daily rental of concrete mixer. |
| | Minimum Charge | Day | | | 147 | 147 | |

| Equipment Rental | | Unit | Material | Labor | Equip. | Total | Specification |
|---|---|---|---|---|---|---|---|
| **Weekly Rental** | | | | | | | |
| | Install | Week | | | 735 | 735 | Weekly rental of concrete mixer. |
| | Minimum Charge | Day | | | 147 | 147 | |
| **Monthly Rental** | | | | | | | |
| | Install | Month | | | 2925 | 2925 | Monthly rental of concrete mixer. |
| | Minimum Charge | Day | | | 147 | 147 | |
| **Concrete Bucket** | | | | | | | |
| **Daily Rental** | | | | | | | |
| | Install | Day | | | 23.50 | 23.50 | Daily rental of concrete bucket. |
| | Minimum Charge | Day | | | 23.50 | 23.50 | |
| **Weekly Rental** | | | | | | | |
| | Install | Week | | | 118 | 118 | Weekly rental of concrete bucket. |
| | Minimum Charge | Day | | | 23.50 | 23.50 | |
| **Monthly Rental** | | | | | | | |
| | Install | Month | | | 470 | 470 | Monthly rental of concrete bucket. |
| | Minimum Charge | Day | | | 23.50 | 23.50 | |
| **Concrete Pump** | | | | | | | |
| **Daily Rental** | | | | | | | |
| | Install | Day | | | 970 | 970 | Daily rental of concrete pump. |
| | Minimum Charge | Day | | | 970 | 970 | |
| **Weekly Rental** | | | | | | | |
| | Install | Week | | | 4850 | 4850 | Weekly rental of concrete pump. |
| | Minimum Charge | Day | | | 970 | 970 | |
| **Monthly Rental** | | | | | | | |
| | Install | Month | | | 19400 | 19400 | Monthly rental of concrete pump. |
| | Minimum Charge | Day | | | 970 | 970 | |
| **Generator** | | | | | | | |
| **Daily Rental** | | | | | | | |
| | Install | Day | | | 31.50 | 31.50 | Daily rental of generator. |
| | Minimum Charge | Day | | | 31.50 | 31.50 | |
| **Weekly Rental** | | | | | | | |
| | Install | Week | | | 156 | 156 | Weekly rental of generator. |
| | Minimum Charge | Day | | | 31.50 | 31.50 | |
| **Monthly Rental** | | | | | | | |
| | Install | Month | | | 625 | 625 | Monthly rental of generator. |
| | Minimum Charge | Day | | | 31.50 | 31.50 | |
| **Welder** | | | | | | | |
| **Daily Rental** | | | | | | | |
| | Install | Day | | | 46 | 46 | Daily rental of welder. |
| | Minimum Charge | Day | | | 46 | 46 | |
| **Weekly Rental** | | | | | | | |
| | Install | Week | | | 229 | 229 | Weekly rental of welder. |
| | Minimum Charge | Day | | | 46 | 46 | |
| **Monthly Rental** | | | | | | | |
| | Install | Month | | | 915 | 915 | Monthly rental of welder. |
| | Minimum Charge | Day | | | 46 | 46 | |
| **Sandblaster** | | | | | | | |
| **Daily Rental** | | | | | | | |
| | Install | Day | | | 23 | 23 | Daily rental of sandblaster. |
| | Minimum Charge | Day | | | 23 | 23 | |
| **Weekly Rental** | | | | | | | |
| | Install | Week | | | 116 | 116 | Weekly rental of sandblaster. |
| | Minimum Charge | Day | | | 23 | 23 | |

## Equipment Rental

| | | Unit | Material | Labor | Equip. | Total | Specification |
|---|---|---|---|---|---|---|---|
| **Monthly Rental** | | | | | | | |
| | Install | Month | | | 460 | 460 | Monthly rental of sandblaster. |
| | Minimum Charge | Day | | | 23 | 23 | |
| **Space Heater** | | | | | | | |
| *Daily Rental* | | | | | | | |
| | Install | Day | | | 18.25 | 18.25 | Daily rental of space heater. |
| | Minimum Charge | Day | | | 23.50 | 23.50 | |
| *Weekly Rental* | | | | | | | |
| | Install | Day | | | 91 | 91 | Weekly rental of space heater. |
| | Minimum Charge | Day | | | 23.50 | 23.50 | |
| *Monthly Rental* | | | | | | | |
| | Install | Day | | | 365 | 365 | Monthly rental of space heater. |
| | Minimum Charge | Day | | | 23.50 | 23.50 | |

## Move / Reset Contents

| | | Unit | Material | Labor | Equip. | Total | Specification |
|---|---|---|---|---|---|---|---|
| **Small Room** | | | | | | | |
| | Install | Room | | 43 | | 43 | Includes cost to remove and reset the |
| | Minimum Charge | Job | | 178 | | 178 | contents of a small room. |
| **Average Room** | | | | | | | |
| | Install | Room | | 52 | | 52 | Includes cost to remove and reset the |
| | Minimum Charge | Job | | 178 | | 178 | contents of an average room. |
| **Large Room** | | | | | | | |
| | Install | Room | | 72 | | 72 | Includes cost to remove and reset the |
| | Minimum Charge | Job | | 178 | | 178 | contents of a large room. |
| **Extra Large Room** | | | | | | | |
| | Install | Room | | 114 | | 114 | Includes cost to remove and reset the |
| | Minimum Charge | Job | | 178 | | 178 | contents of an extra large room. |
| **Store Contents** | | | | | | | |
| *Weekly* | | | | | | | |
| | Install | Week | 298 | | | 298 | Storage fee, per week. |
| *Monthly* | | | | | | | |
| | Install | Month | 895 | | | 895 | Storage fee, per month. |

## Cover and Protect

| | | Unit | Material | Labor | Equip. | Total | Specification |
|---|---|---|---|---|---|---|---|
| **Cover / Protect Walls** | | | | | | | |
| | Install | SF Wall | | .14 | | .14 | Includes labor costs to cover and |
| | Minimum Charge | Job | | 178 | | 178 | protect walls. |
| **Cover / Protect Floors** | | | | | | | |
| | Install | SF Flr. | | .14 | | .14 | Includes labor costs to cover and |
| | Minimum Charge | Job | | 178 | | 178 | protect floors. |

## Temporary Bracing

| | | Unit | Material | Labor | Equip. | Total | Specification |
|---|---|---|---|---|---|---|---|
| **Shim Support Piers** | | | | | | | |
| | Install | Ea. | 4.53 | 137 | | 141.53 | Includes labor and material to install |
| | Minimum Charge | Job | | 410 | | 410 | shims support for piers. |

6

# Job Costs

## Jack and Relevel

| Jack and Relevel | | Unit | Material | Labor | Equip. | Total | Specification |
|---|---|---|---|---|---|---|---|
| Single Story | | | | | | | |
| | Install | S.F. | 1.84 | 1.64 | | 3.48 | Includes labor and equipment to jack |
| | Minimum Charge | Job | | 410 | | 410 | and relevel single-story building. |
| Two-story | | | | | | | |
| | Install | S.F. | 1.84 | 2.74 | | 4.58 | Includes labor and equipment to jack |
| | Minimum Charge | Job | | 410 | | 410 | and relevel two-story building. |

## Dumpster Rental

| Dumpster Rental | | Unit | Material | Labor | Equip. | Total | Specification |
|---|---|---|---|---|---|---|---|
| Small Load | | | | | | | |
| | Install | Week | | | 615 | 615 | Dumpster, weekly rental, 2 dumps, 5 |
| | Minimum Charge | Week | | | 615 | 615 | yard container. |
| Medium Load | | | | | | | |
| | Install | Week | | | 715 | 715 | Dumpster, weekly rental, 2 dumps, 10 |
| | Minimum Charge | Week | | | 715 | 715 | yard container. |
| Large Load | | | | | | | |
| | Install | Week | | | 1075 | 1075 | Dumpster, weekly rental, 2 dumps, 30 |
| | Minimum Charge | Week | | | 1075 | 1075 | yard container. |

## Debris Hauling

| Debris Hauling | | Unit | Material | Labor | Equip. | Total | Specification |
|---|---|---|---|---|---|---|---|
| Per Ton | | | | | | | |
| | Install | Ton | | 112 | 44 | 156 | Includes hauling of debris by trailer or |
| | Minimum Charge | Ton | | 840 | 330 | 1170 | dump truck. |
| Per Cubic Yard | | | | | | | |
| | Install | C.Y. | | 37.50 | 14.75 | 52.25 | Includes hauling of debris by trailer or |
| | Minimum Charge | Job | | 840 | 330 | 1170 | dump truck. |
| Per Pick-up Truck Load | | | | | | | |
| | Install | Ea. | | 197 | | 197 | Includes hauling of debris by pick-up |
| | Minimum Charge | Ea. | | 197 | | 197 | truck. |

## Dump Fees

| Dump Fees | | Unit | Material | Labor | Equip. | Total | Specification |
|---|---|---|---|---|---|---|---|
| Per Cubic Yard | | | | | | | |
| | Install | C.Y. | 49.50 | | | 49.50 | Dump charges, building materials. |
| | Minimum Charge | C.Y. | 49.50 | | | 49.50 | |
| Per Ton | | | | | | | |
| | Install | Ton | 141 | | | 141 | Dump charges, building materials. |
| | Minimum Charge | Ton | 141 | | | 141 | |

## Scaffolding

| Scaffolding | | Unit | Material | Labor | Equip. | Total | Specification |
|---|---|---|---|---|---|---|---|
| 4' to 6' High | | | | | | | |
| | Install | Month | 66 | | | 66 | Includes monthly rental of scaffolding, |
| | Minimum Charge | Month | 66 | | | 66 | steel tubular, 30" wide, 7' long, 5' high. |
| 7' to 11' High | | | | | | | |
| | Install | Month | 77.50 | | | 77.50 | Includes monthly rental of scaffolding, |
| | Minimum Charge | Month | 77.50 | | | 77.50 | steel tubular, 30" wide, 7' long, 10' high. |

**For customer support on your Contractor's Pricing Guide: Residential Repair & Remodeling, call 800.448.8182.**

## Scaffolding

| Scaffolding | | Unit | Material | Labor | Equip. | Total | Specification |
|---|---|---|---|---|---|---|---|
| **12' to 16' High** | | | | | | | |
| | Install | Month | 89 | | | 89 | Includes monthly rental of scaffolding, steel tubular, 30" wide, 7' long, 15' high. |
| | Minimum Charge | Month | 89 | | | 89 | |
| **17' to 21' High** | | | | | | | |
| | Install | Month | 114 | | | 114 | Includes monthly rental of scaffolding, steel tubular, 30" wide, 7' long, 20' high. |
| | Minimum Charge | Month | 114 | | | 114 | |
| **22' to 26' High** | | | | | | | |
| | Install | Month | 127 | | | 127 | Includes monthly rental of scaffolding, steel tubular, 30" wide, 7' long, 25' high. |
| | Minimum Charge | Month | 127 | | | 127 | |
| **27' to 30' High** | | | | | | | |
| | Install | Month | 138 | | | 138 | Includes monthly rental of scaffolding, steel tubular, 30" wide, 7' long, 30' high. |
| | Minimum Charge | Month | 138 | | | 138 | |

| Construction Clean-up | | Unit | Material | Labor | Equip. | Total | Specification |
|---|---|---|---|---|---|---|---|
| **Initial** | | | | | | | |
| | Clean | S.F. | | .44 | | .44 | Includes labor for general clean-up. |
| | Minimum Charge | Job | | 178 | | 178 | |
| **Progressive** | | | | | | | |
| | Clean | S.F. | | .22 | | .22 | Includes labor for progressive site/job clean-up on a weekly basis. |
| | Minimum Charge | Job | | 178 | | 178 | |

| Building Demolition | | Unit | Material | Labor | Equip. | Total | Specification |
|---|---|---|---|---|---|---|---|
| **Wood-framed Building** | | | | | | | |
| 1 Story Demolition | | | | | | | |
| | Demolish | S.F. | | 1.70 | 1.77 | 3.47 | Includes demolition of 1 floor only |
| | Minimum Charge | Job | | 2700 | 2800 | 5500 | light wood-framed building, including hauling of debris. Dump fees not included. |
| 2nd Story Demolition | | | | | | | |
| | Demolish | S.F. | | 1.73 | 1.80 | 3.53 | Includes demolition of 2nd floor only |
| | Minimum Charge | Job | | 2700 | 2800 | 5500 | light wood-framed building, including hauling of debris. Dump fees not included. |
| 3rd Story Demolition | | | | | | | |
| | Demolish | S.F. | | 2.25 | 2.34 | 4.59 | Includes demolition of 3rd floor only |
| | Minimum Charge | Job | | 2700 | 2800 | 5500 | light wood-framed building, including hauling of debris. Dump fees not included. |
| **Masonry Building** | | | | | | | |
| | Demolish | S.F. | | 2.42 | 2.52 | 4.94 | Includes demolition of one story |
| | Minimum Charge | Job | | 2700 | 2800 | 5500 | masonry building, including hauling of debris. Dump fees not included. |
| **Concrete Building** | | | | | | | |
| Unreinforced | | | | | | | |
| | Demolish | S.F. | | 4.92 | 1.10 | 6.02 | Includes demolition of a one story |
| | Minimum Charge | Job | | 2700 | 2800 | 5500 | non-reinforced concrete building, including hauling of debris. Dump fees not included. |
| Reinforced | | | | | | | |
| | Demolish | S.F. | | 5.65 | 1.26 | 6.91 | Includes demolition of a one story |
| | Minimum Charge | Job | | 2700 | 2800 | 5500 | reinforced concrete building, including hauling of debris. Dump fees not included. |
| **Steel Building** | | | | | | | |
| | Demolish | S.F. | | 3.23 | 3.36 | 6.59 | Includes demolition of a metal |
| | Minimum Charge | Job | | 2700 | 2800 | 5500 | building, including hauling of debris. Dump fees not included. No salvage value assumed for scrap metal. |

9

# Job Preparation

## General Clean-up

| General Clean-up | | Unit | Material | Labor | Equip. | Total | Specification |
|---|---|---|---|---|---|---|---|
| **Remove Debris** | | | | | | | |
| | Clean | S.F. | | 1.13 | .21 | 1.34 | Includes removal of debris caused by the loss including contents items and loose debris. |
| | Minimum Charge | Job | | 565 | 104 | 669 | |
| **Muck-out** | | | | | | | |
| | Clean | S.F. | | .69 | | .69 | Includes labor to remove 2" of muck and mud after a flood loss. |
| | Minimum Charge | Job | | 178 | | 178 | |
| **Deodorize / Disinfect** | | | | | | | |
| | Clean | SF Flr. | | .24 | | .24 | Includes deodorizing building with application of mildicide and disinfectant to floors and walls - will vary depending on the extent of deodorization required. |
| | Minimum Charge | Job | | 178 | | 178 | |
| **Clean Walls** | | | | | | | |
| Heavy | | | | | | | |
| | Clean | S.F. | | .15 | | .15 | Includes labor and material for heavy cleaning (multiple applications) with detergent and solvent. |
| | Minimum Charge | Job | | 178 | | 178 | |
| **Clean Ceiling** | | | | | | | |
| Light | | | | | | | |
| | Clean | S.F. | | .26 | | .26 | Includes labor and material to clean light to moderate smoke and soot. |
| | Minimum Charge | Job | | 178 | | 178 | |
| Heavy | | | | | | | |
| | Clean | S.F. | | .39 | | .39 | Includes labor and material to clean heavy smoke and soot. |
| | Minimum Charge | Job | | 178 | | 178 | |

## Flood Clean-up

| Flood Clean-up | | Unit | Material | Labor | Equip. | Total | Specification |
|---|---|---|---|---|---|---|---|
| **Emergency Service Call** | | | | | | | |
| After Hours / Weekend | | | | | | | |
| | Minimum Charge | Job | | 237 | | 237 | Includes minimum charges for deflooding clean-up after hours or on weekends. |
| Grey Water / Sewage | | | | | | | |
| | Minimum Charge | Job | | 178 | | 178 | Includes minimum charges for deflooding clean-up. |
| **Water Extraction** | | | | | | | |
| Clean (non-grey) | | | | | | | |
| | Clean | S.F. | .03 | .71 | | .74 | Includes labor to remove clean uncontaminated water from the loss. |
| | Minimum Charge | Job | | 178 | | 178 | |
| Grey (non-solids) | | | | | | | |
| | Clean | S.F. | .08 | 1.42 | | 1.50 | Includes labor to remove non-solid grey water from the loss. |
| | Minimum Charge | Job | | 178 | | 178 | |
| Sewage / Solids | | | | | | | |
| | Clean | SF Flr. | .12 | 1.78 | | 1.90 | Includes labor to remove water from the loss that may contain sewage or solids. |
| | Minimum Charge | Job | | 178 | | 178 | |

10

# Job Preparation

## Flood Clean-up

| | | Unit | Material | Labor | Equip. | Total | Specification |
|---|---|---|---|---|---|---|---|
| **Mildicide Walls** | | | | | | | |
| Topographical | | | | | | | |
| | Clean | SF Wall | | .32 | | .32 | Clean-up of deflooding including |
| | Minimum Charge | Job | | 178 | | 178 | topographical mildicide application. |
| Injection Treatments | | | | | | | |
| | Clean | SF Flr. | 46 | 5.45 | | 51.45 | Clean-up of deflooding including |
| | Minimum Charge | Job | | 178 | | 178 | mildicide injection. |
| **Carpet Cleaning** | | | | | | | |
| Uncontaminated | | | | | | | |
| | Clean | SF Flr. | | .22 | | .22 | Includes labor to clean carpet. |
| | Minimum Charge | Job | | 178 | | 178 | |
| Contaminated | | | | | | | |
| | Clean | SF Flr. | | .36 | | .36 | Includes labor to clean contaminated |
| | Minimum Charge | Job | | 178 | | 178 | carpet. |
| Stairway | | | | | | | |
| | Clean | Ea. | .09 | 4.18 | | 4.27 | Includes labor to clean contaminated |
| | Minimum Charge | Job | | 178 | | 178 | carpet on a stairway. |
| **Carpet Treatment** | | | | | | | |
| Disinfect / Deodorize | | | | | | | |
| | Clean | SF Flr. | | .22 | | .22 | Includes disinfecting and deodorizing |
| | Minimum Charge | Job | | 178 | | 178 | of carpet. |
| Disinfect For Sewage | | | | | | | |
| | Clean | SF Flr. | | .57 | | .57 | Includes labor to disinfect and |
| | Minimum Charge | Job | | 178 | | 178 | deodorize carpet exposed to sewage. |
| Mildicide | | | | | | | |
| | Clean | SF Flr. | .02 | .22 | | .24 | Includes labor and materials to apply |
| | Minimum Charge | Job | | 178 | | 178 | mildicide to carpet. |
| **Thermal Fog Area** | | | | | | | |
| | Clean | SF Flr. | .01 | .55 | | .56 | Includes labor and materials to |
| | Minimum Charge | Job | | 178 | | 178 | perform thermal fogging during |
| | | | | | | | deflooding clean-up. |
| **Wet Fog / ULV Area** | | | | | | | |
| | Clean | SF Flr. | .03 | .31 | | .34 | Includes labor and materials to |
| | Minimum Charge | Job | | 178 | | 178 | perform wet fogging / ULV during |
| | | | | | | | deflooding clean-up. |
| **Steam Clean Fixtures** | | | | | | | |
| | Clean | Ea. | .01 | 35.50 | | 35.51 | Includes labor and materials to steam |
| | Minimum Charge | Job | | 178 | | 178 | clean fixtures during deflooding |
| | | | | | | | clean-up. |
| **Pressure Wash Sewage** | | | | | | | |
| | Clean | SF Flr. | | 3.23 | | 3.23 | Includes labor to pressure wash |
| | Minimum Charge | Job | | 178 | | 178 | sewage contamination. |
| **Air Mover** | | | | | | | |
| | Minimum Charge | Day | | | 47 | 47 | Daily rental of air mover. |
| **Dehumidifier** | | | | | | | |
| Medium Size | | | | | | | |
| | Minimum Charge | Day | | | 58 | 58 | Daily rental of dehumidifier, medium. |
| Large Size | | | | | | | |
| | Minimum Charge | Day | | | 110 | 110 | Daily rental of dehumidifier, large. |

11

# Job Preparation

| Flood Clean-up | | Unit | Material | Labor | Equip. | Total | Specification |
|---|---|---|---|---|---|---|---|
| **Air Purifier** | | | | | | | |
| HEPA Filter-Bacteria | | | | | | | |
| | Minimum Charge | Ea. | 375 | | | 375 | HEPA filter for air purifier. |

| Asbestos Removal | | Unit | Material | Labor | Equip. | Total | Specification |
|---|---|---|---|---|---|---|---|
| **Asbestos Analysis** | | | | | | | |
| | Minimum Charge | Job | | 271 | | 271 | Minimum charge for testing. |
| **Asbestos Pipe Insulation** | | | | | | | |
| 1/2" to 3/4" Diameter | | | | | | | |
| | Demolish | L.F. | .29 | 7 | | 7.29 | Minimum charge to remove asbestos. |
| | Minimum Charge | Job | | 4350 | | 4350 | Includes full Tyvek suits for workers changed four times per eight-hour shift with respirators, air monitoring and supervision by qualified professionals. |
| 1" to 3" Diameter | | | | | | | |
| | Demolish | L.F. | .89 | 21.50 | | 22.39 | Minimum charge to remove asbestos. |
| | Minimum Charge | Job | | 4350 | | 4350 | Includes full Tyvek suits for workers changed four times per eight-hour shift with respirators, air monitoring and supervision by qualified professionals. |
| **Asbestos-based Siding** | | | | | | | |
| | Demolish | S.F. | 1.99 | 48.50 | | 50.49 | Minimum charge to remove asbestos. |
| | Minimum Charge | Job | | 4350 | | 4350 | Includes full Tyvek suits for workers changed four times per eight-hour shift with respirators, air monitoring and supervision by qualified professionals. |
| **Asbestos-based Plaster** | | | | | | | |
| | Demolish | S.F. | 1.53 | 37 | | 38.53 | Minimum charge to remove asbestos. |
| | Minimum Charge | Job | | 4350 | | 4350 | Includes full Tyvek suits for workers changed four times per eight-hour shift with respirators, air monitoring and supervision by qualified professionals. |
| **Encapsulate Asbestos Ceiling** | | | | | | | |
| | Install | S.F. | .44 | .21 | | .65 | Includes encapsulation with penetrating sealant sprayed onto ceiling and walls with airless sprayer. |
| | Minimum Charge | Job | | 4350 | | 4350 | |
| **Scrape Acoustical Ceiling** | | | | | | | |
| | Demolish | S.F. | .06 | 1.22 | | 1.28 | Minimum charge to remove asbestos. |
| | Minimum Charge | Job | | 4350 | | 4350 | Includes full Tyvek suits for workers changed four times per eight-hour shift with respirators, air monitoring and supervision by qualified professionals. |

| Asbestos Removal | | Unit | Material | Labor | Equip. | Total | Specification |
|---|---|---|---|---|---|---|---|
| **Seal Ceiling Asbestos** | | | | | | | |
| | Install | S.F. | 2.08 | 4.09 | | 6.17 | Includes attaching plastic cover to |
| | Minimum Charge | Job | | 4350 | | 4350 | ceiling area and sealing where asbestos exists, taping all seams, caulk or foam insulation in cracks and negative air system. |
| **Seal Wall Asbestos** | | | | | | | |
| | Install | S.F. | 2.08 | 3.63 | | 5.71 | Includes attaching plastic cover to wall |
| | Minimum Charge | Job | | 4350 | | 4350 | area and sealing where asbestos exists, taping all seams, caulk or foam insulation in cracks and negative air system. |
| **Seal Floor Asbestos** | | | | | | | |
| | Install | S.F. | 3.32 | 7.25 | | 10.57 | Includes attaching plastic cover to |
| | Minimum Charge | Job | | 4350 | | 4350 | floor area and sealing where asbestos exists, taping all seams, caulk or foam insulation in cracks and negative air system. |
| **Negative Air Vent System** | | | | | | | |
| | Install | Day | | | 45.50 | 45.50 | Daily rental of negative air vent |
| | Minimum Charge | Week | | | 227 | 227 | system. |
| **HEPA Vacuum Cleaner** | | | | | | | |
| | Install | Day | | | 19.85 | 19.85 | Daily rental of HEPA vacuum, 16 |
| | Minimum Charge | Week | | | 99 | 99 | gallon. |
| **Airless Sprayer** | | | | | | | |
| | Install | Day | | | 74 | 74 | Daily rental of airless sprayer. |
| | Minimum Charge | Week | | | 370 | 370 | |
| **Decontamination Unit** | | | | | | | |
| | Install | Day | | | 131 | 131 | Daily rental of decontamination unit. |
| | Minimum Charge | Week | | | 655 | 655 | |
| **Light Stand** | | | | | | | |
| | Install | Day | | | 18.55 | 18.55 | Daily rental of floodlight w/tripod. |
| | Minimum Charge | Week | | | 93 | 93 | |

| Room Demolition | | Unit | Material | Labor | Equip. | Total | Specification |
|---|---|---|---|---|---|---|---|
| **Strip Typical Room** | | | | | | | |
| | Demolish | S.F. | | 3.55 | | 3.55 | Includes 4 labor hours for general |
| | Minimum Charge | Job | | 197 | | 197 | demolition. |
| **Strip Typical Bathroom** | | | | | | | |
| | Demolish | S.F. | | 4.74 | | 4.74 | Includes 4 labor hours for general |
| | Minimum Charge | Job | | 197 | | 197 | demolition. |
| **Strip Typical Kitchen** | | | | | | | |
| | Demolish | S.F. | | 4.55 | | 4.55 | Includes 4 labor hours for general |
| | Minimum Charge | Job | | 197 | | 197 | demolition. |
| **Strip Typical Laundry** | | | | | | | |
| | Demolish | S.F. | | 4.06 | | 4.06 | Includes 4 labor hours for general |
| | Minimum Charge | Job | | 197 | | 197 | demolition. |

| Selective Demolition | | Unit | Material | Labor | Equip. | Total | Specification |
|---|---|---|---|---|---|---|---|
| **Concrete Slab** | | | | | | | |
| **4" Unreinforced** | | | | | | | |
| | Demolish | S.F. | | .24 | .53 | .77 | Includes labor and equipment to |
| | Minimum Charge | Job | | 545 | 1200 | 1745 | remove unreinforced concrete slab and haul debris to truck or dumpster. |
| **6" Unreinforced** | | | | | | | |
| | Demolish | S.F. | | .27 | .60 | .87 | Includes labor and equipment to |
| | Minimum Charge | Job | | 545 | 1200 | 1745 | remove unreinforced concrete slab and haul debris to truck or dumpster. |
| **8" Unreinforced** | | | | | | | |
| | Demolish | S.F. | | .31 | .68 | .99 | Includes labor and equipment to |
| | Minimum Charge | Job | | 545 | 1200 | 1745 | remove unreinforced concrete slab and haul debris to truck or dumpster. |
| **4" Reinforced** | | | | | | | |
| | Demolish | S.F. | | .24 | .53 | .77 | Includes labor and equipment to |
| | Minimum Charge | Job | | 545 | 1200 | 1745 | remove reinforced concrete slab and haul debris to truck or dumpster. |
| **6" Reinforced** | | | | | | | |
| | Demolish | S.F. | | .27 | .60 | .87 | Includes labor and equipment to |
| | Minimum Charge | Job | | 545 | 1200 | 1745 | remove reinforced concrete slab and haul debris to truck or dumpster. |
| **8" Reinforced** | | | | | | | |
| | Demolish | S.F. | | .31 | .68 | .99 | Includes labor and equipment to |
| | Minimum Charge | Job | | 545 | 1200 | 1745 | remove reinforced concrete slab and haul debris to truck or dumpster. |
| **Brick Wall** | | | | | | | |
| **4" Thick** | | | | | | | |
| | Demolish | S.F. | | 3.28 | .47 | 3.75 | Includes labor and equipment to |
| | Minimum Charge | Job | | 1000 | 143 | 1143 | remove brick wall and haul debris to a central location for disposal. |
| **8" Thick** | | | | | | | |
| | Demolish | S.F. | | 4.51 | .64 | 5.15 | Includes labor and equipment to |
| | Minimum Charge | Job | | 1000 | 143 | 1143 | remove brick wall and haul debris to a central location for disposal. |
| **12" Thick** | | | | | | | |
| | Demolish | S.F. | | 6 | .86 | 6.86 | Includes labor and equipment to |
| | Minimum Charge | Job | | 1000 | 143 | 1143 | remove brick wall and haul debris to a central location for disposal. |
| **Concrete Masonry Wall** | | | | | | | |
| **4" Thick** | | | | | | | |
| | Demolish | S.F. | | 3.64 | .52 | 4.16 | Includes minimum labor and |
| | Minimum Charge | Job | | 1000 | 143 | 1143 | equipment to remove 4" block wall and haul debris to a central location for disposal. |
| **6" Thick** | | | | | | | |
| | Demolish | S.F. | | 4.51 | .64 | 5.15 | Includes minimum labor and |
| | Minimum Charge | Job | | 1000 | 143 | 1143 | equipment to remove 6" block wall and haul debris to a central location for disposal. |

# Job Preparation

| Selective Demolition | | Unit | Material | Labor | Equip. | Total | Specification |
|---|---|---|---|---|---|---|---|
| **8" Thick** | | | | | | | |
| | Demolish | S.F. | | 5.55 | .79 | 6.34 | Includes minimum labor and |
| | Minimum Charge | Job | | 1000 | 143 | 1143 | equipment to remove 8" block wall and haul debris to a central location for disposal. |
| **12" Thick** | | | | | | | |
| | Demolish | S.F. | | 8 | 1.14 | 9.14 | Includes minimum labor and |
| | Minimum Charge | Job | | 1000 | 143 | 1143 | equipment to remove 12" block wall and haul debris to a central location for disposal. |
| **Tree Removal** | | | | | | | |
| **Small, 4" - 8"** | | | | | | | |
| | Demolish | Ea. | | 127 | 103 | 230 | Minimum charge, common labor |
| | Minimum Charge | Job | | 178 | | 178 | trade. |
| **Medium, 9" - 12"** | | | | | | | |
| | Demolish | Ea. | | 190 | 155 | 345 | Minimum charge, common labor |
| | Minimum Charge | Job | | 178 | | 178 | trade. |
| **Large, 13" - 18"** | | | | | | | |
| | Demolish | Ea. | | 228 | 186 | 414 | Minimum charge, tree removal. |
| | Minimum Charge | Job | | 590 | 213 | 803 | |
| **Very Large, 19" - 24"** | | | | | | | |
| | Demolish | Ea. | | 286 | 232 | 518 | Minimum charge, tree removal. |
| | Minimum Charge | Job | | 590 | 213 | 803 | |
| **Stump Removal** | | | | | | | |
| | Demolish | Ea. | | 83 | 136 | 219 | Includes minimum charges for a site |
| | Minimum Charge | Job | | 325 | 95.50 | 420.50 | demolition crew and backhoe / loader. |

**Backhoe/Loader - Wheel Type**    **Tractor Loader - Wheel Type**    **Backhoe - Crawler Type**

| Excavation | | Unit | Material | Labor | Equip. | Total | Specification |
|---|---|---|---|---|---|---|---|
| **Footings** | | | | | | | |
| Hand Trenching | | | | | | | |
| | Install | L.F. | | 2.22 | | 2.22 | Includes hand excavation for footing |
| | Minimum Charge | Job | | 89 | | 89 | or trench up to 10" deep and 12" wide. |
| Machine Trenching | | | | | | | |
| | Install | L.F. | | .84 | .34 | 1.18 | Includes machine excavation for |
| | Minimum Charge | Job | | 420 | 172 | 592 | footing or trench up to 10" deep and 12" wide. |
| **Trenching w / Backfill** | | | | | | | |
| Hand | | | | | | | |
| | Install | L.F. | | 3.55 | | 3.55 | Includes hand excavation, backfill and |
| | Minimum Charge | Job | | 89 | | 89 | compaction. |
| Machine | | | | | | | |
| | Install | L.F. | | 4.19 | 11.25 | 15.44 | Includes machine excavation, backfill |
| | Minimum Charge | Job | | 178 | 34 | 212 | and compaction. |

| Backfill | | Unit | Material | Labor | Equip. | Total | Specification |
|---|---|---|---|---|---|---|---|
| **Hand (no compaction)** | | | | | | | |
| | Install | L.C.Y. | | 29.50 | | 29.50 | Includes backfill by hand in average |
| | Minimum Charge | Job | | 178 | | 178 | soil without compaction. |
| **Hand (w / compaction)** | | | | | | | |
| | Install | E.C.Y. | | 42 | | 42 | Includes backfill by hand in average |
| | Minimum Charge | Job | | 178 | | 178 | soil with compaction. |
| **Machine (no compaction)** | | | | | | | |
| | Install | L.C.Y. | | 4.19 | 1.72 | 5.91 | Includes backfill, by 55 HP wheel |
| | Minimum Charge | Job | | 420 | 172 | 592 | loader, of trenches from loose material piled adjacent without compaction. |
| **Machine (w / compaction)** | | | | | | | |
| | Install | L.C.Y. | | 11.15 | 4.58 | 15.73 | Includes backfill, by 55 HP wheel |
| | Minimum Charge | Job | | 420 | 172 | 592 | loader, of trenches from loose material piled adjacent with compaction by vibrating tampers. |

**Concrete Footing and Formwork**

**Wood Pile**

| Piles | Unit | Material | Labor | Equip. | Total | Specification |
|---|---|---|---|---|---|---|
| **Treated Wood, 12″ Butt** | | | | | | |
| **To 30′** | | | | | | |
| Demolish | L.F. | | 5.70 | 1.66 | 7.36 | Includes material, labor and |
| Install | V.L.F. | 14.70 | 5.30 | 3.39 | 23.39 | equipment to install up to 30′ long |
| Demolish and Install | V.L.F. | 14.70 | 11 | 5.05 | 30.75 | treated wood piling. |
| Paint | L.F. | .54 | 2.21 | | 2.75 | |
| Minimum Charge | Job | | 3300 | 2125 | 5425 | |
| **30′ to 40′** | | | | | | |
| Demolish | L.F. | | 5.70 | 1.66 | 7.36 | Includes material, labor and |
| Install | V.L.F. | 15.95 | 4.72 | 3.03 | 23.70 | equipment to install 30′ to 40′ long |
| Demolish and Install | V.L.F. | 15.95 | 10.42 | 4.69 | 31.06 | treated wood piling. |
| Paint | L.F. | .54 | 2.21 | | 2.75 | |
| Minimum Charge | Job | | 3300 | 2125 | 5425 | |
| **40′ to 50′** | | | | | | |
| Demolish | L.F. | | 5.70 | 1.66 | 7.36 | Includes material, labor and |
| Install | V.L.F. | 16.90 | 4.59 | 2.95 | 24.44 | equipment to install 40′ to 50′ long |
| Demolish and Install | V.L.F. | 16.90 | 10.29 | 4.61 | 31.80 | treated wood piling. |
| Paint | L.F. | .54 | 2.21 | | 2.75 | |
| Minimum Charge | Job | | 3300 | 2125 | 5425 | |
| **50′ to 60′** | | | | | | |
| Demolish | L.F. | | 5.70 | 1.66 | 7.36 | Includes material, labor and |
| Install | V.L.F. | 19.70 | 4.13 | 2.65 | 26.48 | equipment to install 50′ to 60′ long |
| Demolish and Install | V.L.F. | 19.70 | 9.83 | 4.31 | 33.84 | treated wood piling. |
| Paint | L.F. | .54 | 2.21 | | 2.75 | |
| Minimum Charge | Job | | 3300 | 2125 | 5425 | |
| **60′ to 80′** | | | | | | |
| Demolish | L.F. | | 5.70 | 1.66 | 7.36 | Includes material, labor and |
| Install | V.L.F. | 19.85 | 3.93 | 3.16 | 26.94 | equipment to install 60′ to 80′ long |
| Demolish and Install | V.L.F. | 19.85 | 9.63 | 4.82 | 34.30 | treated wood piling. |
| Paint | L.F. | .54 | 2.21 | | 2.75 | |
| Minimum Charge | Job | | 3300 | 2125 | 5425 | |
| **Untreated Wood, 12″ Butt** | | | | | | |
| **To 30′** | | | | | | |
| Demolish | L.F. | | 5.70 | 1.66 | 7.36 | Includes material, labor and |
| Install | V.L.F. | 12.60 | 5.30 | 3.39 | 21.29 | equipment to install up to 30′ long |
| Demolish and Install | V.L.F. | 12.60 | 11 | 5.05 | 28.65 | untreated wood piling. |
| Paint | L.F. | .54 | 2.21 | | 2.75 | |
| Minimum Charge | Job | | 3300 | 2125 | 5425 | |

| Piles | | Unit | Material | Labor | Equip. | Total | Specification |
|---|---|---|---|---|---|---|---|
| **30' to 40'** | | | | | | | |
| | Demolish | L.F. | | 5.70 | 1.66 | 7.36 | Includes material, labor and |
| | Install | V.L.F. | 12.60 | 4.72 | 3.03 | 20.35 | equipment to install 30' to 40' long |
| | Demolish and Install | V.L.F. | 12.60 | 10.42 | 4.69 | 27.71 | untreated wood piling. |
| | Paint | L.F. | .54 | 2.21 | | 2.75 | |
| | Minimum Charge | Job | | 3300 | 2125 | 5425 | |
| **40' to 50'** | | | | | | | |
| | Demolish | L.F. | | 5.70 | 1.66 | 7.36 | Includes material, labor and |
| | Install | V.L.F. | 12.60 | 4.59 | 2.95 | 20.14 | equipment to install 40' to 50' long |
| | Demolish and Install | V.L.F. | 12.60 | 10.29 | 4.61 | 27.50 | untreated wood piling. |
| | Paint | L.F. | .54 | 2.21 | | 2.75 | |
| | Minimum Charge | Job | | 3300 | 2125 | 5425 | |
| **50' to 60'** | | | | | | | |
| | Demolish | L.F. | | 5.70 | 1.66 | 7.36 | Includes material, labor and |
| | Install | V.L.F. | 14.95 | 4.13 | 2.65 | 21.73 | equipment to install 50' to 60' long |
| | Demolish and Install | V.L.F. | 14.95 | 9.83 | 4.31 | 29.09 | untreated wood piling. |
| | Paint | L.F. | .54 | 2.21 | | 2.75 | |
| | Minimum Charge | Job | | 3300 | 2125 | 5425 | |
| **60' to 80'** | | | | | | | |
| | Demolish | L.F. | | 5.70 | 1.66 | 7.36 | Includes material, labor and |
| | Install | V.L.F. | 18.70 | 3.93 | 2.53 | 25.16 | equipment to install 60' to 80' long |
| | Demolish and Install | V.L.F. | 18.70 | 9.63 | 4.19 | 32.52 | untreated wood piling. |
| | Paint | L.F. | .54 | 2.21 | | 2.75 | |
| | Minimum Charge | Job | | 3300 | 2125 | 5425 | |
| **Precast Concrete** | | | | | | | |
| **10" Square** | | | | | | | |
| | Demolish | L.F. | | 5.80 | 1.70 | 7.50 | Includes material, labor and |
| | Install | V.L.F. | 15.75 | 4.72 | 3.03 | 23.50 | equipment to install a precast, |
| | Demolish and Install | V.L.F. | 15.75 | 10.52 | 4.73 | 31 | prestressed, 40' long, 10" square |
| | Paint | L.F. | .54 | 2.21 | | 2.75 | concrete pile. |
| | Minimum Charge | Job | | 3300 | 2125 | 5425 | |
| **12" Square** | | | | | | | |
| | Demolish | L.F. | | 6.25 | 1.82 | 8.07 | Includes material, labor and |
| | Install | V.L.F. | 23.50 | 4.86 | 3.12 | 31.48 | equipment to install a precast, |
| | Demolish and Install | V.L.F. | 23.50 | 11.11 | 4.94 | 39.55 | prestressed, 40' long, 12" square |
| | Paint | L.F. | .54 | 2.21 | | 2.75 | concrete pile. |
| | Minimum Charge | Job | | 3300 | 2125 | 5425 | |
| **14" Square** | | | | | | | |
| | Demolish | L.F. | | 6.40 | 1.86 | 8.26 | Includes material, labor and |
| | Install | V.L.F. | 26.50 | 5.50 | 3.54 | 35.54 | equipment to install a precast, |
| | Demolish and Install | V.L.F. | 26.50 | 11.90 | 5.40 | 43.80 | prestressed, 40' long, 14" square |
| | Paint | L.F. | .54 | 2.21 | | 2.75 | concrete pile. |
| | Minimum Charge | Job | | 3300 | 2125 | 5425 | |
| **16" Square** | | | | | | | |
| | Demolish | L.F. | | 6.70 | 1.96 | 8.66 | Includes material, labor and |
| | Install | V.L.F. | 33.50 | 5.90 | 3.79 | 43.19 | equipment to install a precast, |
| | Demolish and Install | V.L.F. | 33.50 | 12.60 | 5.75 | 51.85 | prestressed, 40' long, 16" square |
| | Paint | L.F. | .54 | 2.21 | | 2.75 | concrete pile. |
| | Minimum Charge | Job | | 3300 | 2125 | 5425 | |
| **18" Square** | | | | | | | |
| | Demolish | L.F. | | 7.25 | 2.12 | 9.37 | Includes material, labor and |
| | Install | V.L.F. | 45.50 | 6.35 | 5.10 | 56.95 | equipment to install a precast, |
| | Demolish and Install | V.L.F. | 45.50 | 13.60 | 7.22 | 66.32 | prestressed, 40' long, 18" square |
| | Paint | L.F. | .54 | 2.21 | | 2.75 | concrete pile. |
| | Minimum Charge | Job | | 3300 | 2125 | 5425 | |

# Foundation

## Piles

| | Unit | Material | Labor | Equip. | Total | Specification |
|---|---|---|---|---|---|---|
| **Cross Bracing** | | | | | | |
| **Treated 2″ x 6″** | | | | | | |
| Demolish | L.F. | | 1.16 | | 1.16 | Cost includes material and labor to |
| Install | L.F. | .81 | 6.65 | | 7.46 | install 2″ x 6″ lumber for cross-bracing |
| Demolish and Install | L.F. | .81 | 7.81 | | 8.62 | of foundation piling or posts. |
| Reinstall | L.F. | | 5.31 | | 5.31 | |
| Clean | S.F. | | .44 | | .44 | |
| Paint | L.F. | .34 | 1.49 | | 1.83 | |
| Minimum Charge | Job | | 820 | | 820 | |
| **Treated 2″ x 8″** | | | | | | |
| Demolish | L.F. | | 1.16 | | 1.16 | Cost includes material and labor to |
| Install | L.F. | 1.23 | 6.85 | | 8.08 | install 2″ x 8″ lumber for cross-bracing |
| Demolish and Install | L.F. | 1.23 | 8.01 | | 9.24 | of foundation piling or posts. |
| Reinstall | L.F. | | 5.48 | | 5.48 | |
| Clean | S.F. | | .44 | | .44 | |
| Paint | L.F. | .45 | 1.55 | | 2 | |
| Minimum Charge | Job | | 820 | | 820 | |
| **Treated 2″ x 10″** | | | | | | |
| Demolish | L.F. | | 1.20 | | 1.20 | Cost includes material and labor to |
| Install | L.F. | 1.54 | 7.10 | | 8.64 | install 2″ x 10″ lumber for |
| Demolish and Install | L.F. | 1.54 | 8.30 | | 9.84 | cross-bracing of foundation piling or |
| Reinstall | L.F. | | 5.67 | | 5.67 | posts. |
| Clean | S.F. | | .44 | | .44 | |
| Paint | L.F. | .54 | 1.58 | | 2.12 | |
| Minimum Charge | Job | | 820 | | 820 | |
| **Treated 2″ x 12″** | | | | | | |
| Demolish | L.F. | | 1.44 | | 1.44 | Cost includes material and labor to |
| Install | L.F. | 2.23 | 7.35 | | 9.58 | install 2″ x 12″ lumber for |
| Demolish and Install | L.F. | 2.23 | 8.79 | | 11.02 | cross-bracing of foundation piling or |
| Reinstall | L.F. | | 5.87 | | 5.87 | posts. |
| Clean | S.F. | | .44 | | .44 | |
| Paint | L.F. | .64 | 1.61 | | 2.25 | |
| Minimum Charge | Job | | 820 | | 820 | |
| **Metal Pipe** | | | | | | |
| **6″ Concrete Filled** | | | | | | |
| Demolish | L.F. | | 8.05 | 2.36 | 10.41 | Includes material, labor and |
| Install | V.L.F. | 22.50 | 6.90 | 4.42 | 33.82 | equipment to install concrete filled |
| Demolish and Install | V.L.F. | 22.50 | 14.95 | 6.78 | 44.23 | steel pipe pile. |
| Paint | L.F. | .36 | 2.21 | | 2.57 | |
| Minimum Charge | Job | | 3300 | 2125 | 5425 | |
| **6″ Unfilled** | | | | | | |
| Demolish | L.F. | | 8.05 | 2.36 | 10.41 | Includes material, labor and |
| Install | V.L.F. | 18.30 | 6.30 | 4.04 | 28.64 | equipment to install unfilled steel pipe |
| Demolish and Install | V.L.F. | 18.30 | 14.35 | 6.40 | 39.05 | pile. |
| Paint | L.F. | .36 | 2.21 | | 2.57 | |
| Minimum Charge | Job | | 3300 | 2125 | 5425 | |
| **8″ Concrete Filled** | | | | | | |
| Demolish | L.F. | | 8.05 | 2.36 | 10.41 | Includes material, labor and |
| Install | V.L.F. | 26.50 | 7.20 | 4.61 | 38.31 | equipment to install concrete filled |
| Demolish and Install | V.L.F. | 26.50 | 15.25 | 6.97 | 48.72 | steel pipe pile. |
| Paint | L.F. | .36 | 2.21 | | 2.57 | |
| Minimum Charge | Job | | 3300 | 2125 | 5425 | |
| **8″ Unfilled** | | | | | | |
| Demolish | L.F. | | 8.05 | 2.36 | 10.41 | Includes material, labor and |
| Install | V.L.F. | 23.50 | 6.60 | 4.24 | 34.34 | equipment to install unfilled steel pipe |
| Demolish and Install | V.L.F. | 23.50 | 14.65 | 6.60 | 44.75 | pile. |
| Paint | L.F. | .36 | 2.21 | | 2.57 | |
| Minimum Charge | Job | | 3300 | 2125 | 5425 | |

| Piles | Unit | Material | Labor | Equip. | Total | Specification |
|---|---|---|---|---|---|---|
| **10" Concrete Filled** | | | | | | |
| Demolish | L.F. | | 8.35 | 2.43 | 10.78 | Includes material, labor and |
| Install | V.L.F. | 31.50 | 7.35 | 4.71 | 43.56 | equipment to install concrete filled |
| Demolish and Install | V.L.F. | 31.50 | 15.70 | 7.14 | 54.34 | steel pipe pile. |
| Paint | L.F. | .36 | 2.21 | | 2.57 | |
| Minimum Charge | Job | | 3300 | 2125 | 5425 | |
| **10" Unfilled** | | | | | | |
| Demolish | L.F. | | 8.35 | 2.43 | 10.78 | Includes material, labor and |
| Install | V.L.F. | 26.50 | 6.60 | 4.24 | 37.34 | equipment to install unfilled steel pipe |
| Demolish and Install | V.L.F. | 26.50 | 14.95 | 6.67 | 48.12 | pile. |
| Paint | L.F. | .36 | 2.21 | | 2.57 | |
| Minimum Charge | Job | | 3300 | 2125 | 5425 | |
| **12" Concrete Filled** | | | | | | |
| Demolish | L.F. | | 8.35 | 2.43 | 10.78 | Includes material, labor and |
| Install | V.L.F. | 38.50 | 7.95 | 5.10 | 51.55 | equipment to install concrete filled |
| Demolish and Install | V.L.F. | 38.50 | 16.30 | 7.53 | 62.33 | steel pipe pile. |
| Paint | L.F. | .36 | 2.21 | | 2.57 | |
| Minimum Charge | Job | | 3300 | 2125 | 5425 | |
| **12" Unfilled** | | | | | | |
| Demolish | L.F. | | 8.35 | 2.43 | 10.78 | Includes material, labor and |
| Install | V.L.F. | 32.50 | 6.95 | 4.47 | 43.92 | equipment to install unfilled steel pipe |
| Demolish and Install | V.L.F. | 32.50 | 15.30 | 6.90 | 54.70 | pile. |
| Paint | L.F. | .36 | 2.21 | | 2.57 | |
| Minimum Charge | Job | | 3300 | 2125 | 5425 | |
| **Steel H Section** | | | | | | |
| **H Section 8 x 8 x 36#** | | | | | | |
| Demolish | L.F. | | 5.65 | 1.65 | 7.30 | Includes material, labor and |
| Install | V.L.F. | 18.75 | 5.15 | 3.31 | 27.21 | equipment to install steel H section |
| Demolish and Install | V.L.F. | 18.75 | 10.80 | 4.96 | 34.51 | pile. |
| Paint | S.F. | .11 | .33 | | .44 | |
| Minimum Charge | Job | | 3300 | 2125 | 5425 | |
| **H Section 10 x 10 x 57#** | | | | | | |
| Demolish | L.F. | | 6.15 | 1.80 | 7.95 | Includes material, labor and |
| Install | V.L.F. | 29.50 | 5.40 | 3.48 | 38.38 | equipment to install steel H section |
| Demolish and Install | V.L.F. | 29.50 | 11.55 | 5.28 | 46.33 | pile. |
| Paint | S.F. | .11 | .33 | | .44 | |
| Minimum Charge | Job | | 3300 | 2125 | 5425 | |
| **H Section 12 x 12 x 74#** | | | | | | |
| Demolish | L.F. | | 6.40 | 1.86 | 8.26 | Includes material, labor and |
| Install | V.L.F. | 38.50 | 5.60 | 4.50 | 48.60 | equipment to install steel H section |
| Demolish and Install | V.L.F. | 38.50 | 12 | 6.36 | 56.86 | pile. |
| Paint | S.F. | .11 | .33 | | .44 | |
| Minimum Charge | Job | | 3300 | 2125 | 5425 | |
| **H Section 14 x 14 x 89#** | | | | | | |
| Demolish | L.F. | | 6.70 | 1.96 | 8.66 | Includes material, labor and |
| Install | V.L.F. | 46 | 6.10 | 4.91 | 57.01 | equipment to install steel H section |
| Demolish and Install | V.L.F. | 46 | 12.80 | 6.87 | 65.67 | pile. |
| Paint | S.F. | .11 | .33 | | .44 | |
| Minimum Charge | Job | | 3300 | 2125 | 5425 | |

| Foundation Post | | Unit | Material | Labor | Equip. | Total | Specification |
|---|---|---|---|---|---|---|---|
| **Wood Foundation Post** | | | | | | | |
| 4" x 4" | | | | | | | |
| | Demolish | L.F. | | .91 | .27 | 1.18 | Cost includes material and labor to |
| | Install | L.F. | 1.38 | 5.90 | | 7.28 | install 4" x 4" treated wood post by |
| | Demolish and Install | L.F. | 1.38 | 6.81 | .27 | 8.46 | hand in normal soil conditions. |
| | Reinstall | L.F. | | 4.74 | | 4.74 | |
| | Minimum Charge | Job | | 820 | | 820 | |
| 4" x 6" | | | | | | | |
| | Demolish | L.F. | | .91 | .27 | 1.18 | Cost includes material and labor to |
| | Install | L.F. | 2.04 | 5.90 | | 7.94 | install 4" x 6" treated wood post by |
| | Demolish and Install | L.F. | 2.04 | 6.81 | .27 | 9.12 | hand in normal soil conditions. |
| | Reinstall | L.F. | | 4.74 | | 4.74 | |
| | Minimum Charge | Job | | 820 | | 820 | |
| 6" x 6" | | | | | | | |
| | Demolish | L.F. | | .91 | .27 | 1.18 | Cost includes material and labor to |
| | Install | L.F. | 2.98 | 5.90 | | 8.88 | install 6" x 6" treated wood post by |
| | Demolish and Install | L.F. | 2.98 | 6.81 | .27 | 10.06 | hand in normal soil conditions. |
| | Reinstall | L.F. | | 4.74 | | 4.74 | |
| | Minimum Charge | Job | | 820 | | 820 | |
| Water Jet | | | | | | | |
| | Demolish | Ea. | | 105 | | 105 | Includes material, labor and |
| | Install | Ea. | 75 | 810 | | 885 | equipment to water jet a 6" x 6" wood |
| | Demolish and Install | Ea. | 75 | 915 | | 990 | post under an existing building. Cost |
| | Reinstall | Ea. | | 811.89 | | 811.89 | may vary depending upon location |
| | Minimum Charge | Job | | 710 | | 710 | and access. |
| Shim | | | | | | | |
| | Install | Ea. | 1.83 | 82 | | 83.83 | Includes labor and material to trim or |
| | Minimum Charge | Job | | 820 | | 820 | readjust existing foundation posts |
| | | | | | | | under an existing building. Access of |
| | | | | | | | 2 - 4 foot assumed. |
| **Cross-bracing** | | | | | | | |
| Treated 2" x 6" | | | | | | | |
| | Demolish | L.F. | | 1.16 | | 1.16 | Cost includes material and labor to |
| | Install | L.F. | .81 | 6.65 | | 7.46 | install 2" x 6" lumber for cross-bracing |
| | Demolish and Install | L.F. | .81 | 7.81 | | 8.62 | of foundation piling or posts. |
| | Reinstall | L.F. | | 5.31 | | 5.31 | |
| | Clean | S.F. | | .44 | | .44 | |
| | Paint | L.F. | .34 | 1.49 | | 1.83 | |
| | Minimum Charge | Job | | 820 | | 820 | |
| Treated 2" x 8" | | | | | | | |
| | Demolish | L.F. | | 1.16 | | 1.16 | Cost includes material and labor to |
| | Install | L.F. | 1.23 | 6.85 | | 8.08 | install 2" x 8" lumber for cross-bracing |
| | Demolish and Install | L.F. | 1.23 | 8.01 | | 9.24 | of foundation piling or posts. |
| | Reinstall | L.F. | | 5.48 | | 5.48 | |
| | Clean | S.F. | | .44 | | .44 | |
| | Paint | L.F. | .45 | 1.55 | | 2 | |
| | Minimum Charge | Job | | 820 | | 820 | |
| Treated 2" x 10" | | | | | | | |
| | Demolish | L.F. | | 1.20 | | 1.20 | Cost includes material and labor to |
| | Install | L.F. | 1.54 | 7.10 | | 8.64 | install 2" x 10" lumber for |
| | Demolish and Install | L.F. | 1.54 | 8.30 | | 9.84 | cross-bracing of foundation piling or |
| | Reinstall | L.F. | | 5.67 | | 5.67 | posts. |
| | Clean | S.F. | | .44 | | .44 | |
| | Paint | L.F. | .54 | 1.58 | | 2.12 | |
| | Minimum Charge | Job | | 820 | | 820 | |

# Foundation

## Foundation Post

| Treated 2″ x 12″ | Unit | Material | Labor | Equip. | Total | Specification |
|---|---|---|---|---|---|---|
| Demolish | L.F. | | 1.44 | | 1.44 | Cost includes material and labor to |
| Install | L.F. | 2.23 | 7.35 | | 9.58 | install 2″ x 12″ lumber for |
| Demolish and Install | L.F. | 2.23 | 8.79 | | 11.02 | cross-bracing of foundation piling or |
| Reinstall | L.F. | | 5.87 | | 5.87 | posts. |
| Clean | S.F. | | .44 | | .44 | |
| Paint | L.F. | .64 | 1.61 | | 2.25 | |
| Minimum Charge | Job | | 820 | | 820 | |

## Concrete Footing

| | Unit | Material | Labor | Equip. | Total | Specification |
|---|---|---|---|---|---|---|
| **Continuous** | | | | | | |
| **12″ w x 6″ d** | | | | | | |
| Demolish | L.F. | | .60 | 1.33 | 1.93 | Includes material and labor to install |
| Install | L.F. | 4.79 | 3.87 | .02 | 8.68 | continuous reinforced concrete footing |
| Demolish and Install | L.F. | 4.79 | 4.47 | 1.35 | 10.61 | poured by chute cast against earth |
| Minimum Charge | Job | | 930 | | 930 | including reinforcing, forming and |
| | | | | | | finishing. |
| **12″ w x 12″ d** | | | | | | |
| Demolish | L.F. | | .72 | 1.59 | 2.31 | Includes material and labor to install |
| Install | L.F. | 10.75 | 7.75 | .04 | 18.54 | continuous reinforced concrete footing |
| Demolish and Install | L.F. | 10.75 | 8.47 | 1.63 | 20.85 | poured by chute cast against earth |
| Minimum Charge | Job | | 930 | | 930 | including reinforcing, forming and |
| | | | | | | finishing. |
| **18″ w x 10″ d** | | | | | | |
| Demolish | L.F. | | .72 | 1.59 | 2.31 | Includes material and labor to install |
| Install | L.F. | 10.05 | 9.50 | .04 | 19.59 | continuous reinforced concrete footing |
| Demolish and Install | L.F. | 10.05 | 10.22 | 1.63 | 21.90 | poured by chute cast against earth |
| Minimum Charge | Job | | 930 | | 930 | including reinforcing, forming and |
| | | | | | | finishing. |
| **24″ w x 24″ d** | | | | | | |
| Demolish | L.F. | | 1.09 | 2.39 | 3.48 | Includes material and labor to install |
| Install | L.F. | 30.50 | 31 | .14 | 61.64 | continuous reinforced concrete footing |
| Demolish and Install | L.F. | 30.50 | 32.09 | 2.53 | 65.12 | poured by chute cast against earth |
| Minimum Charge | Job | | 930 | | 930 | including reinforcing, forming and |
| | | | | | | finishing. |
| **Stem Wall** | | | | | | |
| **Single Story** | | | | | | |
| Demolish | L.F. | | 15.90 | 13 | 28.90 | Includes material and labor to install |
| Install | L.F. | 25 | 84 | 69 | 178 | stem wall 24″ above and 18″ below |
| Demolish and Install | L.F. | 25 | 99.90 | 82 | 206.90 | grade for a single story structure. |
| Minimum Charge | Job | | 930 | | 930 | Includes 12″ wide and 8″ deep |
| | | | | | | footing. |
| **Two Story** | | | | | | |
| Demolish | L.F. | | 15.90 | 13 | 28.90 | Includes material and labor to install |
| Install | L.F. | 31 | 121 | 99 | 251 | stem wall 24″ above and 18″ below |
| Demolish and Install | L.F. | 31 | 136.90 | 112 | 279.90 | grade typical for a two story structure, |
| Minimum Charge | Job | | 930 | | 930 | includes 18″ wide and 10″ deep |
| | | | | | | footing. |

| Concrete Slab | | Unit | Material | Labor | Equip. | Total | Specification |
|---|---|---|---|---|---|---|---|
| **Reinforced** | | | | | | | |
| **4"** | Demolish | S.F. | | .24 | .53 | .77 | Cost includes material and labor to |
| | Install | S.F. | 2.55 | .46 | | 3.01 | install 4" reinforced concrete |
| | Demolish and Install | S.F. | 2.55 | .70 | .53 | 3.78 | slab-on-grade poured by chute |
| | Clean | S.F. | .09 | .39 | | .48 | including forms, vapor barrier, wire |
| | Paint | S.F. | .19 | .85 | | 1.04 | mesh, 3000 PSI concrete, float finish, |
| | Minimum Charge | Job | | 225 | | 225 | and curing. |
| **6"** | Demolish | S.F. | | .27 | .60 | .87 | Cost includes material and labor to |
| | Install | S.F. | 3.50 | .69 | .01 | 4.20 | install 6" reinforced concrete |
| | Demolish and Install | S.F. | 3.50 | .96 | .61 | 5.07 | slab-on-grade poured by chute |
| | Clean | S.F. | .09 | .39 | | .48 | including forms, vapor barrier, wire |
| | Paint | S.F. | .19 | .85 | | 1.04 | mesh, 3000 PSI concrete, float finish, |
| | Minimum Charge | Job | | 225 | | 225 | and curing. |
| **8"** | Demolish | S.F. | | .31 | .68 | .99 | Cost includes material and labor to |
| | Install | S.F. | 4.57 | .90 | .01 | 5.48 | install 8" reinforced concrete |
| | Demolish and Install | S.F. | 4.57 | 1.21 | .69 | 6.47 | slab-on-grade poured by chute |
| | Clean | S.F. | .09 | .39 | | .48 | including forms, vapor barrier, wire |
| | Paint | S.F. | .19 | .85 | | 1.04 | mesh, 3000 PSI concrete, float finish, |
| | Minimum Charge | Job | | 225 | | 225 | and curing. |
| **Unreinforced** | | | | | | | |
| **4"** | Demolish | S.F. | | .24 | .53 | .77 | Cost includes material and labor to |
| | Install | S.F. | 1.77 | 1.11 | .01 | 2.89 | install 4" concrete slab-on-grade |
| | Demolish and Install | S.F. | 1.77 | 1.35 | .54 | 3.66 | poured by chute, including forms, |
| | Clean | S.F. | .09 | .39 | | .48 | vapor barrier, 3000 PSI concrete, float |
| | Paint | S.F. | .19 | .85 | | 1.04 | finish, and curing. |
| | Minimum Charge | Job | | 225 | | 225 | |
| **6"** | Demolish | S.F. | | .27 | .60 | .87 | Cost includes material and labor to |
| | Install | S.F. | 2.60 | 1.13 | .01 | 3.74 | install 6" concrete slab-on-grade |
| | Demolish and Install | S.F. | 2.60 | 1.40 | .61 | 4.61 | poured by chute, including forms, |
| | Clean | S.F. | .09 | .39 | | .48 | vapor barrier, 3000 PSI concrete, float |
| | Paint | S.F. | .19 | .85 | | 1.04 | finish, and curing. |
| | Minimum Charge | Job | | 225 | | 225 | |
| **8"** | Demolish | S.F. | | .31 | .68 | .99 | Cost includes material and labor to |
| | Install | S.F. | 3.55 | 1.19 | .01 | 4.75 | install 8" concrete slab-on-grade |
| | Demolish and Install | S.F. | 3.55 | 1.50 | .69 | 5.74 | poured by chute, including forms, |
| | Clean | S.F. | .09 | .39 | | .48 | vapor barrier, 3000 PSI concrete, float |
| | Paint | S.F. | .19 | .85 | | 1.04 | finish, and curing. |
| | Minimum Charge | Job | | 225 | | 225 | |
| **Lightweight** | Demolish | S.F. | | 1.05 | .23 | 1.28 | Includes material and labor to install |
| | Install | S.F. | 1.55 | 3.35 | | 4.90 | lightweight concrete, 3" - 4" thick. |
| | Demolish and Install | S.F. | 1.55 | 4.40 | .23 | 6.18 | |
| | Clean | S.F. | .09 | .39 | | .48 | |
| | Paint | S.F. | .19 | .85 | | 1.04 | |
| | Minimum Charge | Job | | 225 | | 225 | |

| Concrete Slab | Unit | Material | Labor | Equip. | Total | Specification |
|---|---|---|---|---|---|---|
| **Gunite** | | | | | | |
| Demolish | S.F. | | 1.31 | .38 | 1.69 | Includes material and labor to install |
| Install | S.F. | .40 | 1.42 | .21 | 2.03 | gunite on flat plane per inch of |
| Demolish and Install | S.F. | .40 | 2.73 | .59 | 3.72 | thickness using 1″ as base price. No |
| Clean | S.F. | .09 | .39 | | .48 | forms or reinforcing are included. |
| Paint | S.F. | .19 | .85 | | 1.04 | |
| Minimum Charge | Job | | 1250 | | 1250 | |
| **Epoxy Inject** | | | | | | |
| Install | L.F. | 6.15 | 7.50 | | 13.65 | Includes labor and material to inject |
| Minimum Charge | Job | | 225 | | 225 | epoxy into cracks. |
| **Slabjacking** | | | | | | |
| Install | S.F. | .54 | 1.64 | .12 | 2.30 | Includes labor, material and |
| Minimum Charge | Job | | 225 | | 225 | equipment to install slabjacking under |
| | | | | | | an existing building. |
| **Scrape and Paint** | | | | | | |
| Paint | S.F. | .23 | .97 | | 1.20 | Includes labor and materials to scrape |
| Minimum Charge | Job | | 193 | | 193 | and paint a concrete floor. |
| **Vapor Barrier** | | | | | | |
| Install | S.F. | .03 | .13 | | .16 | Cost includes material and labor to |
| Minimum Charge | Job | | 232 | | 232 | install 4 mil polyethylene vapor barrier |
| | | | | | | 10′ wide sheets with 6″ overlaps. |

**Rough Framing**

| Beam / Girder | | Unit | Material | Labor | Equip. | Total | Specification |
|---|---|---|---|---|---|---|---|
| **Solid Wood Beam** | | | | | | | |
| **2″ x 8″** | | | | | | | |
| | Demolish | L.F. | | .96 | | .96 | Cost includes material and labor to |
| | Install | L.F. | .98 | 1.43 | | 2.41 | install 2″ x 8″ beam. |
| | Demolish and Install | L.F. | .98 | 2.39 | | 3.37 | |
| | Clean | L.F. | .04 | .32 | | .36 | |
| | Paint | L.F. | .14 | 2.09 | | 2.23 | |
| | Minimum Charge | Job | | 232 | | 232 | |
| **2″ x 10″** | | | | | | | |
| | Demolish | L.F. | | 1.19 | | 1.19 | Cost includes material and labor to |
| | Install | L.F. | 1.55 | 1.55 | | 3.10 | install 2″ x 10″ beam. |
| | Demolish and Install | L.F. | 1.55 | 2.74 | | 4.29 | |
| | Clean | L.F. | .06 | .38 | | .44 | |
| | Paint | L.F. | .19 | 1.29 | | 1.48 | |
| | Minimum Charge | Job | | 232 | | 232 | |
| **2″ x 12″** | | | | | | | |
| | Demolish | L.F. | | 1.44 | | 1.44 | Cost includes material and labor to |
| | Install | L.F. | 1.97 | 1.69 | | 3.66 | install 2″ x 12″ beam. |
| | Demolish and Install | L.F. | 1.97 | 3.13 | | 5.10 | |
| | Clean | L.F. | .07 | .44 | | .51 | |
| | Paint | L.F. | .22 | 3.09 | | 3.31 | |
| | Minimum Charge | Job | | 232 | | 232 | |
| **4″ x 8″** | | | | | | | |
| | Demolish | L.F. | | 2.94 | | 2.94 | Cost includes material and labor to |
| | Install | L.F. | 4.57 | 2.13 | .55 | 7.25 | install 4″ x 8″ beam. |
| | Demolish and Install | L.F. | 4.57 | 5.07 | .55 | 10.19 | |
| | Clean | L.F. | .06 | .38 | | .44 | |
| | Paint | L.F. | .19 | 1.29 | | 1.48 | |
| | Minimum Charge | Job | | 232 | | 232 | |
| **4″ x 10″** | | | | | | | |
| | Demolish | L.F. | | 3.68 | | 3.68 | Cost includes material and labor to |
| | Install | L.F. | 5.45 | 2.25 | .57 | 8.27 | install 4″ x 10″ beam. |
| | Demolish and Install | L.F. | 5.45 | 5.93 | .57 | 11.95 | |
| | Clean | L.F. | .07 | .44 | | .51 | |
| | Paint | L.F. | .22 | 1.50 | | 1.72 | |
| | Minimum Charge | Job | | 232 | | 232 | |
| **4″ x 12″** | | | | | | | |
| | Demolish | L.F. | | 4.41 | | 4.41 | Cost includes material and labor to |
| | Install | L.F. | 6.05 | 2.37 | .61 | 9.03 | install 4″ x 12″ beam. |
| | Demolish and Install | L.F. | 6.05 | 6.78 | .61 | 13.44 | |
| | Clean | L.F. | .08 | .51 | | .59 | |
| | Paint | L.F. | .22 | 3.09 | | 3.31 | |
| | Minimum Charge | Job | | 232 | | 232 | |

# Rough Frame / Structure

| Beam / Girder | | Unit | Material | Labor | Equip. | Total | Specification |
|---|---|---|---|---|---|---|---|
| **6" x 8"** | | | | | | | |
| | Demolish | L.F. | | 4.41 | | 4.41 | Cost includes material and labor to |
| | Install | L.F. | 8.70 | 4.06 | 1.04 | 13.80 | install 6" x 8" beam. |
| | Demolish and Install | L.F. | 8.70 | 8.47 | 1.04 | 18.21 | |
| | Clean | L.F. | .07 | .44 | | .51 | |
| | Paint | L.F. | .22 | 1.50 | | 1.72 | |
| | Minimum Charge | Job | | 232 | | 232 | |
| **6" x 10"** | | | | | | | |
| | Demolish | L.F. | | 5.50 | | 5.50 | Cost includes material and labor to |
| | Install | L.F. | 4.65 | 1.86 | | 6.51 | install 6" x 10" triple beam. |
| | Demolish and Install | L.F. | 4.65 | 7.36 | | 12.01 | |
| | Clean | L.F. | .08 | .51 | | .59 | |
| | Paint | L.F. | .25 | 1.72 | | 1.97 | |
| | Minimum Charge | Job | | 232 | | 232 | |
| **6" x 12"** | | | | | | | |
| | Demolish | L.F. | | 6.55 | | 6.55 | Cost includes material and labor to |
| | Install | L.F. | 5.95 | 1.96 | | 7.91 | install 6" x 12" triple beam. |
| | Demolish and Install | L.F. | 5.95 | 8.51 | | 14.46 | |
| | Clean | L.F. | .09 | .57 | | .66 | |
| | Paint | L.F. | .28 | 1.93 | | 2.21 | |
| | Minimum Charge | Job | | 232 | | 232 | |
| **Steel I-type** | | | | | | | |
| **W8 x 31** | | | | | | | |
| | Demolish | L.F. | | 10.20 | 2.89 | 13.09 | Includes material and labor to install |
| | Install | L.F. | 50.50 | 3.03 | 1.72 | 55.25 | W shaped steel beam / girder. |
| | Demolish and Install | L.F. | 50.50 | 13.23 | 4.61 | 68.34 | |
| | Minimum Charge | Job | | 550 | | 550 | |
| **W8 x 48** | | | | | | | |
| | Demolish | L.F. | | 10.20 | 2.89 | 13.09 | Includes material and labor to install |
| | Install | L.F. | 78 | 3.17 | 1.80 | 82.97 | W shaped steel beam / girder. |
| | Demolish and Install | L.F. | 78 | 13.37 | 4.69 | 96.06 | |
| | Minimum Charge | Job | | 550 | | 550 | |
| **W8 x 67** | | | | | | | |
| | Demolish | L.F. | | 10.20 | 2.89 | 13.09 | Includes material and labor to install |
| | Install | L.F. | 109 | 3.32 | 1.89 | 114.21 | W shaped steel beam / girder. |
| | Demolish and Install | L.F. | 109 | 13.52 | 4.78 | 127.30 | |
| | Minimum Charge | Job | | 550 | | 550 | |
| **W10 x 45** | | | | | | | |
| | Demolish | L.F. | | 10.20 | 2.89 | 13.09 | Includes material and labor to install |
| | Install | L.F. | 73 | 3.17 | 1.80 | 77.97 | W shaped steel beam / girder. |
| | Demolish and Install | L.F. | 73 | 13.37 | 4.69 | 91.06 | |
| | Minimum Charge | Job | | 550 | | 550 | |
| **W10 x 68** | | | | | | | |
| | Demolish | L.F. | | 10.20 | 2.89 | 13.09 | Includes material and labor to install |
| | Install | L.F. | 111 | 3.32 | 1.89 | 116.21 | W shaped steel beam / girder. |
| | Demolish and Install | L.F. | 111 | 13.52 | 4.78 | 129.30 | |
| | Minimum Charge | Job | | 550 | | 550 | |
| **W12 x 50** | | | | | | | |
| | Demolish | L.F. | | 10.20 | 2.89 | 13.09 | Includes material and labor to install |
| | Install | L.F. | 81.50 | 3.17 | 1.80 | 86.47 | W shaped steel beam / girder. |
| | Demolish and Install | L.F. | 81.50 | 13.37 | 4.69 | 99.56 | |
| | Minimum Charge | Job | | 550 | | 550 | |
| **W12 x 87** | | | | | | | |
| | Demolish | L.F. | | 14.95 | 4.24 | 19.19 | Includes material and labor to install |
| | Install | L.F. | 142 | 3.32 | 1.89 | 147.21 | W shaped steel beam / girder. |
| | Demolish and Install | L.F. | 142 | 18.27 | 6.13 | 166.40 | |
| | Minimum Charge | Job | | 550 | | 550 | |

# Rough Frame / Structure

## Beam / Girder

| | Unit | Material | Labor | Equip. | Total | Specification |
|---|---|---|---|---|---|---|
| **W12 x 120** | | | | | | |
| Demolish | L.F. | | 14.95 | 4.24 | 19.19 | Includes material and labor to install |
| Install | L.F. | 196 | 3.41 | 1.93 | 201.34 | W shaped steel beam / girder. |
| Demolish and Install | L.F. | 196 | 18.36 | 6.17 | 220.53 | |
| Minimum Charge | Job | | 550 | | 550 | |
| **W14 x 74** | | | | | | |
| Demolish | L.F. | | 14.95 | 4.24 | 19.19 | Includes material and labor to install |
| Install | L.F. | 121 | 3.32 | 1.89 | 126.21 | W shaped steel beam / girder. |
| Demolish and Install | L.F. | 121 | 18.27 | 6.13 | 145.40 | |
| Minimum Charge | Job | | 550 | | 550 | |
| **W14 x 120** | | | | | | |
| Demolish | L.F. | | 14.95 | 4.24 | 19.19 | Includes material and labor to install |
| Install | L.F. | 196 | 3.41 | 1.93 | 201.34 | W shaped steel beam / girder. |
| Demolish and Install | L.F. | 196 | 18.36 | 6.17 | 220.53 | |
| Minimum Charge | Job | | 550 | | 550 | |

## Column

| | Unit | Material | Labor | Equip. | Total | Specification |
|---|---|---|---|---|---|---|
| **Concrete** | | | | | | |
| **Small Diameter** | | | | | | |
| Demolish | L.F. | | 16.70 | 2.39 | 19.09 | Includes material and labor to install |
| Install | L.F. | 211 | 15.30 | 2.60 | 228.90 | small diameter concrete column. |
| Demolish and Install | L.F. | 211 | 32 | 4.99 | 247.99 | |
| Minimum Charge | Job | | 765 | 130 | 895 | |
| **Large Diameter** | | | | | | |
| Demolish | L.F. | | 50 | 7.15 | 57.15 | Includes material and labor to install |
| Install | L.F. | 375 | 18.05 | 3.05 | 396.10 | large diameter concrete column. |
| Demolish and Install | L.F. | 375 | 68.05 | 10.20 | 453.25 | |
| Minimum Charge | Job | | 765 | 130 | 895 | |

## Floor Framing System

| | Unit | Material | Labor | Equip. | Total | Specification |
|---|---|---|---|---|---|---|
| **12" O.C.** | | | | | | |
| **2" x 6" Joists** | | | | | | |
| Demolish | S.F. | | .81 | | .81 | Cost includes material and labor to |
| Install | S.F. | .83 | .74 | | 1.57 | install 2" x 6" joists, 12" O.C. |
| Demolish and Install | S.F. | .83 | 1.55 | | 2.38 | including box or band joist. Blocking |
| Reinstall | S.F. | | .59 | | .59 | or bridging not included. |
| Minimum Charge | Job | | 232 | | 232 | |
| **2" x 8" Joists** | | | | | | |
| Demolish | S.F. | | .84 | | .84 | Cost includes material and labor to |
| Install | S.F. | 1.14 | .85 | | 1.99 | install 2" x 8" joists, 12" O.C. |
| Demolish and Install | S.F. | 1.14 | 1.69 | | 2.83 | including box or band joist. Blocking |
| Reinstall | S.F. | | .68 | | .68 | or bridging not included. |
| Minimum Charge | Job | | 232 | | 232 | |
| **2" x 10" Joists** | | | | | | |
| Demolish | S.F. | | .87 | | .87 | Cost includes material and labor to |
| Install | S.F. | 1.82 | 1.03 | | 2.85 | install 2" x 10" joists, 12" O.C. |
| Demolish and Install | S.F. | 1.82 | 1.90 | | 3.72 | including box or band joist. Blocking |
| Reinstall | S.F. | | .83 | | .83 | or bridging not included. |
| Minimum Charge | Job | | 232 | | 232 | |

| Floor Framing System | | Unit | Material | Labor | Equip. | Total | Specification |
|---|---|---|---|---|---|---|---|
| **2" x 12" Joists** | | | | | | | Cost includes material and labor to install 2" x 12" joists, 12" O.C. including box or band joist. Blocking or bridging not included. |
| | Demolish | S.F. | | .90 | | .90 | |
| | Install | S.F. | 2.31 | 1.06 | | 3.37 | |
| | Demolish and Install | S.F. | 2.31 | 1.96 | | 4.27 | |
| | Reinstall | S.F. | | .85 | | .85 | |
| | Minimum Charge | Job | | 232 | | 232 | |
| **Block / Bridge** | | | | | | | Includes material and labor to install set of cross bridging or per block of solid bridging for 2" x 10" joists cut to size on site. |
| | Demolish | Ea. | | 1.23 | | 1.23 | |
| | Install | Ea. | 2.07 | 2.61 | | 4.68 | |
| | Demolish and Install | Ea. | 2.07 | 3.84 | | 5.91 | |
| | Minimum Charge | Job | | 232 | | 232 | |
| **16" O.C.** | | | | | | | |
| **2" x 6" Joists** | | | | | | | Cost includes material and labor to install 2" x 6" joists including box or band joist installed 16" O.C. Does not include beams, blocking or bridging. |
| | Demolish | S.F. | | .61 | | .61 | |
| | Install | S.F. | .64 | .56 | | 1.20 | |
| | Demolish and Install | S.F. | .64 | 1.17 | | 1.81 | |
| | Reinstall | S.F. | | .45 | | .45 | |
| | Minimum Charge | Job | | 232 | | 232 | |
| **2" x 8" Joists** | | | | | | | Cost includes material and labor to install 2" x 8" joists including box or band joist installed 16" O.C. Does not include beams, blocking or bridging. |
| | Demolish | S.F. | | .63 | | .63 | |
| | Install | S.F. | .88 | .63 | | 1.51 | |
| | Demolish and Install | S.F. | .88 | 1.26 | | 2.14 | |
| | Reinstall | S.F. | | .51 | | .51 | |
| | Minimum Charge | Job | | 232 | | 232 | |
| **2" x 10" Joists** | | | | | | | Cost includes material and labor to install 2" x 10" joists including box or band joist 16" O.C. Does not include beams, blocking or bridging. |
| | Demolish | S.F. | | .65 | | .65 | |
| | Install | S.F. | 1.40 | .77 | | 2.17 | |
| | Demolish and Install | S.F. | 1.40 | 1.42 | | 2.82 | |
| | Reinstall | S.F. | | .62 | | .62 | |
| | Minimum Charge | Job | | 232 | | 232 | |
| **2" x 12" Joists** | | | | | | | Cost includes material and labor to install 2" x 12" joists including box or band joist installed 16" O.C. Does not include beams, blocking or bridging. |
| | Demolish | S.F. | | .67 | | .67 | |
| | Install | S.F. | 1.77 | .80 | | 2.57 | |
| | Demolish and Install | S.F. | 1.77 | 1.47 | | 3.24 | |
| | Reinstall | S.F. | | .64 | | .64 | |
| | Minimum Charge | Job | | 232 | | 232 | |
| **Block / Bridge** | | | | | | | Includes material and labor to install set of cross bridging or per block of solid bridging for 2" x 10" joists cut to size on site. |
| | Demolish | Ea. | | 1.23 | | 1.23 | |
| | Install | Ea. | 2.07 | 2.61 | | 4.68 | |
| | Demolish and Install | Ea. | 2.07 | 3.84 | | 5.91 | |
| | Minimum Charge | Job | | 232 | | 232 | |
| **24" O.C.** | | | | | | | |
| **2" x 6" Joists** | | | | | | | Cost includes material and labor to install 2" x 6" joists including box or band joist installed 24" O.C. Does not include beams, blocking or bridging. |
| | Demolish | S.F. | | .41 | | .41 | |
| | Install | S.F. | .45 | .37 | | .82 | |
| | Demolish and Install | S.F. | .45 | .78 | | 1.23 | |
| | Reinstall | S.F. | | .30 | | .30 | |
| | Minimum Charge | Job | | 232 | | 232 | |
| **2" x 8" Joists** | | | | | | | Cost includes material and labor to install 2" x 8" joists including box or band joist installed 24" O.C. Does not include beams, blocking or bridging. |
| | Demolish | S.F. | | .42 | | .42 | |
| | Install | S.F. | .63 | .42 | | 1.05 | |
| | Demolish and Install | S.F. | .63 | .84 | | 1.47 | |
| | Reinstall | S.F. | | .34 | | .34 | |
| | Minimum Charge | Job | | 232 | | 232 | |

| Floor Framing System | | Unit | Material | Labor | Equip. | Total | Specification |
|---|---|---|---|---|---|---|---|
| **2" x 10" Joists** | | | | | | | |
| | Demolish | S.F. | | .43 | | .43 | Cost includes material and labor to |
| | Install | S.F. | .99 | .52 | | 1.51 | install 2" x 10" joists including box or |
| | Demolish and Install | S.F. | .99 | .95 | | 1.94 | band joist installed 24" O.C. Does not |
| | Reinstall | S.F. | | .41 | | .41 | include beams, blocking or bridging. |
| | Minimum Charge | Job | | 232 | | 232 | |
| **2" x 12" Joists** | | | | | | | |
| | Demolish | S.F. | | .45 | | .45 | Cost includes material and labor to |
| | Install | S.F. | 1.27 | .53 | | 1.80 | install 2" x 12" joists including box or |
| | Demolish and Install | S.F. | 1.27 | .98 | | 2.25 | band joist installed 24" O.C. Does not |
| | Reinstall | S.F. | | .43 | | .43 | include beams, blocking or bridging. |
| | Minimum Charge | Job | | 232 | | 232 | |
| **Block / Bridge** | | | | | | | |
| | Demolish | Ea. | | 1.23 | | 1.23 | Includes material and labor to install |
| | Install | Ea. | 2.07 | 2.61 | | 4.68 | set of cross bridging or per block of |
| | Demolish and Install | Ea. | 2.07 | 3.84 | | 5.91 | solid bridging for 2" x 10" joists cut to |
| | Minimum Charge | Job | | 232 | | 232 | size on site. |
| **Engineered Lumber, Joist** | | | | | | | |
| **9-1/2"** | | | | | | | |
| | Demolish | S.F. | | .65 | | .65 | Includes material and labor to install |
| | Install | S.F. | 1.75 | .46 | | 2.21 | engineered lumber truss / joists per |
| | Demolish and Install | S.F. | 1.75 | 1.11 | | 2.86 | S.F. of floor area based on joists 16" |
| | Reinstall | S.F. | | .37 | | .37 | O.C. Beams, supports and bridging |
| | Minimum Charge | Job | | 232 | | 232 | are not included. |
| **11-7/8"** | | | | | | | |
| | Demolish | S.F. | | .67 | | .67 | Includes material and labor to install |
| | Install | S.F. | 2.02 | .48 | | 2.50 | engineered lumber truss / joists per |
| | Demolish and Install | S.F. | 2.02 | 1.15 | | 3.17 | S.F. of floor area based on joists 16" |
| | Reinstall | S.F. | | .38 | | .38 | O.C. Beams, supports and bridging |
| | Minimum Charge | Job | | 232 | | 232 | are not included. |
| **14"** | | | | | | | |
| | Demolish | S.F. | | .70 | | .70 | Includes material and labor to install |
| | Install | S.F. | 2.53 | .51 | | 3.04 | engineered lumber truss / joists per |
| | Demolish and Install | S.F. | 2.53 | 1.21 | | 3.74 | S.F. of floor area based on joists 16" |
| | Reinstall | S.F. | | .41 | | .41 | O.C. Beams, supports and bridging |
| | Minimum Charge | Job | | 232 | | 232 | are not included. |
| **16"** | | | | | | | |
| | Demolish | S.F. | | .72 | | .72 | Includes material and labor to install |
| | Install | S.F. | 4.03 | .54 | | 4.57 | engineered lumber truss / joists per |
| | Demolish and Install | S.F. | 4.03 | 1.26 | | 5.29 | S.F. of floor area based on joists 16" |
| | Reinstall | S.F. | | .43 | | .43 | O.C. Beams, supports and bridging |
| | Minimum Charge | Job | | 232 | | 232 | are not included. |
| **Block / Bridge** | | | | | | | |
| | Demolish | Pr. | | 1.23 | | 1.23 | Includes material and labor to install |
| | Install | Pr. | 1.88 | 3.44 | | 5.32 | set of steel, one-nail type cross |
| | Demolish and Install | Pr. | 1.88 | 4.67 | | 6.55 | bridging for trusses placed 16" O.C. |
| | Minimum Charge | Job | | 232 | | 232 | |
| **Block / Bridge** | | | | | | | |
| **2" x 6"** | | | | | | | |
| | Demolish | Ea. | | 1.23 | | 1.23 | Includes material and labor to install |
| | Install | Ea. | .95 | 2.09 | | 3.04 | set of cross bridging or per block of |
| | Demolish and Install | Ea. | .95 | 3.32 | | 4.27 | solid bridging for 2" x 6" joists cut to |
| | Minimum Charge | Job | | 232 | | 232 | size on site. |

| Floor Framing System | | Unit | Material | Labor | Equip. | Total | Specification |
|---|---|---|---|---|---|---|---|
| **2″ x 8″** | | | | | | | |
| | Demolish | Ea. | | 1.23 | | 1.23 | Includes material and labor to install |
| | Install | Ea. | 1.31 | 2.32 | | 3.63 | set of cross bridging or per block of |
| | Demolish and Install | Ea. | 1.31 | 3.55 | | 4.86 | solid bridging for 2″ x 8″ joists cut to |
| | Minimum Charge | Job | | 232 | | 232 | size on site. |
| **2″ x 10″** | | | | | | | |
| | Demolish | Ea. | | 1.23 | | 1.23 | Includes material and labor to install |
| | Install | Ea. | 2.07 | 2.61 | | 4.68 | set of cross bridging or per block of |
| | Demolish and Install | Ea. | 2.07 | 3.84 | | 5.91 | solid bridging for 2″ x 10″ joists cut to |
| | Minimum Charge | Job | | 232 | | 232 | size on site. |
| **2″ x 12″** | | | | | | | |
| | Demolish | Ea. | | 1.23 | | 1.23 | Includes material and labor to install |
| | Install | Ea. | 2.64 | 3.08 | | 5.72 | set of cross bridging or per block of |
| | Demolish and Install | Ea. | 2.64 | 4.31 | | 6.95 | solid bridging for 2″ x 12″ joists cut to |
| | Minimum Charge | Job | | 232 | | 232 | size on site. |
| **Ledger Strips** | | | | | | | |
| **1″ x 2″** | | | | | | | |
| | Demolish | L.F. | | .33 | | .33 | Cost includes material and labor to |
| | Install | L.F. | .30 | 1.35 | | 1.65 | install 1″ x 2″ ledger strip nailed to the |
| | Demolish and Install | L.F. | .30 | 1.68 | | 1.98 | face of studs, beams or joist. |
| | Minimum Charge | Job | | 232 | | 232 | |
| **1″ x 3″** | | | | | | | |
| | Demolish | L.F. | | .33 | | .33 | Includes material and labor to install |
| | Install | L.F. | .56 | 1.55 | | 2.11 | up to 1″ x 4″ ledger strip nailed to the |
| | Demolish and Install | L.F. | .56 | 1.88 | | 2.44 | face of studs, beams or joist. |
| | Minimum Charge | Job | | 232 | | 232 | |
| **1″ x 4″** | | | | | | | |
| | Demolish | L.F. | | .33 | | .33 | Includes material and labor to install |
| | Install | L.F. | .56 | 1.55 | | 2.11 | up to 1″ x 4″ ledger strip nailed to the |
| | Demolish and Install | L.F. | .56 | 1.88 | | 2.44 | face of studs, beams or joist. |
| | Minimum Charge | Job | | 232 | | 232 | |
| **2″ x 2″** | | | | | | | |
| | Demolish | L.F. | | .36 | | .36 | Cost includes material and labor to |
| | Install | L.F. | .46 | 1.41 | | 1.87 | install 2″ x 2″ ledger strip nailed to the |
| | Demolish and Install | L.F. | .46 | 1.77 | | 2.23 | face of studs, beams or joist. |
| | Minimum Charge | Job | | 232 | | 232 | |
| **2″ x 4″** | | | | | | | |
| | Demolish | L.F. | | .39 | | .39 | Cost includes material and labor to |
| | Install | L.F. | .46 | 1.86 | | 2.32 | install 2″ x 4″ ledger strip nailed to the |
| | Demolish and Install | L.F. | .46 | 2.25 | | 2.71 | face of studs, beams or joist. |
| | Reinstall | L.F. | | 1.49 | | 1.49 | |
| | Minimum Charge | Job | | 232 | | 232 | |
| **Ledger Boards** | | | | | | | |
| **2″ x 6″** | | | | | | | |
| | Demolish | L.F. | | .39 | | .39 | Cost includes material and labor to |
| | Install | L.F. | 5.45 | 2.91 | | 8.36 | install 2″ x 6″ ledger board fastened to |
| | Demolish and Install | L.F. | 5.45 | 3.30 | | 8.75 | a wall, joists or studs. |
| | Reinstall | L.F. | | 2.32 | | 2.32 | |
| | Minimum Charge | Job | | 232 | | 232 | |
| **4″ x 6″** | | | | | | | |
| | Demolish | L.F. | | .66 | | .66 | Cost includes material and labor to |
| | Install | L.F. | 9.55 | 3.32 | | 12.87 | install 4″ x 6″ ledger board fastened to |
| | Demolish and Install | L.F. | 9.55 | 3.98 | | 13.53 | a wall, joists or studs. |
| | Reinstall | L.F. | | 2.66 | | 2.66 | |
| | Minimum Charge | Job | | 232 | | 232 | |

| Floor Framing System | | Unit | Material | Labor | Equip. | Total | Specification |
|---|---|---|---|---|---|---|---|
| **4" x 8"** | | | | | | | |
| | Demolish | L.F. | | .88 | | .88 | Cost includes material and labor to |
| | Install | L.F. | 10.80 | 3.87 | | 14.67 | install 4" x 8" ledger board fastened to |
| | Demolish and Install | L.F. | 10.80 | 4.75 | | 15.55 | a wall, joists or studs. |
| | Reinstall | L.F. | | 3.10 | | 3.10 | |
| | Minimum Charge | Job | | 232 | | 232 | |
| **Sill Plate Per L.F.** | | | | | | | |
| **2" x 4"** | | | | | | | |
| | Demolish | L.F. | | .34 | | .34 | Cost includes material and labor to |
| | Install | L.F. | .64 | 1.69 | | 2.33 | install 2" x 4" pressure treated lumber, |
| | Demolish and Install | L.F. | .64 | 2.03 | | 2.67 | drilled and installed on foundation |
| | Reinstall | L.F. | | 1.35 | | 1.35 | bolts 48" O.C. Bolts, nuts and washers |
| | Minimum Charge | Job | | 232 | | 232 | are not included. |
| **2" x 6"** | | | | | | | |
| | Demolish | L.F. | | .51 | | .51 | Cost includes material and labor to |
| | Install | L.F. | .80 | 1.86 | | 2.66 | install 2" x 6" pressure treated lumber, |
| | Demolish and Install | L.F. | .80 | 2.37 | | 3.17 | drilled and installed on foundation |
| | Reinstall | L.F. | | 1.49 | | 1.49 | bolts 48" O.C. Bolts, nuts and washers |
| | Minimum Charge | Job | | 232 | | 232 | not included. |
| **2" x 8"** | | | | | | | |
| | Demolish | L.F. | | .67 | | .67 | Cost includes material and labor to |
| | Install | L.F. | 1.22 | 2.07 | | 3.29 | install 2" x 8" pressure treated lumber, |
| | Demolish and Install | L.F. | 1.22 | 2.74 | | 3.96 | drilled and installed on foundation |
| | Reinstall | L.F. | | 1.65 | | 1.65 | bolts 48" O.C. Bolts, nuts and washers |
| | Minimum Charge | Job | | 232 | | 232 | not included. |
| **Earthquake Strapping** | | | | | | | |
| | Install | Ea. | 2.97 | 2.91 | | 5.88 | Includes labor and material to replace |
| | Minimum Charge | Job | | 232 | | 232 | an earthquake strap. |

| Subflooring | | Unit | Material | Labor | Equip. | Total | Specification |
|---|---|---|---|---|---|---|---|
| **Plywood** | | | | | | | |
| **1/2"** | | | | | | | |
| | Demolish | S.F. | | 1.03 | | 1.03 | Cost includes material and labor to |
| | Install | SF Flr. | .52 | .50 | | 1.02 | install 1/2" plywood subfloor, CD |
| | Demolish and Install | SF Flr. | .52 | 1.53 | | 2.05 | standard interior grade, plugged and |
| | Minimum Charge | Job | | 232 | | 232 | touch sanded. |
| **5/8"** | | | | | | | |
| | Demolish | S.F. | | 1.04 | | 1.04 | Cost includes material and labor to |
| | Install | SF Flr. | .64 | .56 | | 1.20 | install 5/8" plywood subfloor, CD |
| | Demolish and Install | SF Flr. | .64 | 1.60 | | 2.24 | standard interior grade, plugged and |
| | Reinstall | SF Flr. | | .44 | | .44 | touch sanded. |
| | Minimum Charge | Job | | 232 | | 232 | |
| **3/4"** | | | | | | | |
| | Demolish | S.F. | | 1.05 | | 1.05 | Cost includes material and labor to |
| | Install | SF Flr. | .77 | .60 | | 1.37 | install 3/4" plywood subfloor, CD |
| | Demolish and Install | SF Flr. | .77 | 1.65 | | 2.42 | standard interior grade, plugged and |
| | Reinstall | SF Flr. | | .48 | | .48 | touch sanded. |
| | Minimum Charge | Job | | 232 | | 232 | |

33

For customer support on your Contractor's Pricing Guide: Residential Repair & Remodeling, call 800.448.8182.

## Subflooring

| Subflooring | Unit | Material | Labor | Equip. | Total | Specification |
|---|---|---|---|---|---|---|
| **Particle Board** | | | | | | |
| **1/2"** | | | | | | |
| Demolish | S.F. | | .51 | | .51 | Cost includes material and labor to |
| Install | SF Flr. | .47 | .62 | | 1.09 | install 1/2" particle board subfloor. |
| Demolish and Install | SF Flr. | .47 | 1.13 | | 1.60 | |
| Minimum Charge | Job | | 232 | | 232 | |
| **5/8"** | | | | | | |
| Demolish | S.F. | | .52 | | .52 | Cost includes material and labor to |
| Install | SF Flr. | .61 | .68 | | 1.29 | install 5/8" particle board subfloor. |
| Demolish and Install | SF Flr. | .61 | 1.20 | | 1.81 | |
| Minimum Charge | Job | | 232 | | 232 | |
| **3/4"** | | | | | | |
| Demolish | S.F. | | .53 | | .53 | Cost includes material and labor to |
| Install | SF Flr. | .74 | .74 | | 1.48 | install 3/4" particle board subfloor. |
| Demolish and Install | SF Flr. | .74 | 1.27 | | 2.01 | |
| Minimum Charge | Job | | 232 | | 232 | |
| **Plank Board** | | | | | | |
| Demolish | S.F. | | 1.59 | | 1.59 | Cost includes material and labor to |
| Install | SF Flr. | 1.87 | 1.03 | | 2.90 | install 1" x 6" standard grade plank |
| Demolish and Install | SF Flr. | 1.87 | 2.62 | | 4.49 | flooring. |
| Reinstall | SF Flr. | | 1.03 | | 1.03 | |
| Minimum Charge | Job | | 232 | | 232 | |
| **Felt Underlay** | | | | | | |
| Demolish | S.F. | | .09 | | .09 | Includes material and labor to install |
| Install | S.F. | .06 | .13 | | .19 | #15 felt building paper. |
| Demolish and Install | S.F. | .06 | .22 | | .28 | |
| Minimum Charge | Job | | 232 | | 232 | |
| **Prep (for flooring)** | | | | | | |
| Install | S.F. | | .42 | | .42 | Preparation of subflooring for |
| Minimum Charge | Job | | 219 | | 219 | installation of new finished flooring. Cost reflects average time to prep area. |

## Wall Framing System

| Wall Framing System | Unit | Material | Labor | Equip. | Total | Specification |
|---|---|---|---|---|---|---|
| **2" x 4"** | | | | | | |
| **8' High** | | | | | | |
| Demolish | L.F. | | 5.25 | | 5.25 | Cost includes material and labor to |
| Install | L.F. | 4.79 | 7.75 | | 12.54 | install 2" x 4" x 8' high wall system, |
| Demolish and Install | L.F. | 4.79 | 13 | | 17.79 | including studs, treated bottom plate, |
| Reinstall | L.F. | | 6.20 | | 6.20 | double top plate and one row of fire |
| Minimum Charge | Job | | 232 | | 232 | blocking. |
| **9' High** | | | | | | |
| Demolish | L.F. | | 5.90 | | 5.90 | Cost includes material and labor to |
| Install | L.F. | 5.15 | 7.75 | | 12.90 | install 2" x 4" x 9' wall system |
| Demolish and Install | L.F. | 5.15 | 13.65 | | 18.80 | including studs, treated bottom plate, |
| Reinstall | L.F. | | 6.20 | | 6.20 | double top plate and one row of fire |
| Minimum Charge | Job | | 232 | | 232 | blocking. |
| **10' High** | | | | | | |
| Demolish | L.F. | | 6.60 | | 6.60 | Cost includes material and labor to |
| Install | L.F. | 5.50 | 7.75 | | 13.25 | install 2" x 4" x 10' wall system |
| Demolish and Install | L.F. | 5.50 | 14.35 | | 19.85 | including studs, treated bottom plate, |
| Reinstall | L.F. | | 6.20 | | 6.20 | double top plate and one row of fire |
| Minimum Charge | Job | | 232 | | 232 | blocking. |

| Wall Framing System | | Unit | Material | Labor | Equip. | Total | Specification |
|---|---|---|---|---|---|---|---|
| **12′ High** | | | | | | | |
| | Demolish | L.F. | | 7.90 | | 7.90 | Cost includes material and labor to |
| | Install | L.F. | 6.15 | 9.70 | | 15.85 | install 2″ x 4″ x 12′ wall system |
| | Demolish and Install | L.F. | 6.15 | 17.60 | | 23.75 | including studs, treated bottom plate, |
| | Reinstall | L.F. | | 7.75 | | 7.75 | double top plate and one row of fire |
| | Minimum Charge | Job | | 232 | | 232 | blocking. |
| **2″ x 6″** | | | | | | | |
| **8′ High** | | | | | | | |
| | Demolish | L.F. | | 4.93 | | 4.93 | Cost includes material and labor to |
| | Install | L.F. | 7.15 | 8.60 | | 15.75 | install 2″ x 6″ x 8′ wall system |
| | Demolish and Install | L.F. | 7.15 | 13.53 | | 20.68 | including studs, treated bottom plate, |
| | Reinstall | L.F. | | 6.89 | | 6.89 | double top plate and one row of fire |
| | Minimum Charge | Job | | 232 | | 232 | blocking. |
| **9′ High** | | | | | | | |
| | Demolish | L.F. | | 5.55 | | 5.55 | Cost includes material and labor to |
| | Install | L.F. | 7.70 | 8.60 | | 16.30 | install 2″ x 6″ x 9′ wall system |
| | Demolish and Install | L.F. | 7.70 | 14.15 | | 21.85 | including studs, treated bottom plate, |
| | Reinstall | L.F. | | 6.89 | | 6.89 | double top plate and one row of fire |
| | Minimum Charge | Job | | 232 | | 232 | blocking. |
| **10′ High** | | | | | | | |
| | Demolish | L.F. | | 6.05 | | 6.05 | Cost includes material and labor to |
| | Install | L.F. | 8.25 | 8.60 | | 16.85 | install 2″ x 6″ x 10′ wall system |
| | Demolish and Install | L.F. | 8.25 | 14.65 | | 22.90 | including studs, treated bottom plate, |
| | Reinstall | L.F. | | 6.89 | | 6.89 | double top plate and one row of fire |
| | Minimum Charge | Job | | 232 | | 232 | blocking. |
| **12′ High** | | | | | | | |
| | Demolish | L.F. | | 7.20 | | 7.20 | Cost includes material and labor to |
| | Install | L.F. | 9.30 | 10.55 | | 19.85 | install 2″ x 6″ x 12′ wall system |
| | Demolish and Install | L.F. | 9.30 | 17.75 | | 27.05 | including studs, treated bottom plate, |
| | Reinstall | L.F. | | 8.45 | | 8.45 | double top plate and one row of fire |
| | Minimum Charge | Job | | 232 | | 232 | blocking. |
| **Fireblock** | | | | | | | |
| **2″ x 4″, 16″ O.C. System** | | | | | | | |
| | Demolish | L.F. | | .66 | | .66 | Cost includes material and labor to |
| | Install | L.F. | .46 | 1.55 | | 2.01 | install 2″ x 4″ fireblocks in wood frame |
| | Demolish and Install | L.F. | .46 | 2.21 | | 2.67 | walls per L.F. of wall to be blocked. |
| | Reinstall | L.F. | | 1.24 | | 1.24 | |
| | Minimum Charge | Job | | 232 | | 232 | |
| **2″ x 4″, 24″ O.C. System** | | | | | | | |
| | Demolish | L.F. | | .66 | | .66 | Cost includes material and labor to |
| | Install | L.F. | .46 | 1.55 | | 2.01 | install 2″ x 4″ fireblocks in wood frame |
| | Demolish and Install | L.F. | .46 | 2.21 | | 2.67 | walls per L.F. of wall to be blocked. |
| | Reinstall | L.F. | | 1.24 | | 1.24 | |
| | Minimum Charge | Job | | 232 | | 232 | |
| **2″ x 6″, 16″ O.C. System** | | | | | | | |
| | Demolish | L.F. | | .99 | | .99 | Cost includes material and labor to |
| | Install | L.F. | .70 | 1.55 | | 2.25 | install 2″ x 6″ fireblocks in wood frame |
| | Demolish and Install | L.F. | .70 | 2.54 | | 3.24 | walls per L.F. of wall to be blocked. |
| | Reinstall | L.F. | | 1.24 | | 1.24 | |
| | Minimum Charge | Job | | 232 | | 232 | |
| **2″ x 6″, 24″ O.C. System** | | | | | | | |
| | Demolish | L.F. | | .99 | | .99 | Cost includes material and labor to |
| | Install | L.F. | .70 | 1.55 | | 2.25 | install 2″ x 6″ fireblocks in wood frame |
| | Demolish and Install | L.F. | .70 | 2.54 | | 3.24 | walls per L.F. of wall to be blocked. |
| | Reinstall | L.F. | | 1.24 | | 1.24 | |
| | Minimum Charge | Job | | 232 | | 232 | |

| Wall Framing System | | Unit | Material | Labor | Equip. | Total | Specification |
|---|---|---|---|---|---|---|---|
| **Bracing** | | | | | | | |
| **1″ x 3″** | | | | | | | |
| | Demolish | L.F. | | .38 | | .38 | Cost includes material and labor to |
| | Install | L.F. | .47 | 2.91 | | 3.38 | install 1″ x 3″ let-in bracing. |
| | Demolish and Install | L.F. | .47 | 3.29 | | 3.76 | |
| | Minimum Charge | Job | | 232 | | 232 | |
| **1″ x 4″** | | | | | | | |
| | Demolish | L.F. | | .38 | | .38 | Cost includes material and labor to |
| | Install | L.F. | .56 | 2.91 | | 3.47 | install 1″ x 4″ let-in bracing. |
| | Demolish and Install | L.F. | .56 | 3.29 | | 3.85 | |
| | Minimum Charge | Job | | 232 | | 232 | |
| **1″ x 6″** | | | | | | | |
| | Demolish | L.F. | | .38 | | .38 | Cost includes material and labor to |
| | Install | L.F. | .86 | 3.10 | | 3.96 | install 1″ x 6″ let-in bracing. |
| | Demolish and Install | L.F. | .86 | 3.48 | | 4.34 | |
| | Minimum Charge | Job | | 232 | | 232 | |
| **2″ x 3″** | | | | | | | |
| | Demolish | L.F. | | .49 | | .49 | Cost includes material and labor to |
| | Install | L.F. | .46 | 3.10 | | 3.56 | install 2″ x 3″ let-in bracing. |
| | Demolish and Install | L.F. | .46 | 3.59 | | 4.05 | |
| | Reinstall | L.F. | | 2.48 | | 2.48 | |
| | Minimum Charge | Job | | 232 | | 232 | |
| **2″ x 4″** | | | | | | | |
| | Demolish | L.F. | | .49 | | .49 | Cost includes material and labor to |
| | Install | L.F. | .46 | 3.10 | | 3.56 | install 2″ x 4″ let-in bracing. |
| | Demolish and Install | L.F. | .46 | 3.59 | | 4.05 | |
| | Reinstall | L.F. | | 2.48 | | 2.48 | |
| | Minimum Charge | Job | | 232 | | 232 | |
| **2″ x 6″** | | | | | | | |
| | Demolish | L.F. | | .49 | | .49 | Cost includes material and labor to |
| | Install | L.F. | .70 | 3.32 | | 4.02 | install 2″ x 6″ let-in bracing. |
| | Demolish and Install | L.F. | .70 | 3.81 | | 4.51 | |
| | Reinstall | L.F. | | 2.66 | | 2.66 | |
| | Minimum Charge | Job | | 232 | | 232 | |
| **2″ x 8″** | | | | | | | |
| | Demolish | L.F. | | .49 | | .49 | Cost includes material and labor to |
| | Install | L.F. | .98 | 3.32 | | 4.30 | install 2″ x 8″ let-in bracing. |
| | Demolish and Install | L.F. | .98 | 3.81 | | 4.79 | |
| | Reinstall | L.F. | | 2.66 | | 2.66 | |
| | Minimum Charge | Job | | 232 | | 232 | |
| **Earthquake Strapping** | | | | | | | |
| | Install | Ea. | 2.97 | 2.91 | | 5.88 | Includes labor and material to replace |
| | Minimum Charge | Job | | 232 | | 232 | an earthquake strap. |
| **Hurricane Clips** | | | | | | | |
| | Install | Ea. | 1.11 | 3.21 | | 4.32 | Includes labor and material to install a |
| | Minimum Charge | Job | | 232 | | 232 | hurricane clip. |
| **Stud** | | | | | | | |
| **2″ x 4″** | | | | | | | |
| | Demolish | L.F. | | .39 | | .39 | Cost includes material and labor to |
| | Install | L.F. | .45 | .85 | | 1.30 | install 2″ x 4″ wall stud per L.F. of stud. |
| | Demolish and Install | L.F. | .45 | 1.24 | | 1.69 | |
| | Reinstall | L.F. | | .68 | | .68 | |
| | Minimum Charge | Job | | 232 | | 232 | |

| Wall Framing System | | Unit | Material | Labor | Equip. | Total | Specification |
|---|---|---|---|---|---|---|---|
| **2″ x 6″** | | | | | | | |
| | Demolish | L.F. | | .49 | | .49 | Cost includes material and labor to |
| | Install | L.F. | .70 | .93 | | 1.63 | install 2″ x 6″ stud per L.F. of stud. |
| | Demolish and Install | L.F. | .70 | 1.42 | | 2.12 | |
| | Reinstall | L.F. | | .74 | | .74 | |
| | Minimum Charge | Job | | 232 | | 232 | |
| **Plates** | | | | | | | |
| **2″ x 4″** | | | | | | | |
| | Demolish | L.F. | | .36 | | .36 | Cost includes material and labor to |
| | Install | L.F. | .46 | 1.16 | | 1.62 | install 2″ x 4″ plate per L.F. of plate. |
| | Demolish and Install | L.F. | .46 | 1.52 | | 1.98 | |
| | Reinstall | L.F. | | .93 | | .93 | |
| | Minimum Charge | Job | | 232 | | 232 | |
| **2″ x 6″** | | | | | | | |
| | Demolish | L.F. | | .38 | | .38 | Cost includes material and labor to |
| | Install | L.F. | .70 | 1.24 | | 1.94 | install 2″ x 6″ plate per L.F. of plate. |
| | Demolish and Install | L.F. | .70 | 1.62 | | 2.32 | |
| | Reinstall | L.F. | | .99 | | .99 | |
| | Minimum Charge | Job | | 232 | | 232 | |
| **Headers** | | | | | | | |
| **2″ x 6″** | | | | | | | |
| | Demolish | L.F. | | .35 | | .35 | Includes material and labor to install |
| | Install | L.F. | .70 | 2.58 | | 3.28 | header over wall openings and |
| | Demolish and Install | L.F. | .70 | 2.93 | | 3.63 | around floor, ceiling and roof |
| | Clean | L.F. | .01 | .45 | | .46 | openings or flush beam. |
| | Paint | L.F. | .11 | 1.55 | | 1.66 | |
| | Minimum Charge | Job | | 232 | | 232 | |
| **2″ x 8″** | | | | | | | |
| | Demolish | L.F. | | .38 | | .38 | Includes material and labor to install |
| | Install | L.F. | .98 | 2.73 | | 3.71 | header over wall openings and |
| | Demolish and Install | L.F. | .98 | 3.11 | | 4.09 | around floor, ceiling and roof |
| | Clean | L.F. | .01 | .45 | | .46 | openings or flush beam. |
| | Paint | L.F. | .14 | 2.09 | | 2.23 | |
| | Minimum Charge | Job | | 232 | | 232 | |
| **2″ x 10″** | | | | | | | |
| | Demolish | L.F. | | .38 | | .38 | Includes material and labor to install |
| | Install | L.F. | 1.55 | 2.91 | | 4.46 | header over wall openings and |
| | Demolish and Install | L.F. | 1.55 | 3.29 | | 4.84 | around floor, ceiling and roof |
| | Clean | L.F. | .01 | .45 | | .46 | openings or flush beam. |
| | Paint | L.F. | .19 | 2.58 | | 2.77 | |
| | Minimum Charge | Job | | 232 | | 232 | |
| **2″ x 12″** | | | | | | | |
| | Demolish | L.F. | | .39 | | .39 | Includes material and labor to install |
| | Install | L.F. | 1.97 | 3.10 | | 5.07 | header over wall openings and |
| | Demolish and Install | L.F. | 1.97 | 3.49 | | 5.46 | around floor, ceiling and roof |
| | Clean | L.F. | .01 | .46 | | .47 | openings or flush beam. |
| | Paint | L.F. | .22 | 3.09 | | 3.31 | |
| | Minimum Charge | Job | | 232 | | 232 | |
| **4″ x 8″** | | | | | | | |
| | Demolish | L.F. | | .70 | | .70 | Includes material and labor to install |
| | Install | L.F. | 4.57 | 3.58 | | 8.15 | header over wall openings and |
| | Demolish and Install | L.F. | 4.57 | 4.28 | | 8.85 | around floor, ceiling and roof |
| | Clean | L.F. | .01 | .45 | | .46 | openings or flush beam. |
| | Paint | L.F. | .14 | 2.09 | | 2.23 | |
| | Minimum Charge | Job | | 232 | | 232 | |

# Rough Frame / Structure

## Wall Framing System

| | Unit | Material | Labor | Equip. | Total | Specification |
|---|---|---|---|---|---|---|
| **4" x 10"** | | | | | | |
| Demolish | L.F. | | .75 | | .75 | Includes material and labor to install |
| Install | L.F. | 5.45 | 3.87 | | 9.32 | header over wall openings and |
| Demolish and Install | L.F. | 5.45 | 4.62 | | 10.07 | around floor, ceiling and roof |
| Clean | L.F. | .01 | .45 | | .46 | openings or flush beam. |
| Paint | L.F. | .19 | 2.58 | | 2.77 | |
| Minimum Charge | Job | | 232 | | 232 | |
| **4" x 12"** | | | | | | |
| Demolish | L.F. | | .79 | | .79 | Includes material and labor to install |
| Install | L.F. | 6.05 | 4.89 | | 10.94 | header over wall openings and |
| Demolish and Install | L.F. | 6.05 | 5.68 | | 11.73 | around floor, ceiling and roof |
| Clean | L.F. | .01 | .46 | | .47 | openings or flush beam. |
| Paint | L.F. | .22 | 3.09 | | 3.31 | |
| Minimum Charge | Job | | 232 | | 232 | |
| **6" x 8"** | | | | | | |
| Demolish | L.F. | | .70 | | .70 | Includes material and labor to install |
| Install | L.F. | 8.70 | 2.58 | | 11.28 | header over wall openings and |
| Demolish and Install | L.F. | 8.70 | 3.28 | | 11.98 | around floor, ceiling and roof |
| Clean | L.F. | .01 | .45 | | .46 | openings or flush beam. |
| Paint | L.F. | .14 | 2.09 | | 2.23 | |
| Minimum Charge | Job | | 232 | | 232 | |
| **6" x 10"** | | | | | | |
| Demolish | L.F. | | .75 | | .75 | Includes material and labor to install |
| Install | L.F. | 7.60 | 5.65 | | 13.25 | header over wall openings and |
| Demolish and Install | L.F. | 7.60 | 6.40 | | 14 | around floor, ceiling and roof |
| Clean | L.F. | .01 | .45 | | .46 | openings or flush beam. |
| Paint | L.F. | .19 | 2.58 | | 2.77 | |
| Minimum Charge | Job | | 232 | | 232 | |
| **6" x 12"** | | | | | | |
| Demolish | L.F. | | .79 | | .79 | Includes material and labor to install |
| Install | L.F. | 9.55 | 6.65 | | 16.20 | header over wall openings and |
| Demolish and Install | L.F. | 9.55 | 7.44 | | 16.99 | around floor, ceiling and roof |
| Clean | L.F. | .01 | .46 | | .47 | openings or flush beam. |
| Paint | L.F. | .22 | 1.07 | | 1.29 | |
| Minimum Charge | Job | | 232 | | 232 | |

## Rough-in Opening

| | Unit | Material | Labor | Equip. | Total | Specification |
|---|---|---|---|---|---|---|
| **Door w / 2" x 4" Lumber** | | | | | | |
| **3' Wide** | | | | | | |
| Demolish | Ea. | | 12.35 | | 12.35 | Includes material and labor to install |
| Install | Ea. | 19.30 | 14.55 | | 33.85 | header, double studs each side, |
| Demolish and Install | Ea. | 19.30 | 26.90 | | 46.20 | cripples, blocking and nails, up to 3' |
| Reinstall | Ea. | | 11.62 | | 11.62 | opening in 2" x 4" stud wall 8' high. |
| Minimum Charge | Job | | 232 | | 232 | |
| **4' Wide** | | | | | | |
| Demolish | Ea. | | 12.35 | | 12.35 | Includes material and labor to install |
| Install | Ea. | 20.50 | 14.55 | | 35.05 | header, double studs each side, |
| Demolish and Install | Ea. | 20.50 | 26.90 | | 47.40 | cripples, blocking and nails, up to 4' |
| Reinstall | Ea. | | 11.62 | | 11.62 | opening in 2" x 4" stud wall 8' high. |
| Minimum Charge | Job | | 232 | | 232 | |
| **5' Wide** | | | | | | |
| Demolish | Ea. | | 12.35 | | 12.35 | Includes material and labor to install |
| Install | Ea. | 25 | 14.55 | | 39.55 | header, double studs each side, |
| Demolish and Install | Ea. | 25 | 26.90 | | 51.90 | cripples, blocking and nails, up to 5' |
| Reinstall | Ea. | | 11.62 | | 11.62 | opening in 2" x 4" stud wall 8' high. |
| Minimum Charge | Job | | 232 | | 232 | |

| Rough-in Opening | | Unit | Material | Labor | Equip. | Total | Specification |
|---|---|---|---|---|---|---|---|
| **6' Wide** | | | | | | | |
| | Demolish | Ea. | | 12.35 | | 12.35 | Includes material and labor to install |
| | Install | Ea. | 27 | 14.55 | | 41.55 | header, double studs each side, |
| | Demolish and Install | Ea. | 27 | 26.90 | | 53.90 | cripples, blocking and nails, up to 6' |
| | Reinstall | Ea. | | 11.62 | | 11.62 | opening in 2" x 4" stud wall 8' high. |
| | Minimum Charge | Job | | 232 | | 232 | |
| **8' Wide** | | | | | | | |
| | Demolish | Ea. | | 13.15 | | 13.15 | Includes material and labor to install |
| | Install | Ea. | 40 | 15.50 | | 55.50 | header, double studs each side, |
| | Demolish and Install | Ea. | 40 | 28.65 | | 68.65 | cripples, blocking and nails, up to 8' |
| | Reinstall | Ea. | | 12.39 | | 12.39 | opening in 2" x 4" stud wall 8' high. |
| | Minimum Charge | Job | | 232 | | 232 | |
| **10' Wide** | | | | | | | |
| | Demolish | Ea. | | 13.15 | | 13.15 | Includes material and labor to install |
| | Install | Ea. | 55 | 15.50 | | 70.50 | header, double studs each side, |
| | Demolish and Install | Ea. | 55 | 28.65 | | 83.65 | cripples, blocking and nails, up to 10' |
| | Reinstall | Ea. | | 12.39 | | 12.39 | opening in 2" x 4" stud wall 8' high. |
| | Minimum Charge | Job | | 232 | | 232 | |
| **12' Wide** | | | | | | | |
| | Demolish | Ea. | | 13.15 | | 13.15 | Includes material and labor to install |
| | Install | Ea. | 75.50 | 15.50 | | 91 | header, double studs each side, |
| | Demolish and Install | Ea. | 75.50 | 28.65 | | 104.15 | cripples, blocking and nails, up to 12' |
| | Reinstall | Ea. | | 12.39 | | 12.39 | opening in 2" x 4" stud wall 8' high. |
| | Minimum Charge | Job | | 232 | | 232 | |
| **Door w / 2" x 6" Lumber** | | | | | | | |
| **3' Wide** | | | | | | | |
| | Demolish | Ea. | | 12.35 | | 12.35 | Includes material and labor to install |
| | Install | Ea. | 27.50 | 14.55 | | 42.05 | header, double studs each side, |
| | Demolish and Install | Ea. | 27.50 | 26.90 | | 54.40 | cripples, blocking and nails, up to 3' |
| | Reinstall | Ea. | | 11.62 | | 11.62 | opening in 2" x 6" stud wall 8' high. |
| | Minimum Charge | Job | | 232 | | 232 | |
| **4' Wide** | | | | | | | |
| | Demolish | Ea. | | 12.35 | | 12.35 | Includes material and labor to install |
| | Install | Ea. | 28.50 | 14.55 | | 43.05 | header, double studs each side, |
| | Demolish and Install | Ea. | 28.50 | 26.90 | | 55.40 | cripples, blocking and nails, up to 4' |
| | Reinstall | Ea. | | 11.62 | | 11.62 | opening in 2" x 6" stud wall 8' high. |
| | Minimum Charge | Job | | 232 | | 232 | |
| **5' Wide** | | | | | | | |
| | Demolish | Ea. | | 12.35 | | 12.35 | Includes material and labor to install |
| | Install | Ea. | 33 | 14.55 | | 47.55 | header, double studs each side, |
| | Demolish and Install | Ea. | 33 | 26.90 | | 59.90 | cripples, blocking and nails up to 5' |
| | Reinstall | Ea. | | 11.62 | | 11.62 | opening in 2" x 6" stud wall 8' high. |
| | Minimum Charge | Job | | 232 | | 232 | |
| **6' Wide** | | | | | | | |
| | Demolish | Ea. | | 12.35 | | 12.35 | Includes material and labor to install |
| | Install | Ea. | 34.50 | 14.55 | | 49.05 | header, double studs each side, |
| | Demolish and Install | Ea. | 34.50 | 26.90 | | 61.40 | cripples, blocking and nails up to 6' |
| | Reinstall | Ea. | | 11.62 | | 11.62 | opening in 2" x 6" stud wall 8' high. |
| | Minimum Charge | Job | | 232 | | 232 | |
| **8' Wide** | | | | | | | |
| | Demolish | Ea. | | 13.15 | | 13.15 | Includes material and labor to install |
| | Install | Ea. | 48 | 15.50 | | 63.50 | header, double studs each side, |
| | Demolish and Install | Ea. | 48 | 28.65 | | 76.65 | cripples, blocking and nails up to 8' |
| | Reinstall | Ea. | | 12.39 | | 12.39 | opening in 2" x 6" stud wall 8' high. |
| | Minimum Charge | Job | | 232 | | 232 | |
| **10' Wide** | | | | | | | |
| | Demolish | Ea. | | 13.15 | | 13.15 | Includes material and labor to install |
| | Install | Ea. | 62.50 | 15.50 | | 78 | header, double studs each side, |
| | Demolish and Install | Ea. | 62.50 | 28.65 | | 91.15 | cripples, blocking and nails up to 10' |
| | Reinstall | Ea. | | 12.39 | | 12.39 | opening in 2" x 6" stud wall 8' high. |
| | Minimum Charge | Job | | 232 | | 232 | |

# Rough Frame / Structure

## Rough-in Opening

### 12' Wide

| | Unit | Material | Labor | Equip. | Total | Specification |
|---|---|---|---|---|---|---|
| Demolish | Ea. | | 13.15 | | 13.15 | Includes material and labor to install header, double studs each side, cripples, blocking and nails, up to 12' opening in 2" x 6" stud wall 8' high. |
| Install | Ea. | 83.50 | 15.50 | | 99 | |
| Demolish and Install | Ea. | 83.50 | 28.65 | | 112.15 | |
| Reinstall | Ea. | | 12.39 | | 12.39 | |
| Minimum Charge | Job | | 232 | | 232 | |

## Rough-in Opening

### Window w / 2" x 4" Lumber

#### 2' Wide

| | Unit | Material | Labor | Equip. | Total | Specification |
|---|---|---|---|---|---|---|
| Demolish | Ea. | | 16.45 | | 16.45 | Includes material and labor to install header, double studs each side of opening, cripples, blocking, nails and sub-sills for opening up to 2' in a 2" x 4" stud wall 8' high. |
| Install | Ea. | 20.50 | 19.35 | | 39.85 | |
| Demolish and Install | Ea. | 20.50 | 35.80 | | 56.30 | |
| Reinstall | Ea. | | 15.49 | | 15.49 | |
| Minimum Charge | Job | | 232 | | 232 | |

#### 3' Wide

| | Unit | Material | Labor | Equip. | Total | Specification |
|---|---|---|---|---|---|---|
| Demolish | Ea. | | 16.45 | | 16.45 | Includes material and labor to install header, double studs each side of opening, cripples, blocking, nails and sub-sills for opening up to 3' in a 2" x 4" stud wall 8' high. |
| Install | Ea. | 23.50 | 19.35 | | 42.85 | |
| Demolish and Install | Ea. | 23.50 | 35.80 | | 59.30 | |
| Reinstall | Ea. | | 15.49 | | 15.49 | |
| Minimum Charge | Job | | 232 | | 232 | |

#### 4' Wide

| | Unit | Material | Labor | Equip. | Total | Specification |
|---|---|---|---|---|---|---|
| Demolish | Ea. | | 16.45 | | 16.45 | Includes material and labor to install header, double studs each side of opening, cripples, blocking, nails and sub-sills for opening up to 4' in a 2" x 4" stud wall 8' high. |
| Install | Ea. | 26.50 | 19.35 | | 45.85 | |
| Demolish and Install | Ea. | 26.50 | 35.80 | | 62.30 | |
| Reinstall | Ea. | | 15.49 | | 15.49 | |
| Minimum Charge | Job | | 232 | | 232 | |

#### 5' Wide

| | Unit | Material | Labor | Equip. | Total | Specification |
|---|---|---|---|---|---|---|
| Demolish | Ea. | | 16.45 | | 16.45 | Includes material and labor to install header, double studs each side of opening, cripples, blocking, nails and sub-sills for opening up to 5' in a 2" x 4" stud wall 8' high. |
| Install | Ea. | 30.50 | 19.35 | | 49.85 | |
| Demolish and Install | Ea. | 30.50 | 35.80 | | 66.30 | |
| Reinstall | Ea. | | 15.49 | | 15.49 | |
| Minimum Charge | Job | | 232 | | 232 | |

#### 6' Wide

| | Unit | Material | Labor | Equip. | Total | Specification |
|---|---|---|---|---|---|---|
| Demolish | Ea. | | 16.45 | | 16.45 | Includes material and labor to install header, double studs each side of opening, cripples, blocking, nails and sub-sills for opening up to 6' in a 2" x 4" stud wall 8' high. |
| Install | Ea. | 34 | 19.35 | | 53.35 | |
| Demolish and Install | Ea. | 34 | 35.80 | | 69.80 | |
| Reinstall | Ea. | | 15.49 | | 15.49 | |
| Minimum Charge | Job | | 232 | | 232 | |

#### 7' Wide

| | Unit | Material | Labor | Equip. | Total | Specification |
|---|---|---|---|---|---|---|
| Demolish | Ea. | | 16.45 | | 16.45 | Includes material and labor to install header, double studs each side of opening, cripples, blocking, nails and sub-sills for opening up to 7' in a 2" x 4" stud wall 8' high. |
| Install | Ea. | 45 | 19.35 | | 64.35 | |
| Demolish and Install | Ea. | 45 | 35.80 | | 80.80 | |
| Reinstall | Ea. | | 15.49 | | 15.49 | |
| Minimum Charge | Job | | 232 | | 232 | |

#### 8' Wide

| | Unit | Material | Labor | Equip. | Total | Specification |
|---|---|---|---|---|---|---|
| Demolish | Ea. | | 17.95 | | 17.95 | Includes material and labor to install header, double studs each side of opening, cripples, blocking, nails and sub-sills for opening up to 8' in a 2" x 4" stud wall 8' high. |
| Install | Ea. | 50 | 21 | | 71 | |
| Demolish and Install | Ea. | 50 | 38.95 | | 88.95 | |
| Reinstall | Ea. | | 16.90 | | 16.90 | |
| Minimum Charge | Job | | 232 | | 232 | |

# Rough Frame / Structure

| Rough-in Opening | | Unit | Material | Labor | Equip. | Total | Specification |
|---|---|---|---|---|---|---|---|
| **10' Wide** | | | | | | | |
| | Demolish | Ea. | | 17.95 | | 17.95 | Includes material and labor to install |
| | Install | Ea. | 66 | 21 | | 87 | header, double studs each side of |
| | Demolish and Install | Ea. | 66 | 38.95 | | 104.95 | opening, cripples, blocking, nails and |
| | Reinstall | Ea. | | 16.90 | | 16.90 | sub-sills for opening up to 10' in a 2" |
| | Minimum Charge | Job | | 232 | | 232 | x 4" stud wall 8' high. |
| **12' Wide** | | | | | | | |
| | Demolish | Ea. | | 17.95 | | 17.95 | Includes material and labor to install |
| | Install | Ea. | 89 | 21 | | 110 | header, double studs each side of |
| | Demolish and Install | Ea. | 89 | 38.95 | | 127.95 | opening, cripples, blocking, nails and |
| | Reinstall | Ea. | | 16.90 | | 16.90 | sub-sills for opening up to 12' in a 2" |
| | Minimum Charge | Job | | 232 | | 232 | x 4" stud wall 8' high. |
| **Window w / 2" x 6" Lumber** | | | | | | | |
| **2' Wide** | | | | | | | |
| | Demolish | Ea. | | 16.45 | | 16.45 | Includes material and labor to install |
| | Install | Ea. | 30.50 | 19.35 | | 49.85 | header, double studs each side of |
| | Demolish and Install | Ea. | 30.50 | 35.80 | | 66.30 | opening, cripples, blocking, nails and |
| | Reinstall | Ea. | | 15.49 | | 15.49 | sub-sills for opening up to 2' in a 2" x |
| | Minimum Charge | Job | | 232 | | 232 | 6" stud wall 8' high. |
| **3' Wide** | | | | | | | |
| | Demolish | Ea. | | 16.45 | | 16.45 | Includes material and labor to install |
| | Install | Ea. | 34 | 19.35 | | 53.35 | header, double studs each side of |
| | Demolish and Install | Ea. | 34 | 35.80 | | 69.80 | opening, cripples, blocking, nails and |
| | Reinstall | Ea. | | 15.49 | | 15.49 | sub-sills for opening up to 3' in a 2" x |
| | Minimum Charge | Job | | 232 | | 232 | 6" stud wall 8' high. |
| **4' Wide** | | | | | | | |
| | Demolish | Ea. | | 16.45 | | 16.45 | Includes material and labor to install |
| | Install | Ea. | 37 | 19.35 | | 56.35 | header, double studs each side of |
| | Demolish and Install | Ea. | 37 | 35.80 | | 72.80 | opening, cripples, blocking, nails and |
| | Reinstall | Ea. | | 15.49 | | 15.49 | sub-sills for opening up to 4' in a 2" x |
| | Minimum Charge | Job | | 232 | | 232 | 6" stud wall 8' high. |
| **5' Wide** | | | | | | | |
| | Demolish | Ea. | | 16.45 | | 16.45 | Includes material and labor to install |
| | Install | Ea. | 41.50 | 19.35 | | 60.85 | header, double studs each side of |
| | Demolish and Install | Ea. | 41.50 | 35.80 | | 77.30 | opening, cripples, blocking, nails and |
| | Reinstall | Ea. | | 15.49 | | 15.49 | sub-sills for opening up to 5' in a 2" x |
| | Minimum Charge | Job | | 232 | | 232 | 6" stud wall 8' high. |
| **6' Wide** | | | | | | | |
| | Demolish | Ea. | | 16.45 | | 16.45 | Includes material and labor to install |
| | Install | Ea. | 45.50 | 19.35 | | 64.85 | header, double studs each side of |
| | Demolish and Install | Ea. | 45.50 | 35.80 | | 81.30 | opening, cripples, blocking, nails and |
| | Reinstall | Ea. | | 15.49 | | 15.49 | sub-sills for opening up to 6' in a 2" x |
| | Minimum Charge | Job | | 232 | | 232 | 6" stud wall 8' high. |
| **7' Wide** | | | | | | | |
| | Demolish | Ea. | | 16.45 | | 16.45 | Includes material and labor to install |
| | Install | Ea. | 58 | 19.35 | | 77.35 | header, double studs each side of |
| | Demolish and Install | Ea. | 58 | 35.80 | | 93.80 | opening, cripples, blocking, nails and |
| | Reinstall | Ea. | | 15.49 | | 15.49 | sub-sills for opening up to 7' in a 2" x |
| | Minimum Charge | Job | | 232 | | 232 | 6" stud wall 8' high. |

# Rough Frame / Structure

## Rough-in Opening

| | Unit | Material | Labor | Equip. | Total | Specification |
|---|---|---|---|---|---|---|
| **8' Wide** | | | | | | |
| Demolish | Ea. | | 17.95 | | 17.95 | Includes material and labor to install |
| Install | Ea. | 62.50 | 21 | | 83.50 | header, double studs each side of |
| Demolish and Install | Ea. | 62.50 | 38.95 | | 101.45 | opening, cripples, blocking, nails and |
| Reinstall | Ea. | | 16.90 | | 16.90 | sub-sills for opening up to 8' in a 2" x |
| Minimum Charge | Job | | 232 | | 232 | 6" stud wall 8' high. |
| **10' Wide** | | | | | | |
| Demolish | Ea. | | 17.95 | | 17.95 | Includes material and labor to install |
| Install | Ea. | 80 | 21 | | 101 | header, double studs each side of |
| Demolish and Install | Ea. | 80 | 38.95 | | 118.95 | opening, cripples, blocking, nails and |
| Reinstall | Ea. | | 16.90 | | 16.90 | sub-sills for opening up to 10' in a 2" |
| Minimum Charge | Job | | 232 | | 232 | x 6" stud wall 8' high. |
| **12' Wide** | | | | | | |
| Demolish | Ea. | | 17.95 | | 17.95 | Includes material and labor to install |
| Install | Ea. | 105 | 21 | | 126 | header, double studs each side of |
| Demolish and Install | Ea. | 105 | 38.95 | | 143.95 | opening, cripples, blocking, nails and |
| Reinstall | Ea. | | 16.90 | | 16.90 | sub-sills for opening up to 12' in a 2" |
| Minimum Charge | Job | | 232 | | 232 | x 6" stud wall 8' high. |

## Glue-Laminated Beams

| | Unit | Material | Labor | Equip. | Total | Specification |
|---|---|---|---|---|---|---|
| **3-1/2" x 6"** | | | | | | |
| Demolish | L.F. | | 2.94 | | 2.94 | Includes material and labor to install |
| Install | L.F. | 6.10 | 3.56 | .91 | 10.57 | glue-laminated wood beam. |
| Demolish and Install | L.F. | 6.10 | 6.50 | .91 | 13.51 | |
| Reinstall | L.F. | | 2.84 | .73 | 3.57 | |
| Clean | L.F. | .04 | .44 | | .48 | |
| Paint | L.F. | .39 | 1.14 | | 1.53 | |
| Minimum Charge | Job | | 232 | | 232 | |
| **3-1/2" x 9"** | | | | | | |
| Demolish | L.F. | | 2.94 | | 2.94 | Includes material and labor to install |
| Install | L.F. | 8.20 | 3.56 | .91 | 12.67 | glue-laminated wood beam. |
| Demolish and Install | L.F. | 8.20 | 6.50 | .91 | 15.61 | |
| Reinstall | L.F. | | 2.84 | .73 | 3.57 | |
| Clean | L.F. | .07 | .46 | | .53 | |
| Paint | L.F. | .50 | 1.47 | | 1.97 | |
| Minimum Charge | Job | | 232 | | 232 | |
| **3-1/2" x 12"** | | | | | | |
| Demolish | L.F. | | 4.41 | | 4.41 | Includes material and labor to install |
| Install | L.F. | 10.95 | 3.56 | .91 | 15.42 | glue-laminated wood beam. |
| Demolish and Install | L.F. | 10.95 | 7.97 | .91 | 19.83 | |
| Reinstall | L.F. | | 2.84 | .73 | 3.57 | |
| Clean | L.F. | .08 | .47 | | .55 | |
| Paint | L.F. | .62 | 1.84 | | 2.46 | |
| Minimum Charge | Job | | 232 | | 232 | |
| **3-1/2" x 15"** | | | | | | |
| Demolish | L.F. | | 4.41 | | 4.41 | Includes material and labor to install |
| Install | L.F. | 13.70 | 3.68 | .94 | 18.32 | glue-laminated wood beam. |
| Demolish and Install | L.F. | 13.70 | 8.09 | .94 | 22.73 | |
| Reinstall | L.F. | | 2.94 | .75 | 3.69 | |
| Clean | L.F. | .09 | .49 | | .58 | |
| Paint | L.F. | .73 | 2.15 | | 2.88 | |
| Minimum Charge | Job | | 232 | | 232 | |

| Glue-Laminated Beams | Unit | Material | Labor | Equip. | Total | Specification |
|---|---|---|---|---|---|---|
| **3-1/2" x 18"** | | | | | | |
| Demolish | L.F. | | 5.50 | | 5.50 | Includes material and labor to install |
| Install | L.F. | 18.35 | 3.68 | .94 | 22.97 | glue-laminated wood beam. |
| Demolish and Install | L.F. | 18.35 | 9.18 | .94 | 28.47 | |
| Reinstall | L.F. | | 2.94 | .75 | 3.69 | |
| Clean | L.F. | .11 | .51 | | .62 | |
| Paint | L.F. | .85 | 2.49 | | 3.34 | |
| Minimum Charge | Job | | 232 | | 232 | |
| **5-1/8" x 6"** | | | | | | |
| Demolish | L.F. | | 4.41 | | 4.41 | Includes material and labor to install |
| Install | L.F. | 8.95 | 3.56 | .91 | 13.42 | glue-laminated wood beam. |
| Demolish and Install | L.F. | 8.95 | 7.97 | .91 | 17.83 | |
| Reinstall | L.F. | | 2.84 | .73 | 3.57 | |
| Clean | L.F. | .06 | .44 | | .50 | |
| Paint | L.F. | .42 | 1.25 | | 1.67 | |
| Minimum Charge | Job | | 232 | | 232 | |
| **5-1/8" x 9"** | | | | | | |
| Demolish | L.F. | | 5.50 | | 5.50 | Includes material and labor to install |
| Install | L.F. | 13.50 | 3.56 | .91 | 17.97 | glue-laminated wood beam. |
| Demolish and Install | L.F. | 13.50 | 9.06 | .91 | 23.47 | |
| Reinstall | L.F. | | 2.84 | .73 | 3.57 | |
| Clean | L.F. | .07 | .46 | | .53 | |
| Paint | L.F. | .54 | 1.60 | | 2.14 | |
| Minimum Charge | Job | | 232 | | 232 | |
| **5-1/8" x 12"** | | | | | | |
| Demolish | L.F. | | 6.55 | | 6.55 | Includes material and labor to install |
| Install | L.F. | 17.95 | 3.56 | .91 | 22.42 | glue-laminated wood beam. |
| Demolish and Install | L.F. | 17.95 | 10.11 | .91 | 28.97 | |
| Reinstall | L.F. | | 2.84 | .73 | 3.57 | |
| Clean | L.F. | .09 | .47 | | .56 | |
| Paint | L.F. | .66 | 1.93 | | 2.59 | |
| Minimum Charge | Job | | 232 | | 232 | |
| **5-1/8" x 18"** | | | | | | |
| Demolish | L.F. | | 11.05 | | 11.05 | Includes material and labor to install |
| Install | L.F. | 27 | 3.68 | .94 | 31.62 | glue-laminated wood beam. |
| Demolish and Install | L.F. | 27 | 14.73 | .94 | 42.67 | |
| Reinstall | L.F. | | 2.94 | .75 | 3.69 | |
| Clean | L.F. | .11 | .51 | | .62 | |
| Paint | L.F. | .76 | 2.61 | | 3.37 | |
| Minimum Charge | Job | | 232 | | 232 | |
| **6-3/4" x 12"** | | | | | | |
| Demolish | L.F. | | 6.55 | | 6.55 | Includes material and labor to install |
| Install | L.F. | 23.50 | 3.68 | .94 | 28.12 | glue-laminated wood beam. |
| Demolish and Install | L.F. | 23.50 | 10.23 | .94 | 34.67 | |
| Reinstall | L.F. | | 2.94 | .75 | 3.69 | |
| Clean | L.F. | .09 | .49 | | .58 | |
| Paint | L.F. | .73 | 2.15 | | 2.88 | |
| Minimum Charge | Job | | 232 | | 232 | |
| **6-3/4" x 15"** | | | | | | |
| Demolish | L.F. | | 11.05 | | 11.05 | Includes material and labor to install |
| Install | L.F. | 29.50 | 3.68 | .94 | 34.12 | glue-laminated wood beam. |
| Demolish and Install | L.F. | 29.50 | 14.73 | .94 | 45.17 | |
| Reinstall | L.F. | | 2.94 | .75 | 3.69 | |
| Clean | L.F. | .11 | .51 | | .62 | |
| Paint | L.F. | .85 | 2.49 | | 3.34 | |
| Minimum Charge | Job | | 232 | | 232 | |

# Rough Frame / Structure

## Glue-Laminated Beams

| | Unit | Material | Labor | Equip. | Total | Specification |
|---|---|---|---|---|---|---|
| **6-3/4″ x 18″** | | | | | | |
| Demolish | L.F. | | 11.05 | | 11.05 | Includes material and labor to install |
| Install | L.F. | 35 | 3.81 | .97 | 39.78 | glue-laminated wood beam. |
| Demolish and Install | L.F. | 35 | 14.86 | .97 | 50.83 | |
| Reinstall | L.F. | | 3.05 | .78 | 3.83 | |
| Clean | L.F. | .12 | .53 | | .65 | |
| Paint | L.F. | .97 | 2.84 | | 3.81 | |
| Minimum Charge | Job | | 232 | | 232 | |
| **Hardware** | | | | | | |
| **5-1/4″ Glue-Lam Seat** | | | | | | |
| Install | Ea. | 193 | 2.58 | | 195.58 | Includes labor and material to install |
| Minimum Charge | Job | | 232 | | 232 | beam hangers. |
| **6-3/4″ Glue-Lam Seat** | | | | | | |
| Install | Ea. | 227 | 2.58 | | 229.58 | Includes labor and material to install |
| Minimum Charge | Job | | 232 | | 232 | beam hangers. |
| **Earthquake Strapping** | | | | | | |
| Install | Ea. | 2.97 | 2.91 | | 5.88 | Includes labor and material to replace |
| Minimum Charge | Job | | 232 | | 232 | an earthquake strap. |
| **Hurricane Clips** | | | | | | |
| Install | Ea. | 1.11 | 3.21 | | 4.32 | Includes labor and material to install a |
| Minimum Charge | Job | | 232 | | 232 | hurricane clip. |

## Metal Stud Framing

| | Unit | Material | Labor | Equip. | Total | Specification |
|---|---|---|---|---|---|---|
| **16″ O.C. System** | | | | | | |
| **4″, 16 Ga.** | | | | | | |
| Demolish | S.F. | | 1.45 | | 1.45 | Includes material and labor to install |
| Install | S.F. | 1.41 | 1.43 | | 2.84 | load bearing cold rolled metal stud |
| Demolish and Install | S.F. | 1.41 | 2.88 | | 4.29 | walls, to 10′ high, including studs, top |
| Reinstall | S.F. | | 1.14 | | 1.14 | and bottom track and screws. |
| Minimum Charge | Job | | 232 | | 232 | |
| **6″, 16 Ga.** | | | | | | |
| Demolish | S.F. | | 1.45 | | 1.45 | Includes material and labor to install |
| Install | S.F. | 1.78 | 1.45 | | 3.23 | load bearing cold rolled metal stud |
| Demolish and Install | S.F. | 1.78 | 2.90 | | 4.68 | walls, to 10′ high, including studs, top |
| Reinstall | S.F. | | 1.16 | | 1.16 | and bottom track and screws. |
| Minimum Charge | Job | | 232 | | 232 | |
| **4″, 25 Ga.** | | | | | | |
| Demolish | S.F. | | 1 | | 1 | Includes material and labor to install |
| Install | S.F. | .46 | .98 | | 1.44 | cold rolled metal stud walls, to 10′ |
| Demolish and Install | S.F. | .46 | 1.98 | | 2.44 | high, including studs, top and bottom |
| Reinstall | S.F. | | .78 | | .78 | track and screws. |
| Minimum Charge | Job | | 232 | | 232 | |
| **6″, 25 Ga.** | | | | | | |
| Demolish | S.F. | | 1 | | 1 | Includes material and labor to install |
| Install | S.F. | .55 | .99 | | 1.54 | cold rolled metal stud walls, to 10′ |
| Demolish and Install | S.F. | .55 | 1.99 | | 2.54 | high, including studs, top and bottom |
| Reinstall | S.F. | | .79 | | .79 | track and screws. |
| Minimum Charge | Job | | 232 | | 232 | |
| **24″ O.C. System** | | | | | | |
| **4″, 25 Ga.** | | | | | | |
| Demolish | S.F. | | .85 | | .85 | Includes material and labor to install |
| Install | S.F. | .34 | .63 | | .97 | cold rolled metal stud walls, to 10′ |
| Demolish and Install | S.F. | .34 | 1.48 | | 1.82 | high, including studs, top and bottom |
| Reinstall | S.F. | | .50 | | .50 | track and screws. |
| Minimum Charge | Job | | 232 | | 232 | |

**For customer support on your Contractor's Pricing Guide: Residential Repair & Remodeling, call 800.448.8182.**

# Rough Frame / Structure

## Metal Stud Framing

| | Unit | Material | Labor | Equip. | Total | Specification |
|---|---|---|---|---|---|---|
| **6", 25 Ga.** | | | | | | |
| Demolish | S.F. | | .85 | | .85 | Includes material and labor to install |
| Install | S.F. | .41 | .64 | | 1.05 | cold rolled metal stud walls, to 10' |
| Demolish and Install | S.F. | .41 | 1.49 | | 1.90 | high, including studs, top and bottom |
| Reinstall | S.F. | | .51 | | .51 | track and screws. |
| Minimum Charge | Job | | 232 | | 232 | |
| **4", 16 Ga.** | | | | | | |
| Demolish | L.F. | | 9.70 | | 9.70 | Includes material and labor to install |
| Install | S.F. | 1.02 | 1.03 | | 2.05 | load bearing cold rolled metal stud |
| Demolish and Install | S.F. | 1.02 | 10.73 | | 11.75 | walls, to 10' high, including studs, top |
| Reinstall | S.F. | | .83 | | .83 | and bottom track and screws. |
| Minimum Charge | Job | | 232 | | 232 | |
| **6", 16 Ga.** | | | | | | |
| Demolish | L.F. | | 9.70 | | 9.70 | Includes material and labor to install |
| Install | S.F. | 1.30 | 1.06 | | 2.36 | load bearing cold rolled metal stud |
| Demolish and Install | S.F. | 1.30 | 10.76 | | 12.06 | walls, to 10' high, including studs, top |
| Reinstall | S.F. | | .85 | | .85 | and bottom track and screws. |
| Minimum Charge | Job | | 232 | | 232 | |

## Metal Joist

| | Unit | Material | Labor | Equip. | Total | Specification |
|---|---|---|---|---|---|---|
| **Bar Joist** | | | | | | |
| **18K9** | | | | | | |
| Demolish | L.F. | | 1.16 | .33 | 1.49 | Includes material, labor and |
| Install | L.F. | 10.10 | 2.46 | 1.04 | 13.60 | equipment to install open web joist. |
| Demolish and Install | L.F. | 10.10 | 3.62 | 1.37 | 15.09 | |
| Reinstall | L.F. | | 1.97 | .83 | 2.80 | |
| Minimum Charge | Job | | 2125 | 545 | 2670 | |
| **16K6** | | | | | | |
| Demolish | L.F. | | 1.16 | .33 | 1.49 | Includes material, labor and |
| Install | L.F. | 5.45 | 1.65 | .08 | 7.18 | equipment to install open web joist. |
| Demolish and Install | L.F. | 5.45 | 2.81 | .41 | 8.67 | |
| Reinstall | L.F. | | 1.32 | .06 | 1.38 | |
| Minimum Charge | Job | | 2125 | 545 | 2670 | |

## Exterior Sheathing

| | Unit | Material | Labor | Equip. | Total | Specification |
|---|---|---|---|---|---|---|
| **CDX Plywood** | | | | | | |
| **5/16"** | | | | | | |
| Demolish | S.F. | | .67 | | .67 | Cost includes material and labor to |
| Install | S.F. | .65 | .58 | | 1.23 | install 5/16" CDX plywood sheathing. |
| Demolish and Install | S.F. | .65 | 1.25 | | 1.90 | |
| Minimum Charge | Job | | 232 | | 232 | |
| **3/8"** | | | | | | |
| Demolish | S.F. | | .69 | | .69 | Cost includes material and labor to |
| Install | S.F. | .66 | .77 | | 1.43 | install 3/8" CDX plywood sheathing. |
| Demolish and Install | S.F. | .66 | 1.46 | | 2.12 | |
| Minimum Charge | Job | | 232 | | 232 | |
| **1/2"** | | | | | | |
| Demolish | S.F. | | .70 | | .70 | Cost includes material and labor to |
| Install | S.F. | .68 | .83 | | 1.51 | install 1/2" CDX plywood sheathing. |
| Demolish and Install | S.F. | .68 | 1.53 | | 2.21 | |
| Minimum Charge | Job | | 232 | | 232 | |

# Rough Frame / Structure

## Exterior Sheathing

| Exterior Sheathing | Unit | Material | Labor | Equip. | Total | Specification |
|---|---|---|---|---|---|---|
| **5/8″** | | | | | | |
| Demolish | S.F. | | .72 | | .72 | Cost includes material and labor to |
| Install | S.F. | .85 | .89 | | 1.74 | install 5/8″ CDX plywood sheathing. |
| Demolish and Install | S.F. | .85 | 1.61 | | 2.46 | |
| Minimum Charge | Job | | 232 | | 232 | |
| **3/4″** | | | | | | |
| Demolish | S.F. | | .73 | | .73 | Cost includes material and labor to |
| Install | S.F. | 1.01 | .95 | | 1.96 | install 3/4″ CDX plywood sheathing. |
| Demolish and Install | S.F. | 1.01 | 1.68 | | 2.69 | |
| Minimum Charge | Job | | 232 | | 232 | |
| **OSB** | | | | | | |
| **1/2″** | | | | | | |
| Demolish | S.F. | | .70 | | .70 | Cost includes material and labor to |
| Install | S.F. | .53 | .66 | | 1.19 | install 4′ x 8′ x 1/2″ OSB sheathing. |
| Demolish and Install | S.F. | .53 | 1.36 | | 1.89 | |
| Minimum Charge | Job | | 232 | | 232 | |
| **5/8″** | | | | | | |
| Demolish | S.F. | | .72 | | .72 | Cost includes material and labor to |
| Install | S.F. | .53 | .54 | | 1.07 | install 4′ x 8′ x 5/8″ OSB sheathing. |
| Demolish and Install | S.F. | .53 | 1.26 | | 1.79 | |
| Minimum Charge | Job | | 232 | | 232 | |
| **Vapor Barrier** | | | | | | |
| **Black Paper** | | | | | | |
| Install | S.F. | .06 | .13 | | .19 | Includes material and labor to install |
| Minimum Charge | Job | | 232 | | 232 | #15 felt building paper. |

## Plywood Sheathing

| Plywood Sheathing | Unit | Material | Labor | Equip. | Total | Specification |
|---|---|---|---|---|---|---|
| **Finish Plywood** | | | | | | |
| **5/16″** | | | | | | |
| Demolish | S.F. | | .67 | | .67 | Includes material and labor to install |
| Install | S.F. | 1.14 | .29 | | 1.43 | exterior 5/16″ AC plywood on walls. |
| Demolish and Install | S.F. | 1.14 | .96 | | 2.10 | |
| Minimum Charge | Job | | 232 | | 232 | |
| **3/8″** | | | | | | |
| Demolish | S.F. | | .69 | | .69 | Includes material and labor to install |
| Install | S.F. | 1.14 | .39 | | 1.53 | exterior 3/8″ AC plywood on walls. |
| Demolish and Install | S.F. | 1.14 | 1.08 | | 2.22 | |
| Minimum Charge | Job | | 232 | | 232 | |
| **1/2″** | | | | | | |
| Demolish | S.F. | | .70 | | .70 | Includes material and labor to install |
| Install | S.F. | 1.35 | .41 | | 1.76 | exterior 1/2″ AC plywood on walls. |
| Demolish and Install | S.F. | 1.35 | 1.11 | | 2.46 | |
| Minimum Charge | Job | | 232 | | 232 | |
| **5/8″** | | | | | | |
| Demolish | S.F. | | .72 | | .72 | Includes material and labor to install |
| Install | S.F. | 1.50 | .41 | | 1.91 | exterior 5/8″ AC plywood on walls. |
| Demolish and Install | S.F. | 1.50 | 1.13 | | 2.63 | |
| Reinstall | S.F. | | .41 | | .41 | |
| Minimum Charge | Job | | 232 | | 232 | |
| **3/4″** | | | | | | |
| Demolish | S.F. | | .73 | | .73 | Includes material and labor to install |
| Install | S.F. | 1.62 | .41 | | 2.03 | exterior 3/4″ AC plywood on walls. |
| Demolish and Install | S.F. | 1.62 | 1.14 | | 2.76 | |
| Reinstall | S.F. | | .41 | | .41 | |
| Minimum Charge | Job | | 232 | | 232 | |

# Rough Frame / Structure

## Stairs

| Stairs | | Unit | Material | Labor | Equip. | Total | Specification |
|---|---|---|---|---|---|---|---|
| **Job-built** | | | | | | | |
| **Treads and Risers** | | | | | | | |
| | Demolish | Ea. | | 49.50 | | 49.50 | Includes material and labor to install |
| | Install | Ea. | 770 | 465 | | 1235 | three 2" x 12" stringers, treads and |
| | Demolish and Install | Ea. | 770 | 514.50 | | 1284.50 | risers of 3/4" CDX plywood, installed |
| | Clean | Flight | 1.47 | 30 | | 31.47 | in a straight or "L" shaped run |
| | Paint | Ea. | .12 | 77.50 | | 77.62 | including carpet. |
| | Minimum Charge | Job | | 232 | | 232 | |
| **Landing** | | | | | | | |
| | Demolish | S.F. | | .39 | | .39 | Includes material and labor to install |
| | Install | S.F. | 77.50 | 3.23 | | 80.73 | landing framing, 3/4" CDX plywood |
| | Demolish and Install | S.F. | 77.50 | 3.62 | | 81.12 | surface and carpet. |
| | Clean | S.F. | | .44 | | .44 | |
| | Paint | S.F. | .48 | .95 | | 1.43 | |
| | Minimum Charge | Job | | 232 | | 232 | |

## Ceiling Framing

| Ceiling Framing | | Unit | Material | Labor | Equip. | Total | Specification |
|---|---|---|---|---|---|---|---|
| **16" O.C. System** | | | | | | | |
| **2" x 6" Joists** | | | | | | | |
| | Demolish | S.F. | | .61 | | .61 | Cost includes material and labor to |
| | Install | S.F. | .54 | .76 | | 1.30 | install 2" x 6" joists including end and |
| | Demolish and Install | S.F. | .54 | 1.37 | | 1.91 | header joist installed 16" O.C. Does |
| | Reinstall | S.F. | | .61 | | .61 | not include beams, ledger strips, |
| | Minimum Charge | Job | | 232 | | 232 | blocking or bridging. |
| **2" x 8" Joists** | | | | | | | |
| | Demolish | S.F. | | .63 | | .63 | Cost includes material and labor to |
| | Install | S.F. | .75 | .94 | | 1.69 | install 2" x 8" joists including end and |
| | Demolish and Install | S.F. | .75 | 1.57 | | 2.32 | header joist installed 16" O.C. Does |
| | Reinstall | S.F. | | .75 | | .75 | not include beams, ledger strips, |
| | Minimum Charge | Job | | 232 | | 232 | blocking or bridging. |
| **2" x 10" Joists** | | | | | | | |
| | Demolish | S.F. | | .65 | | .65 | Cost includes material and labor to |
| | Install | S.F. | 1.18 | 1.11 | | 2.29 | install 2" x 10" joists including end |
| | Demolish and Install | S.F. | 1.18 | 1.76 | | 2.94 | and header joist installed 16" O.C. |
| | Reinstall | S.F. | | .89 | | .89 | Does not include beams, ledger strips, |
| | Minimum Charge | Job | | 232 | | 232 | blocking or bridging. |
| **2" x 12" Joists** | | | | | | | |
| | Demolish | S.F. | | .67 | | .67 | Cost includes material and labor to |
| | Install | S.F. | 1.50 | 1.36 | | 2.86 | install 2" x 12" joists including end |
| | Demolish and Install | S.F. | 1.50 | 2.03 | | 3.53 | and header joist installed 16" O.C. |
| | Reinstall | S.F. | | 1.09 | | 1.09 | Does not include beams, ledger strips, |
| | Minimum Charge | Job | | 232 | | 232 | blocking or bridging. |
| **Block / Bridge** | | | | | | | |
| | Demolish | Ea. | | 1.23 | | 1.23 | Includes material and labor to install |
| | Install | Ea. | 2.07 | 2.61 | | 4.68 | set of cross bridging or per block of |
| | Demolish and Install | Ea. | 2.07 | 3.84 | | 5.91 | solid bridging for 2" x 10" joists cut to |
| | Minimum Charge | Job | | 232 | | 232 | size on site. |

# Rough Frame / Structure

| Ceiling Framing | Unit | Material | Labor | Equip. | Total | Specification |
|---|---|---|---|---|---|---|
| **24″ O.C. System** | | | | | | |
| **2″ x 6″ Joists** | | | | | | |
| Demolish | S.F. | | .41 | | .41 | Cost includes material and labor to |
| Install | S.F. | .36 | .51 | | .87 | install 2″ x 6″ joists including end and |
| Demolish and Install | S.F. | .36 | .92 | | 1.28 | header joist installed 24″ O.C. Does |
| Reinstall | S.F. | | .41 | | .41 | not include beams, ledger strips, |
| Minimum Charge | Job | | 232 | | 232 | blocking or bridging. |
| **2″ x 8″ Joists** | | | | | | |
| Demolish | S.F. | | .42 | | .42 | Cost includes material and labor to |
| Install | S.F. | .51 | .63 | | 1.14 | install 2″ x 8″ joists including end and |
| Demolish and Install | S.F. | .51 | 1.05 | | 1.56 | header joist installed 24″ O.C. Does |
| Reinstall | S.F. | | .50 | | .50 | not include beams, ledger strips, |
| Minimum Charge | Job | | 232 | | 232 | blocking or bridging. |
| **2″ x 10″ Joists** | | | | | | |
| Demolish | S.F. | | .43 | | .43 | Cost includes material and labor to |
| Install | S.F. | .79 | .74 | | 1.53 | install 2″ x 10″ joists including end |
| Demolish and Install | S.F. | .79 | 1.17 | | 1.96 | and header joist installed 24″ O.C. |
| Reinstall | S.F. | | .59 | | .59 | Does not include beams, ledger strips, |
| Minimum Charge | Job | | 232 | | 232 | blocking or bridging. |
| **2″ x 12″ Joists** | | | | | | |
| Demolish | S.F. | | .45 | | .45 | Cost includes material and labor to |
| Install | S.F. | 1.01 | .91 | | 1.92 | install 2″ x 12″ joists including end |
| Demolish and Install | S.F. | 1.01 | 1.36 | | 2.37 | and header joist installed 24″ O.C. |
| Reinstall | S.F. | | .72 | | .72 | Does not include beams, ledger strips, |
| Minimum Charge | Job | | 232 | | 232 | blocking or bridging. |
| **Block / Bridge** | | | | | | |
| Demolish | Ea. | | 1.23 | | 1.23 | Includes material and labor to install |
| Install | Ea. | 2.07 | 2.61 | | 4.68 | set of cross bridging or per block of |
| Demolish and Install | Ea. | 2.07 | 3.84 | | 5.91 | solid bridging for 2″ x 10″ joists cut to |
| Minimum Charge | Job | | 232 | | 232 | size on site. |
| **Ledger Strips** | | | | | | |
| **1″ x 2″** | | | | | | |
| Demolish | L.F. | | .33 | | .33 | Cost includes material and labor to |
| Install | L.F. | .30 | 1.35 | | 1.65 | install 1″ x 2″ ledger strip nailed to the |
| Demolish and Install | L.F. | .30 | 1.68 | | 1.98 | face of studs, beams or joist. |
| Minimum Charge | Job | | 232 | | 232 | |
| **1″ x 3″** | | | | | | |
| Demolish | L.F. | | .33 | | .33 | Includes material and labor to install |
| Install | L.F. | .56 | 1.55 | | 2.11 | up to 1″ x 4″ ledger strip nailed to the |
| Demolish and Install | L.F. | .56 | 1.88 | | 2.44 | face of studs, beams or joist. |
| Minimum Charge | Job | | 232 | | 232 | |
| **1″ x 4″** | | | | | | |
| Demolish | L.F. | | .33 | | .33 | Includes material and labor to install |
| Install | L.F. | .56 | 1.55 | | 2.11 | up to 1″ x 4″ ledger strip nailed to the |
| Demolish and Install | L.F. | .56 | 1.88 | | 2.44 | face of studs, beams or joist. |
| Minimum Charge | Job | | 232 | | 232 | |
| **2″ x 2″** | | | | | | |
| Demolish | L.F. | | .36 | | .36 | Cost includes material and labor to |
| Install | L.F. | .46 | 1.41 | | 1.87 | install 2″ x 2″ ledger strip nailed to the |
| Demolish and Install | L.F. | .46 | 1.77 | | 2.23 | face of studs, beams or joist. |
| Minimum Charge | Job | | 232 | | 232 | |
| **2″ x 4″** | | | | | | |
| Demolish | L.F. | | .39 | | .39 | Cost includes material and labor to |
| Install | L.F. | .46 | 1.86 | | 2.32 | install 2″ x 4″ ledger strip nailed to the |
| Demolish and Install | L.F. | .46 | 2.25 | | 2.71 | face of studs, beams or joist. |
| Reinstall | L.F. | | 1.49 | | 1.49 | |
| Minimum Charge | Job | | 232 | | 232 | |

# Rough Frame / Structure

## Ceiling Framing

| Ceiling Framing | Unit | Material | Labor | Equip. | Total | Specification |
|---|---|---|---|---|---|---|
| **Ledger Boards** | | | | | | |
| **2" x 4"** | | | | | | |
| Demolish | L.F. | | .39 | | .39 | Cost includes material and labor to |
| Install | L.F. | 5.20 | 2.58 | | 7.78 | install 2" x 4" ledger board fastened to |
| Demolish and Install | L.F. | 5.20 | 2.97 | | 8.17 | a wall, joists or studs. |
| Reinstall | L.F. | | 2.07 | | 2.07 | |
| Minimum Charge | Job | | 232 | | 232 | |
| **4" x 6"** | | | | | | |
| Demolish | L.F. | | .66 | | .66 | Cost includes material and labor to |
| Install | L.F. | 9.55 | 3.32 | | 12.87 | install 4" x 6" ledger board fastened to |
| Demolish and Install | L.F. | 9.55 | 3.98 | | 13.53 | a wall, joists or studs. |
| Reinstall | L.F. | | 2.66 | | 2.66 | |
| Minimum Charge | Job | | 232 | | 232 | |
| **4" x 8"** | | | | | | |
| Demolish | L.F. | | .88 | | .88 | Cost includes material and labor to |
| Install | L.F. | 10.80 | 3.87 | | 14.67 | install 4" x 8" ledger board fastened to |
| Demolish and Install | L.F. | 10.80 | 4.75 | | 15.55 | a wall, joists or studs. |
| Reinstall | L.F. | | 3.10 | | 3.10 | |
| Minimum Charge | Job | | 232 | | 232 | |
| **Earthquake Strapping** | | | | | | |
| Install | Ea. | 2.97 | 2.91 | | 5.88 | Includes labor and material to replace |
| Minimum Charge | Job | | 232 | | 232 | an earthquake strap. |
| **Hurricane Clips** | | | | | | |
| Install | Ea. | 1.11 | 3.21 | | 4.32 | Includes labor and material to install a |
| Minimum Charge | Job | | 232 | | 232 | hurricane clip. |

## Roof Framing

| Roof Framing | Unit | Material | Labor | Equip. | Total | Specification |
|---|---|---|---|---|---|---|
| **16" O.C. System** | | | | | | |
| **2" x 4" Rafters** | | | | | | |
| Demolish | S.F. | | .90 | | .90 | Cost includes material and labor to |
| Install | S.F. | .34 | .83 | | 1.17 | install 2" x 4", 16" O.C. rafter framing |
| Demolish and Install | S.F. | .34 | 1.73 | | 2.07 | for flat, shed or gable roofs, 25' span, |
| Reinstall | S.F. | | .66 | | .66 | up to 5/12 slope, per S.F. of roof. |
| Minimum Charge | Job | | 232 | | 232 | |
| **2" x 6" Rafters** | | | | | | |
| Demolish | S.F. | | .94 | | .94 | Cost includes material and labor to |
| Install | S.F. | .53 | .87 | | 1.40 | install 2" x 6", 16" O.C. rafter framing |
| Demolish and Install | S.F. | .53 | 1.81 | | 2.34 | for flat, shed or gable roofs, 25' span, |
| Reinstall | S.F. | | .70 | | .70 | up to 5/12 slope, per S.F. of roof. |
| Minimum Charge | Job | | 232 | | 232 | |
| **2" x 8" Rafters** | | | | | | |
| Demolish | S.F. | | .96 | | .96 | Cost includes material and labor to |
| Install | S.F. | .74 | .93 | | 1.67 | install 2" x 8", 16" O.C. rafter framing |
| Demolish and Install | S.F. | .74 | 1.89 | | 2.63 | for flat, shed or gable roofs, 25' span, |
| Reinstall | S.F. | | .74 | | .74 | up to 5/12 slope, per S.F. of roof |
| Minimum Charge | Job | | 232 | | 232 | |
| **2" x 10" Rafters** | | | | | | |
| Demolish | S.F. | | .96 | | .96 | Cost includes material and labor to |
| Install | S.F. | 1.17 | 1.41 | | 2.58 | install 2" x 10", 16" O.C. rafter |
| Demolish and Install | S.F. | 1.17 | 2.37 | | 3.54 | framing for flat, shed or gable roofs, |
| Reinstall | S.F. | | 1.13 | | 1.13 | 25' span, up to 5/12 slope, per S.F. |
| Minimum Charge | Job | | 232 | | 232 | of roof. |

| Roof Framing | | Unit | Material | Labor | Equip. | Total | Specification |
|---|---|---|---|---|---|---|---|
| **2" x 12" Rafters** | | | | | | | |
| | Demolish | S.F. | | .97 | | .97 | Cost includes material and labor to |
| | Install | S.F. | 1.49 | 1.53 | | 3.02 | install 2" x 12", 16" O.C. rafter |
| | Demolish and Install | S.F. | 1.49 | 2.50 | | 3.99 | framing for flat, shed or gable roofs, |
| | Reinstall | S.F. | | 1.23 | | 1.23 | 25' span, up to 5/12 slope, per S.F. |
| | Minimum Charge | Job | | 232 | | 232 | of roof. |
| **24" O.C. System** | | | | | | | |
| **2" x 4" Rafters** | | | | | | | |
| | Demolish | S.F. | | .67 | | .67 | Cost includes material and labor to |
| | Install | S.F. | .23 | .62 | | .85 | install 2" x 4", 24" O.C. rafter framing |
| | Demolish and Install | S.F. | .23 | 1.29 | | 1.52 | for flat, shed or gable roofs, 25' span, |
| | Reinstall | S.F. | | .50 | | .50 | up to 5/12 slope, per S.F. of roof. |
| | Minimum Charge | Job | | 232 | | 232 | |
| **2" x 6" Rafters** | | | | | | | |
| | Demolish | S.F. | | .71 | | .71 | Cost includes material and labor to |
| | Install | S.F. | .35 | .66 | | 1.01 | install 2" x 6", 24" O.C. rafter framing |
| | Demolish and Install | S.F. | .35 | 1.37 | | 1.72 | for flat, shed or gable roofs, 25' span, |
| | Reinstall | S.F. | | .52 | | .52 | up to 5/12 slope, per S.F. of roof. |
| | Minimum Charge | Job | | 232 | | 232 | |
| **2" x 8" Rafters** | | | | | | | |
| | Demolish | S.F. | | .72 | | .72 | Cost includes material and labor to |
| | Install | S.F. | .50 | .70 | | 1.20 | install 2" x 8", 24" O.C. rafter framing |
| | Demolish and Install | S.F. | .50 | 1.42 | | 1.92 | for flat, shed or gable roofs, 25' span, |
| | Reinstall | S.F. | | .56 | | .56 | up to 5/12 slope, per S.F. of roof. |
| | Minimum Charge | Job | | 232 | | 232 | |
| **2" x 10" Rafters** | | | | | | | |
| | Demolish | S.F. | | .72 | | .72 | Cost includes material and labor to |
| | Install | S.F. | .77 | 1.06 | | 1.83 | install 2" x 10", 24" O.C. rafter |
| | Demolish and Install | S.F. | .77 | 1.78 | | 2.55 | framing for flat, shed or gable roofs, |
| | Reinstall | S.F. | | .85 | | .85 | 25' span, up to 5/12 slope, per S.F. |
| | Minimum Charge | Job | | 232 | | 232 | of roof. |
| **2" x 12" Rafters** | | | | | | | |
| | Demolish | S.F. | | .73 | | .73 | Cost includes material and labor to |
| | Install | S.F. | .99 | 1.15 | | 2.14 | install 2" x 12", 24" O.C. rafter |
| | Demolish and Install | S.F. | .99 | 1.88 | | 2.87 | framing for flat, shed or gable roofs, |
| | Reinstall | S.F. | | .92 | | .92 | 25' span, up to 5/12 slope, per S.F. |
| | Minimum Charge | Job | | 232 | | 232 | of roof. |
| **Rafter** | | | | | | | |
| **2" x 4"** | | | | | | | |
| | Demolish | L.F. | | .92 | | .92 | Cost includes material and labor to |
| | Install | L.F. | .46 | 1.16 | | 1.62 | install 2" x 4" rafters for flat, shed or |
| | Demolish and Install | L.F. | .46 | 2.08 | | 2.54 | gable roofs, up to 5/12 slope, 25' |
| | Reinstall | L.F. | | .93 | | .93 | span, per L.F. |
| | Minimum Charge | Job | | 232 | | 232 | |
| **2" x 6"** | | | | | | | |
| | Demolish | L.F. | | .93 | | .93 | Cost includes material and labor to |
| | Install | L.F. | .70 | 1.16 | | 1.86 | install 2" x 6" rafters for flat, shed or |
| | Demolish and Install | L.F. | .70 | 2.09 | | 2.79 | gable roofs, up to 5/12 slope, 25' |
| | Reinstall | L.F. | | .93 | | .93 | span, per L.F. |
| | Minimum Charge | Job | | 232 | | 232 | |
| **2" x 8"** | | | | | | | |
| | Demolish | L.F. | | .94 | | .94 | Cost includes material and labor to |
| | Install | L.F. | .98 | 1.24 | | 2.22 | install 2" x 8" rafters for flat, shed or |
| | Demolish and Install | L.F. | .98 | 2.18 | | 3.16 | gable roofs, up to 5/12 slope, 25' |
| | Reinstall | L.F. | | .99 | | .99 | span, per L.F. |
| | Minimum Charge | Job | | 232 | | 232 | |

| Roof Framing | | Unit | Material | Labor | Equip. | Total | Specification |
|---|---|---|---|---|---|---|---|
| **2" x 10"** | | | | | | | Cost includes material and labor to |
| | Demolish | L.F. | | .96 | | .96 | install 2" x 10" rafters for flat, shed or |
| | Install | L.F. | 1.55 | 1.88 | | 3.43 | gable roofs, up to 5/12 slope, 25' |
| | Demolish and Install | L.F. | 1.55 | 2.84 | | 4.39 | span, per L.F. |
| | Reinstall | L.F. | | 1.50 | | 1.50 | |
| | Minimum Charge | Job | | 232 | | 232 | |
| **2" x 12"** | | | | | | | Cost includes material and labor to |
| | Demolish | L.F. | | .97 | | .97 | install 2" x 12" rafters for flat, shed or |
| | Install | L.F. | 1.97 | 2.04 | | 4.01 | gable roofs, up to 5/12 slope, 25' |
| | Demolish and Install | L.F. | 1.97 | 3.01 | | 4.98 | span, per L.F. |
| | Reinstall | L.F. | | 1.63 | | 1.63 | |
| | Minimum Charge | Job | | 232 | | 232 | |
| **2" x 4" Valley / Jack** | | | | | | | Includes material and labor to install |
| | Demolish | L.F. | | .92 | | .92 | up to 2" x 6" valley/jack rafters for |
| | Install | L.F. | .70 | 1.96 | | 2.66 | flat, shed or gable roofs, up to 5/12 |
| | Demolish and Install | L.F. | .70 | 2.88 | | 3.58 | slope, 25' span, per L.F. |
| | Reinstall | L.F. | | 1.57 | | 1.57 | |
| | Minimum Charge | Job | | 232 | | 232 | |
| **2" x 6" Valley / Jack** | | | | | | | Includes material and labor to install |
| | Demolish | L.F. | | .93 | | .93 | up to 2" x 6" valley/jack rafters for |
| | Install | L.F. | .70 | 1.96 | | 2.66 | flat, shed or gable roofs, up to 5/12 |
| | Demolish and Install | L.F. | .70 | 2.89 | | 3.59 | slope, 25' span, per L.F. |
| | Reinstall | L.F. | | 1.57 | | 1.57 | |
| | Minimum Charge | Job | | 232 | | 232 | |
| **Ridgeboard** | | | | | | | |
| **2" x 4"** | | | | | | | Cost includes material and labor to |
| | Demolish | L.F. | | .88 | | .88 | install 2" x 4" ridgeboard for flat, shed |
| | Install | L.F. | .46 | 1.69 | | 2.15 | or gable roofs, up to 5/12 slope, 25' |
| | Demolish and Install | L.F. | .46 | 2.57 | | 3.03 | span, per L.F. |
| | Reinstall | L.F. | | 1.35 | | 1.35 | |
| | Minimum Charge | Job | | 232 | | 232 | |
| **2" x 6"** | | | | | | | Cost includes material and labor to |
| | Demolish | L.F. | | .90 | | .90 | install 2" x 6" ridgeboard for flat, shed |
| | Install | L.F. | .70 | 1.86 | | 2.56 | or gable roofs, up to 5/12 slope, 25' |
| | Demolish and Install | L.F. | .70 | 2.76 | | 3.46 | span, per L.F. |
| | Reinstall | L.F. | | 1.49 | | 1.49 | |
| | Minimum Charge | Job | | 232 | | 232 | |
| **2" x 8"** | | | | | | | Cost includes material and labor to |
| | Demolish | L.F. | | .93 | | .93 | install 2" x 8" ridgeboard for flat, shed |
| | Install | L.F. | .98 | 2.07 | | 3.05 | or gable roofs, up to 5/12 slope, 25' |
| | Demolish and Install | L.F. | .98 | 3 | | 3.98 | span, per L.F. |
| | Reinstall | L.F. | | 1.65 | | 1.65 | |
| | Minimum Charge | Job | | 232 | | 232 | |
| **2" x 10"** | | | | | | | Cost includes material and labor to |
| | Demolish | L.F. | | .96 | | .96 | install 2" x 10" ridgeboard for flat, |
| | Install | L.F. | 1.55 | 2.32 | | 3.87 | shed or gable roofs, up to 5/12 |
| | Demolish and Install | L.F. | 1.55 | 3.28 | | 4.83 | slope, 25' span, per L.F. |
| | Reinstall | L.F. | | 1.86 | | 1.86 | |
| | Minimum Charge | Job | | 232 | | 232 | |
| **2" x 12"** | | | | | | | Cost includes material and labor to |
| | Demolish | L.F. | | .99 | | .99 | install 2" x 12" ridgeboard for flat, |
| | Install | L.F. | 1.97 | 1.33 | | 3.30 | shed or gable roofs, up to 5/12 |
| | Demolish and Install | L.F. | 1.97 | 2.32 | | 4.29 | slope, 25' span, per L.F. |
| | Reinstall | L.F. | | 1.06 | | 1.06 | |
| | Minimum Charge | Job | | 232 | | 232 | |

**For customer support on your Contractor's Pricing Guide: Residential Repair & Remodeling, call 800.448.8182.**

# Rough Frame / Structure

## Roof Framing

| | Unit | Material | Labor | Equip. | Total | Specification |
|---|---|---|---|---|---|---|
| **Collar Beam** | | | | | | |
| **1" x 6"** | | | | | | |
| Demolish | L.F. | | .70 | | .70 | Cost includes material and labor to |
| Install | L.F. | .86 | .58 | | 1.44 | install 1" x 6" collar beam or tie for |
| Demolish and Install | L.F. | .86 | 1.28 | | 2.14 | roof framing. |
| Reinstall | L.F. | | .46 | | .46 | |
| Minimum Charge | Job | | 232 | | 232 | |
| **2" x 6"** | | | | | | |
| Demolish | L.F. | | .81 | | .81 | Cost includes material and labor to |
| Install | L.F. | .70 | .58 | | 1.28 | install 2" x 6" collar beam or tie for |
| Demolish and Install | L.F. | .70 | 1.39 | | 2.09 | roof framing. |
| Reinstall | L.F. | | .46 | | .46 | |
| Minimum Charge | Job | | 232 | | 232 | |
| **Purlins** | | | | | | |
| **2" x 6"** | | | | | | |
| Demolish | L.F. | | .79 | | .79 | Cost includes material and labor to |
| Install | L.F. | .70 | .52 | | 1.22 | install 2" x 6" purlins below roof |
| Demolish and Install | L.F. | .70 | 1.31 | | 2.01 | rafters. |
| Reinstall | L.F. | | .41 | | .41 | |
| Minimum Charge | Job | | 232 | | 232 | |
| **2" x 8"** | | | | | | |
| Demolish | L.F. | | .80 | | .80 | Cost includes material and labor to |
| Install | L.F. | .98 | .52 | | 1.50 | install 2" x 8" purlins below roof |
| Demolish and Install | L.F. | .98 | 1.32 | | 2.30 | rafters. |
| Reinstall | L.F. | | .42 | | .42 | |
| Minimum Charge | Job | | 232 | | 232 | |
| **2" x 10"** | | | | | | |
| Demolish | L.F. | | .81 | | .81 | Cost includes material and labor to |
| Install | L.F. | 1.55 | .53 | | 2.08 | install 2" x 10" purlins below roof |
| Demolish and Install | L.F. | 1.55 | 1.34 | | 2.89 | rafters. |
| Reinstall | L.F. | | .42 | | .42 | |
| Minimum Charge | Job | | 232 | | 232 | |
| **2" x 12"** | | | | | | |
| Demolish | L.F. | | .81 | | .81 | Cost includes material and labor to |
| Install | L.F. | 1.97 | .53 | | 2.50 | install 2" x 12" purlins below roof |
| Demolish and Install | L.F. | 1.97 | 1.34 | | 3.31 | rafters. |
| Reinstall | L.F. | | .43 | | .43 | |
| Minimum Charge | Job | | 232 | | 232 | |
| **4" x 6"** | | | | | | |
| Demolish | L.F. | | .81 | | .81 | Cost includes material and labor to |
| Install | L.F. | 3.37 | .53 | | 3.90 | install 4" x 6" purlins below roof |
| Demolish and Install | L.F. | 3.37 | 1.34 | | 4.71 | rafters. |
| Reinstall | L.F. | | .43 | | .43 | |
| Minimum Charge | Job | | 232 | | 232 | |
| **4" x 8"** | | | | | | |
| Demolish | L.F. | | .82 | | .82 | Cost includes material and labor to |
| Install | L.F. | 4.57 | .55 | | 5.12 | install 4" x 8" purlins below roof |
| Demolish and Install | L.F. | 4.57 | 1.37 | | 5.94 | rafters. |
| Reinstall | L.F. | | .44 | | .44 | |
| Minimum Charge | Job | | 232 | | 232 | |
| **Ledger Board** | | | | | | |
| **2" x 6"** | | | | | | |
| Demolish | L.F. | | .39 | | .39 | Cost includes material and labor to |
| Install | L.F. | .70 | 2.09 | | 2.79 | install 2" x 6" ledger board nailed to |
| Demolish and Install | L.F. | .70 | 2.48 | | 3.18 | the face of a rafter. |
| Reinstall | L.F. | | 1.67 | | 1.67 | |
| Minimum Charge | Job | | 232 | | 232 | |

**For customer support on your Contractor's Pricing Guide: Residential Repair & Remodeling, call 800.448.8182.**

# Rough Frame / Structure

## Roof Framing

| | Unit | Material | Labor | Equip. | Total | Specification |
|---|---|---|---|---|---|---|
| **4" x 6"** | | | | | | |
| Demolish | L.F. | | .66 | | .66 | Cost includes material and labor to |
| Install | L.F. | 3.37 | 3.08 | | 6.45 | install 4" x 6" ledger board nailed to |
| Demolish and Install | L.F. | 3.37 | 3.74 | | 7.11 | the face of a rafter. |
| Reinstall | L.F. | | 2.46 | | 2.46 | |
| Minimum Charge | Job | | 232 | | 232 | |
| **4" x 8"** | | | | | | |
| Demolish | L.F. | | .88 | | .88 | Cost includes material and labor to |
| Install | L.F. | 4.57 | 3.55 | | 8.12 | install 4" x 8" ledger board nailed to |
| Demolish and Install | L.F. | 4.57 | 4.43 | | 9 | the face of a rafter. |
| Reinstall | L.F. | | 2.84 | | 2.84 | |
| Minimum Charge | Job | | 232 | | 232 | |

## Outriggers

| | Unit | Material | Labor | Equip. | Total | Specification |
|---|---|---|---|---|---|---|
| **2" x 4"** | | | | | | |
| Demolish | L.F. | | .27 | | .27 | Cost includes material and labor to |
| Install | L.F. | .46 | 1.86 | | 2.32 | install 2" x 4" outrigger rafters for flat, |
| Demolish and Install | L.F. | .46 | 2.13 | | 2.59 | shed or gabled roofs, up to 5/12 |
| Reinstall | L.F. | | 1.49 | | 1.49 | slope, per L.F. |
| Minimum Charge | Job | | 232 | | 232 | |
| **2" x 6"** | | | | | | |
| Demolish | L.F. | | .41 | | .41 | Cost includes material and labor to |
| Install | L.F. | .70 | 1.86 | | 2.56 | install 2" x 6" outrigger rafters for flat, |
| Demolish and Install | L.F. | .70 | 2.27 | | 2.97 | shed or gabled roofs, up to 5/12 |
| Reinstall | L.F. | | 1.49 | | 1.49 | slope, per L.F. |
| Minimum Charge | Job | | 232 | | 232 | |

## Lookout Rafter

| | Unit | Material | Labor | Equip. | Total | Specification |
|---|---|---|---|---|---|---|
| **2" x 4"** | | | | | | |
| Demolish | L.F. | | .92 | | .92 | Cost includes material and labor to |
| Install | L.F. | .46 | .77 | | 1.23 | install 2" x 4" lookout rafters for flat, |
| Demolish and Install | L.F. | .46 | 1.69 | | 2.15 | shed or gabled roofs, up to 5/12 |
| Reinstall | L.F. | | .62 | | .62 | slope, per L.F. |
| Minimum Charge | Job | | 232 | | 232 | |
| **2" x 6"** | | | | | | |
| Demolish | L.F. | | .93 | | .93 | Cost includes material and labor to |
| Install | L.F. | .70 | .79 | | 1.49 | install 2" x 6" lookout rafters for flat, |
| Demolish and Install | L.F. | .70 | 1.72 | | 2.42 | shed or gabled roofs, up to 5/12 |
| Reinstall | L.F. | | .63 | | .63 | slope, per L.F. |
| Minimum Charge | Job | | 232 | | 232 | |

## Fly Rafter

| | Unit | Material | Labor | Equip. | Total | Specification |
|---|---|---|---|---|---|---|
| **2" x 4"** | | | | | | |
| Demolish | L.F. | | .92 | | .92 | Cost includes material and labor to |
| Install | L.F. | .46 | .77 | | 1.23 | install 2" x 4" fly rafters for flat, shed |
| Demolish and Install | L.F. | .46 | 1.69 | | 2.15 | or gabled roofs, up to 5/12 slope, |
| Reinstall | L.F. | | .62 | | .62 | per L.F. |
| Minimum Charge | Job | | 232 | | 232 | |
| **2" x 6"** | | | | | | |
| Demolish | L.F. | | .93 | | .93 | Cost includes material and labor to |
| Install | L.F. | .69 | .79 | | 1.48 | install 2" x 6" fly rafters for flat, shed |
| Demolish and Install | L.F. | .69 | 1.72 | | 2.41 | or gabled roofs, up to 5/12 slope, |
| Reinstall | L.F. | | .63 | | .63 | per L.F. |
| Minimum Charge | Job | | 232 | | 232 | |

For customer support on your Contractor's Pricing Guide: Residential Repair & Remodeling, call 800.448.8182.

| Residential Trusses | Unit | Material | Labor | Equip. | Total | Specification |
|---|---|---|---|---|---|---|
| **W or Fink Truss** | | | | | | |
| **24' Span** | | | | | | |
| Demolish | Ea. | | 36 | 9.20 | 45.20 | Includes material, labor and |
| Install | Ea. | 87 | 35.50 | 9.10 | 131.60 | equipment to install wood gang-nailed |
| Demolish and Install | Ea. | 87 | 71.50 | 18.30 | 176.80 | residential truss, up to 24' span. |
| Reinstall | Ea. | | 35.47 | 9.09 | 44.56 | |
| Minimum Charge | Job | | 232 | | 232 | |
| **28' Span** | | | | | | |
| Demolish | Ea. | | 38 | 9.80 | 47.80 | Includes material, labor and |
| Install | Ea. | 95 | 40 | 10.30 | 145.30 | equipment to install wood gang-nailed |
| Demolish and Install | Ea. | 95 | 78 | 20.10 | 193.10 | residential truss, up to 28' span. |
| Reinstall | Ea. | | 40.16 | 10.29 | 50.45 | |
| Minimum Charge | Job | | 232 | | 232 | |
| **32' Span** | | | | | | |
| Demolish | Ea. | | 42.50 | 10.85 | 53.35 | Includes material, labor and |
| Install | Ea. | 124 | 42.50 | 10.90 | 177.40 | equipment to install wood gang-nailed |
| Demolish and Install | Ea. | 124 | 85 | 21.75 | 230.75 | residential truss, up to 32' span. |
| Reinstall | Ea. | | 42.57 | 10.91 | 53.48 | |
| Minimum Charge | Job | | 232 | | 232 | |
| **36' Span** | | | | | | |
| Demolish | Ea. | | 45.50 | 11.65 | 57.15 | Includes material, labor and |
| Install | Ea. | 156 | 46.50 | 11.85 | 214.35 | equipment to install wood gang-nailed |
| Demolish and Install | Ea. | 156 | 92 | 23.50 | 271.50 | residential truss, up to 36' span. |
| Reinstall | Ea. | | 46.27 | 11.86 | 58.13 | |
| Minimum Charge | Job | | 232 | | 232 | |
| **Gable End** | | | | | | |
| **24' Span** | | | | | | |
| Demolish | Ea. | | 36 | 9.20 | 45.20 | Includes material, labor and |
| Install | Ea. | 95 | 38 | 9.75 | 142.75 | equipment to install wood gang-nailed |
| Demolish and Install | Ea. | 95 | 74 | 18.95 | 187.95 | residential gable end truss, up to 24' |
| Reinstall | Ea. | | 38.01 | 9.74 | 47.75 | span. |
| Clean | Ea. | .62 | 47.50 | | 48.12 | |
| Paint | Ea. | 4.83 | 51.50 | | 56.33 | |
| Minimum Charge | Job | | 232 | | 232 | |
| **28' Span** | | | | | | |
| Demolish | Ea. | | 38 | 9.80 | 47.80 | Includes material, labor and |
| Install | Ea. | 113 | 45.50 | 11.60 | 170.10 | equipment to install wood gang-nailed |
| Demolish and Install | Ea. | 113 | 83.50 | 21.40 | 217.90 | residential gable end truss, up to 30' |
| Reinstall | Ea. | | 45.29 | 11.61 | 56.90 | span. |
| Clean | Ea. | .64 | 64.50 | | 65.14 | |
| Paint | Ea. | 5.05 | 70.50 | | 75.55 | |
| Minimum Charge | Job | | 232 | | 232 | |
| **32' Span** | | | | | | |
| Demolish | Ea. | | 42.50 | 10.85 | 53.35 | Includes material, labor and |
| Install | Ea. | 121 | 46.50 | 11.85 | 179.35 | equipment to install wood gang-nailed |
| Demolish and Install | Ea. | 121 | 89 | 22.70 | 232.70 | residential gable end truss, up to 32' |
| Reinstall | Ea. | | 46.27 | 11.86 | 58.13 | span. |
| Clean | Ea. | .67 | 84.50 | | 85.17 | |
| Paint | Ea. | 5.25 | 92 | | 97.25 | |
| Minimum Charge | Job | | 232 | | 232 | |
| **36' Span** | | | | | | |
| Demolish | Ea. | | 45.50 | 11.65 | 57.15 | Includes material, labor and |
| Install | Ea. | 155 | 53 | 13.65 | 221.65 | equipment to install wood gang-nailed |
| Demolish and Install | Ea. | 155 | 98.50 | 25.30 | 278.80 | residential gable end truss, up to 36' |
| Reinstall | Ea. | | 53.21 | 13.64 | 66.85 | span. |
| Clean | Ea. | .70 | 107 | | 107.70 | |
| Paint | Ea. | 5.50 | 116 | | 121.50 | |
| Minimum Charge | Job | | 232 | | 232 | |

# Rough Frame / Structure

## Residential Trusses

| Residential Trusses | Unit | Material | Labor | Equip. | Total | Specification |
|---|---|---|---|---|---|---|
| **Truss Hardware** | | | | | | |
| **5-1/4" Glue-Lam Seat** | | | | | | |
| Install | Ea. | 193 | 2.58 | | 195.58 | Includes labor and material to install |
| Minimum Charge | Job | | 232 | | 232 | beam hangers. |
| **6-3/4" Glue-Lam Seat** | | | | | | |
| Install | Ea. | 227 | 2.58 | | 229.58 | Includes labor and material to install |
| Minimum Charge | Job | | 232 | | 232 | beam hangers. |
| **2" x 4" Joist Hanger** | | | | | | |
| Install | Ea. | .83 | 2.66 | | 3.49 | Includes labor and material to install |
| Minimum Charge | Job | | 232 | | 232 | joist and beam hangers. |
| **2" x 12" Joist Hanger** | | | | | | |
| Install | Ea. | 1.40 | 2.82 | | 4.22 | Cost includes labor and material to |
| Minimum Charge | Job | | 232 | | 232 | install 2" x 12" joist hanger. |

## Commercial Trusses

| Commercial Trusses | Unit | Material | Labor | Equip. | Total | Specification |
|---|---|---|---|---|---|---|
| **Engineered Lumber, Truss / Joist** | | | | | | |
| **9-1/2"** | | | | | | |
| Demolish | S.F. | | .65 | | .65 | Includes material and labor to install |
| Install | S.F. | 1.75 | .46 | | 2.21 | engineered lumber truss / joists per |
| Demolish and Install | S.F. | 1.75 | 1.11 | | 2.86 | S.F. of floor area based on joists 16" |
| Reinstall | S.F. | | .37 | | .37 | O.C. Beams, supports and bridging |
| Minimum Charge | Job | | 232 | | 232 | are not included. |
| **11-7/8"** | | | | | | |
| Demolish | S.F. | | .67 | | .67 | Includes material and labor to install |
| Install | S.F. | 2.02 | .48 | | 2.50 | engineered lumber truss / joists per |
| Demolish and Install | S.F. | 2.02 | 1.15 | | 3.17 | S.F. of floor area based on joists 16" |
| Reinstall | S.F. | | .38 | | .38 | O.C. Beams, supports and bridging |
| Minimum Charge | Job | | 232 | | 232 | are not included. |
| **14"** | | | | | | |
| Demolish | S.F. | | .70 | | .70 | Includes material and labor to install |
| Install | S.F. | 2.53 | .51 | | 3.04 | engineered lumber truss / joists per |
| Demolish and Install | S.F. | 2.53 | 1.21 | | 3.74 | S.F. of floor area based on joists 16" |
| Reinstall | S.F. | | .41 | | .41 | O.C. Beams, supports and bridging |
| Minimum Charge | Job | | 232 | | 232 | are not included. |
| **16"** | | | | | | |
| Demolish | S.F. | | .72 | | .72 | Includes material and labor to install |
| Install | S.F. | 4.03 | .54 | | 4.57 | engineered lumber truss / joists per |
| Demolish and Install | S.F. | 4.03 | 1.26 | | 5.29 | S.F. of floor area based on joists 16" |
| Reinstall | S.F. | | .43 | | .43 | O.C. Beams, supports and bridging |
| Minimum Charge | Job | | 232 | | 232 | are not included. |
| **Bowstring Truss** | | | | | | |
| **100' Clear Span** | | | | | | |
| Demolish | SF Flr. | | .66 | .46 | 1.12 | Includes material, labor and |
| Install | SF Flr. | 8.40 | .53 | .27 | 9.20 | equipment to install bow string truss |
| Demolish and Install | SF Flr. | 8.40 | 1.19 | .73 | 10.32 | system up to 100' clear span per S.F. |
| Minimum Charge | Job | | 232 | | 232 | of floor area. |
| **120' Clear Span** | | | | | | |
| Demolish | SF Flr. | | .73 | .52 | 1.25 | Includes material, labor and |
| Install | SF Flr. | 8.90 | .59 | .30 | 9.79 | equipment to install bow string truss |
| Demolish and Install | SF Flr. | 8.90 | 1.32 | .82 | 11.04 | system up to 120' clear span per S.F. |
| Minimum Charge | Job | | 232 | | 232 | of floor area. |

# Rough Frame / Structure

| Roof Sheathing | | Unit | Material | Labor | Equip. | Total | Specification |
|---|---|---|---|---|---|---|---|
| **Plywood** | | | | | | | |
| **3/8"** | | | | | | | |
| | Demolish | S.F. | | .51 | | .51 | Cost includes material and labor to |
| | Install | S.F. | .66 | .61 | | 1.27 | install 3/8" CDX plywood roof |
| | Demolish and Install | S.F. | .66 | 1.12 | | 1.78 | sheathing. |
| | Reinstall | S.F. | | .61 | | .61 | |
| | Minimum Charge | Job | | 232 | | 232 | |
| **1/2"** | | | | | | | |
| | Demolish | S.F. | | .51 | | .51 | Cost includes material and labor to |
| | Install | S.F. | .68 | .66 | | 1.34 | install 1/2" CDX plywood roof |
| | Demolish and Install | S.F. | .68 | 1.17 | | 1.85 | sheathing. |
| | Minimum Charge | Job | | 232 | | 232 | |
| **5/8"** | | | | | | | |
| | Demolish | S.F. | | .51 | | .51 | Cost includes material and labor to |
| | Install | S.F. | .85 | .72 | | 1.57 | install 5/8" CDX plywood roof |
| | Demolish and Install | S.F. | .85 | 1.23 | | 2.08 | sheathing. |
| | Minimum Charge | Job | | 232 | | 232 | |
| **3/4"** | | | | | | | |
| | Demolish | S.F. | | .51 | | .51 | Cost includes material and labor to |
| | Install | S.F. | 1.01 | .77 | | 1.78 | install 3/4" CDX plywood roof |
| | Demolish and Install | S.F. | 1.01 | 1.28 | | 2.29 | sheathing. |
| | Minimum Charge | Job | | 232 | | 232 | |
| **OSB** | | | | | | | |
| **1/2"** | | | | | | | |
| | Demolish | S.F. | | .51 | | .51 | Cost includes material and labor to |
| | Install | S.F. | .53 | .66 | | 1.19 | install 4' x 8' x 1/2" OSB sheathing. |
| | Demolish and Install | S.F. | .53 | 1.17 | | 1.70 | |
| | Minimum Charge | Job | | 232 | | 232 | |
| **5/8"** | | | | | | | |
| | Demolish | S.F. | | .51 | | .51 | Cost includes material and labor to |
| | Install | S.F. | .53 | .54 | | 1.07 | install 4' x 8' x 5/8" OSB sheathing. |
| | Demolish and Install | S.F. | .53 | 1.05 | | 1.58 | |
| | Minimum Charge | Job | | 232 | | 232 | |
| **Plank** | | | | | | | |
| | Demolish | S.F. | | .51 | | .51 | Cost includes material and labor to |
| | Install | S.F. | 2.28 | 1.28 | | 3.56 | install 1" x 6" or 1" x 8" utility T&G |
| | Demolish and Install | S.F. | 2.28 | 1.79 | | 4.07 | board sheathing laid diagonal. |
| | Reinstall | S.F. | | 1.28 | | 1.28 | |
| | Minimum Charge | Job | | 232 | | 232 | |
| **Skip-type** | | | | | | | |
| **1" x 4"** | | | | | | | |
| | Demolish | S.F. | | .79 | | .79 | Includes material and labor to install |
| | Install | S.F. | .69 | .39 | | 1.08 | sheathing board material 1" x 4", 7" |
| | Demolish and Install | S.F. | .69 | 1.18 | | 1.87 | O.C. |
| | Minimum Charge | Job | | 232 | | 232 | |
| **1" x 6"** | | | | | | | |
| | Demolish | S.F. | | .79 | | .79 | Includes material and labor to install |
| | Install | S.F. | .86 | .32 | | 1.18 | sheathing board material 1" x 6", 7" |
| | Demolish and Install | S.F. | .86 | 1.11 | | 1.97 | or 9" O.C. |
| | Minimum Charge | Job | | 232 | | 232 | |

## Wood Deck

| Wood Deck | Unit | Material | Labor | Equip. | Total | Specification |
|---|---|---|---|---|---|---|
| **Stairs** | | | | | | |
| Demolish | Ea. | | 455 | | 455 | Includes material and labor to install |
| Install | Ea. | 156 | 820 | | 976 | wood steps. 5 - 7 risers included. |
| Demolish and Install | Ea. | 156 | 1275 | | 1431 | |
| Reinstall | Ea. | | 657.02 | | 657.02 | |
| Clean | Flight | 1.47 | 30 | | 31.47 | |
| Paint | Flight | 21 | 51.50 | | 72.50 | |
| Minimum Charge | Job | | 232 | | 232 | |
| **Bench Seating** | | | | | | |
| Demolish | L.F. | | 6.60 | | 6.60 | Includes material and labor to install |
| Install | L.F. | 2.27 | 23 | | 25.27 | pressure treated bench seating. |
| Demolish and Install | L.F. | 2.27 | 29.60 | | 31.87 | |
| Reinstall | L.F. | | 18.59 | | 18.59 | |
| Clean | L.F. | .06 | 1.78 | | 1.84 | |
| Paint | L.F. | .69 | 3.09 | | 3.78 | |
| Minimum Charge | Job | | 232 | | 232 | |
| **Privacy Wall** | | | | | | |
| Demolish | SF Flr. | | 1.58 | | 1.58 | Includes material, labor and |
| Install | SF Flr. | 1.71 | .85 | .16 | 2.72 | equipment to install privacy wall. |
| Demolish and Install | SF Flr. | 1.71 | 2.43 | .16 | 4.30 | |
| Reinstall | SF Flr. | | .68 | .13 | .81 | |
| Clean | SF Flr. | .03 | .36 | | .39 | |
| Paint | S.F. | .11 | .30 | | .41 | |
| Minimum Charge | Job | | 232 | | 232 | |
| **Treated Column or Post** | | | | | | |
| **4" x 4"** | | | | | | |
| Demolish | L.F. | | .99 | | .99 | Includes material and labor to install |
| Install | L.F. | 1.32 | 2.38 | | 3.70 | 4" x 4" treated column or post. |
| Demolish and Install | L.F. | 1.32 | 3.37 | | 4.69 | |
| Reinstall | L.F. | | 1.91 | | 1.91 | |
| Clean | L.F. | .04 | .32 | | .36 | |
| Paint | L.F. | .18 | .97 | | 1.15 | |
| Minimum Charge | Job | | 232 | | 232 | |
| **4" x 6"** | | | | | | |
| Demolish | L.F. | | 1.44 | | 1.44 | Includes material and labor to install |
| Install | L.F. | 2.13 | 3.38 | | 5.51 | 4" x 6" treated column or post. |
| Demolish and Install | L.F. | 2.13 | 4.82 | | 6.95 | |
| Reinstall | L.F. | | 2.70 | | 2.70 | |
| Clean | L.F. | .07 | .44 | | .51 | |
| Paint | L.F. | .18 | .97 | | 1.15 | |
| Minimum Charge | Job | | 232 | | 232 | |
| **Cedar Columns or Posts** | | | | | | |
| **4" x 4"** | | | | | | |
| Demolish | L.F. | | .99 | | .99 | Includes material and labor to install |
| Install | L.F. | 4.06 | 2.38 | | 6.44 | 4" x 4" cedar columns or posts. |
| Demolish and Install | L.F. | 4.06 | 3.37 | | 7.43 | |
| Reinstall | L.F. | | 1.91 | | 1.91 | |
| Clean | L.F. | .04 | .32 | | .36 | |
| Paint | L.F. | .18 | .97 | | 1.15 | |
| Minimum Charge | Job | | 232 | | 232 | |
| **4" x 6"** | | | | | | |
| Demolish | L.F. | | 1.44 | | 1.44 | Includes material and labor to install |
| Install | L.F. | 7.60 | 3.38 | | 10.98 | 4" x 6" cedar columns or posts. |
| Demolish and Install | L.F. | 7.60 | 4.82 | | 12.42 | |
| Reinstall | L.F. | | 2.70 | | 2.70 | |
| Clean | L.F. | .07 | .44 | | .51 | |
| Paint | L.F. | .18 | .97 | | 1.15 | |
| Minimum Charge | Job | | 232 | | 232 | |

**For customer support on your Contractor's Pricing Guide: Residential Repair & Remodeling, call 800.448.8182.**

## Wood Deck

| Wood Deck | Unit | Material | Labor | Equip. | Total | Specification |
|---|---|---|---|---|---|---|
| **Redwood Columns or Posts** | | | | | | |
| **4″ x 4″** | | | | | | |
| Demolish | L.F. | | .99 | | .99 | Includes material and labor to install |
| Install | L.F. | 7.10 | 2.38 | | 9.48 | 4″ x 4″ redwood columns or posts. |
| Demolish and Install | L.F. | 7.10 | 3.37 | | 10.47 | |
| Reinstall | L.F. | | 1.91 | | 1.91 | |
| Clean | L.F. | .04 | .32 | | .36 | |
| Paint | L.F. | .18 | .97 | | 1.15 | |
| Minimum Charge | Job | | 232 | | 232 | |
| **4″ x 6″** | | | | | | |
| Demolish | L.F. | | 1.44 | | 1.44 | Includes material and labor to install |
| Install | L.F. | 13.85 | 3.38 | | 17.23 | 4″ x 6″ redwood columns or posts. |
| Demolish and Install | L.F. | 13.85 | 4.82 | | 18.67 | |
| Reinstall | L.F. | | 2.70 | | 2.70 | |
| Clean | L.F. | .07 | .44 | | .51 | |
| Paint | L.F. | .18 | .97 | | 1.15 | |
| Minimum Charge | Job | | 232 | | 232 | |
| **Treated Girder - Single** | | | | | | |
| **4″ x 8″** | | | | | | |
| Demolish | L.F. | | 2.94 | | 2.94 | Includes material and labor to install a |
| Install | L.F. | 4.36 | 1.77 | | 6.13 | treated lumber girder consisting of one |
| Demolish and Install | L.F. | 4.36 | 4.71 | | 9.07 | 4″ x 8″ for porch or deck construction. |
| Reinstall | L.F. | | 1.42 | | 1.42 | |
| Clean | L.F. | .06 | .38 | | .44 | |
| Minimum Charge | Job | | 232 | | 232 | |
| **Treated Girder - Double** | | | | | | |
| **2″ x 8″** | | | | | | |
| Demolish | L.F. | | 2.94 | | 2.94 | Includes material and labor to install a |
| Install | L.F. | 2.53 | 1.62 | | 4.15 | treated lumber girder consisting of two |
| Demolish and Install | L.F. | 2.53 | 4.56 | | 7.09 | 2″ x 8″s for porch or deck |
| Reinstall | L.F. | | 1.29 | | 1.29 | construction. |
| Clean | L.F. | .06 | .38 | | .44 | |
| Minimum Charge | Job | | 232 | | 232 | |
| **2″ x 10″** | | | | | | |
| Demolish | L.F. | | 3.68 | | 3.68 | Includes material and labor to install a |
| Install | L.F. | 3.17 | 1.69 | | 4.86 | treated lumber girder consisting of two |
| Demolish and Install | L.F. | 3.17 | 5.37 | | 8.54 | 2″ x 10″s for porch or deck |
| Reinstall | L.F. | | 1.35 | | 1.35 | construction. |
| Clean | L.F. | .07 | .44 | | .51 | |
| Minimum Charge | Job | | 232 | | 232 | |
| **2″ x 12″** | | | | | | |
| Demolish | L.F. | | 4.41 | | 4.41 | Includes material and labor to install a |
| Install | L.F. | 4.57 | 1.77 | | 6.34 | treated lumber girder consisting of two |
| Demolish and Install | L.F. | 4.57 | 6.18 | | 10.75 | 2″ x 12″s for porch or deck |
| Reinstall | L.F. | | 1.42 | | 1.42 | construction. |
| Clean | L.F. | .08 | .51 | | .59 | |
| Minimum Charge | Job | | 232 | | 232 | |
| **Treated Girder - Triple** | | | | | | |
| **2″ x 8″** | | | | | | |
| Demolish | L.F. | | 4.41 | | 4.41 | Includes material and labor to install a |
| Install | L.F. | 3.80 | 1.77 | | 5.57 | treated lumber girder consisting of |
| Demolish and Install | L.F. | 3.80 | 6.18 | | 9.98 | three 2″ x 8″s for porch or deck |
| Reinstall | L.F. | | 1.42 | | 1.42 | construction. |
| Clean | L.F. | .07 | .44 | | .51 | |
| Minimum Charge | Job | | 232 | | 232 | |

**For customer support on your Contractor's Pricing Guide: Residential Repair & Remodeling, call 800.448.8182.**

# Rough Frame / Structure

## Wood Deck

| | Unit | Material | Labor | Equip. | Total | Specification |
|---|---|---|---|---|---|---|
| **2″ x 10″** | | | | | | |
| Demolish | L.F. | | 5.50 | | 5.50 | Includes material and labor to install a |
| Install | L.F. | 4.75 | 1.86 | | 6.61 | treated lumber girder consisting of |
| Demolish and Install | L.F. | 4.75 | 7.36 | | 12.11 | three 2″ x 10″s for porch or deck |
| Reinstall | L.F. | | 1.49 | | 1.49 | construction. |
| Clean | L.F. | .08 | .51 | | .59 | |
| Minimum Charge | Job | | 232 | | 232 | |
| **2″ x 12″** | | | | | | |
| Demolish | L.F. | | 6.55 | | 6.55 | Includes material and labor to install a |
| Install | L.F. | 6.80 | 1.96 | | 8.76 | treated lumber girder consisting of |
| Demolish and Install | L.F. | 6.80 | 8.51 | | 15.31 | three 2″ x 12″s for porch or deck |
| Reinstall | L.F. | | 1.57 | | 1.57 | construction. |
| Clean | L.F. | .09 | .57 | | .66 | |
| Minimum Charge | Job | | 232 | | 232 | |

## Treated Ledger - Bolted

| | Unit | Material | Labor | Equip. | Total | Specification |
|---|---|---|---|---|---|---|
| **2″ x 8″** | | | | | | |
| Demolish | L.F. | | .49 | | .49 | Includes material and labor to install a |
| Install | L.F. | 1.40 | 2.38 | | 3.78 | treated lumber 2″ x 8″ ledger with |
| Demolish and Install | L.F. | 1.40 | 2.87 | | 4.27 | carriage bolts 4′ on center for porch |
| Reinstall | L.F. | | 1.91 | | 1.91 | or deck construction. |
| Clean | L.F. | .01 | .25 | | .26 | |
| Minimum Charge | Job | | 232 | | 232 | |
| **2″ x 10″** | | | | | | |
| Demolish | L.F. | | .49 | | .49 | Includes material and labor to install a |
| Install | L.F. | 1.71 | 2.41 | | 4.12 | treated lumber 2″ x 10″ ledger with |
| Demolish and Install | L.F. | 1.71 | 2.90 | | 4.61 | carriage bolts 4′ on center for porch |
| Reinstall | L.F. | | 1.93 | | 1.93 | or deck construction. |
| Clean | L.F. | .02 | .30 | | .32 | |
| Minimum Charge | Job | | 232 | | 232 | |
| **2″ x 12″** | | | | | | |
| Demolish | L.F. | | .49 | | .49 | Includes material and labor to install a |
| Install | L.F. | 2.39 | 2.45 | | 4.84 | treated lumber 2″ x 12″ ledger with |
| Demolish and Install | L.F. | 2.39 | 2.94 | | 5.33 | carriage bolts 4′ on center for porch |
| Reinstall | L.F. | | 1.96 | | 1.96 | or deck construction. |
| Clean | L.F. | .02 | .30 | | .32 | |
| Minimum Charge | Job | | 232 | | 232 | |

## Treated Joists

| | Unit | Material | Labor | Equip. | Total | Specification |
|---|---|---|---|---|---|---|
| **2″ x 8″** | | | | | | |
| Demolish | L.F. | | .84 | | .84 | Includes material and labor to install |
| Install | L.F. | 1.27 | .85 | | 2.12 | treated lumber 2″ x 8″ deck or floor |
| Demolish and Install | L.F. | 1.27 | 1.69 | | 2.96 | joists for porch or deck construction. |
| Reinstall | L.F. | | .68 | | .68 | |
| Clean | L.F. | .04 | .32 | | .36 | |
| Minimum Charge | Job | | 232 | | 232 | |
| **2″ x 10″** | | | | | | |
| Demolish | L.F. | | .87 | | .87 | Includes material and labor to install |
| Install | L.F. | 1.58 | 1.03 | | 2.61 | treated lumber 2″ x 10″ deck or floor |
| Demolish and Install | L.F. | 1.58 | 1.90 | | 3.48 | joists for porch or deck construction. |
| Reinstall | L.F. | | .83 | | .83 | |
| Clean | L.F. | .06 | .38 | | .44 | |
| Minimum Charge | Job | | 232 | | 232 | |
| **2″ x 12″** | | | | | | |
| Demolish | L.F. | | .90 | | .90 | Includes material and labor to install |
| Install | L.F. | 1.90 | 1.06 | | 2.96 | treated lumber 2″ x 12″ deck or floor |
| Demolish and Install | L.F. | 1.90 | 1.96 | | 3.86 | joists for porch or deck construction. |
| Reinstall | L.F. | | .85 | | .85 | |
| Clean | L.F. | .07 | .44 | | .51 | |
| Minimum Charge | Job | | 232 | | 232 | |

| Wood Deck | Unit | Material | Labor | Equip. | Total | Specification |
|---|---|---|---|---|---|---|
| **Treated Railings and Trim** | | | | | | |
| **2″ x 2″** | | | | | | |
| Demolish | L.F. | | .33 | | .33 | Includes material and labor to install |
| Install | L.F. | .47 | 1.55 | | 2.02 | treated lumber 2″ x 2″ railings and |
| Demolish and Install | L.F. | .47 | 1.88 | | 2.35 | trim for porch or deck construction. |
| Reinstall | L.F. | | 1.24 | | 1.24 | |
| Clean | L.F. | | .24 | | .24 | |
| Paint | L.F. | .18 | .97 | | 1.15 | |
| Minimum Charge | Job | | 232 | | 232 | |
| **2″ x 4″** | | | | | | |
| Demolish | L.F. | | .33 | | .33 | Includes material and labor to install |
| Install | L.F. | .65 | 1.55 | | 2.20 | treated lumber 2″ x 4″ railings and |
| Demolish and Install | L.F. | .65 | 1.88 | | 2.53 | trim for porch or deck construction. |
| Reinstall | L.F. | | 1.24 | | 1.24 | |
| Clean | L.F. | .01 | .25 | | .26 | |
| Paint | L.F. | .18 | .97 | | 1.15 | |
| Minimum Charge | Job | | 232 | | 232 | |
| **2″ x 6″** | | | | | | |
| Demolish | L.F. | | .33 | | .33 | Includes material and labor to install |
| Install | L.F. | .81 | 1.55 | | 2.36 | treated lumber 2″ x 6″ railings and |
| Demolish and Install | L.F. | .81 | 1.88 | | 2.69 | trim for porch or deck construction. |
| Reinstall | L.F. | | 1.24 | | 1.24 | |
| Clean | L.F. | .01 | .30 | | .31 | |
| Paint | L.F. | .18 | .97 | | 1.15 | |
| Minimum Charge | Job | | 232 | | 232 | |
| **Cedar Railings and Trim** | | | | | | |
| **1″ x 4″** | | | | | | |
| Demolish | L.F. | | .33 | | .33 | Includes material and labor to install |
| Install | L.F. | 3.03 | 1.55 | | 4.58 | cedar 1″ x 4″ railings and trim for |
| Demolish and Install | L.F. | 3.03 | 1.88 | | 4.91 | porch or deck construction. |
| Reinstall | L.F. | | 1.24 | | 1.24 | |
| Clean | L.F. | | .24 | | .24 | |
| Paint | L.F. | .18 | .97 | | 1.15 | |
| Minimum Charge | Job | | 232 | | 232 | |
| **2″ x 4″** | | | | | | |
| Demolish | L.F. | | .33 | | .33 | Includes material and labor to install |
| Install | L.F. | 6.15 | 1.55 | | 7.70 | cedar 2″ x 4″ railings and trim for |
| Demolish and Install | L.F. | 6.15 | 1.88 | | 8.03 | porch or deck construction. |
| Reinstall | L.F. | | 1.24 | | 1.24 | |
| Clean | L.F. | .01 | .25 | | .26 | |
| Paint | L.F. | .18 | .97 | | 1.15 | |
| Minimum Charge | Job | | 232 | | 232 | |
| **2″ x 6″** | | | | | | |
| Demolish | L.F. | | .33 | | .33 | Includes material and labor to install |
| Install | L.F. | 11.15 | 1.55 | | 12.70 | cedar 2″ x 6″ railings and trim for |
| Demolish and Install | L.F. | 11.15 | 1.88 | | 13.03 | porch or deck construction. |
| Reinstall | L.F. | | 1.24 | | 1.24 | |
| Clean | L.F. | .01 | .30 | | .31 | |
| Paint | L.F. | .18 | .97 | | 1.15 | |
| Minimum Charge | Job | | 232 | | 232 | |
| **Redwood Railings and Trim** | | | | | | |
| **1″ x 4″** | | | | | | |
| Demolish | L.F. | | .33 | | .33 | Includes material and labor to install |
| Install | L.F. | 1.29 | 1.55 | | 2.84 | redwood 1″ x 4″ railings and trim for |
| Demolish and Install | L.F. | 1.29 | 1.88 | | 3.17 | porch or deck construction. |
| Reinstall | L.F. | | 1.24 | | 1.24 | |
| Clean | L.F. | | .24 | | .24 | |
| Paint | L.F. | .18 | .97 | | 1.15 | |
| Minimum Charge | Job | | 232 | | 232 | |

## Wood Deck

| Wood Deck | Unit | Material | Labor | Equip. | Total | Specification |
|---|---|---|---|---|---|---|
| **2″ x 6″** | | | | | | |
| Demolish | L.F. | | .33 | | .33 | Includes material and labor to install |
| Install | L.F. | 8.25 | 1.55 | | 9.80 | redwood 2″ x 6″ railings and trim for |
| Demolish and Install | L.F. | 8.25 | 1.88 | | 10.13 | porch or deck construction. |
| Reinstall | L.F. | | 1.24 | | 1.24 | |
| Clean | L.F. | .01 | .30 | | .31 | |
| Paint | L.F. | .18 | .97 | | 1.15 | |
| Minimum Charge | Job | | 232 | | 232 | |

## Treated Decking

| Treated Decking | Unit | Material | Labor | Equip. | Total | Specification |
|---|---|---|---|---|---|---|
| **1″ x 4″** | | | | | | |
| Demolish | L.F. | | .48 | | .48 | Includes material and labor to install |
| Install | S.F. | 3.04 | 1.69 | | 4.73 | treated lumber 1″ x 4″ decking for |
| Demolish and Install | S.F. | 3.04 | 2.17 | | 5.21 | porch or deck construction. |
| Reinstall | S.F. | | 1.35 | | 1.35 | |
| Clean | S.F. | .03 | .36 | | .39 | |
| Paint | S.F. | .23 | .77 | | 1 | |
| Minimum Charge | Job | | 232 | | 232 | |
| **2″ x 4″** | | | | | | |
| Demolish | L.F. | | .48 | | .48 | Includes material and labor to install |
| Install | S.F. | 2.20 | 1.55 | | 3.75 | treated lumber 2″ x 4″ decking for |
| Demolish and Install | S.F. | 2.20 | 2.03 | | 4.23 | porch or deck construction. |
| Reinstall | S.F. | | 1.24 | | 1.24 | |
| Clean | S.F. | .03 | .36 | | .39 | |
| Paint | S.F. | .23 | .77 | | 1 | |
| Minimum Charge | Job | | 232 | | 232 | |
| **2″ x 6″** | | | | | | |
| Demolish | L.F. | | .48 | | .48 | Includes material and labor to install |
| Install | S.F. | 1.76 | 1.45 | | 3.21 | treated lumber 2″ x 6″ decking for |
| Demolish and Install | S.F. | 1.76 | 1.93 | | 3.69 | porch or deck construction. |
| Reinstall | S.F. | | 1.16 | | 1.16 | |
| Clean | S.F. | .03 | .36 | | .39 | |
| Paint | S.F. | .23 | .77 | | 1 | |
| Minimum Charge | Job | | 232 | | 232 | |
| **5/4″ x 6″** | | | | | | |
| Demolish | L.F. | | .48 | | .48 | Includes material and labor to install |
| Install | S.F. | 2.37 | 1.45 | | 3.82 | treated lumber 5/4″ x 6″ decking for |
| Demolish and Install | S.F. | 2.37 | 1.93 | | 4.30 | porch or deck construction. |
| Reinstall | S.F. | | 1.16 | | 1.16 | |
| Clean | S.F. | .03 | .36 | | .39 | |
| Paint | S.F. | .23 | .77 | | 1 | |
| Minimum Charge | Job | | 232 | | 232 | |

## Wood / Plastic Composite Decking

| Wood / Plastic Composite Decking | Unit | Material | Labor | Equip. | Total | Specification |
|---|---|---|---|---|---|---|
| **5/4″ x 6″** | | | | | | |
| Demolish | L.F. | | .48 | | .48 | Includes material and labor to install |
| Install | L.F. | 3.60 | 1.45 | | 5.05 | wood/plastic composite 5/4″ x 6″ |
| Demolish and Install | L.F. | 3.60 | 1.93 | | 5.53 | decking for porch or deck |
| Reinstall | L.F. | | 1.16 | | 1.16 | construction. |
| Clean | S.F. | .03 | .36 | | .39 | |
| Paint | S.F. | .23 | .77 | | 1 | |
| Minimum Charge | Job | | 232 | | 232 | |

## Square Edge Fir Decking

| Square Edge Fir Decking | Unit | Material | Labor | Equip. | Total | Specification |
|---|---|---|---|---|---|---|
| **1″ x 4″** | | | | | | |
| Demolish | L.F. | | .48 | | .48 | Includes material and labor to install |
| Install | L.F. | 3.05 | 1.69 | | 4.74 | square edge fir 1″ x 4″ decking for |
| Demolish and Install | L.F. | 3.05 | 2.17 | | 5.22 | porch or deck construction. |
| Reinstall | L.F. | | 1.35 | | 1.35 | |
| Clean | S.F. | .03 | .36 | | .39 | |
| Paint | S.F. | .23 | .77 | | 1 | |
| Minimum Charge | Job | | 232 | | 232 | |

| Wood Deck | Unit | Material | Labor | Equip. | Total | Specification |
|---|---|---|---|---|---|---|
| **Tongue and Groove Fir Decking** | | | | | | |
| **1″ x 4″** | | | | | | |
| Demolish | L.F. | | .48 | | .48 | Includes material and labor to install |
| Install | L.F. | 1.66 | 2.07 | | 3.73 | treated lumber 1″ x 4″ decking for |
| Demolish and Install | L.F. | 1.66 | 2.55 | | 4.21 | porch or deck construction. |
| Reinstall | L.F. | | 1.65 | | 1.65 | |
| Clean | S.F. | .03 | .36 | | .39 | |
| Paint | S.F. | .23 | .77 | | 1 | |
| Minimum Charge | Job | | 232 | | 232 | |
| **Mahogany Decking** | | | | | | |
| **1″ x 4″** | | | | | | |
| Demolish | L.F. | | .48 | | .48 | Includes material and labor to install |
| Install | L.F. | 2.28 | 1.69 | | 3.97 | treated lumber 1″ x 4″ decking for |
| Demolish and Install | L.F. | 2.28 | 2.17 | | 4.45 | porch or deck construction. |
| Reinstall | L.F. | | 1.35 | | 1.35 | |
| Clean | S.F. | .03 | .36 | | .39 | |
| Paint | S.F. | .23 | .77 | | 1 | |
| Minimum Charge | Job | | 232 | | 232 | |
| **PVC Decking** | | | | | | |
| **5/4″ x 6″** | | | | | | |
| Demolish | L.F. | | .48 | | .48 | Includes material and labor to install |
| Install | L.F. | 3.70 | 1.69 | | 5.39 | PVC 5/4″ x 6″ decking for porch or |
| Demolish and Install | L.F. | 3.70 | 2.17 | | 5.87 | deck construction. |
| Reinstall | L.F. | | 1.35 | | 1.35 | |
| Clean | S.F. | .03 | .36 | | .39 | |
| Minimum Charge | Job | | 232 | | 232 | |
| **PVC Handrails and Balusters** | | | | | | |
| **Straight** | | | | | | |
| Demolish | L.F. | | 3.29 | | 3.29 | Includes material and labor to install |
| Install | L.F. | 32 | 4.84 | | 36.84 | straight PVC handrails and balusters |
| Demolish and Install | L.F. | 32 | 8.13 | | 40.13 | for porch or deck construction. |
| Reinstall | L.F. | | 3.87 | | 3.87 | |
| Clean | L.F. | .30 | 2.78 | | 3.08 | |
| Minimum Charge | Job | | 232 | | 232 | |
| **Angled** | | | | | | |
| Demolish | L.F. | | 3.29 | | 3.29 | Includes material and labor to install |
| Install | L.F. | 20.50 | 6.45 | | 26.95 | angled PVC handrails and balusters |
| Demolish and Install | L.F. | 20.50 | 9.74 | | 30.24 | for porch or deck construction. |
| Reinstall | L.F. | | 5.16 | | 5.16 | |
| Clean | L.F. | .30 | 2.78 | | 3.08 | |
| Minimum Charge | Job | | 232 | | 232 | |
| **PVC Post Sleeve** | | | | | | |
| **for 4″ x 4″ post** | | | | | | |
| Demolish | L.F. | | .99 | | .99 | Includes material and labor to install |
| Install | L.F. | 16.15 | 4.84 | | 20.99 | PVC post sleeves over 4″ x 4″ porch |
| Demolish and Install | L.F. | 16.15 | 5.83 | | 21.98 | or deck posts. |
| Reinstall | L.F. | | 3.87 | | 3.87 | |
| Clean | L.F. | .04 | .32 | | .36 | |
| Minimum Charge | Job | | 232 | | 232 | |
| **Post Cap** | | | | | | |
| Demolish | Ea. | | 4.93 | | 4.93 | Includes material and labor to install |
| Install | Ea. | 28 | 9.70 | | 37.70 | PVC post caps for porch or deck |
| Demolish and Install | Ea. | 28 | 14.63 | | 42.63 | construction. |
| Reinstall | Ea. | | 7.75 | | 7.75 | |
| Clean | L.F. | .04 | .32 | | .36 | |
| Minimum Charge | Job | | 232 | | 232 | |

# Rough Frame / Structure

| Porch | | Unit | Material | Labor | Equip. | Total | Specification |
|---|---|---|---|---|---|---|---|
| **Stairs** | | | | | | | |
| | Demolish | Ea. | | 455 | | 455 | Includes material and labor to install |
| | Install | Ea. | 156 | 820 | | 976 | wood steps. 5 - 7 risers included. |
| | Demolish and Install | Ea. | 156 | 1275 | | 1431 | |
| | Reinstall | Ea. | | 657.02 | | 657.02 | |
| | Clean | Flight | 1.47 | 30 | | 31.47 | |
| | Paint | Flight | 21 | 51.50 | | 72.50 | |
| | Minimum Charge | Job | | 232 | | 232 | |
| **Bench Seating** | | | | | | | |
| | Demolish | L.F. | | 6.60 | | 6.60 | Includes material and labor to install |
| | Install | L.F. | 2.27 | 23 | | 25.27 | pressure treated bench seating. |
| | Demolish and Install | L.F. | 2.27 | 29.60 | | 31.87 | |
| | Reinstall | L.F. | | 18.59 | | 18.59 | |
| | Clean | L.F. | .06 | 1.78 | | 1.84 | |
| | Paint | L.F. | .69 | 3.09 | | 3.78 | |
| | Minimum Charge | Job | | 232 | | 232 | |
| **Privacy Wall** | | | | | | | |
| | Demolish | SF Flr. | | 1.58 | | 1.58 | Includes material, labor and |
| | Install | SF Flr. | 1.71 | .85 | .16 | 2.72 | equipment to install privacy wall. |
| | Demolish and Install | SF Flr. | 1.71 | 2.43 | .16 | 4.30 | |
| | Reinstall | SF Flr. | | .68 | .13 | .81 | |
| | Clean | SF Flr. | .03 | .36 | | .39 | |
| | Paint | S.F. | .11 | .30 | | .41 | |
| | Minimum Charge | Job | | 232 | | 232 | |
| **Wood Board Ceiling** | | | | | | | |
| | Demolish | S.F. | | .66 | | .66 | Includes material and labor to install |
| | Install | S.F. | 3.63 | 2.07 | | 5.70 | wood board on ceiling. |
| | Demolish and Install | S.F. | 3.63 | 2.73 | | 6.36 | |
| | Clean | S.F. | .01 | .21 | | .22 | |
| | Paint | S.F. | .19 | .48 | | .67 | |
| | Minimum Charge | Job | | 232 | | 232 | |
| **Re-screen** | | | | | | | |
| | Install | S.F. | 1.83 | 1.86 | | 3.69 | Includes labor and material to |
| | Minimum Charge | Job | | 232 | | 232 | re-screen wood frame. |

**Exterior Trim**

| Exterior Trim | | Unit | Material | Labor | Equip. | Total | Specification |
|---|---|---|---|---|---|---|---|
| **Stock Lumber** | | | | | | | |
| **1" x 2"** | | | | | | | |
| | Demolish | L.F. | | .33 | | .33 | Cost includes material and labor to |
| | Install | L.F. | .31 | 1.41 | | 1.72 | install 1" x 2" pine trim. |
| | Demolish and Install | L.F. | .31 | 1.74 | | 2.05 | |
| | Clean | L.F. | | .24 | | .24 | |
| | Paint | L.F. | .02 | .60 | | .62 | |
| | Minimum Charge | Job | | 232 | | 232 | |
| **1" x 3"** | | | | | | | |
| | Demolish | L.F. | | .33 | | .33 | Cost includes material and labor to |
| | Install | L.F. | .33 | 1.60 | | 1.93 | install 1" x 3" pine trim. |
| | Demolish and Install | L.F. | .33 | 1.93 | | 2.26 | |
| | Clean | L.F. | | .24 | | .24 | |
| | Paint | L.F. | .02 | .60 | | .62 | |
| | Minimum Charge | Job | | 232 | | 232 | |
| **1" x 4"** | | | | | | | |
| | Demolish | L.F. | | .33 | | .33 | Cost includes material and labor to |
| | Install | L.F. | .59 | 1.86 | | 2.45 | install 1" x 4" pine trim. |
| | Demolish and Install | L.F. | .59 | 2.19 | | 2.78 | |
| | Clean | L.F. | | .24 | | .24 | |
| | Paint | L.F. | .02 | .60 | | .62 | |
| | Minimum Charge | Job | | 232 | | 232 | |
| **1" x 6"** | | | | | | | |
| | Demolish | L.F. | | .33 | | .33 | Cost includes material and labor to |
| | Install | L.F. | .90 | 1.86 | | 2.76 | install 1" x 6" pine trim. |
| | Demolish and Install | L.F. | .90 | 2.19 | | 3.09 | |
| | Clean | L.F. | .01 | .24 | | .25 | |
| | Paint | L.F. | .09 | .60 | | .69 | |
| | Minimum Charge | Job | | 232 | | 232 | |
| **1" x 8"** | | | | | | | |
| | Demolish | L.F. | | .36 | | .36 | Cost includes material and labor to |
| | Install | L.F. | 1.46 | 2.32 | | 3.78 | install 1" x 8" pine trim. |
| | Demolish and Install | L.F. | 1.46 | 2.68 | | 4.14 | |
| | Reinstall | L.F. | | 1.86 | | 1.86 | |
| | Clean | L.F. | .01 | .25 | | .26 | |
| | Paint | L.F. | .09 | .60 | | .69 | |
| | Minimum Charge | Job | | 232 | | 232 | |

# Exterior Trim

| Exterior Trim | Unit | Material | Labor | Equip. | Total | Specification |
|---|---|---|---|---|---|---|
| **Stock Lumber** | | | | | | |
| **1" x 10"** | | | | | | |
| Demolish | L.F. | | .55 | | .55 | Cost includes material and labor to |
| Install | L.F. | 1.91 | 2.58 | | 4.49 | install 1" x 10" pine trim. |
| Demolish and Install | L.F. | 1.91 | 3.13 | | 5.04 | |
| Minimum Charge | Job | | 232 | | 232 | |
| **1" x 12"** | | | | | | |
| Demolish | L.F. | | .55 | | .55 | Cost includes material and labor to |
| Install | L.F. | 2.34 | 2.58 | | 4.92 | install 1" x 12" pine trim. |
| Demolish and Install | L.F. | 2.34 | 3.13 | | 5.47 | |
| Minimum Charge | Job | | 232 | | 232 | |
| **PVC** | | | | | | |
| **1" x 4"** | | | | | | |
| Demolish | L.F. | | .33 | | .33 | Includes material and labor to install |
| Install | L.F. | 1.53 | 1.86 | | 3.39 | 1" x 4" PVC exterior trim. |
| Demolish and Install | L.F. | 1.53 | 2.19 | | 3.72 | |
| Reinstall | L.F. | | 1.49 | | 1.49 | |
| Clean | L.F. | | .24 | | .24 | |
| Paint | L.F. | .02 | .60 | | .62 | |
| Minimum Charge | Job | | 232 | | 232 | |
| **1" x 6"** | | | | | | |
| Demolish | L.F. | | .33 | | .33 | Includes material and labor to install |
| Install | L.F. | 2.27 | 1.86 | | 4.13 | 1" x 6" PVC exterior trim. |
| Demolish and Install | L.F. | 2.27 | 2.19 | | 4.46 | |
| Reinstall | L.F. | | 1.49 | | 1.49 | |
| Clean | L.F. | .01 | .24 | | .25 | |
| Paint | L.F. | .09 | .60 | | .69 | |
| Minimum Charge | Job | | 232 | | 232 | |
| **1" x 8"** | | | | | | |
| Demolish | L.F. | | .36 | | .36 | Includes material and labor to install |
| Install | L.F. | 2.99 | 2.07 | | 5.06 | 1" x 8" PVC exterior trim. |
| Demolish and Install | L.F. | 2.99 | 2.43 | | 5.42 | |
| Reinstall | L.F. | | 1.65 | | 1.65 | |
| Clean | L.F. | .01 | .25 | | .26 | |
| Paint | L.F. | .09 | .60 | | .69 | |
| Minimum Charge | Job | | 232 | | 232 | |
| **1" x 10"** | | | | | | |
| Demolish | L.F. | | .38 | | .38 | Includes material and labor to install |
| Install | L.F. | 3.78 | 2.07 | | 5.85 | 1" x 10" PVC exterior trim. |
| Demolish and Install | L.F. | 3.78 | 2.45 | | 6.23 | |
| Reinstall | L.F. | | 1.65 | | 1.65 | |
| Clean | L.F. | .02 | .30 | | .32 | |
| Paint | L.F. | .09 | .60 | | .69 | |
| Minimum Charge | Job | | 232 | | 232 | |
| **1" x 12"** | | | | | | |
| Demolish | L.F. | | .39 | | .39 | Includes material and labor to install |
| Install | L.F. | 4.64 | 2.32 | | 6.96 | 1" x 12" PVC exterior trim. |
| Demolish and Install | L.F. | 4.64 | 2.71 | | 7.35 | |
| Reinstall | L.F. | | 1.86 | | 1.86 | |
| Clean | L.F. | .02 | .30 | | .32 | |
| Paint | L.F. | .09 | .60 | | .69 | |
| Minimum Charge | Job | | 232 | | 232 | |

# Exterior Trim

## Fascia

| | Unit | Material | Labor | Equip. | Total | Specification |
|---|---|---|---|---|---|---|
| **Aluminum** | | | | | | |
| Demolish | L.F. | | .69 | | .69 | Includes material and labor to install |
| Install | S.F. | 2.28 | 2.21 | | 4.49 | aluminum fascia. |
| Demolish and Install | S.F. | 2.28 | 2.90 | | 5.18 | |
| Reinstall | S.F. | | 1.77 | | 1.77 | |
| Clean | L.F. | | .44 | | .44 | |
| Minimum Charge | Job | | 232 | | 232 | |
| **Vinyl** | | | | | | |
| Demolish | L.F. | | .69 | | .69 | Cost includes material and labor to |
| Install | L.F. | 4.99 | 2.66 | | 7.65 | install 6″ vinyl fascia with 12″ vinyl |
| Demolish and Install | L.F. | 4.99 | 3.35 | | 8.34 | soffit and J-channel. |
| Reinstall | L.F. | | 2.12 | | 2.12 | |
| Clean | L.F. | | .44 | | .44 | |
| Minimum Charge | Job | | 232 | | 232 | |
| **Plywood** | | | | | | |
| Demolish | L.F. | | .68 | | .68 | Cost includes material and labor to |
| Install | L.F. | 1.68 | 1.03 | | 2.71 | install 12″ wide plywood fascia. |
| Demolish and Install | L.F. | 1.68 | 1.71 | | 3.39 | |
| Reinstall | L.F. | | .83 | | .83 | |
| Clean | L.F. | .02 | .30 | | .32 | |
| Paint | L.F. | .18 | .97 | | 1.15 | |
| Minimum Charge | Job | | 232 | | 232 | |
| **Cedar** | | | | | | |
| Demolish | L.F. | | .68 | | .68 | Cost includes material and labor to |
| Install | L.F. | 4.72 | 1.03 | | 5.75 | install 12″ wide cedar fascia. |
| Demolish and Install | L.F. | 4.72 | 1.71 | | 6.43 | |
| Reinstall | L.F. | | .83 | | .83 | |
| Clean | L.F. | .02 | .30 | | .32 | |
| Paint | L.F. | .18 | .97 | | 1.15 | |
| Minimum Charge | Job | | 232 | | 232 | |
| **Redwood** | | | | | | |
| **1″ x 6″** | | | | | | |
| Demolish | L.F. | | .53 | | .53 | Cost includes material and labor to |
| Install | L.F. | 1.97 | 1.86 | | 3.83 | install 1″ x 6″ redwood fascia board. |
| Demolish and Install | L.F. | 1.97 | 2.39 | | 4.36 | |
| Reinstall | L.F. | | 1.49 | | 1.49 | |
| Minimum Charge | Job | | 232 | | 232 | |
| **1″ x 8″** | | | | | | |
| Demolish | L.F. | | .59 | | .59 | Cost includes material and labor to |
| Install | L.F. | 2.61 | 2.02 | | 4.63 | install 1″ x 8″ redwood fascia board. |
| Demolish and Install | L.F. | 2.61 | 2.61 | | 5.22 | |
| Reinstall | L.F. | | 1.62 | | 1.62 | |
| Minimum Charge | Job | | 232 | | 232 | |
| **2″ x 6″** | | | | | | |
| Demolish | L.F. | | .53 | | .53 | Cost includes material and labor to |
| Install | L.F. | .70 | 3.72 | | 4.42 | install 2″ x 6″ exterior trim board. |
| Demolish and Install | L.F. | .70 | 4.25 | | 4.95 | |
| Reinstall | L.F. | | 2.97 | | 2.97 | |
| Minimum Charge | Job | | 232 | | 232 | |
| **2″ x 8″** | | | | | | |
| Demolish | L.F. | | .59 | | .59 | Cost includes material and labor to |
| Install | L.F. | .98 | 4.13 | | 5.11 | install 2″ x 8″ exterior trim board. |
| Demolish and Install | L.F. | .98 | 4.72 | | 5.70 | |
| Reinstall | L.F. | | 3.31 | | 3.31 | |
| Minimum Charge | Job | | 232 | | 232 | |

| Fascia | Unit | Material | Labor | Equip. | Total | Specification |
|---|---|---|---|---|---|---|
| **Hem-fir Std & Better** | | | | | | |
| **2" x 6"** | | | | | | |
| Demolish | L.F. | | .53 | | .53 | Cost includes material and labor to |
| Install | L.F. | .70 | 3.72 | | 4.42 | install 2" x 6" exterior trim board. |
| Demolish and Install | L.F. | .70 | 4.25 | | 4.95 | |
| Reinstall | L.F. | | 2.97 | | 2.97 | |
| Clean | L.F. | .01 | .30 | | .31 | |
| Paint | L.F. | .18 | .97 | | 1.15 | |
| Minimum Charge | Job | | 232 | | 232 | |
| **2" x 8"** | | | | | | |
| Demolish | L.F. | | .59 | | .59 | Cost includes material and labor to |
| Install | L.F. | .98 | 4.13 | | 5.11 | install 2" x 8" exterior trim board. |
| Demolish and Install | L.F. | .98 | 4.72 | | 5.70 | |
| Reinstall | L.F. | | 3.31 | | 3.31 | |
| Clean | L.F. | .01 | .30 | | .31 | |
| Paint | L.F. | .18 | .97 | | 1.15 | |
| Minimum Charge | Job | | 232 | | 232 | |
| **2" x 10"** | | | | | | |
| Demolish | L.F. | | .68 | | .68 | Cost includes material and labor to |
| Install | L.F. | 1.55 | 5.15 | | 6.70 | install 2" x 10" exterior trim board. |
| Demolish and Install | L.F. | 1.55 | 5.83 | | 7.38 | |
| Reinstall | L.F. | | 4.13 | | 4.13 | |
| Clean | L.F. | .01 | .32 | | .33 | |
| Paint | L.F. | .18 | .97 | | 1.15 | |
| Minimum Charge | Job | | 232 | | 232 | |
| **Barge Rafter** | | | | | | |
| **2" x 6"** | | | | | | |
| Demolish | L.F. | | 1.05 | | 1.05 | Cost includes material and labor to |
| Install | L.F. | .70 | .52 | | 1.22 | install 2" x 6" barge rafters. |
| Demolish and Install | L.F. | .70 | 1.57 | | 2.27 | |
| Reinstall | L.F. | | .41 | | .41 | |
| Clean | L.F. | .01 | .30 | | .31 | |
| Paint | L.F. | .18 | .97 | | 1.15 | |
| Minimum Charge | Job | | 232 | | 232 | |
| **2" x 8"** | | | | | | |
| Demolish | L.F. | | 1.07 | | 1.07 | Cost includes material and labor to |
| Install | L.F. | .98 | .55 | | 1.53 | install 2" x 8" barge rafters. |
| Demolish and Install | L.F. | .98 | 1.62 | | 2.60 | |
| Reinstall | L.F. | | .44 | | .44 | |
| Clean | L.F. | .01 | .30 | | .31 | |
| Paint | L.F. | .18 | .97 | | 1.15 | |
| Minimum Charge | Job | | 232 | | 232 | |
| **2" x 10"** | | | | | | |
| Demolish | L.F. | | 1.09 | | 1.09 | Cost includes material and labor to |
| Install | L.F. | 1.55 | .56 | | 2.11 | install 2" x 10" barge rafters. |
| Demolish and Install | L.F. | 1.55 | 1.65 | | 3.20 | |
| Reinstall | L.F. | | .45 | | .45 | |
| Clean | L.F. | .01 | .32 | | .33 | |
| Paint | L.F. | .18 | .97 | | 1.15 | |
| Minimum Charge | Job | | 232 | | 232 | |

| Soffit | | Unit | Material | Labor | Equip. | Total | Specification |
|---|---|---|---|---|---|---|---|
| **Aluminum** | | | | | | | |
| **12″ w / Fascia** | | | | | | | |
| | Demolish | L.F. | | .69 | | .69 | Cost includes material and labor to |
| | Install | L.F. | 5.05 | 2.66 | | 7.71 | install 12″ aluminum soffit with 6″ |
| | Demolish and Install | L.F. | 5.05 | 3.35 | | 8.40 | fascia and J-channel. |
| | Reinstall | L.F. | | 2.12 | | 2.12 | |
| | Clean | L.F. | | .44 | | .44 | |
| | Paint | L.F. | .22 | .97 | | 1.19 | |
| | Minimum Charge | Job | | 232 | | 232 | |
| **18″ w / Fascia** | | | | | | | |
| | Demolish | L.F. | | .69 | | .69 | Cost includes material and labor to |
| | Install | L.F. | 6.10 | 3.10 | | 9.20 | install 18″ aluminum soffit with 6″ |
| | Demolish and Install | L.F. | 6.10 | 3.79 | | 9.89 | fascia and J-channel. |
| | Reinstall | L.F. | | 2.48 | | 2.48 | |
| | Clean | L.F. | | .44 | | .44 | |
| | Paint | L.F. | .22 | .97 | | 1.19 | |
| | Minimum Charge | Job | | 232 | | 232 | |
| **24″ w / Fascia** | | | | | | | |
| | Demolish | L.F. | | .69 | | .69 | Cost includes material and labor to |
| | Install | L.F. | 7.25 | 3.44 | | 10.69 | install 24″ aluminum soffit with 6″ |
| | Demolish and Install | L.F. | 7.25 | 4.13 | | 11.38 | fascia and J-channel. |
| | Reinstall | L.F. | | 2.75 | | 2.75 | |
| | Clean | L.F. | | .65 | | .65 | |
| | Paint | L.F. | .22 | .97 | | 1.19 | |
| | Minimum Charge | Job | | 232 | | 232 | |
| **Vinyl** | | | | | | | |
| | Demolish | L.F. | | .69 | | .69 | Cost includes material and labor to |
| | Install | L.F. | 4.99 | 2.66 | | 7.65 | install 6″ vinyl fascia with 12″ vinyl |
| | Demolish and Install | L.F. | 4.99 | 3.35 | | 8.34 | soffit and J-channel. |
| | Reinstall | L.F. | | 2.12 | | 2.12 | |
| | Clean | L.F. | | .44 | | .44 | |
| | Paint | L.F. | .22 | .97 | | 1.19 | |
| | Minimum Charge | Job | | 232 | | 232 | |
| **Plywood** | | | | | | | |
| | Demolish | L.F. | | .76 | | .76 | Cost includes material and labor to |
| | Install | L.F. | 1.62 | 1.33 | | 2.95 | install 12″ wide plywood soffit. |
| | Demolish and Install | L.F. | 1.62 | 2.09 | | 3.71 | |
| | Reinstall | L.F. | | 1.06 | | 1.06 | |
| | Clean | L.F. | .02 | .30 | | .32 | |
| | Paint | L.F. | .18 | .97 | | 1.15 | |
| | Minimum Charge | Job | | 232 | | 232 | |
| **Redwood** | | | | | | | |
| | Demolish | L.F. | | .76 | | .76 | Cost includes material and labor to |
| | Install | L.F. | 5.20 | 1.33 | | 6.53 | install 12″ wide redwood soffit. |
| | Demolish and Install | L.F. | 5.20 | 2.09 | | 7.29 | |
| | Reinstall | L.F. | | 1.06 | | 1.06 | |
| | Clean | L.F. | .02 | .30 | | .32 | |
| | Paint | L.F. | .18 | .97 | | 1.15 | |
| | Minimum Charge | Job | | 232 | | 232 | |
| **Cedar** | | | | | | | |
| | Demolish | L.F. | | .76 | | .76 | Cost includes material and labor to |
| | Install | L.F. | 4.65 | 1.33 | | 5.98 | install 12″ wide cedar soffit. |
| | Demolish and Install | L.F. | 4.65 | 2.09 | | 6.74 | |
| | Reinstall | L.F. | | 1.06 | | 1.06 | |
| | Clean | L.F. | .02 | .30 | | .32 | |
| | Paint | L.F. | .18 | .97 | | 1.15 | |
| | Minimum Charge | Job | | 232 | | 232 | |

69

For customer support on your Contractor's Pricing Guide: Residential Repair & Remodeling, call 800.448.8182.

# Exterior Trim

## Soffit

| Soffit | Unit | Material | Labor | Equip. | Total | Specification |
|---|---|---|---|---|---|---|
| **Douglas Fir** | | | | | | |
| Demolish | L.F. | | .76 | | .76 | Cost includes material and labor to |
| Install | L.F. | 2.28 | 1.33 | | 3.61 | install 12" wide Douglas fir soffit. |
| Demolish and Install | L.F. | 2.28 | 2.09 | | 4.37 | |
| Reinstall | L.F. | | 1.06 | | 1.06 | |
| Clean | L.F. | .02 | .30 | | .32 | |
| Paint | L.F. | .18 | .97 | | 1.15 | |
| Minimum Charge | Job | | 232 | | 232 | |

**Roofing and Flashing**    **Tile Roofing**

| Metal Flashing | Unit | Material | Labor | Equip. | Total | Specification |
|---|---|---|---|---|---|---|
| **Galvanized** | | | | | | |
| **6″** | | | | | | |
| Demolish | L.F. | | .26 | | .26 | Includes material and labor to install |
| Install | L.F. | .73 | 1.09 | | 1.82 | flat galvanized steel flashing, 6″ wide. |
| Demolish and Install | L.F. | .73 | 1.35 | | 2.08 | |
| Paint | L.F. | .12 | .48 | | .60 | |
| Minimum Charge | Job | | 218 | | 218 | |
| **14″** | | | | | | |
| Demolish | L.F. | | .26 | | .26 | Includes material and labor to install |
| Install | L.F. | 1.65 | 1.09 | | 2.74 | flat galvanized steel flashing, 14″ |
| Demolish and Install | L.F. | 1.65 | 1.35 | | 3 | wide. |
| Paint | L.F. | .12 | .48 | | .60 | |
| Minimum Charge | Job | | 218 | | 218 | |
| **20″** | | | | | | |
| Demolish | L.F. | | .26 | | .26 | Includes material and labor to install |
| Install | L.F. | 2.48 | 1.09 | | 3.57 | flat galvanized steel flashing, 20″ |
| Demolish and Install | L.F. | 2.48 | 1.35 | | 3.83 | wide. |
| Paint | L.F. | .12 | .48 | | .60 | |
| Minimum Charge | Job | | 218 | | 218 | |
| **Aluminum** | | | | | | |
| **6″** | | | | | | |
| Demolish | L.F. | | .26 | | .26 | Includes material and labor to install |
| Install | L.F. | .43 | 1.09 | | 1.52 | flat aluminum flashing, 6″ wide. |
| Demolish and Install | L.F. | .43 | 1.35 | | 1.78 | |
| Paint | L.F. | .12 | .48 | | .60 | |
| Minimum Charge | Job | | 218 | | 218 | |
| **14″** | | | | | | |
| Demolish | L.F. | | .26 | | .26 | Includes material and labor to install |
| Install | L.F. | .84 | 1.09 | | 1.93 | flat aluminum flashing, 14″ wide. |
| Demolish and Install | L.F. | .84 | 1.35 | | 2.19 | |
| Paint | L.F. | .12 | .48 | | .60 | |
| Minimum Charge | Job | | 218 | | 218 | |
| **20″** | | | | | | |
| Demolish | L.F. | | .26 | | .26 | Includes material and labor to install |
| Install | L.F. | 1.14 | 1.09 | | 2.23 | flat aluminum flashing, 20″ wide. |
| Demolish and Install | L.F. | 1.14 | 1.35 | | 2.49 | |
| Paint | L.F. | .12 | .48 | | .60 | |
| Minimum Charge | Job | | 218 | | 218 | |

| Metal Flashing | Unit | Material | Labor | Equip. | Total | Specification |
|---|---|---|---|---|---|---|
| **Copper** | | | | | | |
| **6"** | | | | | | |
| Demolish | L.F. | | .26 | | .26 | Includes material and labor to install |
| Install | L.F. | 4.46 | 1.45 | | 5.91 | flat copper flashing, 6" wide. |
| Demolish and Install | L.F. | 4.46 | 1.71 | | 6.17 | |
| Minimum Charge | Job | | 255 | | 255 | |
| **14"** | | | | | | |
| Demolish | L.F. | | .26 | | .26 | Includes material and labor to install |
| Install | L.F. | 10.45 | 1.45 | | 11.90 | flat copper flashing, 14" wide. |
| Demolish and Install | L.F. | 10.45 | 1.71 | | 12.16 | |
| Minimum Charge | Job | | 255 | | 255 | |
| **18"** | | | | | | |
| Demolish | L.F. | | .26 | | .26 | Includes material and labor to install |
| Install | L.F. | 14.90 | 1.45 | | 16.35 | flat copper flashing, 18" wide. |
| Demolish and Install | L.F. | 14.90 | 1.71 | | 16.61 | |
| Minimum Charge | Job | | 255 | | 255 | |
| **Aluminum Valley** | | | | | | |
| Demolish | L.F. | | .26 | | .26 | Includes material and labor to install |
| Install | L.F. | 4.37 | 2.73 | | 7.10 | aluminum valley flashing, .024" thick. |
| Demolish and Install | L.F. | 4.37 | 2.99 | | 7.36 | |
| Paint | L.F. | .12 | .48 | | .60 | |
| Minimum Charge | Job | | 232 | | 232 | |
| **Gravel Stop** | | | | | | |
| Demolish | L.F. | | .42 | | .42 | Includes material and labor to install |
| Install | L.F. | 7.20 | 3.51 | | 10.71 | aluminum gravel stop, .050" thick, 4" |
| Demolish and Install | L.F. | 7.20 | 3.93 | | 11.13 | high, mill finish. |
| Reinstall | L.F. | | 2.81 | | 2.81 | |
| Paint | L.F. | .19 | .48 | | .67 | |
| Minimum Charge | Job | | 232 | | 232 | |
| **Drip Edge** | | | | | | |
| Demolish | L.F. | | .39 | | .39 | Includes material and labor to install |
| Install | L.F. | .64 | 1.16 | | 1.80 | aluminum drip edge, .016" thick, 5" |
| Demolish and Install | L.F. | .64 | 1.55 | | 2.19 | wide, mill finish. |
| Reinstall | L.F. | | .93 | | .93 | |
| Paint | L.F. | .19 | .48 | | .67 | |
| Minimum Charge | Job | | 232 | | 232 | |
| **End Wall** | | | | | | |
| Demolish | L.F. | | 1.21 | | 1.21 | Includes material and labor to install |
| Install | L.F. | 2.50 | 2.73 | | 5.23 | aluminum end wall flashing, .024" |
| Demolish and Install | L.F. | 2.50 | 3.94 | | 6.44 | thick. |
| Minimum Charge | Job | | 232 | | 232 | |
| **Side Wall** | | | | | | |
| Demolish | L.F. | | .24 | | .24 | Includes material and labor to install |
| Install | L.F. | 2.44 | 2.73 | | 5.17 | aluminum side wall flashing, .024" |
| Demolish and Install | L.F. | 2.44 | 2.97 | | 5.41 | thick. |
| Minimum Charge | Job | | 232 | | 232 | |
| **Coping and Wall Cap** | | | | | | |
| Demolish | L.F. | | 1.64 | | 1.64 | Includes material and labor to install |
| Install | L.F. | 2.52 | 3.01 | | 5.53 | aluminum flashing, mill finish, 0.40 |
| Demolish and Install | L.F. | 2.52 | 4.65 | | 7.17 | thick, 12" wide. |
| Clean | L.F. | .03 | 1.11 | | 1.14 | |
| Minimum Charge | Job | | 218 | | 218 | |

# Roofing

| Roof Shingles | | Unit | Material | Labor | Equip. | Total | Specification |
|---|---|---|---|---|---|---|---|
| **Underlayment** | | | | | | | |
| 15# | | | | | | | |
| | Demolish | Sq. | | 13.15 | | 13.15 | Includes material and labor to install |
| | Install | Sq. | 5.90 | 6.80 | | 12.70 | #15 felt underlayment. |
| | Demolish and Install | Sq. | 5.90 | 19.95 | | 25.85 | |
| | Minimum Charge | Job | | 218 | | 218 | |
| 30# | | | | | | | |
| | Demolish | Sq. | | 13.15 | | 13.15 | Includes material and labor to install |
| | Install | Sq. | 11.35 | 7.50 | | 18.85 | #30 felt underlayment. |
| | Demolish and Install | Sq. | 11.35 | 20.65 | | 32 | |
| | Minimum Charge | Job | | 218 | | 218 | |
| **Self Adhering** | | | | | | | |
| | Demolish | Sq. | | 33 | | 33 | Includes material and labor to install |
| | Install | Sq. | 86 | 19.80 | | 105.80 | self adhering ice barrier roofing |
| | Demolish and Install | Sq. | 86 | 52.80 | | 138.80 | underlayment. |
| | Minimum Charge | Job | | 218 | | 218 | |

| Rolled Roofing | | Unit | Material | Labor | Equip. | Total | Specification |
|---|---|---|---|---|---|---|---|
| **90 lb** | | | | | | | |
| | Demolish | Sq. | | 33 | | 33 | Cost includes material and labor to |
| | Install | Sq. | 39.50 | 36.50 | | 76 | install 90 lb mineral surface rolled |
| | Demolish and Install | Sq. | 39.50 | 69.50 | | 109 | roofing. |
| | Minimum Charge | Job | | 218 | | 218 | |

| Composition Shingles | | Unit | Material | Labor | Equip. | Total | Specification |
|---|---|---|---|---|---|---|---|
| **Fiberglass (3-tab)** | | | | | | | |
| 25 Year | | | | | | | |
| | Demolish | Sq. | | 36 | | 36 | Cost includes material and labor to |
| | Install | Sq. | 76.50 | 48.50 | | 125 | install 25 year fiberglass shingles. |
| | Demolish and Install | Sq. | 76.50 | 84.50 | | 161 | |
| | Minimum Charge | Job | | 218 | | 218 | |
| 30 Year | | | | | | | |
| | Demolish | Sq. | | 36 | | 36 | Cost includes material and labor to |
| | Install | Sq. | 97.50 | 54.50 | | 152 | install 30 year fiberglass shingles. |
| | Demolish and Install | Sq. | 97.50 | 90.50 | | 188 | |
| | Minimum Charge | Job | | 218 | | 218 | |
| **Emergency Repairs** | | | | | | | |
| | Install | Ea. | 173 | 218 | | 391 | Includes labor and material for |
| | Minimum Charge | Job | | 218 | | 218 | emergency repairs. |
| **Architectural** | | | | | | | |
| Laminated, 25 Year | | | | | | | |
| | Demolish | Sq. | | 36 | | 36 | Cost includes material and labor to |
| | Install | Sq. | 113 | 60.50 | | 173.50 | install 25 year fiberglass roof shingles. |
| | Demolish and Install | Sq. | 113 | 96.50 | | 209.50 | |
| | Minimum Charge | Job | | 218 | | 218 | |
| Laminated, 30 Year | | | | | | | |
| | Demolish | Sq. | | 36 | | 36 | Cost includes material and labor to |
| | Install | Sq. | 147 | 78 | | 225 | install 30 year laminated roof |
| | Demolish and Install | Sq. | 147 | 114 | | 261 | shingles. |
| | Minimum Charge | Job | | 218 | | 218 | |

For customer support on your Contractor's Pricing Guide: Residential Repair & Remodeling, call 800.448.8182.

| Composition Shingles | | Unit | Material | Labor | Equip. | Total | Specification |
|---|---|---|---|---|---|---|---|
| **Laminated, 40 Year** | | | | | | | |
| | Demolish | Sq. | | 36 | | 36 | Cost includes material and labor to |
| | Install | Sq. | 266 | 90.50 | | 356.50 | install 40 year laminated roof |
| | Demolish and Install | Sq. | 266 | 126.50 | | 392.50 | shingles. |
| | Minimum Charge | Job | | 218 | | 218 | |
| **Laminated Shake, 40 Year** | | | | | | | |
| | Demolish | Sq. | | 36 | | 36 | Cost includes material and labor to |
| | Install | Sq. | 266 | 90.50 | | 356.50 | install 40 year laminated roof |
| | Demolish and Install | Sq. | 266 | 126.50 | | 392.50 | shingles. |
| | Minimum Charge | Job | | 218 | | 218 | |
| **Emergency Repairs** | | | | | | | |
| | Install | Ea. | 173 | 218 | | 391 | Includes labor and material for |
| | Minimum Charge | Job | | 218 | | 218 | emergency repairs. |
| **Roof Jack** | | | | | | | |
| | Demolish | Ea. | | 7.20 | | 7.20 | Includes material and labor to install |
| | Install | Ea. | 27 | 46.50 | | 73.50 | residential roof jack, w/bird screen, |
| | Demolish and Install | Ea. | 27 | 53.70 | | 80.70 | backdraft damper, 3" & 4" dia. round |
| | Paint | Ea. | .07 | 4.83 | | 4.90 | duct. |
| | Minimum Charge | Job | | 263 | | 263 | |
| **Ridge Vent** | | | | | | | |
| | Demolish | L.F. | | 1.27 | | 1.27 | Includes material and labor to install a |
| | Install | L.F. | 4.32 | 3.29 | | 7.61 | mill finish aluminum ridge vent strip. |
| | Demolish and Install | L.F. | 4.32 | 4.56 | | 8.88 | |
| | Minimum Charge | Job | | 232 | | 232 | |
| **Shingle Molding** | | | | | | | |
| **1" x 2"** | | | | | | | |
| | Demolish | L.F. | | .25 | | .25 | Cost includes material and labor to |
| | Install | L.F. | .31 | 1.41 | | 1.72 | install 1" x 2" pine trim. |
| | Demolish and Install | L.F. | .31 | 1.66 | | 1.97 | |
| | Paint | L.F. | .02 | .60 | | .62 | |
| | Minimum Charge | Job | | 232 | | 232 | |
| **1" x 3"** | | | | | | | |
| | Demolish | L.F. | | .25 | | .25 | Cost includes material and labor to |
| | Install | L.F. | .33 | 1.60 | | 1.93 | install 1" x 3" pine trim. |
| | Demolish and Install | L.F. | .33 | 1.85 | | 2.18 | |
| | Paint | L.F. | .02 | .60 | | .62 | |
| | Minimum Charge | Job | | 232 | | 232 | |
| **1" x 4"** | | | | | | | |
| | Demolish | L.F. | | .25 | | .25 | Cost includes material and labor to |
| | Install | L.F. | .59 | 1.86 | | 2.45 | install 1" x 4" pine trim. |
| | Demolish and Install | L.F. | .59 | 2.11 | | 2.70 | |
| | Paint | L.F. | .02 | .60 | | .62 | |
| | Minimum Charge | Job | | 232 | | 232 | |
| **Add for Steep Pitch** | | | | | | | |
| | Install | Sq. | | 27.50 | | 27.50 | Includes labor and material added for |
| **Add for Additional Story** | | | | | | | 2nd story or steep roofs. |
| | Install | Sq. | | 27.50 | | 27.50 | Includes labor and material added for |
| | | | | | | | 2nd story or steep roofs. |

# Roofing

| Wood Shingles | | Unit | Material | Labor | Equip. | Total | Specification |
|---|---|---|---|---|---|---|---|
| **Cedar Wood Shingles** | | | | | | | |
| 16" Red Label (#2) | | | | | | | |
| | Demolish | Sq. | | 87.50 | | 87.50 | Includes material and labor to install |
| | Install | Sq. | 176 | 186 | | 362 | wood shingles/shakes. |
| | Demolish and Install | Sq. | 176 | 273.50 | | 449.50 | |
| | Minimum Charge | Job | | 232 | | 232 | |
| 16" Blue Label (#1) | | | | | | | |
| | Demolish | Sq. | | 87.50 | | 87.50 | Includes material and labor to install |
| | Install | Sq. | 325 | 143 | | 468 | wood shingles/shakes. |
| | Demolish and Install | Sq. | 325 | 230.50 | | 555.50 | |
| | Minimum Charge | Job | | 232 | | 232 | |
| 18" Red Label (#2) | | | | | | | |
| | Demolish | Sq. | | 87.50 | | 87.50 | Includes material and labor to install |
| | Install | Sq. | 320 | 119 | | 439 | wood shingles/shakes. |
| | Demolish and Install | Sq. | 320 | 206.50 | | 526.50 | |
| | Minimum Charge | Job | | 232 | | 232 | |
| 18" Blue Label (#1) | | | | | | | |
| | Demolish | Sq. | | 87.50 | | 87.50 | Includes material and labor to install |
| | Install | Sq. | 278 | 130 | | 408 | wood shingles/shakes. |
| | Demolish and Install | Sq. | 278 | 217.50 | | 495.50 | |
| | Minimum Charge | Job | | 232 | | 232 | |
| **Replace Shingle** | | | | | | | |
| | Install | Ea. | 80.50 | 11.60 | | 92.10 | Includes labor and material to replace |
| | Minimum Charge | Job | | 232 | | 232 | individual wood shingle/shake. Cost is per shingle. |
| **Cedar Shake** | | | | | | | |
| 18" Medium Handsplit | | | | | | | |
| | Demolish | Sq. | | 87.50 | | 87.50 | Includes material and labor to install |
| | Install | Sq. | 305 | 179 | | 484 | wood shingles/shakes. |
| | Demolish and Install | Sq. | 305 | 266.50 | | 571.50 | |
| | Minimum Charge | Job | | 232 | | 232 | |
| 18" Heavy Handsplit | | | | | | | |
| | Demolish | Sq. | | 87.50 | | 87.50 | Includes material and labor to install |
| | Install | Sq. | 305 | 199 | | 504 | wood shingles/shakes. |
| | Demolish and Install | Sq. | 305 | 286.50 | | 591.50 | |
| | Minimum Charge | Job | | 232 | | 232 | |
| 24" Heavy Handsplit | | | | | | | |
| | Demolish | Sq. | | 87.50 | | 87.50 | Includes material and labor to install |
| | Install | Sq. | 330 | 159 | | 489 | wood shingles/shakes. |
| | Demolish and Install | Sq. | 330 | 246.50 | | 576.50 | |
| | Minimum Charge | Job | | 232 | | 232 | |
| 24" Medium Handsplit | | | | | | | |
| | Demolish | Sq. | | 87.50 | | 87.50 | Includes material and labor to install |
| | Install | Sq. | 330 | 143 | | 473 | wood shingles/shakes. |
| | Demolish and Install | Sq. | 330 | 230.50 | | 560.50 | |
| | Minimum Charge | Job | | 232 | | 232 | |
| 24" Straight Split | | | | | | | |
| | Demolish | Sq. | | 87.50 | | 87.50 | Includes material and labor to install |
| | Install | Sq. | 345 | 211 | | 556 | wood shingles/shakes. |
| | Demolish and Install | Sq. | 345 | 298.50 | | 643.50 | |
| | Minimum Charge | Job | | 232 | | 232 | |
| 24" Tapersplit | | | | | | | |
| | Demolish | Sq. | | 87.50 | | 87.50 | Includes material and labor to install |
| | Install | Sq. | 299 | 211 | | 510 | wood shingles/shakes. |
| | Demolish and Install | Sq. | 299 | 298.50 | | 597.50 | |
| | Minimum Charge | Job | | 232 | | 232 | |

# Roofing

## Wood Shingles

| Wood Shingles | | Unit | Material | Labor | Equip. | Total | Specification |
|---|---|---|---|---|---|---|---|
| **18" Straight Split** | | | | | | | |
| | Demolish | Sq. | | 87.50 | | 87.50 | Includes material and labor to install |
| | Install | Sq. | 325 | 211 | | 536 | wood shingles/shakes. |
| | Demolish and Install | Sq. | 325 | 298.50 | | 623.50 | |
| | Minimum Charge | Job | | 232 | | 232 | |
| **Replace Shingle** | | | | | | | |
| | Install | Ea. | 80.50 | 11.60 | | 92.10 | Includes labor and material to replace |
| | Minimum Charge | Job | | 232 | | 232 | individual wood shingle/shake. Cost is per shingle. |

## Slate Tile Roofing

| Slate Tile Roofing | | Unit | Material | Labor | Equip. | Total | Specification |
|---|---|---|---|---|---|---|---|
| **Unfading Green** | | | | | | | |
| | Demolish | Sq. | | 79 | | 79 | Includes material and labor to install |
| | Install | Sq. | 555 | 251 | | 806 | clear Vermont slate tile. |
| | Demolish and Install | Sq. | 555 | 330 | | 885 | |
| | Reinstall | Sq. | | 200.41 | | 200.41 | |
| | Minimum Charge | Job | | 219 | | 219 | |
| **Unfading Purple** | | | | | | | |
| | Demolish | Sq. | | 79 | | 79 | Includes material and labor to install |
| | Install | Sq. | 480 | 251 | | 731 | clear Vermont slate tile. |
| | Demolish and Install | Sq. | 480 | 330 | | 810 | |
| | Reinstall | Sq. | | 200.41 | | 200.41 | |
| | Minimum Charge | Job | | 219 | | 219 | |
| **Replace Slate Tiles** | | | | | | | |
| | Install | Ea. | 7.75 | 23 | | 30.75 | Includes labor and material to replace |
| | Minimum Charge | Job | | 219 | | 219 | individual slate tiles. Cost is per tile. |
| **Variegated Purple** | | | | | | | |
| | Demolish | Sq. | | 79 | | 79 | Includes material and labor to install |
| | Install | Sq. | 470 | 251 | | 721 | clear Vermont slate tile. |
| | Demolish and Install | Sq. | 470 | 330 | | 800 | |
| | Reinstall | Sq. | | 200.41 | | 200.41 | |
| | Minimum Charge | Job | | 219 | | 219 | |
| **Unfading Grey / Black** | | | | | | | |
| | Demolish | Sq. | | 79 | | 79 | Includes material and labor to install |
| | Install | Sq. | 535 | 251 | | 786 | clear Vermont slate tile. |
| | Demolish and Install | Sq. | 535 | 330 | | 865 | |
| | Reinstall | Sq. | | 200.41 | | 200.41 | |
| | Minimum Charge | Job | | 219 | | 219 | |
| **Unfading Red** | | | | | | | |
| | Demolish | Sq. | | 79 | | 79 | Includes material and labor to install |
| | Install | Sq. | 1300 | 251 | | 1551 | clear Vermont slate tile. |
| | Demolish and Install | Sq. | 1300 | 330 | | 1630 | |
| | Reinstall | Sq. | | 200.41 | | 200.41 | |
| | Minimum Charge | Job | | 219 | | 219 | |
| **Weathering Black** | | | | | | | |
| | Demolish | Sq. | | 79 | | 79 | Includes material and labor to install |
| | Install | Sq. | 545 | 251 | | 796 | clear Pennsylvania slate tile. |
| | Demolish and Install | Sq. | 545 | 330 | | 875 | |
| | Reinstall | Sq. | | 200.41 | | 200.41 | |
| | Minimum Charge | Job | | 219 | | 219 | |
| **Weathering Green** | | | | | | | |
| | Demolish | Sq. | | 79 | | 79 | Includes material and labor to install |
| | Install | Sq. | 395 | 251 | | 646 | clear Vermont slate tile. |
| | Demolish and Install | Sq. | 395 | 330 | | 725 | |
| | Reinstall | Sq. | | 200.41 | | 200.41 | |
| | Minimum Charge | Job | | 219 | | 219 | |

**For customer support on your Contractor's Pricing Guide: Residential Repair & Remodeling, call 800.448.8182.**

## Flat Clay Tile Roofing

| Flat Clay Tile Roofing | Unit | Material | Labor | Equip. | Total | Specification |
|---|---|---|---|---|---|---|
| **Glazed** | | | | | | |
| Demolish | Sq. | | 87.50 | | 87.50 | Includes material and labor to install |
| Install | Sq. | 570 | 219 | | 789 | tile roofing. Cost based on terra cotta |
| Demolish and Install | Sq. | 570 | 306.50 | | 876.50 | red flat clay tile and includes ridge |
| Reinstall | Sq. | | 175.36 | | 175.36 | and rake tiles. |
| Minimum Charge | Job | | 219 | | 219 | |
| **Terra Cotta Red** | | | | | | |
| Demolish | Sq. | | 87.50 | | 87.50 | Includes material and labor to install |
| Install | Sq. | 570 | 219 | | 789 | tile roofing. Cost based on terra cotta |
| Demolish and Install | Sq. | 570 | 306.50 | | 876.50 | red flat clay tile and includes ridge |
| Reinstall | Sq. | | 175.36 | | 175.36 | and rake tiles. |
| Minimum Charge | Job | | 219 | | 219 | |

## Mission Tile Roofing

| Mission Tile Roofing | Unit | Material | Labor | Equip. | Total | Specification |
|---|---|---|---|---|---|---|
| **Glazed Red** | | | | | | |
| Demolish | Sq. | | 87.50 | | 87.50 | Includes material and labor to install |
| Install | Sq. | 605 | 239 | | 844 | tile roofing. Cost based on glazed red |
| Demolish and Install | Sq. | 605 | 326.50 | | 931.50 | tile and includes birdstop, ridge and |
| Reinstall | Sq. | | 191.30 | | 191.30 | rake tiles. |
| Minimum Charge | Job | | 219 | | 219 | |
| **Unglazed Red** | | | | | | |
| Demolish | Sq. | | 87.50 | | 87.50 | Includes material and labor to install |
| Install | Sq. | 460 | 239 | | 699 | tile roofing. Cost based on unglazed |
| Demolish and Install | Sq. | 460 | 326.50 | | 786.50 | red tile and includes birdstop, ridge |
| Reinstall | Sq. | | 191.30 | | 191.30 | and rake tiles. |
| Minimum Charge | Job | | 219 | | 219 | |
| **Glazed Blue** | | | | | | |
| Demolish | Sq. | | 87.50 | | 87.50 | Includes material and labor to install |
| Install | Sq. | 415 | 239 | | 654 | tile roofing. Cost based on glazed |
| Demolish and Install | Sq. | 415 | 326.50 | | 741.50 | blue tile and includes birdstop, ridge |
| Reinstall | Sq. | | 191.30 | | 191.30 | and rake tiles. |
| Minimum Charge | Job | | 219 | | 219 | |
| **Glazed White** | | | | | | |
| Demolish | Sq. | | 87.50 | | 87.50 | Includes material and labor to install |
| Install | Sq. | 440 | 239 | | 679 | tile roofing. Cost based on glazed |
| Demolish and Install | Sq. | 440 | 326.50 | | 766.50 | white tile and includes birdstop, ridge |
| Reinstall | Sq. | | 191.30 | | 191.30 | and rake tiles. |
| Minimum Charge | Job | | 219 | | 219 | |
| **Unglazed White** | | | | | | |
| Demolish | Sq. | | 87.50 | | 87.50 | Includes material and labor to install |
| Install | Sq. | 365 | 239 | | 604 | tile roofing. Cost based on unglazed |
| Demolish and Install | Sq. | 365 | 326.50 | | 691.50 | white tile and includes birdstop, ridge |
| Reinstall | Sq. | | 191.30 | | 191.30 | and rake tiles. |
| Minimum Charge | Job | | 219 | | 219 | |
| **Color Blend** | | | | | | |
| Demolish | Sq. | | 87.50 | | 87.50 | Includes material and labor to install |
| Install | Sq. | 370 | 239 | | 609 | tile roofing. Cost based on color blend |
| Demolish and Install | Sq. | 370 | 326.50 | | 696.50 | tile and includes birdstop, ridge and |
| Reinstall | Sq. | | 191.30 | | 191.30 | rake tiles. |
| Minimum Charge | Job | | 219 | | 219 | |

## Mission Tile Roofing

| Mission Tile Roofing | | Unit | Material | Labor | Equip. | Total | Specification |
|---|---|---|---|---|---|---|---|
| **Accessories / Extras** | | | | | | | |
| Add for Cap Furring Strips | Install | Sq. | 99 | 274 | | 373 | Includes labor and material to install vertically placed wood furring strips under cap tiles. |
| Add for Tile Adhesive | Install | Sq. | 23.50 | 16 | | 39.50 | Includes material and labor for adhesive material placed between tiles. |
| Add for Wire Attachment | Install | Sq. | 15.40 | 191 | | 206.40 | Includes labor and material to install single wire anchors for attaching tile. |
| Add for Braided Runners | Install | Sq. | 38.50 | 183 | | 221.50 | Includes labor and material to install braided wire runners for secure attachment in high wind areas. |
| Add for Stainless Steel Nails | Install | Sq. | 25.50 | | | 25.50 | Includes material only for stainless steel nails. |
| Add for Brass Nails | Install | Sq. | 44 | | | 44 | Includes material only for brass nails. |
| Add for Hurricane Clips | Install | Sq. | 23.50 | 258 | | 281.50 | Includes material and labor for hurricane or wind clips for high wind areas. |
| Add for Copper Nails | Install | Sq. | 44 | | | 44 | Includes material only for copper nails. |
| Add for Vertical Furring Strips | Install | Sq. | 19.80 | 51 | | 70.80 | Includes labor and material to install vertically placed wood furring strips under tiles. |
| Add for Horizontal Furring Strips | Install | Sq. | 13.20 | 23 | | 36.20 | Includes labor and material to install horizontally placed wood furring strips under tiles. |

## Spanish Tile Roofing

| Spanish Tile Roofing | | Unit | Material | Labor | Equip. | Total | Specification |
|---|---|---|---|---|---|---|---|
| **Glazed Red** | Demolish | Sq. | | 87.50 | | 87.50 | Includes material and labor to install |
| | Install | Sq. | 605 | 183 | | 788 | tile roofing. Cost based on glazed red |
| | Demolish and Install | Sq. | 605 | 270.50 | | 875.50 | tile and includes birdstop, ridge and |
| | Reinstall | Sq. | | 146.13 | | 146.13 | rake tiles. |
| | Minimum Charge | Job | | 219 | | 219 | |
| **Glazed White** | Demolish | Sq. | | 87.50 | | 87.50 | Includes material and labor to install |
| | Install | Sq. | 395 | 244 | | 639 | tile roofing. Cost based on glazed |
| | Demolish and Install | Sq. | 395 | 331.50 | | 726.50 | white tile and includes birdstop, ridge |
| | Reinstall | Sq. | | 194.84 | | 194.84 | and rake tiles. |
| | Minimum Charge | Job | | 219 | | 219 | |

| Spanish Tile Roofing | | Unit | Material | Labor | Equip. | Total | Specification |
|---|---|---|---|---|---|---|---|
| **Glazed Blue** | | | | | | | |
| | Demolish | Sq. | | 87.50 | | 87.50 | Includes material and labor to install |
| | Install | Sq. | 395 | 183 | | 578 | tile roofing. Cost based on glazed |
| | Demolish and Install | Sq. | 395 | 270.50 | | 665.50 | blue tile and includes birdstop, ridge |
| | Reinstall | Sq. | | 146.13 | | 146.13 | and rake tiles. |
| | Minimum Charge | Job | | 219 | | 219 | |
| **Unglazed Red** | | | | | | | |
| | Demolish | Sq. | | 87.50 | | 87.50 | Includes material and labor to install |
| | Install | Sq. | 625 | 183 | | 808 | tile roofing. Cost based on unglazed |
| | Demolish and Install | Sq. | 625 | 270.50 | | 895.50 | red tile and includes birdstop, ridge |
| | Reinstall | Sq. | | 146.13 | | 146.13 | and rake tiles. |
| | Minimum Charge | Job | | 219 | | 219 | |
| **Unglazed White** | | | | | | | |
| | Demolish | Sq. | | 87.50 | | 87.50 | Includes material and labor to install |
| | Install | Sq. | 550 | 183 | | 733 | tile roofing. Cost based on unglazed |
| | Demolish and Install | Sq. | 550 | 270.50 | | 820.50 | white tile and includes birdstop, ridge |
| | Reinstall | Sq. | | 146.13 | | 146.13 | and rake tiles. |
| | Minimum Charge | Job | | 219 | | 219 | |
| **Color Blend** | | | | | | | |
| | Demolish | Sq. | | 87.50 | | 87.50 | Includes material and labor to install |
| | Install | Sq. | 670 | 183 | | 853 | tile roofing. Cost based on color blend |
| | Demolish and Install | Sq. | 670 | 270.50 | | 940.50 | tile and includes birdstop, ridge and |
| | Reinstall | Sq. | | 146.13 | | 146.13 | rake tiles. |
| | Minimum Charge | Job | | 219 | | 219 | |
| **Accessories / Extras** | | | | | | | |
| **Add for Vertical Furring Strips** | | | | | | | |
| | Install | Sq. | 19.80 | 51 | | 70.80 | Includes labor and material to install vertically placed wood furring strips under tiles. |
| **Add for Wire Attachment** | | | | | | | |
| | Install | Sq. | 8.25 | 146 | | 154.25 | Includes labor and material to install single wire ties. |
| **Add for Braided Runners** | | | | | | | |
| | Install | Sq. | 30.50 | 125 | | 155.50 | Includes labor and material to install braided wire runners for secure attachment in high wind areas. |
| **Add for Stainless Steel Nails** | | | | | | | |
| | Install | Sq. | 21 | | | 21 | Includes material only for stainless steel nails. |
| **Add for Copper Nails** | | | | | | | |
| | Install | Sq. | 33 | | | 33 | Includes material only for copper nails. |
| **Add for Brass Nails** | | | | | | | |
| | Install | Sq. | 33 | | | 33 | Includes material only for brass nails. |
| **Add for Hurricane Clips** | | | | | | | |
| | Install | Sq. | 12.40 | 151 | | 163.40 | Includes material and labor to install hurricane or wind clips for high wind areas. |
| **Add for Mortar Set Tiles** | | | | | | | |
| | Install | Sq. | 51.50 | 219 | | 270.50 | Includes material and labor to install setting tiles by using mortar material placed between tiles. |

| Spanish Tile Roofing | | Unit | Material | Labor | Equip. | Total | Specification |
|---|---|---|---|---|---|---|---|
| **Add for Tile Adhesive** | Install | Sq. | 13.50 | 14.25 | | 27.75 | Includes material and labor for adhesive placed between tiles. |
| **Add for Cap Furring Strips** | Install | Sq. | 99 | 274 | | 373 | Includes labor and material to install vertically placed wood furring strips under cap tiles. |
| **Add for Horizontal Furring Strips** | Install | Sq. | 13.20 | 23 | | 36.20 | Includes labor and material to install horizontally placed wood furring strips under tiles. |

| Concrete Tile Roofing | | Unit | Material | Labor | Equip. | Total | Specification |
|---|---|---|---|---|---|---|---|
| **Corrugated Red / Brown** | Demolish | Sq. | | 87.50 | | 87.50 | Includes material and labor to install tile roofing. Cost includes birdstop, booster, ridge and rake tiles. |
| | Install | Sq. | 116 | 325 | | 441 | |
| | Demolish and Install | Sq. | 116 | 412.50 | | 528.50 | |
| | Reinstall | Sq. | | 259.79 | | 259.79 | |
| | Minimum Charge | Job | | 219 | | 219 | |
| **Corrugated Gray** | Demolish | Sq. | | 87.50 | | 87.50 | Includes material and labor to install tile roofing. Cost includes birdstop, booster, ridge and rake tiles. |
| | Install | Sq. | 116 | 325 | | 441 | |
| | Demolish and Install | Sq. | 116 | 412.50 | | 528.50 | |
| | Reinstall | Sq. | | 259.79 | | 259.79 | |
| | Minimum Charge | Job | | 219 | | 219 | |
| **Corrugated Bright Red** | Demolish | Sq. | | 87.50 | | 87.50 | Includes material and labor to install tile roofing. Cost includes birdstop, booster, ridge and rake tiles. |
| | Install | Sq. | 117 | 325 | | 442 | |
| | Demolish and Install | Sq. | 117 | 412.50 | | 529.50 | |
| | Reinstall | Sq. | | 259.79 | | 259.79 | |
| | Minimum Charge | Job | | 219 | | 219 | |
| **Corrugated Black** | Demolish | Sq. | | 87.50 | | 87.50 | Includes material and labor to install tile roofing. Cost includes birdstop, booster, ridge and rake tiles. |
| | Install | Sq. | 117 | 325 | | 442 | |
| | Demolish and Install | Sq. | 117 | 412.50 | | 529.50 | |
| | Reinstall | Sq. | | 259.79 | | 259.79 | |
| | Minimum Charge | Job | | 219 | | 219 | |
| **Corrugated Green** | Demolish | Sq. | | 87.50 | | 87.50 | Includes material and labor to install tile roofing. Cost includes birdstop, booster, ridge and rake tiles. |
| | Install | Sq. | 117 | 325 | | 442 | |
| | Demolish and Install | Sq. | 117 | 412.50 | | 529.50 | |
| | Reinstall | Sq. | | 259.79 | | 259.79 | |
| | Minimum Charge | Job | | 219 | | 219 | |
| **Corrugated Blue** | Demolish | Sq. | | 87.50 | | 87.50 | Includes material and labor to install tile roofing. Cost includes birdstop, booster, ridge and rake tiles. |
| | Install | Sq. | 117 | 325 | | 442 | |
| | Demolish and Install | Sq. | 117 | 412.50 | | 529.50 | |
| | Reinstall | Sq. | | 259.79 | | 259.79 | |
| | Minimum Charge | Job | | 219 | | 219 | |
| **Flat Natural Gray** | Demolish | Sq. | | 87.50 | | 87.50 | Includes material and labor to install tile roofing. Cost includes birdstop, booster, ridge and rake tiles. |
| | Install | Sq. | 138 | 325 | | 463 | |
| | Demolish and Install | Sq. | 138 | 412.50 | | 550.50 | |
| | Reinstall | Sq. | | 259.79 | | 259.79 | |
| | Minimum Charge | Job | | 219 | | 219 | |

| Concrete Tile Roofing | | Unit | Material | Labor | Equip. | Total | Specification |
|---|---|---|---|---|---|---|---|
| **Flat Red / Brown** | | | | | | | |
| | Demolish | Sq. | | 87.50 | | 87.50 | Includes material and labor to install |
| | Install | Sq. | 138 | 325 | | 463 | tile roofing. Cost includes birdstop, |
| | Demolish and Install | Sq. | 138 | 412.50 | | 550.50 | booster, ridge and rake tiles. |
| | Reinstall | Sq. | | 259.79 | | 259.79 | |
| | Minimum Charge | Job | | 219 | | 219 | |
| **Flat Bright Red** | | | | | | | |
| | Demolish | Sq. | | 87.50 | | 87.50 | Includes material and labor to install |
| | Install | Sq. | 138 | 325 | | 463 | tile roofing. Cost includes birdstop, |
| | Demolish and Install | Sq. | 138 | 412.50 | | 550.50 | booster, ridge and rake tiles. |
| | Reinstall | Sq. | | 259.79 | | 259.79 | |
| | Minimum Charge | Job | | 219 | | 219 | |
| **Flat Green** | | | | | | | |
| | Demolish | Sq. | | 87.50 | | 87.50 | Includes material and labor to install |
| | Install | Sq. | 138 | 325 | | 463 | tile roofing. Cost includes birdstop, |
| | Demolish and Install | Sq. | 138 | 412.50 | | 550.50 | booster, ridge and rake tiles. |
| | Minimum Charge | Job | | 219 | | 219 | |
| **Flat Blue** | | | | | | | |
| | Demolish | Sq. | | 87.50 | | 87.50 | Includes material and labor to install |
| | Install | Sq. | 138 | 325 | | 463 | tile roofing. Cost includes birdstop, |
| | Demolish and Install | Sq. | 138 | 412.50 | | 550.50 | booster, ridge and rake tiles. |
| | Reinstall | Sq. | | 259.79 | | 259.79 | |
| | Minimum Charge | Job | | 219 | | 219 | |
| **Flat Black** | | | | | | | |
| | Demolish | Sq. | | 87.50 | | 87.50 | Includes material and labor to install |
| | Install | Sq. | 138 | 325 | | 463 | tile roofing. Cost includes birdstop, |
| | Demolish and Install | Sq. | 138 | 412.50 | | 550.50 | booster, ridge and rake tiles. |
| | Reinstall | Sq. | | 259.79 | | 259.79 | |
| | Minimum Charge | Job | | 219 | | 219 | |

| Aluminum Sheet | | Unit | Material | Labor | Equip. | Total | Specification |
|---|---|---|---|---|---|---|---|
| **Corrugated** | | | | | | | |
| **.019" Thick Natural** | | | | | | | |
| | Demolish | S.F. | | .93 | | .93 | Includes material and labor to install |
| | Install | S.F. | 1.36 | 2.12 | | 3.48 | aluminum sheet metal roofing. |
| | Demolish and Install | S.F. | 1.36 | 3.05 | | 4.41 | |
| | Reinstall | S.F. | | 1.70 | | 1.70 | |
| | Minimum Charge | Job | | 510 | | 510 | |
| **.016" Thick Natural** | | | | | | | |
| | Demolish | S.F. | | .93 | | .93 | Includes material and labor to install |
| | Install | S.F. | 1.10 | 2.12 | | 3.22 | aluminum sheet metal roofing. |
| | Demolish and Install | S.F. | 1.10 | 3.05 | | 4.15 | |
| | Reinstall | S.F. | | 1.70 | | 1.70 | |
| | Minimum Charge | Job | | 510 | | 510 | |
| **.016" Colored Finish** | | | | | | | |
| | Demolish | S.F. | | .93 | | .93 | Includes material and labor to install |
| | Install | S.F. | 1.60 | 2.12 | | 3.72 | aluminum sheet metal roofing. |
| | Demolish and Install | S.F. | 1.60 | 3.05 | | 4.65 | |
| | Reinstall | S.F. | | 1.70 | | 1.70 | |
| | Minimum Charge | Job | | 510 | | 510 | |
| **.019" Colored Finish** | | | | | | | |
| | Demolish | S.F. | | .93 | | .93 | Includes material and labor to install |
| | Install | S.F. | 1.71 | 2.12 | | 3.83 | aluminum sheet metal roofing. |
| | Demolish and Install | S.F. | 1.71 | 3.05 | | 4.76 | |
| | Reinstall | S.F. | | 1.70 | | 1.70 | |
| | Minimum Charge | Job | | 510 | | 510 | |

# Roofing

| Aluminum Sheet | Unit | Material | Labor | Equip. | Total | Specification |
|---|---|---|---|---|---|---|
| **Ribbed** | | | | | | |
| **.019″ Thick Natural** | | | | | | |
| Demolish | S.F. | | .93 | | .93 | Includes material and labor to install |
| Install | S.F. | 1.36 | 2.12 | | 3.48 | aluminum sheet metal roofing. |
| Demolish and Install | S.F. | 1.36 | 3.05 | | 4.41 | |
| Reinstall | S.F. | | 1.70 | | 1.70 | |
| Minimum Charge | Job | | 510 | | 510 | |
| **.016″ Thick Natural** | | | | | | |
| Demolish | S.F. | | .93 | | .93 | Includes material and labor to install |
| Install | S.F. | 1.10 | 2.12 | | 3.22 | aluminum sheet metal roofing. |
| Demolish and Install | S.F. | 1.10 | 3.05 | | 4.15 | |
| Reinstall | S.F. | | 1.70 | | 1.70 | |
| Minimum Charge | Job | | 510 | | 510 | |
| **.050″ Colored Finish** | | | | | | |
| Demolish | S.F. | | .93 | | .93 | Includes material and labor to install |
| Install | S.F. | 5.25 | 2.12 | | 7.37 | aluminum sheet metal roofing. |
| Demolish and Install | S.F. | 5.25 | 3.05 | | 8.30 | |
| Reinstall | S.F. | | 1.70 | | 1.70 | |
| Minimum Charge | Job | | 510 | | 510 | |
| **.032″ Thick Natural** | | | | | | |
| Demolish | S.F. | | .93 | | .93 | Includes material and labor to install |
| Install | S.F. | 2.88 | 2.12 | | 5 | aluminum sheet metal roofing. |
| Demolish and Install | S.F. | 2.88 | 3.05 | | 5.93 | |
| Reinstall | S.F. | | 1.70 | | 1.70 | |
| Minimum Charge | Job | | 510 | | 510 | |
| **.040″ Thick Natural** | | | | | | |
| Demolish | S.F. | | .93 | | .93 | Includes material and labor to install |
| Install | S.F. | 4.20 | 2.12 | | 6.32 | aluminum sheet metal roofing. |
| Demolish and Install | S.F. | 4.20 | 3.05 | | 7.25 | |
| Reinstall | S.F. | | 1.70 | | 1.70 | |
| Minimum Charge | Job | | 510 | | 510 | |
| **.050″ Thick Natural** | | | | | | |
| Demolish | S.F. | | .93 | | .93 | Includes material and labor to install |
| Install | S.F. | 4.35 | 2.12 | | 6.47 | aluminum sheet metal roofing. |
| Demolish and Install | S.F. | 4.35 | 3.05 | | 7.40 | |
| Reinstall | S.F. | | 1.70 | | 1.70 | |
| Minimum Charge | Job | | 510 | | 510 | |
| **.016″ Colored Finish** | | | | | | |
| Demolish | S.F. | | .93 | | .93 | Includes material and labor to install |
| Install | S.F. | 1.60 | 2.12 | | 3.72 | aluminum sheet metal roofing. |
| Demolish and Install | S.F. | 1.60 | 3.05 | | 4.65 | |
| Reinstall | S.F. | | 1.70 | | 1.70 | |
| Minimum Charge | Job | | 510 | | 510 | |
| **.019″ Colored Finish** | | | | | | |
| Demolish | S.F. | | .93 | | .93 | Includes material and labor to install |
| Install | S.F. | 1.71 | 2.12 | | 3.83 | aluminum sheet metal roofing. |
| Demolish and Install | S.F. | 1.71 | 3.05 | | 4.76 | |
| Reinstall | S.F. | | 1.70 | | 1.70 | |
| Minimum Charge | Job | | 510 | | 510 | |
| **.032″ Colored Finish** | | | | | | |
| Demolish | S.F. | | .93 | | .93 | Includes material and labor to install |
| Install | S.F. | 4.30 | 2.12 | | 6.42 | aluminum sheet metal roofing. |
| Demolish and Install | S.F. | 4.30 | 3.05 | | 7.35 | |
| Reinstall | S.F. | | 1.70 | | 1.70 | |
| Minimum Charge | Job | | 510 | | 510 | |

# Roofing

## Aluminum Sheet

| .040" Colored Finish | Unit | Material | Labor | Equip. | Total | Specification |
|---|---|---|---|---|---|---|
| Demolish | S.F. | | .93 | | .93 | Includes material and labor to install |
| Install | S.F. | 5.05 | 2.12 | | 7.17 | aluminum sheet metal roofing. |
| Demolish and Install | S.F. | 5.05 | 3.05 | | 8.10 | |
| Reinstall | S.F. | | 1.70 | | 1.70 | |
| Minimum Charge | Job | | 510 | | 510 | |

## Fiberglass Sheet

| 12 Ounce Corrugated | Unit | Material | Labor | Equip. | Total | Specification |
|---|---|---|---|---|---|---|
| Demolish | S.F. | | 1.67 | | 1.67 | Cost includes material and labor to |
| Install | S.F. | 4.97 | 1.73 | | 6.70 | install 12 ounce corrugated fiberglass |
| Demolish and Install | S.F. | 4.97 | 3.40 | | 8.37 | sheet roofing. |
| Reinstall | S.F. | | 1.38 | | 1.38 | |
| Minimum Charge | Job | | 218 | | 218 | |
| **8 Ounce Corrugated** | | | | | | |
| Demolish | S.F. | | 1.67 | | 1.67 | Cost includes material and labor to |
| Install | S.F. | 2.72 | 1.73 | | 4.45 | install 8 ounce corrugated fiberglass |
| Demolish and Install | S.F. | 2.72 | 3.40 | | 6.12 | roofing. |
| Reinstall | S.F. | | 1.38 | | 1.38 | |
| Minimum Charge | Job | | 218 | | 218 | |

## Galvanized Steel

| Corrugated | Unit | Material | Labor | Equip. | Total | Specification |
|---|---|---|---|---|---|---|
| **24 Gauge** | | | | | | |
| Demolish | S.F. | | .93 | | .93 | Includes material and labor to install |
| Install | S.F. | 3.30 | 1.82 | | 5.12 | galvanized steel sheet metal roofing. |
| Demolish and Install | S.F. | 3.30 | 2.75 | | 6.05 | Cost based on 24 gauge corrugated |
| Reinstall | S.F. | | 1.46 | | 1.46 | or ribbed steel roofing. |
| Minimum Charge | Job | | 218 | | 218 | |
| **26 Gauge** | | | | | | |
| Demolish | S.F. | | .93 | | .93 | Includes material and labor to install |
| Install | S.F. | 2.22 | 1.73 | | 3.95 | galvanized steel sheet metal roofing. |
| Demolish and Install | S.F. | 2.22 | 2.66 | | 4.88 | Cost based on 26 gauge corrugated |
| Reinstall | S.F. | | 1.38 | | 1.38 | or ribbed steel roofing. |
| Minimum Charge | Job | | 218 | | 218 | |
| **28 Gauge** | | | | | | |
| Demolish | S.F. | | .93 | | .93 | Includes material and labor to install |
| Install | S.F. | 1.58 | 1.65 | | 3.23 | galvanized steel sheet metal roofing. |
| Demolish and Install | S.F. | 1.58 | 2.58 | | 4.16 | Cost based on 28 gauge corrugated |
| Reinstall | S.F. | | 1.32 | | 1.32 | or ribbed steel roofing. |
| Minimum Charge | Job | | 218 | | 218 | |
| **30 Gauge** | | | | | | |
| Demolish | S.F. | | .93 | | .93 | Includes material and labor to install |
| Install | S.F. | 1.84 | 1.57 | | 3.41 | galvanized steel sheet metal roofing. |
| Demolish and Install | S.F. | 1.84 | 2.50 | | 4.34 | Cost based on 30 gauge corrugated |
| Reinstall | S.F. | | 1.26 | | 1.26 | or ribbed steel roofing. |
| Minimum Charge | Job | | 218 | | 218 | |

| Galvanized Steel | Unit | Material | Labor | Equip. | Total | Specification |
|---|---|---|---|---|---|---|
| **Ribbed** | | | | | | |
| **24 Gauge** | | | | | | |
| Demolish | S.F. | | .93 | | .93 | Includes material and labor to install |
| Install | S.F. | 3.30 | 1.82 | | 5.12 | galvanized steel sheet metal roofing. |
| Demolish and Install | S.F. | 3.30 | 2.75 | | 6.05 | Cost based on 24 gauge corrugated |
| Reinstall | S.F. | | 1.46 | | 1.46 | or ribbed steel roofing. |
| Minimum Charge | Job | | 218 | | 218 | |
| **26 Gauge** | | | | | | |
| Demolish | S.F. | | .93 | | .93 | Includes material and labor to install |
| Install | S.F. | 2.22 | 1.73 | | 3.95 | galvanized steel sheet metal roofing. |
| Demolish and Install | S.F. | 2.22 | 2.66 | | 4.88 | Cost based on 26 gauge corrugated |
| Reinstall | S.F. | | 1.38 | | 1.38 | or ribbed steel roofing. |
| Minimum Charge | Job | | 218 | | 218 | |
| **28 Gauge** | | | | | | |
| Demolish | S.F. | | .93 | | .93 | Includes material and labor to install |
| Install | S.F. | 1.58 | 1.65 | | 3.23 | galvanized steel sheet metal roofing. |
| Demolish and Install | S.F. | 1.58 | 2.58 | | 4.16 | Cost based on 28 gauge corrugated |
| Reinstall | S.F. | | 1.32 | | 1.32 | or ribbed steel roofing. |
| Minimum Charge | Job | | 218 | | 218 | |
| **30 Gauge** | | | | | | |
| Demolish | S.F. | | .93 | | .93 | Includes material and labor to install |
| Install | S.F. | 1.84 | 1.57 | | 3.41 | galvanized steel sheet metal roofing. |
| Demolish and Install | S.F. | 1.84 | 2.50 | | 4.34 | Cost based on 30 gauge corrugated |
| Reinstall | S.F. | | 1.26 | | 1.26 | or ribbed steel roofing. |
| Minimum Charge | Job | | 218 | | 218 | |

| Standing Seam | Unit | Material | Labor | Equip. | Total | Specification |
|---|---|---|---|---|---|---|
| **Copper** | | | | | | |
| **16 Ounce** | | | | | | |
| Demolish | Sq. | | 93 | | 93 | Includes material and labor to install |
| Install | Sq. | 1125 | 390 | | 1515 | copper standing seam metal roofing. |
| Demolish and Install | Sq. | 1125 | 483 | | 1608 | Cost based on 16 ounce copper |
| Reinstall | Sq. | | 313.60 | | 313.60 | roofing. |
| Minimum Charge | Job | | 255 | | 255 | |
| **18 Ounce** | | | | | | |
| Demolish | Sq. | | 93 | | 93 | Includes material and labor to install |
| Install | Sq. | 1250 | 425 | | 1675 | copper standing seam metal roofing. |
| Demolish and Install | Sq. | 1250 | 518 | | 1768 | Cost based on 18 ounce copper |
| Reinstall | Sq. | | 339.73 | | 339.73 | roofing. |
| Minimum Charge | Job | | 255 | | 255 | |
| **20 Ounce** | | | | | | |
| Demolish | Sq. | | 93 | | 93 | Includes material and labor to install |
| Install | Sq. | 1475 | 465 | | 1940 | copper standing seam metal roofing. |
| Demolish and Install | Sq. | 1475 | 558 | | 2033 | Cost based on 20 ounce copper |
| Reinstall | Sq. | | 370.62 | | 370.62 | roofing. |
| **Stainless Steel** | | | | | | |
| **26 Gauge** | | | | | | |
| Demolish | Sq. | | 93 | | 93 | Cost includes material and labor to |
| Install | Sq. | 830 | 445 | | 1275 | install 26 gauge stainless steel |
| Demolish and Install | Sq. | 830 | 538 | | 1368 | standing seam metal roofing. |
| Reinstall | Sq. | | 354.50 | | 354.50 | |
| Minimum Charge | Job | | 255 | | 255 | |

# Roofing

## Standing Seam

| | Unit | Material | Labor | Equip. | Total | Specification |
|---|---|---|---|---|---|---|
| **28 Gauge** | | | | | | Includes material and labor to install |
| Demolish | Sq. | | 93 | | 93 | standing seam metal roofing. Cost |
| Install | Sq. | 720 | 425 | | 1145 | based on 28 gauge stainless steel |
| Demolish and Install | Sq. | 720 | 518 | | 1238 | roofing. |
| Reinstall | Sq. | | 339.73 | | 339.73 | |
| Minimum Charge | Job | | 255 | | 255 | |

## Elastomeric Roofing

| | Unit | Material | Labor | Equip. | Total | Specification |
|---|---|---|---|---|---|---|
| **Neoprene** | | | | | | |
| **1/16"** | | | | | | Includes material and labor to install |
| Demolish | Sq. | | 32.50 | | 32.50 | neoprene membrane, fully adhered, |
| Install | Sq. | 245 | 208 | 41 | 494 | 1/16" thick. |
| Demolish and Install | Sq. | 245 | 240.50 | 41 | 526.50 | |
| Minimum Charge | Job | | 995 | 99.50 | 1094.50 | |
| **GRM Membrane** | | | | | | |
| **1/16"** | | | | | | Includes material and labor to install |
| Demolish | Sq. | | 32.50 | | 32.50 | coal tar based elastomeric roofing |
| Install | Sq. | 83.50 | 330 | 33 | 446.50 | membrane, polyester fiber reinforced. |
| Demolish and Install | Sq. | 83.50 | 362.50 | 33 | 479 | |
| Minimum Charge | Job | | 995 | 99.50 | 1094.50 | |
| **Cant Strips** | | | | | | |
| **3"** | | | | | | Includes material and labor to install |
| Demolish | L.F. | | 1 | | 1 | cants 3" x 3", treated timber, cut |
| Install | L.F. | .98 | 1.34 | | 2.32 | diagonally. |
| Demolish and Install | L.F. | .98 | 2.34 | | 3.32 | |
| Minimum Charge | Job | | 109 | | 109 | |
| **4"** | | | | | | Includes material and labor to install |
| Demolish | L.F. | | 1 | | 1 | cants 4" x 4", treated timber, cut |
| Install | L.F. | 2.10 | 1.34 | | 3.44 | diagonally. |
| Demolish and Install | L.F. | 2.10 | 2.34 | | 4.44 | |
| Minimum Charge | Job | | 109 | | 109 | |
| **Polyvinyl Chloride (PVC)** | | | | | | |
| **45 mil (loose laid)** | | | | | | Cost includes material and labor to |
| Demolish | Sq. | | 32.50 | | 32.50 | install 45 mil PVC elastomeric roofing, |
| Install | Sq. | 101 | 39 | 3.91 | 143.91 | loose-laid with stone. |
| Demolish and Install | Sq. | 101 | 71.50 | 3.91 | 176.41 | |
| Minimum Charge | Job | | 995 | 99.50 | 1094.50 | |
| **48 mil (loose laid)** | | | | | | Cost includes material and labor to |
| Demolish | Sq. | | 32.50 | | 32.50 | install 48 mil PVC elastomeric roofing, |
| Install | Sq. | 124 | 39 | 3.91 | 166.91 | loose-laid with stone. |
| Demolish and Install | Sq. | 124 | 71.50 | 3.91 | 199.41 | |
| Minimum Charge | Job | | 995 | 99.50 | 1094.50 | |
| **60 mil (loose laid)** | | | | | | Cost includes material and labor to |
| Demolish | Sq. | | 32.50 | | 32.50 | install 60 mil PVC elastomeric roofing, |
| Install | Sq. | 125 | 39 | 3.91 | 167.91 | loose-laid with stone. |
| Demolish and Install | Sq. | 125 | 71.50 | 3.91 | 200.41 | |
| Minimum Charge | Job | | 995 | 99.50 | 1094.50 | |
| **45 mil (attached)** | | | | | | Cost includes material and labor to |
| Demolish | Sq. | | 51.50 | | 51.50 | install 45 mil PVC elastomeric roofing |
| Install | Sq. | 91 | 76.50 | 7.65 | 175.15 | fully attached to roof deck. |
| Demolish and Install | Sq. | 91 | 128 | 7.65 | 226.65 | |
| Minimum Charge | Job | | 995 | 99.50 | 1094.50 | |

| Elastomeric Roofing | | Unit | Material | Labor | Equip. | Total | Specification |
|---|---|---|---|---|---|---|---|
| **48 mil (attached)** | | | | | | | Cost includes material and labor to |
| | Demolish | Sq. | | 51.50 | | 51.50 | install 48 mil PVC elastomeric roofing |
| | Install | Sq. | 168 | 76.50 | 7.65 | 252.15 | fully attached to roof deck. |
| | Demolish and Install | Sq. | 168 | 128 | 7.65 | 303.65 | |
| | Minimum Charge | Job | | 995 | 99.50 | 1094.50 | |
| **60 mil (attached)** | | | | | | | Cost includes material and labor to |
| | Demolish | Sq. | | 51.50 | | 51.50 | install 60 mil PVC elastomeric roofing |
| | Install | Sq. | 169 | 76.50 | 7.65 | 253.15 | fully attached to roof deck. |
| | Demolish and Install | Sq. | 169 | 128 | 7.65 | 304.65 | |
| | Minimum Charge | Job | | 995 | 99.50 | 1094.50 | |
| **CSPE Type** | | | | | | | |
| **45 mil (loose laid w / stone)** | | | | | | | Cost includes material and labor to |
| | Demolish | Sq. | | 32.50 | | 32.50 | install 45 mil CSPE (chlorosulfonated |
| | Install | Sq. | 277 | 76.50 | 7.65 | 361.15 | polyethylene-hypalon) roofing material |
| | Demolish and Install | Sq. | 277 | 109 | 7.65 | 393.65 | loose laid on deck and ballasted with |
| | Minimum Charge | Job | | 995 | 99.50 | 1094.50 | stone. |
| **45 mil (attached at seams)** | | | | | | | Cost includes material and labor to |
| | Demolish | Sq. | | 38.50 | | 38.50 | install 45 mil CSPE (chlorosulfonated |
| | Install | Sq. | 340 | 57 | 5.70 | 402.70 | polyethylene-hypalon) roofing material |
| | Demolish and Install | Sq. | 340 | 95.50 | 5.70 | 441.20 | attached to deck at seams with batten |
| | Minimum Charge | Job | | 995 | 99.50 | 1094.50 | strips. |
| **45 mil (fully attached)** | | | | | | | Cost includes material and labor to |
| | Demolish | Sq. | | 51.50 | | 51.50 | install 45 mil CSPE (chlorosulfonated |
| | Install | Sq. | 265 | 57 | 5.70 | 327.70 | polyethylene-hypalon) roofing material |
| | Demolish and Install | Sq. | 265 | 108.50 | 5.70 | 379.20 | fully attached to deck. |
| | Minimum Charge | Job | | 995 | 99.50 | 1094.50 | |
| **EPDM Type** | | | | | | | |
| **45 mil (attached at seams)** | | | | | | | Cost includes material and labor to |
| | Demolish | Sq. | | 38.50 | | 38.50 | install 45 mil EPDM (ethylene |
| | Install | Sq. | 87 | 57 | 5.70 | 149.70 | propylene diene monomer) roofing |
| | Demolish and Install | Sq. | 87 | 95.50 | 5.70 | 188.20 | material attached to deck at seams. |
| | Minimum Charge | Job | | 995 | 99.50 | 1094.50 | |
| **45 mil (fully attached)** | | | | | | | Cost includes material and labor to |
| | Demolish | Sq. | | 51.50 | | 51.50 | install 45 mil EPDM (ethylene |
| | Install | Sq. | 123 | 76.50 | 7.65 | 207.15 | propylene diene monomer) roofing |
| | Demolish and Install | Sq. | 123 | 128 | 7.65 | 258.65 | material fully attached to deck. |
| | Minimum Charge | Job | | 995 | 99.50 | 1094.50 | |
| **45 mil (loose laid w / stone)** | | | | | | | Cost includes material and labor to |
| | Demolish | Sq. | | 32.50 | | 32.50 | install 45 mil EPDM (ethylene |
| | Install | Sq. | 93.50 | 39 | 3.91 | 136.41 | propylene diene monomer) roofing |
| | Demolish and Install | Sq. | 93.50 | 71.50 | 3.91 | 168.91 | material loose laid on deck and |
| | Minimum Charge | Job | | 995 | 99.50 | 1094.50 | ballasted with stones (approx. 1/2 ton |
| | | | | | | | of stone per square). |
| **60 mil (attached at seams)** | | | | | | | Cost includes material and labor to |
| | Demolish | Sq. | | 38.50 | | 38.50 | install 60 mil EPDM (ethylene |
| | Install | Sq. | 103 | 57 | 5.70 | 165.70 | propylene diene monomer) roofing |
| | Demolish and Install | Sq. | 103 | 95.50 | 5.70 | 204.20 | material attached to deck at seams. |
| | Minimum Charge | Job | | 995 | 99.50 | 1094.50 | |

86

# Roofing

## Elastomeric Roofing

| Elastomeric Roofing | Unit | Material | Labor | Equip. | Total | Specification |
|---|---|---|---|---|---|---|
| **60 mil (fully attached)** | | | | | | Cost includes material and labor to |
| Demolish | Sq. | | 51.50 | | 51.50 | install 60 mil EPDM (ethylene |
| Install | Sq. | 140 | 76.50 | 7.65 | 224.15 | propylene diene monomer) roofing |
| Demolish and Install | Sq. | 140 | 128 | 7.65 | 275.65 | material fully attached to deck. |
| Minimum Charge | Job | | 995 | 99.50 | 1094.50 | |
| **60 mil (loose laid w / stone)** | | | | | | Cost includes material and labor to |
| Demolish | Sq. | | 32.50 | | 32.50 | install 60 mil EPDM (ethylene |
| Install | Sq. | 111 | 39 | 3.91 | 153.91 | propylene diene monomer) roofing |
| Demolish and Install | Sq. | 111 | 71.50 | 3.91 | 186.41 | material loose laid on deck and |
| Minimum Charge | Job | | 995 | 99.50 | 1094.50 | ballasted with stones (approx. 1/2 ton |
| | | | | | | of stone per square). |

## Modified Bitumen Roofing

| Modified Bitumen Roofing | Unit | Material | Labor | Equip. | Total | Specification |
|---|---|---|---|---|---|---|
| **Fully Attached** | | | | | | |
| **120 mil** | | | | | | |
| Demolish | Sq. | | 77 | | 77 | Cost includes material and labor to |
| Install | Sq. | 71.50 | 9.70 | 1.92 | 83.12 | install 120 mil modified bitumen |
| Demolish and Install | Sq. | 71.50 | 86.70 | 1.92 | 160.12 | roofing, fully attached to deck by |
| Minimum Charge | Job | | 109 | | 109 | torch. |
| **150 mil** | | | | | | |
| Demolish | Sq. | | 77 | | 77 | Cost includes material and labor to |
| Install | Sq. | 93.50 | 24 | 4.71 | 122.21 | install 150 mil modified bitumen |
| Demolish and Install | Sq. | 93.50 | 101 | 4.71 | 199.21 | roofing, fully attached to deck by |
| Minimum Charge | Job | | 109 | | 109 | torch. |
| **160 mil** | | | | | | |
| Demolish | Sq. | | 77 | | 77 | Cost includes material and labor to |
| Install | Sq. | 103 | 25 | 4.92 | 132.92 | install 160 mil modified bitumen |
| Demolish and Install | Sq. | 103 | 102 | 4.92 | 209.92 | roofing, fully attached to deck by |
| Minimum Charge | Job | | 109 | | 109 | torch. |

## Built-up Roofing

| Built-up Roofing | Unit | Material | Labor | Equip. | Total | Specification |
|---|---|---|---|---|---|---|
| **Emergency Repairs** | | | | | | |
| Install | Ea. | 224 | 2050 | 405 | 2679 | Minimum charge for work requiring |
| Minimum Charge | Job | | 1425 | 283 | 1708 | hot asphalt. |
| **Asphalt** | | | | | | |
| **3-ply Asphalt** | | | | | | |
| Demolish | Sq. | | 80 | | 80 | Cost includes material and labor to |
| Install | Sq. | 82.50 | 119 | 23.50 | 225 | install 3-ply asphalt built-up roof |
| Demolish and Install | Sq. | 82.50 | 199 | 23.50 | 305 | including one 30 lb and two 15 lb |
| Minimum Charge | Job | | 1425 | 283 | 1708 | layers of felt. |
| **4-ply Asphalt** | | | | | | |
| Demolish | Sq. | | 95.50 | | 95.50 | Cost includes material and labor to |
| Install | Sq. | 114 | 130 | 25.50 | 269.50 | install 4-ply asphalt built-up roof |
| Demolish and Install | Sq. | 114 | 225.50 | 25.50 | 365 | including one 30 lb and three 15 lb |
| Minimum Charge | Job | | 1425 | 283 | 1708 | layers of felt. |

| Built-up Roofing | Unit | Material | Labor | Equip. | Total | Specification |
|---|---|---|---|---|---|---|
| **5-ply Asphalt** | | | | | | Cost includes material and labor to install 5-ply asphalt built-up roof including one 30 lb and four 15 lb layers of felt. |
| Demolish | Sq. | | 125 | | 125 | |
| Install | Sq. | 156 | 143 | 28.50 | 327.50 | |
| Demolish and Install | Sq. | 156 | 268 | 28.50 | 452.50 | |
| Minimum Charge | Job | | 1425 | 283 | 1708 | |
| **Components / Accessories** | | | | | | |
| **Flood Coat** | | | | | | Includes labor and material to install flood coat. |
| Install | Sq. | 32 | 67 | | 99 | |
| **R / R Gravel and Flood Coat** | | | | | | Includes labor and material to install flood coat and gravel. |
| Install | Sq. | 37.50 | 218 | | 255.50 | |
| **Aluminum UV Coat** | | | | | | Includes labor and material to install aluminum-based UV coating. |
| Install | Sq. | 34.50 | 249 | 25 | 308.50 | |
| **Gravel Coat** | | | | | | Includes labor and material to install gravel coat. |
| Demolish | Sq. | | 40 | | 40 | |
| Install | Sq. | 5.60 | 36.50 | | 42.10 | |
| Demolish and Install | Sq. | 5.60 | 76.50 | | 82.10 | |
| **Mineral Surface Cap Sheet** | | | | | | Cost includes labor and material to install mineral surfaced cap sheet. |
| Install | Sq. | 91.50 | 7.50 | | 99 | |
| **Fibrated Aluminum UV Coating** | | | | | | Includes labor and material to install fibered aluminum-based UV coating. |
| Install | Sq. | 26 | 67 | | 93 | |
| **Impregnated Walk** | | | | | | |
| **1"** | | | | | | Cost includes labor and material to install 1" layer of asphalt coating. |
| Install | S.F. | 3.22 | 2.33 | | 5.55 | |
| **1/2"** | | | | | | Cost includes labor and material to install 1/2" layer of asphalt coating. |
| Install | S.F. | 2.33 | 1.84 | | 4.17 | |
| **3/4"** | | | | | | Cost includes labor and material to install 3/4" layer of asphalt coating. |
| Install | S.F. | 2.88 | 1.94 | | 4.82 | |

| Doors | Roof Window | Double Hung Window |
|-------|-------------|--------------------|

| Exterior Type Door | Unit | Material | Labor | Equip. | Total | Specification |
|--------------------|------|----------|-------|--------|-------|---------------|
| **Entry** | | | | | | |
| **Flush (1-3/4″)** | | | | | | |
| Demolish | Ea. | | 43 | | 43 | Cost includes material and labor to |
| Install | Ea. | 264 | 29 | | 293 | install 3′ x 6′ 8″ pre-hung 1-3/4″ |
| Demolish and Install | Ea. | 264 | 72 | | 336 | door, including casing and stop, |
| Reinstall | Ea. | | 29.05 | | 29.05 | hinges, jamb, aluminum sill, |
| Clean | Ea. | 1.47 | 14.80 | | 16.27 | weatherstripped. |
| Paint | Ea. | 10.60 | 55 | | 65.60 | |
| Minimum Charge | Job | | 232 | | 232 | |
| **Raised Panel** | | | | | | |
| Demolish | Ea. | | 43 | | 43 | Cost includes material and labor to |
| Install | Ea. | 505 | 26 | | 531 | install 3′ x 6′ 8″ pre-hung 1-3/4″ |
| Demolish and Install | Ea. | 505 | 69 | | 574 | door, jamb, including casing and |
| Reinstall | Ea. | | 25.82 | | 25.82 | stop, hinges, aluminum sill, |
| Clean | Ea. | 1.47 | 14.80 | | 16.27 | weatherstripping. |
| Paint | Ea. | 10.60 | 55 | | 65.60 | |
| Minimum Charge | Job | | 232 | | 232 | |
| **Entry w / Lights** | | | | | | |
| **Good Quality** | | | | | | |
| Demolish | Ea. | | 43 | | 43 | Cost includes material and labor to |
| Install | Ea. | 365 | 58 | | 423 | install 3′ x 6′ 8″ pre-hung 1-3/4″ |
| Demolish and Install | Ea. | 365 | 101 | | 466 | door, solid wood lower panel, with up |
| Reinstall | Ea. | | 58.10 | | 58.10 | to 9-light glass panels, casing, sill, |
| Clean | Ea. | 1.47 | 14.80 | | 16.27 | weatherstripped. |
| Paint | Ea. | 7 | 26 | | 33 | |
| Minimum Charge | Job | | 232 | | 232 | |
| **Better Quality** | | | | | | |
| Demolish | Ea. | | 43 | | 43 | Cost includes material and labor to |
| Install | Ea. | 385 | 58 | | 443 | install 3′ x 6′ 8″ pre-hung 1-3/4″ Ash |
| Demolish and Install | Ea. | 385 | 101 | | 486 | door, solid wood, raised panel door |
| Reinstall | Ea. | | 58.10 | | 58.10 | without glass panels. |
| Clean | Ea. | 1.47 | 14.80 | | 16.27 | |
| Paint | Ea. | 10.60 | 55 | | 65.60 | |
| Minimum Charge | Job | | 232 | | 232 | |
| **Premium Quality** | | | | | | |
| Demolish | Ea. | | 43 | | 43 | Cost includes material and labor to |
| Install | Ea. | 1000 | 27.50 | | 1027.50 | install 3′ x 6′ 8″ pre-hung 1-3/4″ Oak |
| Demolish and Install | Ea. | 1000 | 70.50 | | 1070.50 | door, solid wood, raised panel door |
| Reinstall | Ea. | | 27.34 | | 27.34 | without glass panels. |
| Clean | Ea. | 1.47 | 14.80 | | 16.27 | |
| Paint | Ea. | 10.60 | 55 | | 65.60 | |
| Minimum Charge | Job | | 232 | | 232 | |

| Exterior Type Door | | Unit | Material | Labor | Equip. | Total | Specification |
|---|---|---|---|---|---|---|---|
| **Custom Quality** | | | | | | | |
| | Demolish | Ea. | | 43 | | 43 | Cost includes material and labor to |
| | Install | Ea. | 540 | 62 | | 602 | install 3' x 6' 8" pre-hung 1-3/4" |
| | Demolish and Install | Ea. | 540 | 105 | | 645 | hand-made, solid wood, raised panel |
| | Reinstall | Ea. | | 61.97 | | 61.97 | door with custom stained glass. |
| | Clean | Ea. | .10 | 14.80 | | 14.90 | |
| | Paint | Ea. | 2.64 | 129 | | 131.64 | |
| | Minimum Charge | Job | | 232 | | 232 | |
| **Miami Style** | | | | | | | |
| | Demolish | Ea. | | 43 | | 43 | Cost includes material and labor to |
| | Install | Ea. | 243 | 27.50 | | 270.50 | install 3' x 6' 8" pre-hung 1-3/4" |
| | Demolish and Install | Ea. | 243 | 70.50 | | 313.50 | lauan door of solid wood. |
| | Reinstall | Ea. | | 27.34 | | 27.34 | |
| | Clean | Ea. | .10 | 14.80 | | 14.90 | |
| | Paint | Ea. | 10.60 | 55 | | 65.60 | |
| | Minimum Charge | Job | | 232 | | 232 | |
| **French Style** | | | | | | | |
| **1-Light** | | | | | | | |
| | Demolish | Ea. | | 43 | | 43 | Cost includes material and labor to |
| | Install | Ea. | 680 | 77.50 | | 757.50 | install 3' x 6' 8" pre-hung 1-3/4" |
| | Demolish and Install | Ea. | 680 | 120.50 | | 800.50 | French style door with 1 light. |
| | Reinstall | Ea. | | 77.47 | | 77.47 | |
| | Clean | Ea. | .92 | 22 | | 22.92 | |
| | Paint | Ea. | 11.20 | 96.50 | | 107.70 | |
| | Minimum Charge | Job | | 232 | | 232 | |
| **9-Light** | | | | | | | |
| | Demolish | Ea. | | 43 | | 43 | Cost includes material and labor to |
| | Install | Ea. | 775 | 33 | | 808 | install 3' x 6' 8" pre-hung 1-3/4" |
| | Demolish and Install | Ea. | 775 | 76 | | 851 | French style door with 9 lights. |
| | Reinstall | Ea. | | 33.20 | | 33.20 | |
| | Clean | Ea. | .20 | 29.50 | | 29.70 | |
| | Paint | Ea. | 11.20 | 96.50 | | 107.70 | |
| | Minimum Charge | Job | | 232 | | 232 | |
| **15-Light** | | | | | | | |
| | Demolish | Ea. | | 49.50 | | 49.50 | Cost includes material and labor to |
| | Install | Ea. | 810 | 66.50 | | 876.50 | install 3' x 6' 8" pre-hung 1-3/4" |
| | Demolish and Install | Ea. | 810 | 116 | | 926 | French style door with 15 lights. |
| | Reinstall | Ea. | | 66.40 | | 66.40 | |
| | Clean | Ea. | .20 | 29.50 | | 29.70 | |
| | Paint | Ea. | 11.20 | 96.50 | | 107.70 | |
| | Minimum Charge | Job | | 232 | | 232 | |
| **Jamb** | | | | | | | |
| **Wood** | | | | | | | |
| | Demolish | Ea. | | 32.50 | | 32.50 | Includes material and labor to install |
| | Install | Ea. | 116 | 12.55 | | 128.55 | flat pine jamb with square cut heads |
| | Demolish and Install | Ea. | 116 | 45.05 | | 161.05 | and rabbeted sides for 6' 8" high and |
| | Clean | Ea. | .04 | 8.90 | | 8.94 | 3-9/16" door including trim sets for |
| | Paint | Ea. | 7 | 26 | | 33 | both sides. |
| | Minimum Charge | Job | | 232 | | 232 | |
| **Metal 6' 8" High** | | | | | | | |
| | Demolish | Ea. | | 26 | | 26 | Cost includes material and labor to |
| | Install | Ea. | 141 | 29 | | 170 | install 18 gauge hollow metal door |
| | Demolish and Install | Ea. | 141 | 55 | | 196 | frame to fit a 4-1/2" jamb 6' 8" high |
| | Reinstall | Ea. | | 29.05 | | 29.05 | and up to 3' 6" opening. |
| | Clean | Ea. | .31 | 7.40 | | 7.71 | |
| | Paint | Ea. | 1.71 | 7.75 | | 9.46 | |
| | Minimum Charge | Job | | 232 | | 232 | |

# Exterior Doors and Windows

| Exterior Type Door | | Unit | Material | Labor | Equip. | Total | Specification |
|---|---|---|---|---|---|---|---|
| **Metal 7' High** | | | | | | | |
| | Demolish | Ea. | | 26 | | 26 | Cost includes material and labor to |
| | Install | Ea. | 146 | 31 | | 177 | install 18 gauge hollow metal door |
| | Demolish and Install | Ea. | 146 | 57 | | 203 | frame to fit a 4-1/2" jamb 7' high and |
| | Reinstall | Ea. | | 30.99 | | 30.99 | up to 3' 6" opening. |
| | Clean | Ea. | .31 | 7.40 | | 7.71 | |
| | Paint | Ea. | 1.71 | 7.75 | | 9.46 | |
| | Minimum Charge | Job | | 232 | | 232 | |
| **Metal 8' High** | | | | | | | |
| | Demolish | Ea. | | 26 | | 26 | Cost includes material and labor to |
| | Install | Ea. | 178 | 38.50 | | 216.50 | install 18 gauge hollow metal door |
| | Demolish and Install | Ea. | 178 | 64.50 | | 242.50 | frame to fit a 4-1/2" jamb 8' high and |
| | Reinstall | Ea. | | 38.73 | | 38.73 | up to 3' 6" opening. |
| | Clean | Ea. | .43 | 8.90 | | 9.33 | |
| | Paint | Ea. | 1.90 | 8.60 | | 10.50 | |
| | Minimum Charge | Job | | 232 | | 232 | |
| **Metal 9' High** | | | | | | | |
| | Demolish | Ea. | | 26 | | 26 | Cost includes material and labor to |
| | Install | Ea. | 224 | 46.50 | | 270.50 | install 18 gauge hollow metal door |
| | Demolish and Install | Ea. | 224 | 72.50 | | 296.50 | frame to fit a 4-1/2" jamb 9' high and |
| | Reinstall | Ea. | | 46.48 | | 46.48 | up to 3' 6" opening. |
| | Clean | Ea. | .43 | 8.90 | | 9.33 | |
| | Paint | Ea. | 1.90 | 8.60 | | 10.50 | |
| | Minimum Charge | Job | | 232 | | 232 | |
| **Casing Trim** | | | | | | | |
| **Single Width** | | | | | | | |
| | Demolish | Ea. | | 8.20 | | 8.20 | Cost includes material and labor to |
| | Install | Opng. | 26.50 | 38.50 | | 65 | install 11/16" x 2-1/2" pine ranch |
| | Demolish and Install | Opng. | 26.50 | 46.70 | | 73.20 | style casing for one side of a standard |
| | Reinstall | Opng. | | 30.99 | | 30.99 | door opening. |
| | Clean | Opng. | .21 | 3.55 | | 3.76 | |
| | Paint | Ea. | 5 | 7.75 | | 12.75 | |
| | Minimum Charge | Job | | 232 | | 232 | |
| **Double Width** | | | | | | | |
| | Demolish | Ea. | | 9.75 | | 9.75 | Cost includes material and labor to |
| | Install | Opng. | 31 | 46.50 | | 77.50 | install 11/16" x 2-1/2" pine ranch |
| | Demolish and Install | Opng. | 31 | 56.25 | | 87.25 | style door casing for one side of a |
| | Reinstall | Opng. | | 37.18 | | 37.18 | double door opening. |
| | Clean | Opng. | .24 | 7.10 | | 7.34 | |
| | Paint | Ea. | 5 | 8.80 | | 13.80 | |
| | Minimum Charge | Job | | 232 | | 232 | |

| Residential Metal Door | | Unit | Material | Labor | Equip. | Total | Specification |
|---|---|---|---|---|---|---|---|
| **Pre-hung, Steel-clad** | | | | | | | |
| **2' 6" x 6' 8"** | | | | | | | |
| | Demolish | Ea. | | 43 | | 43 | Cost includes material and labor to |
| | Install | Ea. | 445 | 54.50 | | 499.50 | install 2' 6" x 6' 8" pre-hung 1-3/4" |
| | Demolish and Install | Ea. | 445 | 97.50 | | 542.50 | 24 gauge steel clad door with 4-1/2" |
| | Reinstall | Ea. | | 54.68 | | 54.68 | wood frame, trims, hinges, aluminum |
| | Clean | Ea. | 1.47 | 14.80 | | 16.27 | sill, weatherstripped. |
| | Paint | Ea. | 11.20 | 55 | | 66.20 | |
| | Minimum Charge | Job | | 232 | | 232 | |

91

# Exterior Doors and Windows

## Residential Metal Door

| 3' x 6' 8" | | Unit | Material | Labor | Equip. | Total | Specification |
|---|---|---|---|---|---|---|---|
| | Demolish | Ea. | | 43 | | 43 | Cost includes material and labor to |
| | Install | Ea. | 320 | 62 | | 382 | install 3' x 6' 8" pre-hung 1-3/4" 24 |
| | Demolish and Install | Ea. | 320 | 105 | | 425 | gauge steel clad door with 4-1/2" |
| | Reinstall | Ea. | | 61.97 | | 61.97 | wood frame, trims, hinges, aluminum |
| | Clean | Ea. | 1.47 | 14.80 | | 16.27 | sill, weatherstripped. |
| | Paint | Ea. | 11.20 | 55 | | 66.20 | |
| | Minimum Charge | Job | | 232 | | 232 | |

## Metal Door

| | | Unit | Material | Labor | Equip. | Total | Specification |
|---|---|---|---|---|---|---|---|
| **Steel, Door Only** | | | | | | | |
| **2' 8" x 6' 8", 18 Gauge** | | | | | | | |
| | Demolish | Ea. | | 9.20 | | 9.20 | Cost includes material and labor to |
| | Install | Ea. | 550 | 29 | | 579 | install 18 gauge 2' 8" x 6' 8" steel |
| | Demolish and Install | Ea. | 550 | 38.20 | | 588.20 | door only, 1-3/4" thick, hinges. |
| | Reinstall | Ea. | | 29.05 | | 29.05 | |
| | Clean | Ea. | .62 | 11.10 | | 11.72 | |
| | Paint | Ea. | 6.15 | 32 | | 38.15 | |
| | Minimum Charge | Job | | 232 | | 232 | |
| **3' x 6' 8", 18 Gauge** | | | | | | | |
| | Demolish | Ea. | | 9.20 | | 9.20 | Cost includes material and labor to |
| | Install | Ea. | 550 | 31 | | 581 | install 18 gauge 3' x 6' 8" steel door |
| | Demolish and Install | Ea. | 550 | 40.20 | | 590.20 | only, 1-3/4" thick, hinges. |
| | Reinstall | Ea. | | 30.99 | | 30.99 | |
| | Clean | Ea. | .62 | 11.10 | | 11.72 | |
| | Paint | Ea. | 6.15 | 32 | | 38.15 | |
| | Minimum Charge | Job | | 232 | | 232 | |
| **2' 8" x 7', 18 Gauge** | | | | | | | |
| | Demolish | Ea. | | 9.20 | | 9.20 | Cost includes material and labor to |
| | Install | Ea. | 505 | 31 | | 536 | install 18 gauge 2' 8" x 7' steel door |
| | Demolish and Install | Ea. | 505 | 40.20 | | 545.20 | only, 1-3/4" thick, hinges. |
| | Reinstall | Ea. | | 30.99 | | 30.99 | |
| | Clean | Ea. | .62 | 11.10 | | 11.72 | |
| | Paint | Ea. | 6.15 | 32 | | 38.15 | |
| | Minimum Charge | Job | | 232 | | 232 | |
| **3' x 7', 18 Gauge** | | | | | | | |
| | Demolish | Ea. | | 9.20 | | 9.20 | Cost includes material and labor to |
| | Install | Ea. | 560 | 33 | | 593 | install 18 gauge 3' x 7' steel door |
| | Demolish and Install | Ea. | 560 | 42.20 | | 602.20 | only, 1-3/4" thick, hinges. |
| | Reinstall | Ea. | | 33.20 | | 33.20 | |
| | Clean | Ea. | .62 | 11.10 | | 11.72 | |
| | Paint | Ea. | 6.15 | 32 | | 38.15 | |
| | Minimum Charge | Job | | 232 | | 232 | |
| **3' 6" x 7', 18 Gauge** | | | | | | | |
| | Demolish | Ea. | | 9.20 | | 9.20 | Cost includes material and labor to |
| | Install | Ea. | 655 | 36 | | 691 | install 18 gauge 3' 6" x 7' steel door |
| | Demolish and Install | Ea. | 655 | 45.20 | | 700.20 | only, 1-3/4" thick, hinges. |
| | Reinstall | Ea. | | 35.75 | | 35.75 | |
| | Clean | Ea. | .62 | 11.10 | | 11.72 | |
| | Paint | Ea. | 6.15 | 32 | | 38.15 | |
| | Minimum Charge | Job | | 232 | | 232 | |

| Metal Door | Unit | Material | Labor | Equip. | Total | Specification |
|---|---|---|---|---|---|---|
| **Casing Trim** | | | | | | |
| **Single Width** | | | | | | |
| Demolish | Ea. | | 8.20 | | 8.20 | Cost includes material and labor to |
| Install | Opng. | 26.50 | 38.50 | | 65 | install 11/16" x 2-1/2" pine ranch |
| Demolish and Install | Opng. | 26.50 | 46.70 | | 73.20 | style casing for one side of a standard |
| Reinstall | Opng. | | 30.99 | | 30.99 | door opening. |
| Clean | Opng. | .21 | 3.55 | | 3.76 | |
| Paint | Ea. | 5 | 7.75 | | 12.75 | |
| Minimum Charge | Job | | 232 | | 232 | |
| **Double Width** | | | | | | |
| Demolish | Ea. | | 9.75 | | 9.75 | Cost includes material and labor to |
| Install | Opng. | 31 | 46.50 | | 77.50 | install 11/16" x 2-1/2" pine ranch |
| Demolish and Install | Opng. | 31 | 56.25 | | 87.25 | style door casing for one side of a |
| Reinstall | Opng. | | 37.18 | | 37.18 | double door opening. |
| Clean | Opng. | .24 | 7.10 | | 7.34 | |
| Paint | Ea. | 5 | 8.80 | | 13.80 | |
| Minimum Charge | Job | | 232 | | 232 | |

| Solid Core Door | Unit | Material | Labor | Equip. | Total | Specification |
|---|---|---|---|---|---|---|
| **Exterior Type** | | | | | | |
| **Flush (1-3/4")** | | | | | | |
| Demolish | Ea. | | 43 | | 43 | Cost includes material and labor to |
| Install | Ea. | 445 | 58 | | 503 | install 3' x 6' 8" pre-hung 1-3/4" |
| Demolish and Install | Ea. | 445 | 101 | | 546 | door, including casing and stop, |
| Reinstall | Ea. | | 58.10 | | 58.10 | hinges, jamb, aluminum sill, |
| Clean | Ea. | 1.47 | 14.80 | | 16.27 | weatherstripped. |
| Paint | Ea. | 11.20 | 55 | | 66.20 | |
| Minimum Charge | Job | | 232 | | 232 | |
| **Raised Panel** | | | | | | |
| Demolish | Ea. | | 43 | | 43 | Cost includes material and labor to |
| Install | Ea. | 505 | 26 | | 531 | install 3' x 6' 8" pre-hung 1-3/4" |
| Demolish and Install | Ea. | 505 | 69 | | 574 | door, jamb, including casing and |
| Reinstall | Ea. | | 25.82 | | 25.82 | stop, hinges, aluminum sill, |
| Clean | Ea. | 1.47 | 14.80 | | 16.27 | weatherstripping. |
| Paint | Ea. | 11.20 | 129 | | 140.20 | |
| Minimum Charge | Job | | 232 | | 232 | |
| **Entry w / Lights** | | | | | | |
| Demolish | Ea. | | 43 | | 43 | Cost includes material and labor to |
| Install | Ea. | 365 | 58 | | 423 | install 3' x 6' 8" pre-hung 1-3/4" |
| Demolish and Install | Ea. | 365 | 101 | | 466 | door, solid wood lower panel, with up |
| Reinstall | Ea. | | 58.10 | | 58.10 | to 9-light glass panels, casing, sill, |
| Clean | Ea. | 1.47 | 14.80 | | 16.27 | weatherstripped. |
| Paint | Ea. | 2.64 | 129 | | 131.64 | |
| Minimum Charge | Job | | 232 | | 232 | |
| **Custom Entry** | | | | | | |
| Demolish | Ea. | | 43 | | 43 | Cost includes material and labor to |
| Install | Ea. | 3350 | 93 | | 3443 | install prehung, fir, single glazed with |
| Demolish and Install | Ea. | 3350 | 136 | | 3486 | lead caming, 1-3/4" x 6' 8" x 3' |
| Reinstall | Ea. | | 92.96 | | 92.96 | |
| Clean | Ea. | 1.47 | 14.80 | | 16.27 | |
| Paint | Ea. | 6.15 | 32 | | 38.15 | |
| Minimum Charge | Job | | 232 | | 232 | |

| Solid Core Door | | Unit | Material | Labor | Equip. | Total | Specification |
|---|---|---|---|---|---|---|---|
| **Dutch Style** | | | | | | | |
| | Demolish | Ea. | | 43 | | 43 | Includes material and labor to install |
| | Install | Ea. | 765 | 77.50 | | 842.50 | dutch door, pine, 1-3/4" x 6' 8" x 2' |
| | Demolish and Install | Ea. | 765 | 120.50 | | 885.50 | 8" wide. |
| | Reinstall | Ea. | | 77.47 | | 77.47 | |
| | Clean | Ea. | 1.47 | 14.80 | | 16.27 | |
| | Paint | Ea. | 2.64 | 129 | | 131.64 | |
| | Minimum Charge | Job | | 232 | | 232 | |
| **Miami Style** | | | | | | | |
| | Demolish | Ea. | | 43 | | 43 | Cost includes material and labor to |
| | Install | Ea. | 243 | 27.50 | | 270.50 | install 3' x 6' 8" pre-hung 1-3/4" |
| | Demolish and Install | Ea. | 243 | 70.50 | | 313.50 | lauan door of solid wood. |
| | Reinstall | Ea. | | 27.34 | | 27.34 | |
| | Clean | Ea. | 1.47 | 14.80 | | 16.27 | |
| | Paint | Ea. | 10.60 | 55 | | 65.60 | |
| | Minimum Charge | Job | | 232 | | 232 | |
| **French Style** | | | | | | | |
| | Demolish | Ea. | | 49.50 | | 49.50 | Cost includes material and labor to |
| | Install | Ea. | 810 | 66.50 | | 876.50 | install 3' x 6' 8" pre-hung 1-3/4" |
| | Demolish and Install | Ea. | 810 | 116 | | 926 | French style door with 15 lights. |
| | Reinstall | Ea. | | 66.40 | | 66.40 | |
| | Clean | Ea. | .10 | 14.80 | | 14.90 | |
| | Paint | Ea. | 8.20 | 193 | | 201.20 | |
| | Minimum Charge | Job | | 232 | | 232 | |
| **Double** | | | | | | | |
| | Demolish | Ea. | | 33 | | 33 | Includes material and labor to install a |
| | Install | Ea. | 1525 | 116 | | 1641 | pre-hung double entry door including |
| | Demolish and Install | Ea. | 1525 | 149 | | 1674 | frame and exterior trim. |
| | Clean | Ea. | .06 | 53.50 | | 53.56 | |
| | Paint | Ea. | 22.50 | 193 | | 215.50 | |
| | Minimum Charge | Job | | 116 | | 116 | |
| **Commercial Grade** | | | | | | | |
| **3' x 6' 8"** | | | | | | | |
| | Demolish | Ea. | | 43 | | 43 | Includes material and labor to install |
| | Install | Ea. | 560 | 27.50 | | 587.50 | flush solid core 1-3/4" hardwood |
| | Demolish and Install | Ea. | 560 | 70.50 | | 630.50 | veneer 3' x 6' 8" door only. |
| | Reinstall | Ea. | | 27.34 | | 27.34 | |
| | Clean | Ea. | 1.47 | 14.80 | | 16.27 | |
| | Paint | Ea. | 10.30 | 32 | | 42.30 | |
| | Minimum Charge | Job | | 232 | | 232 | |
| **3' 6" x 6' 8"** | | | | | | | |
| | Demolish | Ea. | | 43 | | 43 | Includes material and labor to install |
| | Install | Ea. | 675 | 27.50 | | 702.50 | flush solid core 1-3/4" hardwood |
| | Demolish and Install | Ea. | 675 | 70.50 | | 745.50 | veneer 3' 6" x 6' 8" door only. |
| | Reinstall | Ea. | | 27.34 | | 27.34 | |
| | Clean | Ea. | 1.47 | 14.80 | | 16.27 | |
| | Paint | Ea. | 10.30 | 32 | | 42.30 | |
| | Minimum Charge | Job | | 232 | | 232 | |
| **3' x 7'** | | | | | | | |
| | Demolish | Ea. | | 43 | | 43 | Includes material and labor to install |
| | Install | Ea. | 320 | 66.50 | | 386.50 | flush solid core 1-3/4" hardwood |
| | Demolish and Install | Ea. | 320 | 109.50 | | 429.50 | veneer 3' x 7' door only. |
| | Reinstall | Ea. | | 66.40 | | 66.40 | |
| | Clean | Ea. | 1.47 | 14.80 | | 16.27 | |
| | Paint | Ea. | 10.30 | 32 | | 42.30 | |
| | Minimum Charge | Job | | 232 | | 232 | |

| Solid Core Door | | Unit | Material | Labor | Equip. | Total | Specification |
|---|---|---|---|---|---|---|---|
| **3' 6" x 7'** | | | | | | | Includes material and labor to install |
| | Demolish | Ea. | | 43 | | 43 | flush solid core 1-3/4" hardwood |
| | Install | Ea. | 780 | 33 | | 813 | veneer 3' 6" x 7' door only. |
| | Demolish and Install | Ea. | 780 | 76 | | 856 | |
| | Reinstall | Ea. | | 33.20 | | 33.20 | |
| | Clean | Ea. | 1.47 | 14.80 | | 16.27 | |
| | Paint | Ea. | 10.30 | 32 | | 42.30 | |
| | Minimum Charge | Job | | 232 | | 232 | |
| **3' x 8'** | | | | | | | Includes material and labor to install |
| | Demolish | Ea. | | 43 | | 43 | flush solid core 1-3/4" hardwood |
| | Install | Ea. | 715 | 58 | | 773 | veneer 3' x 8' door only. |
| | Demolish and Install | Ea. | 715 | 101 | | 816 | |
| | Reinstall | Ea. | | 58.10 | | 58.10 | |
| | Clean | Ea. | 1.77 | 22 | | 23.77 | |
| | Paint | Ea. | 10.30 | 32 | | 42.30 | |
| | Minimum Charge | Job | | 232 | | 232 | |
| **3' 6" x 8'** | | | | | | | Includes material and labor to install |
| | Demolish | Ea. | | 43 | | 43 | flush solid core 1-3/4" hardwood |
| | Install | Ea. | 855 | 58 | | 913 | veneer 3' 6" x 8' door only. |
| | Demolish and Install | Ea. | 855 | 101 | | 956 | |
| | Reinstall | Ea. | | 58.10 | | 58.10 | |
| | Clean | Ea. | 1.77 | 22 | | 23.77 | |
| | Paint | Ea. | 10.30 | 32 | | 42.30 | |
| | Minimum Charge | Job | | 232 | | 232 | |
| **3' x 9'** | | | | | | | Includes material and labor to install |
| | Demolish | Ea. | | 43 | | 43 | flush solid core 1-3/4" hardwood |
| | Install | Ea. | 605 | 66.50 | | 671.50 | veneer 3' x 9' door only. |
| | Demolish and Install | Ea. | 605 | 109.50 | | 714.50 | |
| | Reinstall | Ea. | | 66.40 | | 66.40 | |
| | Clean | Ea. | 1.77 | 22 | | 23.77 | |
| | Paint | Ea. | 10.30 | 32 | | 42.30 | |
| | Minimum Charge | Job | | 232 | | 232 | |
| **Stain** | | | | | | | Includes labor and material to stain |
| | Install | Ea. | 12.40 | 43 | | 55.40 | exterior type door and trim on both |
| | Minimum Charge | Job | | 193 | | 193 | sides. |
| **Door Only (SC)** | | | | | | | |
| **2' 4" x 6' 8"** | | | | | | | Cost includes material and labor to |
| | Demolish | Ea. | | 9.20 | | 9.20 | install 2' 4" x 6' 8" lauan solid core |
| | Install | Ea. | 230 | 26 | | 256 | door slab. |
| | Demolish and Install | Ea. | 230 | 35.20 | | 265.20 | |
| | Reinstall | Ea. | | 25.82 | | 25.82 | |
| | Clean | Ea. | .62 | 11.10 | | 11.72 | |
| | Paint | Ea. | 4.92 | 48.50 | | 53.42 | |
| | Minimum Charge | Job | | 232 | | 232 | |
| **2' 8" x 6' 8"** | | | | | | | Cost includes material and labor to |
| | Demolish | Ea. | | 9.20 | | 9.20 | install 2' 8" x 6' 8" lauan solid core |
| | Install | Ea. | 237 | 26 | | 263 | door slab. |
| | Demolish and Install | Ea. | 237 | 35.20 | | 272.20 | |
| | Reinstall | Ea. | | 25.82 | | 25.82 | |
| | Clean | Ea. | .62 | 11.10 | | 11.72 | |
| | Paint | Ea. | 4.92 | 48.50 | | 53.42 | |
| | Minimum Charge | Job | | 232 | | 232 | |

| Solid Core Door | | Unit | Material | Labor | Equip. | Total | Specification |
|---|---|---|---|---|---|---|---|
| **3' x 6' 8"** | | | | | | | |
| | Demolish | Ea. | | 9.20 | | 9.20 | Cost includes material and labor to |
| | Install | Ea. | 243 | 27.50 | | 270.50 | install 3' x 6' 8" lauan solid core door |
| | Demolish and Install | Ea. | 243 | 36.70 | | 279.70 | slab. |
| | Reinstall | Ea. | | 27.34 | | 27.34 | |
| | Clean | Ea. | .62 | 11.10 | | 11.72 | |
| | Paint | Ea. | 4.92 | 48.50 | | 53.42 | |
| | Minimum Charge | Job | | 232 | | 232 | |
| **3' 6" x 6' 8"** | | | | | | | |
| | Demolish | Ea. | | 9.20 | | 9.20 | Cost includes material and labor to |
| | Install | Ea. | 277 | 29 | | 306 | install 3' 6" x 6' 8" lauan solid core |
| | Demolish and Install | Ea. | 277 | 38.20 | | 315.20 | door slab. |
| | Reinstall | Ea. | | 29.05 | | 29.05 | |
| | Clean | Ea. | .62 | 11.10 | | 11.72 | |
| | Paint | Ea. | 4.92 | 48.50 | | 53.42 | |
| | Minimum Charge | Job | | 232 | | 232 | |
| **Panel** | | | | | | | |
| | Demolish | Ea. | | 9.20 | | 9.20 | Cost includes material and labor to |
| | Install | Ea. | 298 | 54.50 | | 352.50 | install 3' x 6' 8" six (6) panel interior |
| | Demolish and Install | Ea. | 298 | 63.70 | | 361.70 | door. |
| | Reinstall | Ea. | | 54.68 | | 54.68 | |
| | Clean | Ea. | .74 | 11.10 | | 11.84 | |
| | Paint | Ea. | 4.92 | 48.50 | | 53.42 | |
| | Minimum Charge | Job | | 232 | | 232 | |
| **Casing Trim** | | | | | | | |
| **Single Width** | | | | | | | |
| | Demolish | Ea. | | 8.20 | | 8.20 | Cost includes material and labor to |
| | Install | Opng. | 26.50 | 38.50 | | 65 | install 11/16" x 2-1/2" pine ranch |
| | Demolish and Install | Opng. | 26.50 | 46.70 | | 73.20 | style casing for one side of a standard |
| | Reinstall | Opng. | | 30.99 | | 30.99 | door opening. |
| | Clean | Opng. | .21 | 3.55 | | 3.76 | |
| | Paint | Ea. | 5 | 7.75 | | 12.75 | |
| | Minimum Charge | Job | | 232 | | 232 | |
| **Double Width** | | | | | | | |
| | Demolish | Ea. | | 9.75 | | 9.75 | Cost includes material and labor to |
| | Install | Opng. | 31 | 46.50 | | 77.50 | install 11/16" x 2-1/2" pine ranch |
| | Demolish and Install | Opng. | 31 | 56.25 | | 87.25 | style door casing for one side of a |
| | Reinstall | Opng. | | 37.18 | | 37.18 | double door opening. |
| | Clean | Opng. | .24 | 7.10 | | 7.34 | |
| | Paint | Ea. | 5 | 8.80 | | 13.80 | |
| | Minimum Charge | Job | | 232 | | 232 | |

| Components | | Unit | Material | Labor | Equip. | Total | Specification |
|---|---|---|---|---|---|---|---|
| **Hardware** | | | | | | | |
| **Doorknob** | | | | | | | |
| | Demolish | Ea. | | 18.45 | | 18.45 | Includes material and labor to install |
| | Install | Ea. | 42.50 | 29 | | 71.50 | residential passage lockset, keyless |
| | Demolish and Install | Ea. | 42.50 | 47.45 | | 89.95 | bored type. |
| | Reinstall | Ea. | | 23.24 | | 23.24 | |
| | Clean | Ea. | .14 | 11.10 | | 11.24 | |
| | Minimum Charge | Job | | 232 | | 232 | |

# Exterior Doors and Windows

| Components | | Unit | Material | Labor | Equip. | Total | Specification |
|---|---|---|---|---|---|---|---|
| **Doorknob w / Lock** | | | | | | | |
| | Demolish | Ea. | | 18.45 | | 18.45 | Includes material and labor to install |
| | Install | Ea. | 40.50 | 29 | | 69.50 | privacy lockset. |
| | Demolish and Install | Ea. | 40.50 | 47.45 | | 87.95 | |
| | Reinstall | Ea. | | 23.24 | | 23.24 | |
| | Clean | Ea. | .14 | 11.10 | | 11.24 | |
| | Minimum Charge | Job | | 232 | | 232 | |
| **Lever Handle** | | | | | | | |
| | Demolish | Ea. | | 18.45 | | 18.45 | Includes material and labor to install |
| | Install | Ea. | 51 | 46.50 | | 97.50 | residential passage lockset, keyless |
| | Demolish and Install | Ea. | 51 | 64.95 | | 115.95 | bored type with lever handle. |
| | Reinstall | Ea. | | 37.18 | | 37.18 | |
| | Clean | Ea. | .14 | 11.10 | | 11.24 | |
| | Minimum Charge | Job | | 232 | | 232 | |
| **Deadbolt** | | | | | | | |
| | Demolish | Ea. | | 16.15 | | 16.15 | Includes material and labor to install a |
| | Install | Ea. | 69.50 | 33 | | 102.50 | deadbolt. |
| | Demolish and Install | Ea. | 69.50 | 49.15 | | 118.65 | |
| | Reinstall | Ea. | | 26.56 | | 26.56 | |
| | Clean | Ea. | .14 | 11.10 | | 11.24 | |
| | Minimum Charge | Job | | 232 | | 232 | |
| **Grip Handle Entry** | | | | | | | |
| | Demolish | Ea. | | 14.75 | | 14.75 | Includes material and labor to install |
| | Install | Ea. | 105 | 51.50 | | 156.50 | residential keyed entry lock set with |
| | Demolish and Install | Ea. | 105 | 66.25 | | 171.25 | grip type handle. |
| | Reinstall | Ea. | | 41.32 | | 41.32 | |
| | Clean | Ea. | .14 | 11.10 | | 11.24 | |
| | Minimum Charge | Job | | 232 | | 232 | |
| **Panic Device** | | | | | | | |
| | Demolish | Ea. | | 39.50 | | 39.50 | Includes material and labor to install |
| | Install | Ea. | 305 | 77.50 | | 382.50 | touch bar, low profile, exit only. |
| | Demolish and Install | Ea. | 305 | 117 | | 422 | |
| | Reinstall | Ea. | | 61.97 | | 61.97 | |
| | Clean | Ea. | .08 | 19.75 | | 19.83 | |
| | Minimum Charge | Job | | 232 | | 232 | |
| **Closer** | | | | | | | |
| | Demolish | Ea. | | 9.70 | | 9.70 | Includes material and labor to install |
| | Install | Ea. | 220 | 77.50 | | 297.50 | pneumatic heavy duty closer for |
| | Demolish and Install | Ea. | 220 | 87.20 | | 307.20 | exterior type doors. |
| | Reinstall | Ea. | | 61.97 | | 61.97 | |
| | Clean | Ea. | .08 | 14.80 | | 14.88 | |
| | Minimum Charge | Job | | 232 | | 232 | |
| **Peep-hole** | | | | | | | |
| | Demolish | Ea. | | 8.75 | | 8.75 | Includes material and labor to install |
| | Install | Ea. | 9.85 | 14.55 | | 24.40 | peep-hole for door. |
| | Demolish and Install | Ea. | 9.85 | 23.30 | | 33.15 | |
| | Reinstall | Ea. | | 11.62 | | 11.62 | |
| | Clean | Ea. | .01 | 4.10 | | 4.11 | |
| | Minimum Charge | Job | | 232 | | 232 | |
| **Exit Sign** | | | | | | | |
| | Demolish | Ea. | | 26.50 | | 26.50 | Includes material and labor to install a |
| | Install | Ea. | 104 | 65.50 | | 169.50 | wall mounted interior electric exit sign. |
| | Demolish and Install | Ea. | 104 | 92 | | 196 | |
| | Reinstall | Ea. | | 52.52 | | 52.52 | |
| | Clean | Ea. | .01 | 8.35 | | 8.36 | |
| | Minimum Charge | Job | | 232 | | 232 | |

| Components | Unit | Material | Labor | Equip. | Total | Specification |
|---|---|---|---|---|---|---|
| **Kickplate** | | | | | | |
| Demolish | Ea. | | 14.35 | | 14.35 | Includes material and labor to install |
| Install | Ea. | 26 | 31 | | 57 | aluminum kickplate, 10" x 28". |
| Demolish and Install | Ea. | 26 | 45.35 | | 71.35 | |
| Reinstall | Ea. | | 24.79 | | 24.79 | |
| Clean | Ea. | .14 | 11.10 | | 11.24 | |
| Minimum Charge | Job | | 232 | | 232 | |
| **Threshold** | | | | | | |
| Install | L.F. | 3.93 | 4.65 | | 8.58 | Includes labor and material to install |
| Minimum Charge | Job | | 116 | | 116 | oak threshold. |
| **Weatherstripping** | | | | | | |
| Install | L.F. | .65 | 1.16 | | 1.81 | Includes labor and material to install |
| Minimum Charge | Job | | 116 | | 116 | rubber weatherstripping for door. |
| **Metal Jamb** | | | | | | |
| **6' 8" High** | | | | | | |
| Demolish | Ea. | | 26 | | 26 | Cost includes material and labor to |
| Install | Ea. | 141 | 29 | | 170 | install 18 gauge hollow metal door |
| Demolish and Install | Ea. | 141 | 55 | | 196 | frame to fit a 4-1/2" jamb 6' 8" high |
| Reinstall | Ea. | | 29.05 | | 29.05 | and up to 3' 6" opening. |
| Clean | Ea. | .31 | 7.40 | | 7.71 | |
| Paint | Ea. | 1.71 | 7.75 | | 9.46 | |
| Minimum Charge | Job | | 232 | | 232 | |
| **7' High** | | | | | | |
| Demolish | Ea. | | 26 | | 26 | Cost includes material and labor to |
| Install | Ea. | 146 | 31 | | 177 | install 18 gauge hollow metal door |
| Demolish and Install | Ea. | 146 | 57 | | 203 | frame to fit a 4-1/2" jamb 7' high and |
| Reinstall | Ea. | | 30.99 | | 30.99 | up to 3' 6" opening. |
| Clean | Ea. | .31 | 7.40 | | 7.71 | |
| Paint | Ea. | 1.71 | 7.75 | | 9.46 | |
| Minimum Charge | Job | | 232 | | 232 | |
| **8' High** | | | | | | |
| Demolish | Ea. | | 26 | | 26 | Cost includes material and labor to |
| Install | Ea. | 178 | 38.50 | | 216.50 | install 18 gauge hollow metal door |
| Demolish and Install | Ea. | 178 | 64.50 | | 242.50 | frame to fit a 4-1/2" jamb 8' high and |
| Reinstall | Ea. | | 38.73 | | 38.73 | up to 3' 6" opening. |
| Clean | Ea. | .43 | 8.90 | | 9.33 | |
| Paint | Ea. | 1.90 | 8.60 | | 10.50 | |
| Minimum Charge | Job | | 232 | | 232 | |
| **9' High** | | | | | | |
| Demolish | Ea. | | 26 | | 26 | Cost includes material and labor to |
| Install | Ea. | 224 | 46.50 | | 270.50 | install 18 gauge hollow metal door |
| Demolish and Install | Ea. | 224 | 72.50 | | 296.50 | frame to fit a 4-1/2" jamb 9' high and |
| Reinstall | Ea. | | 46.48 | | 46.48 | up to 3' 6" opening. |
| Clean | Ea. | .43 | 8.90 | | 9.33 | |
| Paint | Ea. | 1.90 | 8.60 | | 10.50 | |
| Minimum Charge | Job | | 232 | | 232 | |
| **Wood Jamb** | | | | | | |
| Demolish | Ea. | | 32.50 | | 32.50 | Includes material and labor to install |
| Install | Ea. | 116 | 12.55 | | 128.55 | flat pine jamb with square cut heads |
| Demolish and Install | Ea. | 116 | 45.05 | | 161.05 | and rabbeted sides for 6' 8" high and |
| Clean | Ea. | .04 | 8.90 | | 8.94 | 3-9/16" door including trim sets for |
| Paint | Ea. | 7 | 26 | | 33 | both sides. |
| Minimum Charge | Job | | 232 | | 232 | |
| **Stain** | | | | | | |
| Install | Ea. | 8.55 | 32 | | 40.55 | Includes labor and material to stain |
| Minimum Charge | Job | | 193 | | 193 | single door and trim on both sides. |

98

# Exterior Doors and Windows

| Components | | Unit | Material | Labor | Equip. | Total | Specification |
|---|---|---|---|---|---|---|---|
| **Shave & Refit** | | | | | | | |
| | Install | Ea. | | 38.50 | | 38.50 | Includes labor to shave and rework |
| | Minimum Charge | Job | | 116 | | 116 | door to fit opening at the job site. |

| Sliding Patio Door | | Unit | Material | Labor | Equip. | Total | Specification |
|---|---|---|---|---|---|---|---|
| **Aluminum** | | | | | | | |
| **6' Single Pane** | | | | | | | |
| | Demolish | Ea. | | 64.50 | | 64.50 | Cost includes material and labor to |
| | Install | Ea. | 500 | 232 | | 732 | install 6' x 6' 8" sliding anodized |
| | Demolish and Install | Ea. | 500 | 296.50 | | 796.50 | aluminum patio door with single pane |
| | Reinstall | Ea. | | 185.92 | | 185.92 | tempered safety glass, anodized |
| | Clean | Ea. | .92 | 22 | | 22.92 | frame, screen, weatherstripped. |
| | Minimum Charge | Job | | 232 | | 232 | |
| **8' Single Pane** | | | | | | | |
| | Demolish | Ea. | | 64.50 | | 64.50 | Cost includes material and labor to |
| | Install | Ea. | 635 | 310 | | 945 | install 8' x 6' 8" sliding aluminum |
| | Demolish and Install | Ea. | 635 | 374.50 | | 1009.50 | patio door with single pane tempered |
| | Reinstall | Ea. | | 247.89 | | 247.89 | safety glass, anodized frame, screen, |
| | Clean | Ea. | .92 | 22 | | 22.92 | weatherstripped. |
| | Minimum Charge | Job | | 232 | | 232 | |
| **10' Single Pane** | | | | | | | |
| | Demolish | Ea. | | 64.50 | | 64.50 | Cost includes material and labor to |
| | Install | Ea. | 725 | 465 | | 1190 | install 10' x 6' 8" sliding aluminum |
| | Demolish and Install | Ea. | 725 | 529.50 | | 1254.50 | patio door with single pane tempered |
| | Reinstall | Ea. | | 371.84 | | 371.84 | safety glass, anodized frame, screen, |
| | Clean | Ea. | .92 | 22 | | 22.92 | weatherstripped. |
| | Minimum Charge | Job | | 232 | | 232 | |
| **6' Insulated** | | | | | | | |
| | Demolish | Ea. | | 64.50 | | 64.50 | Cost includes material and labor to |
| | Install | Ea. | 1050 | 232 | | 1282 | install 6' x 6' 8" sliding aluminum |
| | Demolish and Install | Ea. | 1050 | 296.50 | | 1346.50 | patio door with double insulated |
| | Reinstall | Ea. | | 185.92 | | 185.92 | tempered glass, baked-on enamel |
| | Clean | Ea. | .92 | 22 | | 22.92 | finish, screen, weatherstripped. |
| | Minimum Charge | Job | | 232 | | 232 | |
| **8' Insulated** | | | | | | | |
| | Demolish | Ea. | | 64.50 | | 64.50 | Cost includes material and labor to |
| | Install | Ea. | 1250 | 310 | | 1560 | install 8' x 6' 8" sliding aluminum |
| | Demolish and Install | Ea. | 1250 | 374.50 | | 1624.50 | patio door with insulated tempered |
| | Reinstall | Ea. | | 247.89 | | 247.89 | glass, baked-on enamel finish, screen, |
| | Clean | Ea. | .92 | 22 | | 22.92 | weatherstripped. |
| | Minimum Charge | Job | | 232 | | 232 | |
| **10' Insulated** | | | | | | | |
| | Demolish | Ea. | | 64.50 | | 64.50 | Cost includes material and labor to |
| | Install | Ea. | 1550 | 465 | | 2015 | install 10' x 6' 8" sliding aluminum |
| | Demolish and Install | Ea. | 1550 | 529.50 | | 2079.50 | patio door with double insulated |
| | Reinstall | Ea. | | 371.84 | | 371.84 | tempered glass, baked-on enamel |
| | Clean | Ea. | .92 | 22 | | 22.92 | finish, screen, weatherstripped. |
| | Minimum Charge | Job | | 232 | | 232 | |

For customer support on your Contractor's Pricing Guide: Residential Repair & Remodeling, call 800.448.8182.

| Sliding Patio Door | | Unit | Material | Labor | Equip. | Total | Specification |
|---|---|---|---|---|---|---|---|
| **Vinyl** | | | | | | | |
| **6' Single Pane** | | | | | | | |
| | Demolish | Ea. | | 64.50 | | 64.50 | Cost includes material and labor to |
| | Install | Ea. | 475 | 232 | | 707 | install 6' x 6' 8" sliding vinyl patio |
| | Demolish and Install | Ea. | 475 | 296.50 | | 771.50 | door with single pane tempered safety |
| | Reinstall | Ea. | | 185.92 | | 185.92 | glass, anodized frame, screen, |
| | Clean | Ea. | .92 | 22 | | 22.92 | weatherstripped. |
| | Minimum Charge | Job | | 232 | | 232 | |
| **8' Single Pane** | | | | | | | |
| | Demolish | Ea. | | 64.50 | | 64.50 | Cost includes material and labor to |
| | Install | Ea. | 645 | 310 | | 955 | install 8' x 6' 8" sliding vinyl patio |
| | Demolish and Install | Ea. | 645 | 374.50 | | 1019.50 | door with single pane tempered safety |
| | Reinstall | Ea. | | 247.89 | | 247.89 | glass, anodized frame, screen, |
| | Clean | Ea. | .92 | 22 | | 22.92 | weatherstripped. |
| | Minimum Charge | Job | | 232 | | 232 | |
| **10' Single Pane** | | | | | | | |
| | Demolish | Ea. | | 64.50 | | 64.50 | Cost includes material and labor to |
| | Install | Ea. | 700 | 465 | | 1165 | install 10' x 6' 8" sliding vinyl patio |
| | Demolish and Install | Ea. | 700 | 529.50 | | 1229.50 | door with single pane tempered safety |
| | Reinstall | Ea. | | 371.84 | | 371.84 | glass, anodized frame, screen, |
| | Clean | Ea. | .92 | 22 | | 22.92 | weatherstripped. |
| | Minimum Charge | Job | | 232 | | 232 | |
| **6' Insulated** | | | | | | | |
| | Demolish | Ea. | | 64.50 | | 64.50 | Cost includes material and labor to |
| | Install | Ea. | 1300 | 232 | | 1532 | install 6' x 6' 8" sliding vinyl patio |
| | Demolish and Install | Ea. | 1300 | 296.50 | | 1596.50 | door with double insulated tempered |
| | Reinstall | Ea. | | 185.92 | | 185.92 | glass, anodized frame, screen, |
| | Clean | Ea. | .92 | 22 | | 22.92 | weatherstripped. |
| | Minimum Charge | Job | | 232 | | 232 | |
| **8' Insulated** | | | | | | | |
| | Demolish | Ea. | | 64.50 | | 64.50 | Cost includes material and labor to |
| | Install | Ea. | 1075 | 310 | | 1385 | install 8' x 6' 8" sliding vinyl patio |
| | Demolish and Install | Ea. | 1075 | 374.50 | | 1449.50 | door with double insulated tempered |
| | Reinstall | Ea. | | 247.89 | | 247.89 | glass, anodized frame, screen, |
| | Clean | Ea. | .92 | 22 | | 22.92 | weatherstripped. |
| | Minimum Charge | Job | | 232 | | 232 | |
| **10' Insulated** | | | | | | | |
| | Demolish | Ea. | | 64.50 | | 64.50 | Cost includes material and labor to |
| | Install | Ea. | 1850 | 232 | | 2082 | install 10' x 6' 8" sliding vinyl patio |
| | Demolish and Install | Ea. | 1850 | 296.50 | | 2146.50 | door with double insulated tempered |
| | Reinstall | Ea. | | 185.92 | | 185.92 | glass, anodized frame, screen, |
| | Clean | Ea. | .92 | 22 | | 22.92 | weatherstripped. |
| | Minimum Charge | Job | | 232 | | 232 | |
| **Wood Framed** | | | | | | | |
| **6' Premium Insulated** | | | | | | | |
| | Demolish | Ea. | | 64.50 | | 64.50 | Cost includes material and labor to |
| | Install | Ea. | 1600 | 232 | | 1832 | install 6' x 6' 8" sliding wood patio |
| | Demolish and Install | Ea. | 1600 | 296.50 | | 1896.50 | door with pine frame, double insulated |
| | Reinstall | Ea. | | 185.92 | | 185.92 | glass, screen, lock and dead bolt, |
| | Clean | Ea. | .92 | 22 | | 22.92 | weatherstripped. Add for casing and |
| | Minimum Charge | Job | | 232 | | 232 | finishing. |

# Exterior Doors and Windows

| Sliding Patio Door | | Unit | Material | Labor | Equip. | Total | Specification |
|---|---|---|---|---|---|---|---|
| **8' Premium Insulated** | | | | | | | |
| | Demolish | Ea. | | 64.50 | | 64.50 | Cost includes material and labor to |
| | Install | Ea. | 2025 | 310 | | 2335 | install 8' x 6' 8" sliding wood patio |
| | Demolish and Install | Ea. | 2025 | 374.50 | | 2399.50 | door with pine frame, double insulated |
| | Reinstall | Ea. | | 247.89 | | 247.89 | glass, screen, lock and dead bolt, |
| | Clean | Ea. | .92 | 22 | | 22.92 | weatherstripped. Add for casing and |
| | Minimum Charge | Job | | 232 | | 232 | finishing. |
| **9' Premium Insulated** | | | | | | | |
| | Demolish | Ea. | | 64.50 | | 64.50 | Cost includes material and labor to |
| | Install | Ea. | 2025 | 310 | | 2335 | install 9' x 6' 8" sliding wood patio |
| | Demolish and Install | Ea. | 2025 | 374.50 | | 2399.50 | door with pine frame, double insulated |
| | Reinstall | Ea. | | 247.89 | | 247.89 | glass, screen, lock and dead bolt, |
| | Clean | Ea. | .92 | 22 | | 22.92 | weatherstripped. Add for casing and |
| | Minimum Charge | Job | | 232 | | 232 | finishing. |
| **Stain** | | | | | | | |
| | Install | Ea. | 12.40 | 43 | | 55.40 | Includes labor and material to stain |
| | Minimum Charge | Job | | 193 | | 193 | exterior type door and trim on both |
| | | | | | | | sides. |
| **Casing Trim** | | | | | | | |
| | Demolish | Ea. | | 9.75 | | 9.75 | Cost includes material and labor to |
| | Install | Opng. | 31 | 46.50 | | 77.50 | install 11/16" x 2-1/2" pine ranch |
| | Demolish and Install | Opng. | 31 | 56.25 | | 87.25 | style door casing for one side of a |
| | Reinstall | Opng. | | 37.18 | | 37.18 | double door opening. |
| | Clean | Opng. | .24 | 7.10 | | 7.34 | |
| | Paint | Ea. | 5 | 8.80 | | 13.80 | |
| | Minimum Charge | Job | | 232 | | 232 | |
| **Sliding Screen** | | | | | | | |
| | Demolish | Ea. | | 9.20 | | 9.20 | Includes material and labor to install |
| | Install | Ea. | 143 | 23 | | 166 | sliding screen door, 6' 8" x 3'. |
| | Demolish and Install | Ea. | 143 | 32.20 | | 175.20 | |
| | Reinstall | Ea. | | 18.59 | | 18.59 | |
| | Clean | Ea. | .31 | 7.40 | | 7.71 | |
| | Minimum Charge | Job | | 232 | | 232 | |

| Screen Door | | Unit | Material | Labor | Equip. | Total | Specification |
|---|---|---|---|---|---|---|---|
| **Aluminum** | | | | | | | |
| | Demolish | Ea. | | 43 | | 43 | Includes material and labor to install |
| | Install | Ea. | 219 | 66.50 | | 285.50 | aluminum frame screen door with |
| | Demolish and Install | Ea. | 219 | 109.50 | | 328.50 | fiberglass screen cloth, closer, hinges, |
| | Reinstall | Ea. | | 53.12 | | 53.12 | latch. |
| | Clean | Ea. | 1.47 | 14.80 | | 16.27 | |
| | Paint | Ea. | 4.81 | 15.45 | | 20.26 | |
| | Minimum Charge | Job | | 232 | | 232 | |
| **Wood** | | | | | | | |
| | Demolish | Ea. | | 43 | | 43 | Includes material and labor to install |
| | Install | Ea. | 320 | 77.50 | | 397.50 | wood screen door with aluminum cloth |
| | Demolish and Install | Ea. | 320 | 120.50 | | 440.50 | screen. Frame and hardware not |
| | Reinstall | Ea. | | 61.97 | | 61.97 | included. |
| | Clean | Ea. | 1.47 | 14.80 | | 16.27 | |
| | Paint | Ea. | 4.81 | 15.45 | | 20.26 | |
| | Minimum Charge | Job | | 232 | | 232 | |
| **Shave & Refit** | | | | | | | |
| | Install | Ea. | | 38.50 | | 38.50 | Includes labor to shave and rework |
| | Minimum Charge | Job | | 232 | | 232 | door to fit opening at the job site. |

# Exterior Doors and Windows

| Storm Door | | Unit | Material | Labor | Equip. | Total | Specification |
|---|---|---|---|---|---|---|---|
| **Good Grade** | | | | | | | |
| | Demolish | Ea. | | 43 | | 43 | Includes material and labor to install |
| | Install | Ea. | 310 | 66.50 | | 376.50 | aluminum frame storm door with |
| | Demolish and Install | Ea. | 310 | 109.50 | | 419.50 | fiberglass screen cloth, upper glass |
| | Reinstall | Ea. | | 53.12 | | 53.12 | panel, closer, hinges, latch. |
| | Clean | Ea. | 1.47 | 14.80 | | 16.27 | |
| | Paint | Ea. | 4.81 | 15.45 | | 20.26 | |
| | Minimum Charge | Job | | 232 | | 232 | |
| **Better Grade** | | | | | | | |
| | Demolish | Ea. | | 43 | | 43 | Includes material and labor to install |
| | Install | Ea. | 320 | 66.50 | | 386.50 | aluminum frame storm door with |
| | Demolish and Install | Ea. | 320 | 109.50 | | 429.50 | fiberglass screen cloth, upper and |
| | Reinstall | Ea. | | 53.12 | | 53.12 | lower glass panel, closer, latch. |
| | Clean | Ea. | 1.47 | 14.80 | | 16.27 | |
| | Paint | Ea. | 4.81 | 15.45 | | 20.26 | |
| | Minimum Charge | Job | | 232 | | 232 | |
| **Premium Grade** | | | | | | | |
| | Demolish | Ea. | | 43 | | 43 | Includes material and labor to install |
| | Install | Ea. | 320 | 66.50 | | 386.50 | aluminum frame storm door with |
| | Demolish and Install | Ea. | 320 | 109.50 | | 429.50 | fiberglass screen cloth, upper and |
| | Reinstall | Ea. | | 53.12 | | 53.12 | lower glass panel, closer, latch. |
| | Clean | Ea. | 1.47 | 14.80 | | 16.27 | |
| | Paint | Ea. | 4.81 | 15.45 | | 20.26 | |
| | Minimum Charge | Job | | 232 | | 232 | |
| **Insulated** | | | | | | | |
| | Demolish | Ea. | | 43 | | 43 | Includes material and labor to install |
| | Install | Ea. | 310 | 33 | | 343 | aluminum frame storm door with |
| | Demolish and Install | Ea. | 310 | 76 | | 386 | fiberglass screen cloth, upper and |
| | Reinstall | Ea. | | 26.56 | | 26.56 | lower glass panel, closer, latch. |
| | Clean | Ea. | 1.47 | 14.80 | | 16.27 | |
| | Paint | Ea. | 4.81 | 15.45 | | 20.26 | |
| | Minimum Charge | Job | | 232 | | 232 | |

| Garage Door | | Unit | Material | Labor | Equip. | Total | Specification |
|---|---|---|---|---|---|---|---|
| **9' One Piece Metal** | | | | | | | |
| | Demolish | Ea. | | 109 | | 109 | Includes material and labor to install |
| | Install | Ea. | 695 | 116 | | 811 | one piece steel garage door 9' x 7' |
| | Demolish and Install | Ea. | 695 | 225 | | 920 | primed including hardware. |
| | Reinstall | Ea. | | 92.96 | | 92.96 | |
| | Clean | Ea. | 1.84 | 44.50 | | 46.34 | |
| | Paint | Ea. | 10.45 | 96.50 | | 106.95 | |
| | Minimum Charge | Job | | 232 | | 232 | |
| **16' One Piece Metal** | | | | | | | |
| | Demolish | Ea. | | 129 | | 129 | Includes material and labor to install |
| | Install | Ea. | 915 | 155 | | 1070 | one piece steel garage door 16' x 7' |
| | Demolish and Install | Ea. | 915 | 284 | | 1199 | primed including hardware. |
| | Reinstall | Ea. | | 123.95 | | 123.95 | |
| | Clean | Ea. | 3.69 | 59 | | 62.69 | |
| | Paint | Ea. | 21 | 110 | | 131 | |
| | Minimum Charge | Job | | 232 | | 232 | |

**For customer support on your Contractor's Pricing Guide: Residential Repair & Remodeling, call 800.448.8182.**

| Garage Door | Unit | Material | Labor | Equip. | Total | Specification |
|---|---|---|---|---|---|---|
| **9' Metal Sectional** | | | | | | |
| Demolish | Ea. | | 109 | | 109 | Includes material and labor to install |
| Install | Ea. | 975 | 116 | | 1091 | sectional metal garage door 9' x 7' |
| Demolish and Install | Ea. | 975 | 225 | | 1200 | primed including hardware. |
| Reinstall | Ea. | | 92.96 | | 92.96 | |
| Clean | Ea. | 1.88 | 50.50 | | 52.38 | |
| Paint | Ea. | 21 | 110 | | 131 | |
| Minimum Charge | Job | | 232 | | 232 | |
| **16' Metal Sectional** | | | | | | |
| Demolish | Ea. | | 129 | | 129 | Includes material and labor to install |
| Install | Ea. | 1175 | 155 | | 1330 | sectional metal garage door 16' x 7' |
| Demolish and Install | Ea. | 1175 | 284 | | 1459 | primed including hardware. |
| Reinstall | Ea. | | 123.95 | | 123.95 | |
| Clean | Ea. | 3.69 | 59 | | 62.69 | |
| Paint | Ea. | 21 | 110 | | 131 | |
| Minimum Charge | Job | | 232 | | 232 | |
| **Metal / Steel Roll Up** | | | | | | |
| **8' x 8' Steel** | | | | | | |
| Demolish | Ea. | | 148 | | 148 | Cost includes material and labor to |
| Install | Ea. | 800 | 690 | | 1490 | install 8' x 8' steel overhead roll-up, |
| Demolish and Install | Ea. | 800 | 838 | | 1638 | chain hoist operated service door. |
| Reinstall | Ea. | | 550.80 | | 550.80 | |
| Clean | Ea. | 1.88 | 50.50 | | 52.38 | |
| Paint | Ea. | 22.50 | 129 | | 151.50 | |
| Minimum Charge | Job | | 1100 | | 1100 | |
| **10' x 10' Steel** | | | | | | |
| Demolish | Ea. | | 172 | | 172 | Cost includes material and labor to |
| Install | Ea. | 1175 | 785 | | 1960 | install 10' x 10' steel overhead roll-up, |
| Demolish and Install | Ea. | 1175 | 957 | | 2132 | chain hoist operated service door. |
| Reinstall | Ea. | | 629.49 | | 629.49 | |
| Clean | Ea. | 2.95 | 71 | | 73.95 | |
| Paint | Ea. | 34.50 | 129 | | 163.50 | |
| Minimum Charge | Job | | 1100 | | 1100 | |
| **12' x 12' Steel** | | | | | | |
| Demolish | Ea. | | 207 | | 207 | Cost includes material and labor to |
| Install | Ea. | 1325 | 920 | | 2245 | install 12' x 12' steel overhead roll-up, |
| Demolish and Install | Ea. | 1325 | 1127 | | 2452 | chain hoist operated service door. |
| Reinstall | Ea. | | 734.40 | | 734.40 | |
| Clean | Ea. | 4.24 | 79 | | 83.24 | |
| Paint | Ea. | 50 | 155 | | 205 | |
| Minimum Charge | Job | | 1100 | | 1100 | |
| **14' x 14' Steel** | | | | | | |
| Demolish | Ea. | | 258 | | 258 | Cost includes material and labor to |
| Install | Ea. | 3300 | 1375 | | 4675 | install 14' x 14' steel overhead roll-up, |
| Demolish and Install | Ea. | 3300 | 1633 | | 4933 | chain hoist operated service door. |
| Reinstall | Ea. | | 1101.60 | | 1101.60 | |
| Clean | Ea. | 5.80 | 101 | | 106.80 | |
| Paint | Ea. | 58.50 | 193 | | 251.50 | |
| Minimum Charge | Job | | 1100 | | 1100 | |
| **18' x 18' Steel** | | | | | | |
| Demolish | Ea. | | 287 | | 287 | Cost includes material and labor to |
| Install | Opng. | 3075 | 1825 | | 4900 | install 18' x 18' steel overhead roll-up, |
| Demolish and Install | Opng. | 3075 | 2112 | | 5187 | chain hoist operated service door. |
| Reinstall | Opng. | | 1468.80 | | 1468.80 | |
| Clean | Ea. | 5.80 | 59 | | 64.80 | |
| Paint | Ea. | 113 | 385 | | 498 | |
| Minimum Charge | Job | | 1100 | | 1100 | |
| **Automatic Motor** | | | | | | |
| Install | Ea. | 645 | 46.50 | | 691.50 | Includes labor and material to install |
| Minimum Charge | Job | | 232 | | 232 | electric opener motor, 1/3 HP. |

103

# Exterior Doors and Windows

| Garage Door | | Unit | Material | Labor | Equip. | Total | Specification |
|---|---|---|---|---|---|---|---|
| **9' One Piece Wood** | | | | | | | |
| | Demolish | Ea. | | 109 | | 109 | Includes material and labor to install |
| | Install | Ea. | 810 | 116 | | 926 | one piece wood garage door 9' x 7' |
| | Demolish and Install | Ea. | 810 | 225 | | 1035 | primed including hardware. |
| | Reinstall | Ea. | | 92.96 | | 92.96 | |
| | Clean | Ea. | 1.84 | 44.50 | | 46.34 | |
| | Paint | Ea. | 10.45 | 96.50 | | 106.95 | |
| | Minimum Charge | Job | | 232 | | 232 | |
| **9' Wood Sectional** | | | | | | | |
| | Demolish | Ea. | | 109 | | 109 | Includes material and labor to install |
| | Install | Ea. | 1125 | 116 | | 1241 | sectional wood garage door 9' x 7' |
| | Demolish and Install | Ea. | 1125 | 225 | | 1350 | unfinished including hardware. |
| | Reinstall | Ea. | | 92.96 | | 92.96 | |
| | Clean | Ea. | 1.84 | 44.50 | | 46.34 | |
| | Paint | Ea. | 10.45 | 96.50 | | 106.95 | |
| | Minimum Charge | Job | | 232 | | 232 | |
| **16' Wood Sectional** | | | | | | | |
| | Demolish | Ea. | | 129 | | 129 | Includes material and labor to install |
| | Install | Ea. | 1800 | 155 | | 1955 | sectional fiberglass garage door 16' x |
| | Demolish and Install | Ea. | 1800 | 284 | | 2084 | 7' unfinished including hardware. |
| | Reinstall | Ea. | | 123.95 | | 123.95 | |
| | Clean | Ea. | 3.69 | 59 | | 62.69 | |
| | Paint | Ea. | 21 | 110 | | 131 | |
| | Minimum Charge | Job | | 232 | | 232 | |
| **Electric Opener** | | | | | | | |
| | Demolish | Ea. | | 79 | | 79 | Includes material and labor to install |
| | Install | Ea. | 430 | 87 | | 517 | electric opener, 1/3 HP economy. |
| | Demolish and Install | Ea. | 430 | 166 | | 596 | |
| | Reinstall | Ea. | | 69.76 | | 69.76 | |
| | Clean | Ea. | .08 | 19.75 | | 19.83 | |
| | Minimum Charge | Job | | 232 | | 232 | |
| **Springs** | | | | | | | |
| | Install | Pr. | 197 | 116 | | 313 | Includes labor and material to replace |
| | Minimum Charge | Job | | 232 | | 232 | springs for garage door. |

| Commercial Metal Door | | Unit | Material | Labor | Equip. | Total | Specification |
|---|---|---|---|---|---|---|---|
| **Pre-hung Steel Door** | | | | | | | |
| **2' 8" x 6' 8"** | | | | | | | |
| | Demolish | Ea. | | 64.50 | | 64.50 | Cost includes material and labor to |
| | Install | Ea. | 705 | 29 | | 734 | install 2' 8" x 6' 8" pre-hung 1-3/4" |
| | Demolish and Install | Ea. | 705 | 93.50 | | 798.50 | 18 gauge steel door with steel frame, |
| | Reinstall | Ea. | | 29.05 | | 29.05 | hinges, aluminum sill, |
| | Clean | Ea. | 1.47 | 14.80 | | 16.27 | weather-stripped. |
| | Paint | Ea. | 6.15 | 32 | | 38.15 | |
| | Minimum Charge | Job | | 232 | | 232 | |
| **3' x 6' 8"** | | | | | | | |
| | Demolish | Ea. | | 64.50 | | 64.50 | Cost includes material and labor to |
| | Install | Ea. | 705 | 31 | | 736 | install 3' x 6' 8" pre-hung 1-3/4" 18 |
| | Demolish and Install | Ea. | 705 | 95.50 | | 800.50 | gauge steel door with steel frame, |
| | Reinstall | Ea. | | 30.99 | | 30.99 | hinges, aluminum sill, |
| | Clean | Ea. | 1.47 | 14.80 | | 16.27 | weather-stripped. |
| | Paint | Ea. | 6.15 | 32 | | 38.15 | |
| | Minimum Charge | Job | | 232 | | 232 | |

# Exterior Doors and Windows

| Commercial Metal Door | Unit | Material | Labor | Equip. | Total | Specification |
|---|---|---|---|---|---|---|
| **3' 6" x 6' 8"** | | | | | | |
| Demolish | Ea. | | 64.50 | | 64.50 | Cost includes material and labor to |
| Install | Ea. | 775 | 36 | | 811 | install 3' 6" x 6' 8" pre-hung 1-3/4" |
| Demolish and Install | Ea. | 775 | 100.50 | | 875.50 | 18 gauge steel door with steel frame, |
| Reinstall | Ea. | | 35.75 | | 35.75 | hinges, aluminum sill, |
| Clean | Ea. | 1.47 | 14.80 | | 16.27 | weather-stripped. |
| Paint | Ea. | 6.15 | 32 | | 38.15 | |
| Minimum Charge | Job | | 232 | | 232 | |
| **4' x 6' 8"** | | | | | | |
| Demolish | Ea. | | 64.50 | | 64.50 | Cost includes material and labor to |
| Install | Ea. | 845 | 46.50 | | 891.50 | install 4' x 6' 8" pre-hung 1-3/4" 18 |
| Demolish and Install | Ea. | 845 | 111 | | 956 | gauge steel door with steel frame, |
| Reinstall | Ea. | | 46.48 | | 46.48 | hinges, aluminum sill, |
| Clean | Ea. | 1.47 | 14.80 | | 16.27 | weather-stripped. |
| Paint | Ea. | 6.15 | 32 | | 38.15 | |
| Minimum Charge | Job | | 232 | | 232 | |
| **Metal Jamb** | | | | | | |
| **6' 8" High** | | | | | | |
| Demolish | Ea. | | 26 | | 26 | Cost includes material and labor to |
| Install | Ea. | 141 | 29 | | 170 | install 18 gauge hollow metal door |
| Demolish and Install | Ea. | 141 | 55 | | 196 | frame to fit a 4-1/2" jamb 6' 8" high |
| Reinstall | Ea. | | 29.05 | | 29.05 | and up to 3' 6" opening. |
| Clean | Ea. | .31 | 7.40 | | 7.71 | |
| Paint | Ea. | 1.71 | 7.75 | | 9.46 | |
| Minimum Charge | Job | | 232 | | 232 | |
| **7' High** | | | | | | |
| Demolish | Ea. | | 26 | | 26 | Cost includes material and labor to |
| Install | Ea. | 146 | 31 | | 177 | install 18 gauge hollow metal door |
| Demolish and Install | Ea. | 146 | 57 | | 203 | frame to fit a 4-1/2" jamb 7' high and |
| Reinstall | Ea. | | 30.99 | | 30.99 | up to 3' 6" opening. |
| Clean | Ea. | .31 | 7.40 | | 7.71 | |
| Paint | Ea. | 1.71 | 7.75 | | 9.46 | |
| Minimum Charge | Job | | 232 | | 232 | |
| **8' High** | | | | | | |
| Demolish | Ea. | | 26 | | 26 | Cost includes material and labor to |
| Install | Ea. | 178 | 38.50 | | 216.50 | install 18 gauge hollow metal door |
| Demolish and Install | Ea. | 178 | 64.50 | | 242.50 | frame to fit a 4-1/2" jamb 8' high and |
| Reinstall | Ea. | | 38.73 | | 38.73 | up to 3' 6" opening. |
| Clean | Ea. | .43 | 8.90 | | 9.33 | |
| Paint | Ea. | 1.90 | 8.60 | | 10.50 | |
| Minimum Charge | Job | | 232 | | 232 | |
| **9' High** | | | | | | |
| Demolish | Ea. | | 26 | | 26 | Cost includes material and labor to |
| Install | Ea. | 224 | 46.50 | | 270.50 | install 18 gauge hollow metal door |
| Demolish and Install | Ea. | 224 | 72.50 | | 296.50 | frame to fit a 4-1/2" jamb 9' high and |
| Reinstall | Ea. | | 46.48 | | 46.48 | up to 3' 6" opening. |
| Clean | Ea. | .43 | 8.90 | | 9.33 | |
| Paint | Ea. | 1.90 | 8.60 | | 10.50 | |
| Minimum Charge | Job | | 232 | | 232 | |
| **Hardware** | | | | | | |
| **Doorknob w / Lock** | | | | | | |
| Demolish | Ea. | | 18.45 | | 18.45 | Includes material and labor to install |
| Install | Ea. | 209 | 38.50 | | 247.50 | keyed entry lock set, bored type |
| Demolish and Install | Ea. | 209 | 56.95 | | 265.95 | including knob trim, strike and strike |
| Reinstall | Ea. | | 30.99 | | 30.99 | box for standard commercial |
| Clean | Ea. | .14 | 11.10 | | 11.24 | application. |
| Minimum Charge | Job | | 232 | | 232 | |

**For customer support on your Contractor's Pricing Guide: Residential Repair & Remodeling, call 800.448.8182.**

# Exterior Doors and Windows

| Commercial Metal Door | | Unit | Material | Labor | Equip. | Total | Specification |
|---|---|---|---|---|---|---|---|
| **Lever Handle** | | | | | | | |
| | Demolish | Ea. | | 18.45 | | 18.45 | Includes material and labor to install |
| | Install | Ea. | 168 | 46.50 | | 214.50 | passage lockset, keyless bored type |
| | Demolish and Install | Ea. | 168 | 64.95 | | 232.95 | with lever handle, non-locking |
| | Reinstall | Ea. | | 37.18 | | 37.18 | standard commercial latchset. |
| | Clean | Ea. | .14 | 11.10 | | 11.24 | |
| | Minimum Charge | Job | | 232 | | 232 | |
| **Deadbolt** | | | | | | | |
| | Demolish | Ea. | | 16.15 | | 16.15 | Includes material and labor to install a |
| | Install | Ea. | 69.50 | 33 | | 102.50 | deadbolt. |
| | Demolish and Install | Ea. | 69.50 | 49.15 | | 118.65 | |
| | Reinstall | Ea. | | 26.56 | | 26.56 | |
| | Clean | Ea. | .14 | 11.10 | | 11.24 | |
| | Minimum Charge | Job | | 232 | | 232 | |
| **Panic Device** | | | | | | | |
| | Demolish | Ea. | | 39.50 | | 39.50 | Includes material and labor to install |
| | Install | Ea. | 305 | 77.50 | | 382.50 | touch bar, low profile, exit only. |
| | Demolish and Install | Ea. | 305 | 117 | | 422 | |
| | Reinstall | Ea. | | 61.97 | | 61.97 | |
| | Clean | Ea. | .08 | 19.75 | | 19.83 | |
| | Minimum Charge | Job | | 232 | | 232 | |
| **Closer** | | | | | | | |
| | Demolish | Ea. | | 9.70 | | 9.70 | Includes material and labor to install |
| | Install | Ea. | 335 | 77.50 | | 412.50 | pneumatic heavy duty model for |
| | Demolish and Install | Ea. | 335 | 87.20 | | 422.20 | exterior type doors. |
| | Reinstall | Ea. | | 61.97 | | 61.97 | |
| | Clean | Ea. | .08 | 14.80 | | 14.88 | |
| | Minimum Charge | Job | | 232 | | 232 | |
| **Peep-Hole** | | | | | | | |
| | Demolish | Ea. | | 8.75 | | 8.75 | Includes material and labor to install |
| | Install | Ea. | 9.85 | 14.55 | | 24.40 | peep-hole for door. |
| | Demolish and Install | Ea. | 9.85 | 23.30 | | 33.15 | |
| | Reinstall | Ea. | | 11.62 | | 11.62 | |
| | Clean | Ea. | .01 | 4.10 | | 4.11 | |
| | Minimum Charge | Job | | 232 | | 232 | |
| **Exit Sign** | | | | | | | |
| | Demolish | Ea. | | 26.50 | | 26.50 | Includes material and labor to install a |
| | Install | Ea. | 104 | 65.50 | | 169.50 | wall mounted interior electric exit sign. |
| | Demolish and Install | Ea. | 104 | 92 | | 196 | |
| | Reinstall | Ea. | | 52.52 | | 52.52 | |
| | Clean | Ea. | .01 | 8.35 | | 8.36 | |
| | Minimum Charge | Job | | 232 | | 232 | |
| **Kickplate** | | | | | | | |
| | Demolish | Ea. | | 14.35 | | 14.35 | Includes material and labor to install |
| | Install | Ea. | 94 | 31 | | 125 | stainless steel kickplate, 10" x 36". |
| | Demolish and Install | Ea. | 94 | 45.35 | | 139.35 | |
| | Reinstall | Ea. | | 24.79 | | 24.79 | |
| | Clean | Ea. | .14 | 11.10 | | 11.24 | |
| | Minimum Charge | Job | | 232 | | 232 | |

| Aluminum Window | | Unit | Material | Labor | Equip. | Total | Specification |
|---|---|---|---|---|---|---|---|
| **Single Hung Two Light** | | | | | | | |
| **2' x 2'** | | | | | | | |
| | Demolish | Ea. | | 24.50 | | 24.50 | Cost includes material and labor to |
| | Install | Ea. | 195 | 100 | | 295 | install 2' x 2' single hung, two light, |
| | Demolish and Install | Ea. | 195 | 124.50 | | 319.50 | double glazed aluminum window |
| | Reinstall | Ea. | | 80.12 | | 80.12 | including screen. |
| | Clean | Ea. | .02 | 9.55 | | 9.57 | |
| | Minimum Charge | Job | | 275 | | 275 | |

106

| Aluminum Window | Unit | Material | Labor | Equip. | Total | Specification |
|---|---|---|---|---|---|---|
| **2' x 2' 6"** | | | | | | |
| Demolish | Ea. | | 24.50 | | 24.50 | Cost includes material and labor to |
| Install | Ea. | 86.50 | 100 | | 186.50 | install 2' x 2' 6" single hung, two |
| Demolish and Install | Ea. | 86.50 | 124.50 | | 211 | light, double glazed aluminum |
| Reinstall | Ea. | | 80.12 | | 80.12 | window including screen. |
| Clean | Ea. | .02 | 9.55 | | 9.57 | |
| Minimum Charge | Job | | 275 | | 275 | |
| **3' x 1' 6"** | | | | | | |
| Demolish | Ea. | | 24.50 | | 24.50 | Cost includes material and labor to |
| Install | Ea. | 78.50 | 110 | | 188.50 | install 3' x 1' 6" single hung, two |
| Demolish and Install | Ea. | 78.50 | 134.50 | | 213 | light, double glazed aluminum |
| Reinstall | Ea. | | 88.13 | | 88.13 | window including screen. |
| Clean | Ea. | .01 | 8.35 | | 8.36 | |
| Minimum Charge | Job | | 275 | | 275 | |
| **3' x 2'** | | | | | | |
| Demolish | Ea. | | 24.50 | | 24.50 | Cost includes material and labor to |
| Install | Ea. | 95 | 110 | | 205 | install 3' x 2' single hung, two light, |
| Demolish and Install | Ea. | 95 | 134.50 | | 229.50 | double glazed aluminum window |
| Reinstall | Ea. | | 88.13 | | 88.13 | including screen. |
| Clean | Ea. | .01 | 8.35 | | 8.36 | |
| Minimum Charge | Job | | 275 | | 275 | |
| **3' x 2' 6"** | | | | | | |
| Demolish | Ea. | | 24.50 | | 24.50 | Cost includes material and labor to |
| Install | Ea. | 103 | 110 | | 213 | install 3' x 2' 6" single hung, two |
| Demolish and Install | Ea. | 103 | 134.50 | | 237.50 | light, double glazed aluminum |
| Reinstall | Ea. | | 88.13 | | 88.13 | window including screen. |
| Clean | Ea. | .02 | 9.55 | | 9.57 | |
| Minimum Charge | Job | | 275 | | 275 | |
| **3' x 3'** | | | | | | |
| Demolish | Ea. | | 24.50 | | 24.50 | Cost includes material and labor to |
| Install | Ea. | 107 | 110 | | 217 | install 3' x 3' single hung, two light, |
| Demolish and Install | Ea. | 107 | 134.50 | | 241.50 | double glazed aluminum window |
| Reinstall | Ea. | | 88.13 | | 88.13 | including screen. |
| Clean | Ea. | .02 | 9.55 | | 9.57 | |
| Minimum Charge | Job | | 275 | | 275 | |
| **3' x 3' 6"** | | | | | | |
| Demolish | Ea. | | 24.50 | | 24.50 | Cost includes material and labor to |
| Install | Ea. | 268 | 110 | | 378 | install 3' x 3' 6" single hung, two |
| Demolish and Install | Ea. | 268 | 134.50 | | 402.50 | light, double glazed aluminum |
| Reinstall | Ea. | | 88.13 | | 88.13 | window including screen. |
| Clean | Ea. | .02 | 11.10 | | 11.12 | |
| Minimum Charge | Job | | 275 | | 275 | |
| **3' x 4'** | | | | | | |
| Demolish | Ea. | | 24.50 | | 24.50 | Cost includes material and labor to |
| Install | Ea. | 277 | 110 | | 387 | install 3' x 4' single hung, two light, |
| Demolish and Install | Ea. | 277 | 134.50 | | 411.50 | double glazed aluminum window |
| Reinstall | Ea. | | 88.13 | | 88.13 | including screen. |
| Clean | Ea. | .02 | 11.10 | | 11.12 | |
| Minimum Charge | Job | | 275 | | 275 | |
| **3' x 5'** | | | | | | |
| Demolish | Ea. | | 36 | | 36 | Cost includes material and labor to |
| Install | Ea. | 140 | 110 | | 250 | install 3' x 5' single hung, two light, |
| Demolish and Install | Ea. | 140 | 146 | | 286 | double glazed aluminum window |
| Reinstall | Ea. | | 88.13 | | 88.13 | including screen. |
| Clean | Ea. | .03 | 13.35 | | 13.38 | |
| Minimum Charge | Job | | 275 | | 275 | |

| Aluminum Window | Unit | Material | Labor | Equip. | Total | Specification |
|---|---|---|---|---|---|---|
| **3' x 6'** | | | | | | |
| Demolish | Ea. | | 36 | | 36 | Cost includes material and labor to |
| Install | Ea. | 380 | 122 | | 502 | install 3' x 6' single hung, two light, |
| Demolish and Install | Ea. | 380 | 158 | | 538 | double glazed aluminum window |
| Reinstall | Ea. | | 97.92 | | 97.92 | including screen. |
| Clean | Ea. | .03 | 15.70 | | 15.73 | |
| Minimum Charge | Job | | 275 | | 275 | |
| **Double Hung Two Light** | | | | | | |
| **2' x 2'** | | | | | | |
| Demolish | Ea. | | 24.50 | | 24.50 | Cost includes material and labor to |
| Install | Ea. | 129 | 100 | | 229 | install 2' x 2' double hung, two light, |
| Demolish and Install | Ea. | 129 | 124.50 | | 253.50 | double glazed aluminum window |
| Reinstall | Ea. | | 80.12 | | 80.12 | including screen. |
| Clean | Ea. | .01 | 8.35 | | 8.36 | |
| Minimum Charge | Job | | 275 | | 275 | |
| **2' x 2' 6"** | | | | | | |
| Demolish | Ea. | | 24.50 | | 24.50 | Cost includes material and labor to |
| Install | Ea. | 144 | 100 | | 244 | install 2' x 2' 6" double hung, two |
| Demolish and Install | Ea. | 144 | 124.50 | | 268.50 | light, double glazed aluminum |
| Reinstall | Ea. | | 80.12 | | 80.12 | window including screen. |
| Clean | Ea. | .01 | 8.35 | | 8.36 | |
| Minimum Charge | Job | | 275 | | 275 | |
| **3' x 1' 6"** | | | | | | |
| Demolish | Ea. | | 24.50 | | 24.50 | Cost includes material and labor to |
| Install | Ea. | 297 | 110 | | 407 | install 3' x 1' 6" double hung, two |
| Demolish and Install | Ea. | 297 | 134.50 | | 431.50 | light, double glazed aluminum |
| Reinstall | Ea. | | 88.13 | | 88.13 | window including screen. |
| Clean | Ea. | .01 | 8.35 | | 8.36 | |
| Minimum Charge | Job | | 275 | | 275 | |
| **3' x 2'** | | | | | | |
| Demolish | Ea. | | 24.50 | | 24.50 | Cost includes material and labor to |
| Install | Ea. | 345 | 110 | | 455 | install 3' x 2' double hung, two light, |
| Demolish and Install | Ea. | 345 | 134.50 | | 479.50 | double glazed aluminum window |
| Reinstall | Ea. | | 88.13 | | 88.13 | including screen. |
| Clean | Ea. | .01 | 8.35 | | 8.36 | |
| Minimum Charge | Job | | 275 | | 275 | |
| **3' x 2' 6"** | | | | | | |
| Demolish | Ea. | | 24.50 | | 24.50 | Cost includes material and labor to |
| Install | Ea. | 420 | 110 | | 530 | install 3' x 2' 6" double hung, two |
| Demolish and Install | Ea. | 420 | 134.50 | | 554.50 | light, double glazed aluminum |
| Reinstall | Ea. | | 88.13 | | 88.13 | window including screen. |
| Clean | Ea. | .02 | 9.55 | | 9.57 | |
| Minimum Charge | Job | | 275 | | 275 | |
| **3' x 3'** | | | | | | |
| Demolish | Ea. | | 24.50 | | 24.50 | Cost includes material and labor to |
| Install | Ea. | 470 | 110 | | 580 | install 3' x 3' double hung, two light, |
| Demolish and Install | Ea. | 470 | 134.50 | | 604.50 | double glazed aluminum window |
| Reinstall | Ea. | | 88.13 | | 88.13 | including screen. |
| Clean | Ea. | .02 | 9.55 | | 9.57 | |
| Minimum Charge | Job | | 275 | | 275 | |
| **3' x 3' 6"** | | | | | | |
| Demolish | Ea. | | 24.50 | | 24.50 | Cost includes material and labor to |
| Install | Ea. | 490 | 110 | | 600 | install 3' x 3' 6" double hung, two |
| Demolish and Install | Ea. | 490 | 134.50 | | 624.50 | light, double glazed aluminum |
| Reinstall | Ea. | | 88.13 | | 88.13 | window including screen. |
| Clean | Ea. | .02 | 11.10 | | 11.12 | |
| Minimum Charge | Job | | 275 | | 275 | |

| Aluminum Window | Unit | Material | Labor | Equip. | Total | Specification |
|---|---|---|---|---|---|---|
| **3' x 4'** | | | | | | |
| Demolish | Ea. | | 24.50 | | 24.50 | Cost includes material and labor to |
| Install | Ea. | 530 | 110 | | 640 | install 3' x 4' double hung, two light, |
| Demolish and Install | Ea. | 530 | 134.50 | | 664.50 | double glazed aluminum window |
| Reinstall | Ea. | | 88.13 | | 88.13 | including screen. |
| Clean | Ea. | .02 | 11.10 | | 11.12 | |
| Minimum Charge | Job | | 275 | | 275 | |
| **3' x 5'** | | | | | | |
| Demolish | Ea. | | 36 | | 36 | Cost includes material and labor to |
| Install | Ea. | 585 | 110 | | 695 | install 3' x 5' double hung, two light, |
| Demolish and Install | Ea. | 585 | 146 | | 731 | double glazed aluminum window |
| Reinstall | Ea. | | 88.13 | | 88.13 | including screen. |
| Clean | Ea. | .03 | 13.35 | | 13.38 | |
| Minimum Charge | Job | | 275 | | 275 | |
| **3' x 6'** | | | | | | |
| Demolish | Ea. | | 36 | | 36 | Cost includes material and labor to |
| Install | Ea. | 615 | 122 | | 737 | install 3' x 6' double hung, two light, |
| Demolish and Install | Ea. | 615 | 158 | | 773 | double glazed aluminum window |
| Reinstall | Ea. | | 97.92 | | 97.92 | including screen. |
| Clean | Ea. | .03 | 15.70 | | 15.73 | |
| Minimum Charge | Job | | 275 | | 275 | |
| **Sliding Sash** | | | | | | |
| **1' 6" x 3'** | | | | | | |
| Demolish | Ea. | | 24.50 | | 24.50 | Cost includes material and labor to |
| Install | Ea. | 244 | 100 | | 344 | install 1' 6" x 3' sliding sash, two |
| Demolish and Install | Ea. | 244 | 124.50 | | 368.50 | light, double glazed aluminum |
| Reinstall | Ea. | | 80.12 | | 80.12 | window including screen. |
| Clean | Ea. | .01 | 8.35 | | 8.36 | |
| Minimum Charge | Job | | 275 | | 275 | |
| **2' x 2'** | | | | | | |
| Demolish | Ea. | | 24.50 | | 24.50 | Cost includes material and labor to |
| Install | Ea. | 66 | 110 | | 176 | install 2' x 2' sliding sash, two light, |
| Demolish and Install | Ea. | 66 | 134.50 | | 200.50 | double glazed aluminum window |
| Reinstall | Ea. | | 88.13 | | 88.13 | including screen. |
| Clean | Ea. | .01 | 8.35 | | 8.36 | |
| Minimum Charge | Job | | 275 | | 275 | |
| **3' x 2'** | | | | | | |
| Demolish | Ea. | | 24.50 | | 24.50 | Cost includes material and labor to |
| Install | Ea. | 253 | 110 | | 363 | install 3' x 2' sliding sash, two light, |
| Demolish and Install | Ea. | 253 | 134.50 | | 387.50 | double glazed aluminum window |
| Reinstall | Ea. | | 88.13 | | 88.13 | including screen. |
| Clean | Ea. | .01 | 8.35 | | 8.36 | |
| Minimum Charge | Job | | 275 | | 275 | |
| **3' x 2' 6"** | | | | | | |
| Demolish | Ea. | | 24.50 | | 24.50 | Cost includes material and labor to |
| Install | Ea. | 92 | 110 | | 202 | install 3' x 2' 6" sliding sash, two |
| Demolish and Install | Ea. | 92 | 134.50 | | 226.50 | light, double glazed aluminum |
| Reinstall | Ea. | | 88.13 | | 88.13 | window including screen. |
| Clean | Ea. | .02 | 9.55 | | 9.57 | |
| Minimum Charge | Job | | 275 | | 275 | |
| **3' x 3'** | | | | | | |
| Demolish | Ea. | | 24.50 | | 24.50 | Cost includes material and labor to |
| Install | Ea. | 310 | 110 | | 420 | install 3' x 3' sliding sash, two light, |
| Demolish and Install | Ea. | 310 | 134.50 | | 444.50 | double glazed aluminum window |
| Reinstall | Ea. | | 88.13 | | 88.13 | including screen. |
| Clean | Ea. | .02 | 9.55 | | 9.57 | |
| Minimum Charge | Job | | 275 | | 275 | |

| Aluminum Window | | Unit | Material | Labor | Equip. | Total | Specification |
|---|---|---|---|---|---|---|---|
| **3' x 3' 6"** | | | | | | | |
| | Demolish | Ea. | | 24.50 | | 24.50 | Cost includes material and labor to |
| | Install | Ea. | 111 | 110 | | 221 | install 3' x 3' 6" sliding sash, two |
| | Demolish and Install | Ea. | 111 | 134.50 | | 245.50 | light, double glazed aluminum |
| | Reinstall | Ea. | | 88.13 | | 88.13 | window including screen. |
| | Clean | Ea. | .02 | 11.10 | | 11.12 | |
| | Minimum Charge | Job | | 275 | | 275 | |
| **3' x 4'** | | | | | | | |
| | Demolish | Ea. | | 24.50 | | 24.50 | Cost includes material and labor to |
| | Install | Ea. | 118 | 122 | | 240 | install 3' x 4' sliding sash, two light, |
| | Demolish and Install | Ea. | 118 | 146.50 | | 264.50 | double glazed aluminum window |
| | Reinstall | Ea. | | 97.92 | | 97.92 | including screen. |
| | Clean | Ea. | .02 | 11.10 | | 11.12 | |
| | Minimum Charge | Job | | 275 | | 275 | |
| **3' x 5'** | | | | | | | |
| | Demolish | Ea. | | 36 | | 36 | Cost includes material and labor to |
| | Install | Ea. | 152 | 122 | | 274 | install 3' x 5' sliding sash, two light, |
| | Demolish and Install | Ea. | 152 | 158 | | 310 | double glazed aluminum window |
| | Reinstall | Ea. | | 97.92 | | 97.92 | including screen. |
| | Clean | Ea. | .03 | 13.35 | | 13.38 | |
| | Minimum Charge | Job | | 275 | | 275 | |
| **4' x 5'** | | | | | | | |
| | Demolish | Ea. | | 36 | | 36 | Cost includes material and labor to |
| | Install | Ea. | 154 | 157 | | 311 | install 4' x 5' sliding sash, two light, |
| | Demolish and Install | Ea. | 154 | 193 | | 347 | double glazed aluminum window |
| | Reinstall | Ea. | | 125.90 | | 125.90 | including screen. |
| | Clean | Ea. | .04 | 19.05 | | 19.09 | |
| | Minimum Charge | Job | | 275 | | 275 | |
| **5' x 5'** | | | | | | | |
| | Demolish | Ea. | | 36 | | 36 | Cost includes material and labor to |
| | Install | Ea. | 360 | 157 | | 517 | install 5' x 5' sliding sash, two light, |
| | Demolish and Install | Ea. | 360 | 193 | | 553 | double glazed aluminum window |
| | Reinstall | Ea. | | 125.90 | | 125.90 | including screen. |
| | Clean | Ea. | .04 | 19.05 | | 19.09 | |
| | Minimum Charge | Job | | 275 | | 275 | |
| **Awning Sash** | | | | | | | |
| **26" x 26", 2 Light** | | | | | | | |
| | Demolish | Ea. | | 24.50 | | 24.50 | Cost includes material and labor to |
| | Install | Ea. | 330 | 122 | | 452 | install 26" x 26" two light awning, |
| | Demolish and Install | Ea. | 330 | 146.50 | | 476.50 | double glazed aluminum window |
| | Reinstall | Ea. | | 97.92 | | 97.92 | including screen. |
| | Clean | Ea. | .02 | 9.55 | | 9.57 | |
| | Minimum Charge | Job | | 275 | | 275 | |
| **26" x 37", 2 Light** | | | | | | | |
| | Demolish | Ea. | | 24.50 | | 24.50 | Cost includes material and labor to |
| | Install | Ea. | 360 | 122 | | 482 | install 26" x 37" two light awning, |
| | Demolish and Install | Ea. | 360 | 146.50 | | 506.50 | double glazed aluminum window |
| | Reinstall | Ea. | | 97.92 | | 97.92 | including screen. |
| | Clean | Ea. | .02 | 9.55 | | 9.57 | |
| | Minimum Charge | Job | | 275 | | 275 | |
| **38" x 26", 3 Light** | | | | | | | |
| | Demolish | Ea. | | 24.50 | | 24.50 | Cost includes material and labor to |
| | Install | Ea. | 400 | 122 | | 522 | install 38" x 26" three light awning, |
| | Demolish and Install | Ea. | 400 | 146.50 | | 546.50 | double glazed aluminum window |
| | Reinstall | Ea. | | 97.92 | | 97.92 | including screen. |
| | Clean | Ea. | .02 | 11.10 | | 11.12 | |
| | Minimum Charge | Job | | 275 | | 275 | |

| Aluminum Window | | Unit | Material | Labor | Equip. | Total | Specification |
|---|---|---|---|---|---|---|---|
| **38" x 37", 3 Light** | | | | | | | |
| | Demolish | Ea. | | 24.50 | | 24.50 | Cost includes material and labor to |
| | Install | Ea. | 405 | 138 | | 543 | install 38" x 37" three light awning, |
| | Demolish and Install | Ea. | 405 | 162.50 | | 567.50 | double glazed aluminum window |
| | Reinstall | Ea. | | 110.16 | | 110.16 | including screen. |
| | Clean | Ea. | .02 | 11.10 | | 11.12 | |
| | Minimum Charge | Job | | 275 | | 275 | |
| **38" x 53", 3 Light** | | | | | | | |
| | Demolish | Ea. | | 36 | | 36 | Cost includes material and labor to |
| | Install | Ea. | 670 | 138 | | 808 | install 38" x 53" three light awning, |
| | Demolish and Install | Ea. | 670 | 174 | | 844 | double glazed aluminum window |
| | Reinstall | Ea. | | 110.16 | | 110.16 | including screen. |
| | Clean | Ea. | .03 | 13.35 | | 13.38 | |
| | Minimum Charge | Job | | 275 | | 275 | |
| **50" x 37", 4 Light** | | | | | | | |
| | Demolish | Ea. | | 36 | | 36 | Cost includes material and labor to |
| | Install | Ea. | 665 | 138 | | 803 | install 50" x 37" four light awning, |
| | Demolish and Install | Ea. | 665 | 174 | | 839 | double glazed aluminum window |
| | Reinstall | Ea. | | 110.16 | | 110.16 | including screen. |
| | Clean | Ea. | .03 | 13.35 | | 13.38 | |
| | Minimum Charge | Job | | 275 | | 275 | |
| **50" x 53", 4 Light** | | | | | | | |
| | Demolish | Ea. | | 36 | | 36 | Cost includes material and labor to |
| | Install | Ea. | 1100 | 157 | | 1257 | install 50" x 53" four light awning, |
| | Demolish and Install | Ea. | 1100 | 193 | | 1293 | double glazed aluminum window |
| | Reinstall | Ea. | | 125.90 | | 125.90 | including screen. |
| | Clean | Ea. | .03 | 15.70 | | 15.73 | |
| | Minimum Charge | Job | | 275 | | 275 | |
| **63" x 37", 5 Light** | | | | | | | |
| | Demolish | Ea. | | 36 | | 36 | Cost includes material and labor to |
| | Install | Ea. | 865 | 157 | | 1022 | install 63" x 37" five light awning, |
| | Demolish and Install | Ea. | 865 | 193 | | 1058 | double glazed aluminum window |
| | Reinstall | Ea. | | 125.90 | | 125.90 | including screen. |
| | Clean | Ea. | .03 | 15.70 | | 15.73 | |
| | Minimum Charge | Job | | 275 | | 275 | |
| **63" x 53", 5 Light** | | | | | | | |
| | Demolish | Ea. | | 36 | | 36 | Cost includes material and labor to |
| | Install | Ea. | 1200 | 157 | | 1357 | install 63" x 53" five light awning, |
| | Demolish and Install | Ea. | 1200 | 193 | | 1393 | double glazed aluminum window |
| | Reinstall | Ea. | | 125.90 | | 125.90 | including screen. |
| | Clean | Ea. | .04 | 19.05 | | 19.09 | |
| | Minimum Charge | Job | | 275 | | 275 | |
| **Mullion Bars** | | | | | | | |
| | Install | Ea. | 14.25 | 44 | | 58.25 | Includes labor and material to install |
| | Minimum Charge | Job | | 275 | | 275 | mullions for sliding aluminum window. |
| **Bay** | | | | | | | |
| **4' 8" x 6' 4"** | | | | | | | |
| | Demolish | Ea. | | 26.50 | | 26.50 | Cost includes material and labor to |
| | Install | Ea. | 1650 | 116 | | 1766 | install 4' 8" x 6' 4" angle bay double |
| | Demolish and Install | Ea. | 1650 | 142.50 | | 1792.50 | hung unit, pine frame with aluminum |
| | Reinstall | Ea. | | 92.96 | | 92.96 | cladding, picture center sash, |
| | Clean | Ea. | .07 | 22 | | 22.07 | hardware, screens. |
| | Minimum Charge | Job | | 275 | | 275 | |

| Aluminum Window | Unit | Material | Labor | Equip. | Total | Specification |
|---|---|---|---|---|---|---|
| **4' 8" x 7' 2"** | | | | | | |
| Demolish | Ea. | | 26.50 | | 26.50 | Cost includes material and labor to |
| Install | Ea. | 1675 | 116 | | 1791 | install 4' 8" x 7' 2" angle bay double |
| Demolish and Install | Ea. | 1675 | 142.50 | | 1817.50 | hung unit, pine frame with aluminum |
| Reinstall | Ea. | | 92.96 | | 92.96 | cladding, picture center sash, |
| Clean | Ea. | .07 | 22 | | 22.07 | hardware, screens. |
| Minimum Charge | Job | | 275 | | 275 | |
| **4' 8" x 7' 10"** | | | | | | |
| Demolish | Ea. | | 26.50 | | 26.50 | Cost includes material and labor to |
| Install | Ea. | 1700 | 116 | | 1816 | install 4' 8" x 7' 10" angle bay |
| Demolish and Install | Ea. | 1700 | 142.50 | | 1842.50 | double hung unit, pine frame with |
| Reinstall | Ea. | | 92.96 | | 92.96 | aluminum cladding, picture center |
| Clean | Ea. | .07 | 22 | | 22.07 | sash, hardware, screens. |
| Minimum Charge | Job | | 275 | | 275 | |
| **4' 8" x 8' 6"** | | | | | | |
| Demolish | Ea. | | 30.50 | | 30.50 | Cost includes material and labor to |
| Install | Ea. | 1675 | 133 | | 1808 | install 4' 8" x 8' 6" angle bay double |
| Demolish and Install | Ea. | 1675 | 163.50 | | 1838.50 | hung unit, pine frame with aluminum |
| Reinstall | Ea. | | 106.24 | | 106.24 | cladding, picture center sash, |
| Clean | Ea. | .01 | 29.50 | | 29.51 | hardware, screens. |
| Minimum Charge | Job | | 275 | | 275 | |
| **4' 8" x 8' 10"** | | | | | | |
| Demolish | Ea. | | 30.50 | | 30.50 | Cost includes material and labor to |
| Install | Ea. | 1700 | 133 | | 1833 | install 4' 8" x 8' 10" angle bay |
| Demolish and Install | Ea. | 1700 | 163.50 | | 1863.50 | double hung unit, pine frame with |
| Reinstall | Ea. | | 106.24 | | 106.24 | aluminum cladding, picture center |
| Clean | Ea. | .01 | 29.50 | | 29.51 | sash, hardware, screens. |
| Minimum Charge | Job | | 275 | | 275 | |
| **4' 8" x 9' 2"** | | | | | | |
| Demolish | Ea. | | 30.50 | | 30.50 | Cost includes material and labor to |
| Install | Ea. | 1750 | 232 | | 1982 | install 4' 8" x 9' 2" angle bay double |
| Demolish and Install | Ea. | 1750 | 262.50 | | 2012.50 | hung unit, pine frame with aluminum |
| Reinstall | Ea. | | 185.92 | | 185.92 | cladding, picture center sash, |
| Clean | Ea. | .01 | 29.50 | | 29.51 | hardware, screens. |
| Minimum Charge | Job | | 275 | | 275 | |
| **4' 8" x 9' 10"** | | | | | | |
| Demolish | Ea. | | 30.50 | | 30.50 | Cost includes material and labor to |
| Install | Ea. | 1900 | 232 | | 2132 | install 4' 8" x 9' 10" angle bay |
| Demolish and Install | Ea. | 1900 | 262.50 | | 2162.50 | double hung unit, pine frame with |
| Reinstall | Ea. | | 185.92 | | 185.92 | aluminum cladding, picture center |
| Clean | Ea. | .01 | 29.50 | | 29.51 | sash, hardware, screens. |
| Minimum Charge | Job | | 275 | | 275 | |
| **Casing** | | | | | | |
| Demolish | Ea. | | 10.75 | | 10.75 | Cost includes material and labor to |
| Install | Opng. | 35 | 36 | | 71 | install 11/16" x 2-1/2" pine ranch |
| Demolish and Install | Opng. | 35 | 46.75 | | 81.75 | style window casing. |
| Reinstall | Opng. | | 28.60 | | 28.60 | |
| Clean | Opng. | .21 | 3.55 | | 3.76 | |
| Paint | Ea. | 5 | 7.75 | | 12.75 | |
| Minimum Charge | Job | | 275 | | 275 | |
| **Storm Window** | | | | | | |
| **2' x 3' 6"** | | | | | | |
| Demolish | Ea. | | 14.60 | | 14.60 | Includes material and labor to install |
| Install | Ea. | 99 | 31 | | 130 | residential aluminum storm window |
| Demolish and Install | Ea. | 99 | 45.60 | | 144.60 | with mill finish, 2' x 3' 6" high. |
| Reinstall | Ea. | | 24.79 | | 24.79 | |
| Clean | Ea. | .01 | 8.35 | | 8.36 | |
| Minimum Charge | Job | | 275 | | 275 | |

# Exterior Doors and Windows

## Aluminum Window

| | | Unit | Material | Labor | Equip. | Total | Specification |
|---|---|---|---|---|---|---|---|
| **2' 6" x 5'** | | | | | | | Includes material and labor to install |
| | Demolish | Ea. | | 14.60 | | 14.60 | residential aluminum storm window |
| | Install | Ea. | 105 | 33 | | 138 | with mill finish, 2' 6" x 5' high. |
| | Demolish and Install | Ea. | 105 | 47.60 | | 152.60 | |
| | Reinstall | Ea. | | 26.56 | | 26.56 | |
| | Clean | Ea. | .02 | 11.10 | | 11.12 | |
| | Minimum Charge | Job | | 275 | | 275 | |
| **4' x 6'** | | | | | | | Includes material and labor to install |
| | Demolish | Ea. | | 28 | | 28 | residential aluminum storm window, 4' |
| | Install | Ea. | 127 | 37 | | 164 | x 6' high. |
| | Demolish and Install | Ea. | 127 | 65 | | 192 | |
| | Reinstall | Ea. | | 29.75 | | 29.75 | |
| | Clean | Ea. | .04 | 16.70 | | 16.74 | |
| | Minimum Charge | Job | | 275 | | 275 | |

### Hardware
#### Weatherstripping

| | | Unit | Material | Labor | Equip. | Total | Specification |
|---|---|---|---|---|---|---|---|
| | Install | L.F. | .78 | 3.87 | | 4.65 | Includes labor and material to install |
| | Minimum Charge | Job | | 155 | | 155 | vinyl V strip weatherstripping for double-hung window. |

### Trim Set

| | | Unit | Material | Labor | Equip. | Total | Specification |
|---|---|---|---|---|---|---|---|
| | Demolish | Ea. | | 10.75 | | 10.75 | Cost includes material and labor to |
| | Install | Opng. | 35 | 36 | | 71 | install 11/16" x 2-1/2" pine ranch |
| | Demolish and Install | Opng. | 35 | 46.75 | | 81.75 | style window casing. |
| | Reinstall | Opng. | | 28.60 | | 28.60 | |
| | Clean | Opng. | .21 | 3.55 | | 3.76 | |
| | Paint | Ea. | 5 | 7.75 | | 12.75 | |
| | Minimum Charge | Job | | 232 | | 232 | |

## Vinyl Window

| | | Unit | Material | Labor | Equip. | Total | Specification |
|---|---|---|---|---|---|---|---|
| **Double Hung** | | | | | | | |
| **2' x 2' 6"** | | | | | | | Cost includes material and labor to |
| | Demolish | Ea. | | 17.95 | | 17.95 | install 2' x 2' 6" double hung, two |
| | Install | Ea. | 226 | 62 | | 288 | light, vinyl window with screen. |
| | Demolish and Install | Ea. | 226 | 79.95 | | 305.95 | |
| | Reinstall | Ea. | | 49.58 | | 49.58 | |
| | Clean | Ea. | .01 | 8.35 | | 8.36 | |
| | Minimum Charge | Job | | 155 | | 155 | |
| **2' x 3' 6"** | | | | | | | Cost includes material and labor to |
| | Demolish | Ea. | | 17.95 | | 17.95 | install 2' x 3' 6" double hung, two |
| | Install | Ea. | 237 | 66.50 | | 303.50 | light, vinyl window with screen. |
| | Demolish and Install | Ea. | 237 | 84.45 | | 321.45 | |
| | Reinstall | Ea. | | 53.12 | | 53.12 | |
| | Clean | Ea. | .02 | 9.55 | | 9.57 | |
| | Minimum Charge | Job | | 155 | | 155 | |
| **2' 6" x 4' 6"** | | | | | | | Cost includes material and labor to |
| | Demolish | Ea. | | 17.95 | | 17.95 | install 2' 6" x 4' 6" double hung, two |
| | Install | Ea. | 282 | 71.50 | | 353.50 | light, vinyl window with screen. |
| | Demolish and Install | Ea. | 282 | 89.45 | | 371.45 | |
| | Reinstall | Ea. | | 57.21 | | 57.21 | |
| | Clean | Ea. | .02 | 11.10 | | 11.12 | |
| | Minimum Charge | Job | | 155 | | 155 | |
| **3' x 4'** | | | | | | | Includes material and labor to install |
| | Demolish | Ea. | | 17.95 | | 17.95 | medium size double hung, 3' x 4', |
| | Install | Ea. | 250 | 93 | | 343 | two light, vinyl window with screen. |
| | Demolish and Install | Ea. | 250 | 110.95 | | 360.95 | |
| | Reinstall | Ea. | | 74.37 | | 74.37 | |
| | Clean | Ea. | .02 | 11.10 | | 11.12 | |
| | Minimum Charge | Job | | 155 | | 155 | |

113

# Exterior Doors and Windows

| Vinyl Window | | Unit | Material | Labor | Equip. | Total | Specification |
|---|---|---|---|---|---|---|---|
| **3' x 4' 6"** | | | | | | | |
| | Demolish | Ea. | | 22 | | 22 | Cost includes material and labor to |
| | Install | Ea. | 320 | 103 | | 423 | install 3' x 4' 6" double hung, two |
| | Demolish and Install | Ea. | 320 | 125 | | 445 | light, vinyl window with screen. |
| | Reinstall | Ea. | | 82.63 | | 82.63 | |
| | Clean | Ea. | .03 | 13.35 | | 13.38 | |
| | Minimum Charge | Job | | 155 | | 155 | |
| **4' x 4' 6"** | | | | | | | |
| | Demolish | Ea. | | 22 | | 22 | Cost includes material and labor to |
| | Install | Ea. | 340 | 116 | | 456 | install 4' x 4' 6" double hung, two |
| | Demolish and Install | Ea. | 340 | 138 | | 478 | light, vinyl window with screen. |
| | Reinstall | Ea. | | 92.96 | | 92.96 | |
| | Clean | Ea. | .03 | 15.70 | | 15.73 | |
| | Minimum Charge | Job | | 155 | | 155 | |
| **4' x 6'** | | | | | | | |
| | Demolish | Ea. | | 22 | | 22 | Includes material and labor to install |
| | Install | Ea. | 425 | 133 | | 558 | large double hung, 4' x 6', two light, |
| | Demolish and Install | Ea. | 425 | 155 | | 580 | vinyl window with screen. |
| | Reinstall | Ea. | | 106.24 | | 106.24 | |
| | Clean | Ea. | .04 | 19.05 | | 19.09 | |
| | Minimum Charge | Job | | 155 | | 155 | |
| **Trim Set** | | | | | | | |
| | Demolish | Ea. | | 10.75 | | 10.75 | Cost includes material and labor to |
| | Install | Opng. | 35 | 36 | | 71 | install 11/16" x 2-1/2" pine ranch |
| | Demolish and Install | Opng. | 35 | 46.75 | | 81.75 | style window casing. |
| | Reinstall | Opng. | | 28.60 | | 28.60 | |
| | Clean | Opng. | .21 | 3.55 | | 3.76 | |
| | Paint | Ea. | 5 | 7.75 | | 12.75 | |
| | Minimum Charge | Job | | 275 | | 275 | |

| Wood Window | | Unit | Material | Labor | Equip. | Total | Specification |
|---|---|---|---|---|---|---|---|
| **Double Hung** | | | | | | | |
| **2' 2" x 3' 4"** | | | | | | | |
| | Demolish | Ea. | | 17.95 | | 17.95 | Cost includes material and labor to |
| | Install | Ea. | 237 | 62 | | 299 | install 2' 2" x 3' 4" double hung, two |
| | Demolish and Install | Ea. | 237 | 79.95 | | 316.95 | light, double glazed wood window |
| | Reinstall | Ea. | | 49.58 | | 49.58 | with screen. |
| | Clean | Ea. | .02 | 9.55 | | 9.57 | |
| | Paint | Ea. | 3.12 | 16.10 | | 19.22 | |
| | Minimum Charge | Job | | 232 | | 232 | |
| **2' 2" x 4' 4"** | | | | | | | |
| | Demolish | Ea. | | 17.95 | | 17.95 | Cost includes material and labor to |
| | Install | Ea. | 257 | 66.50 | | 323.50 | install 2' 2" x 4' 4" double hung, two |
| | Demolish and Install | Ea. | 257 | 84.45 | | 341.45 | light, double glazed wood window |
| | Reinstall | Ea. | | 53.12 | | 53.12 | with screen. |
| | Clean | Ea. | .02 | 9.55 | | 9.57 | |
| | Paint | Ea. | 3.12 | 16.10 | | 19.22 | |
| | Minimum Charge | Job | | 232 | | 232 | |
| **2' 6" x 3' 4"** | | | | | | | |
| | Demolish | Ea. | | 17.95 | | 17.95 | Cost includes material and labor to |
| | Install | Ea. | 248 | 71.50 | | 319.50 | install 2' 6" x 3' 4" double hung, two |
| | Demolish and Install | Ea. | 248 | 89.45 | | 337.45 | light, double glazed wood window |
| | Reinstall | Ea. | | 57.21 | | 57.21 | with screen. |
| | Clean | Ea. | .02 | 11.10 | | 11.12 | |
| | Paint | Ea. | 4 | 19.30 | | 23.30 | |
| | Minimum Charge | Job | | 232 | | 232 | |

# Exterior Doors and Windows

| Wood Window | | Unit | Material | Labor | Equip. | Total | Specification |
|---|---|---|---|---|---|---|---|
| **2' 6" x 4'** | | | | | | | |
| | Demolish | Ea. | | 17.95 | | 17.95 | Cost includes material and labor to |
| | Install | Ea. | 256 | 77.50 | | 333.50 | install 2' 6" x 4' double hung, two |
| | Demolish and Install | Ea. | 256 | 95.45 | | 351.45 | light, double glazed wood window |
| | Reinstall | Ea. | | 61.97 | | 61.97 | with screen. |
| | Clean | Ea. | .02 | 11.10 | | 11.12 | |
| | Paint | Ea. | 4 | 19.30 | | 23.30 | |
| | Minimum Charge | Job | | 232 | | 232 | |
| **2' 6" x 4' 8"** | | | | | | | |
| | Demolish | Ea. | | 17.95 | | 17.95 | Cost includes material and labor to |
| | Install | Ea. | 278 | 77.50 | | 355.50 | install 2' 6" x 4' 8" double hung, two |
| | Demolish and Install | Ea. | 278 | 95.45 | | 373.45 | light, double glazed wood window |
| | Reinstall | Ea. | | 61.97 | | 61.97 | with screen. |
| | Clean | Ea. | .02 | 11.10 | | 11.12 | |
| | Paint | Ea. | 4 | 19.30 | | 23.30 | |
| | Minimum Charge | Job | | 232 | | 232 | |
| **2' 10" x 3' 4"** | | | | | | | |
| | Demolish | Ea. | | 17.95 | | 17.95 | Cost includes material and labor to |
| | Install | Ea. | 250 | 93 | | 343 | install 2' 10" x 3' 4" double hung, two |
| | Demolish and Install | Ea. | 250 | 110.95 | | 360.95 | light, double glazed wood window |
| | Reinstall | Ea. | | 74.37 | | 74.37 | with screen. |
| | Clean | Ea. | .02 | 11.10 | | 11.12 | |
| | Paint | Ea. | 4 | 19.30 | | 23.30 | |
| | Minimum Charge | Job | | 232 | | 232 | |
| **2' 10" x 4'** | | | | | | | |
| | Demolish | Ea. | | 17.95 | | 17.95 | Cost includes material and labor to |
| | Install | Ea. | 277 | 93 | | 370 | install 2' 10" x 4' double hung, two |
| | Demolish and Install | Ea. | 277 | 110.95 | | 387.95 | light, double glazed wood window |
| | Reinstall | Ea. | | 74.37 | | 74.37 | with screen. |
| | Clean | Ea. | .02 | 11.10 | | 11.12 | |
| | Paint | Ea. | 4 | 19.30 | | 23.30 | |
| | Minimum Charge | Job | | 232 | | 232 | |
| **3' 7" x 3' 4"** | | | | | | | |
| | Demolish | Ea. | | 17.95 | | 17.95 | Cost includes material and labor to |
| | Install | Ea. | 285 | 103 | | 388 | install 3' 7" x 3' 4" double hung, two |
| | Demolish and Install | Ea. | 285 | 120.95 | | 405.95 | light, double glazed wood window |
| | Reinstall | Ea. | | 82.63 | | 82.63 | with screen. |
| | Clean | Ea. | .02 | 11.10 | | 11.12 | |
| | Paint | Ea. | 4 | 19.30 | | 23.30 | |
| | Minimum Charge | Job | | 232 | | 232 | |
| **3' 7" x 5' 4"** | | | | | | | |
| | Demolish | Ea. | | 22 | | 22 | Cost includes material and labor to |
| | Install | Ea. | 325 | 103 | | 428 | install 3' 7" x 5' 4" double hung, two |
| | Demolish and Install | Ea. | 325 | 125 | | 450 | light, double glazed wood window |
| | Reinstall | Ea. | | 82.63 | | 82.63 | with screen. |
| | Clean | Ea. | .04 | 19.05 | | 19.09 | |
| | Paint | Ea. | 6.65 | 32 | | 38.65 | |
| | Minimum Charge | Job | | 232 | | 232 | |
| **4' 3" x 5' 4"** | | | | | | | |
| | Demolish | Ea. | | 22 | | 22 | Cost includes material and labor to |
| | Install | Ea. | 570 | 116 | | 686 | install 4' 3" x 5' 4" double hung, two |
| | Demolish and Install | Ea. | 570 | 138 | | 708 | light, double glazed wood window |
| | Reinstall | Ea. | | 92.96 | | 92.96 | with screen. |
| | Clean | Ea. | .04 | 19.05 | | 19.09 | |
| | Paint | Ea. | 4.83 | 38.50 | | 43.33 | |
| | Minimum Charge | Job | | 232 | | 232 | |

| Wood Window | | Unit | Material | Labor | Equip. | Total | Specification |
|---|---|---|---|---|---|---|---|
| **Casement** | | | | | | | |
| **2' x 3' 4"** | | | | | | | |
| | Demolish | Ea. | | 17.95 | | 17.95 | Cost includes material and labor to |
| | Install | Ea. | 284 | 46.50 | | 330.50 | install 2' x 3' 4 " casement window |
| | Demolish and Install | Ea. | 284 | 64.45 | | 348.45 | with 3/4" insulated glass, screens, |
| | Reinstall | Ea. | | 37.18 | | 37.18 | weatherstripping, hardware. |
| | Clean | Ea. | .02 | 9.55 | | 9.57 | |
| | Paint | Ea. | 3.12 | 16.10 | | 19.22 | |
| | Minimum Charge | Job | | 465 | | 465 | |
| **2' x 4'** | | | | | | | |
| | Demolish | Ea. | | 17.95 | | 17.95 | Cost includes material and labor to |
| | Install | Ea. | 300 | 51.50 | | 351.50 | install 2' x 4' casement window with |
| | Demolish and Install | Ea. | 300 | 69.45 | | 369.45 | 3/4" insulated glass, screens, |
| | Reinstall | Ea. | | 41.32 | | 41.32 | weatherstripping, hardware. |
| | Clean | Ea. | .02 | 9.55 | | 9.57 | |
| | Paint | Ea. | 3.12 | 16.10 | | 19.22 | |
| | Minimum Charge | Job | | 465 | | 465 | |
| **2' x 5'** | | | | | | | |
| | Demolish | Ea. | | 17.95 | | 17.95 | Cost includes material and labor to |
| | Install | Ea. | 345 | 54.50 | | 399.50 | install 2' x 5' casement window with |
| | Demolish and Install | Ea. | 345 | 72.45 | | 417.45 | 3/4" insulated glass, screens, |
| | Reinstall | Ea. | | 43.75 | | 43.75 | weatherstripping, hardware. |
| | Clean | Ea. | .02 | 11.10 | | 11.12 | |
| | Paint | Ea. | 4 | 19.30 | | 23.30 | |
| | Minimum Charge | Job | | 465 | | 465 | |
| **2' x 6'** | | | | | | | |
| | Demolish | Ea. | | 17.95 | | 17.95 | Cost includes material and labor to |
| | Install | Ea. | 345 | 58 | | 403 | install 2' x 6' casement window with |
| | Demolish and Install | Ea. | 345 | 75.95 | | 420.95 | 3/4" insulated glass, screens, |
| | Reinstall | Ea. | | 46.48 | | 46.48 | weatherstripping, hardware. |
| | Clean | Ea. | .02 | 11.10 | | 11.12 | |
| | Paint | Ea. | 4 | 19.30 | | 23.30 | |
| | Minimum Charge | Job | | 465 | | 465 | |
| **4' x 3' 4"** | | | | | | | |
| | Demolish | Ea. | | 22 | | 22 | Cost includes material and labor to |
| | Install | Ea. | 650 | 62 | | 712 | install 4' x 3' 4 " casement windows |
| | Demolish and Install | Ea. | 650 | 84 | | 734 | with 3/4" insulated glass, screens, |
| | Reinstall | Ea. | | 49.58 | | 49.58 | weatherstripping, hardware. |
| | Clean | Ea. | .02 | 11.10 | | 11.12 | |
| | Paint | Ea. | 2.42 | 19.30 | | 21.72 | |
| | Minimum Charge | Job | | 465 | | 465 | |
| **4' x 4'** | | | | | | | |
| | Demolish | Ea. | | 22 | | 22 | Cost includes material and labor to |
| | Install | Ea. | 730 | 62 | | 792 | install 4' x 4' casement windows with |
| | Demolish and Install | Ea. | 730 | 84 | | 814 | 3/4" insulated glass, screens, |
| | Reinstall | Ea. | | 49.58 | | 49.58 | weatherstripping, hardware. |
| | Clean | Ea. | .03 | 15.70 | | 15.73 | |
| | Paint | Ea. | 2.86 | 24 | | 26.86 | |
| | Minimum Charge | Job | | 465 | | 465 | |
| **4' x 5'** | | | | | | | |
| | Demolish | Ea. | | 22 | | 22 | Cost includes material and labor to |
| | Install | Ea. | 825 | 66.50 | | 891.50 | install 4' x 5' casement windows with |
| | Demolish and Install | Ea. | 825 | 88.50 | | 913.50 | 3/4" insulated glass, screens, |
| | Reinstall | Ea. | | 53.12 | | 53.12 | weatherstripping, hardware. |
| | Clean | Ea. | .04 | 19.05 | | 19.09 | |
| | Paint | Ea. | 3.74 | 32 | | 35.74 | |
| | Minimum Charge | Job | | 465 | | 465 | |

# Exterior Doors and Windows

| Wood Window | | Unit | Material | Labor | Equip. | Total | Specification |
|---|---|---|---|---|---|---|---|
| **4' x 6'** | | | | | | | Cost includes material and labor to |
| | Demolish | Ea. | | 22 | | 22 | install 4' x 6' casement windows with |
| | Install | Ea. | 930 | 77.50 | | 1007.50 | 3/4" insulated glass, screens, |
| | Demolish and Install | Ea. | 930 | 99.50 | | 1029.50 | weatherstripping, hardware. |
| | Reinstall | Ea. | | 61.97 | | 61.97 | |
| | Clean | Ea. | .04 | 19.05 | | 19.09 | |
| | Paint | Ea. | 4.39 | 38.50 | | 42.89 | |
| | Minimum Charge | Job | | 465 | | 465 | |
| **Casement Bow** | | | | | | | |
| **4' 8" x 8' 1"** | | | | | | | Includes material and labor to install |
| | Demolish | Ea. | | 30.50 | | 30.50 | casement bow window unit with 4 |
| | Install | Ea. | 1750 | 174 | | 1924 | lights, pine frame, 1/2" insulated |
| | Demolish and Install | Ea. | 1750 | 204.50 | | 1954.50 | glass, weatherstripping, hardware, |
| | Reinstall | Ea. | | 139.44 | | 139.44 | screens. |
| | Clean | Ea. | .07 | 22 | | 22.07 | |
| | Paint | Ea. | 8.15 | 77.50 | | 85.65 | |
| | Minimum Charge | Job | | 465 | | 465 | |
| **4' 8" x 10'** | | | | | | | Includes material and labor to install |
| | Demolish | Ea. | | 30.50 | | 30.50 | casement bow window unit with 5 |
| | Install | Ea. | 2150 | 232 | | 2382 | lights, pine frame, 1/2" insulated |
| | Demolish and Install | Ea. | 2150 | 262.50 | | 2412.50 | glass, weatherstripping, hardware, |
| | Reinstall | Ea. | | 185.92 | | 185.92 | screens. |
| | Clean | Ea. | .01 | 29.50 | | 29.51 | |
| | Paint | Ea. | 10.10 | 96.50 | | 106.60 | |
| | Minimum Charge | Job | | 465 | | 465 | |
| **5' 4" x 8' 1"** | | | | | | | Includes material and labor to install |
| | Demolish | Ea. | | 30.50 | | 30.50 | casement bow window unit with 4 |
| | Install | Ea. | 1900 | 174 | | 2074 | lights, pine frame, 1/2" insulated |
| | Demolish and Install | Ea. | 1900 | 204.50 | | 2104.50 | glass, weatherstripping, hardware, |
| | Reinstall | Ea. | | 139.44 | | 139.44 | screens. |
| | Clean | Ea. | .01 | 29.50 | | 29.51 | |
| | Paint | Ea. | 9.25 | 77.50 | | 86.75 | |
| | Minimum Charge | Job | | 465 | | 465 | |
| **5' 4" x 10'** | | | | | | | Includes material and labor to install |
| | Demolish | Ea. | | 30.50 | | 30.50 | casement bow window unit with 5 |
| | Install | Ea. | 2125 | 232 | | 2357 | lights, pine frame, 1/2" insulated |
| | Demolish and Install | Ea. | 2125 | 262.50 | | 2387.50 | glass, weatherstripping, hardware, |
| | Reinstall | Ea. | | 185.92 | | 185.92 | screens. |
| | Clean | Ea. | .01 | 29.50 | | 29.51 | |
| | Paint | Ea. | 11.65 | 96.50 | | 108.15 | |
| | Minimum Charge | Job | | 465 | | 465 | |
| **6' x 8' 1"** | | | | | | | Includes material and labor to install |
| | Demolish | Ea. | | 30.50 | | 30.50 | casement bow window unit with 5 |
| | Install | Ea. | 2150 | 232 | | 2382 | lights, pine frame, 1/2" insulated |
| | Demolish and Install | Ea. | 2150 | 262.50 | | 2412.50 | glass, weatherstripping, hardware, |
| | Reinstall | Ea. | | 185.92 | | 185.92 | screens. |
| | Clean | Ea. | .01 | 29.50 | | 29.51 | |
| | Paint | Ea. | 10.55 | 96.50 | | 107.05 | |
| | Minimum Charge | Job | | 465 | | 465 | |
| **6' x 10'** | | | | | | | Includes material and labor to install |
| | Demolish | Ea. | | 30.50 | | 30.50 | casement bow window unit with 5 |
| | Install | Ea. | 2250 | 232 | | 2482 | lights, pine frame, 1/2" insulated |
| | Demolish and Install | Ea. | 2250 | 262.50 | | 2512.50 | glass, weatherstripping, hardware, |
| | Reinstall | Ea. | | 185.92 | | 185.92 | screens. |
| | Clean | Ea. | .01 | 29.50 | | 29.51 | |
| | Paint | Ea. | 13.20 | 129 | | 142.20 | |
| | Minimum Charge | Job | | 465 | | 465 | |

**For customer support on your Contractor's Pricing Guide: Residential Repair & Remodeling, call 800.448.8182.**

## Wood Window

| Wood Window | Unit | Material | Labor | Equip. | Total | Specification |
|---|---|---|---|---|---|---|
| **Casing** | | | | | | |
| Demolish | Ea. | | 10.75 | | 10.75 | Cost includes material and labor to |
| Install | Opng. | 35 | 36 | | 71 | install 11/16" x 2-1/2" pine ranch |
| Demolish and Install | Opng. | 35 | 46.75 | | 81.75 | style window casing. |
| Reinstall | Opng. | | 28.60 | | 28.60 | |
| Clean | Opng. | .21 | 3.55 | | 3.76 | |
| Paint | Ea. | 5 | 7.75 | | 12.75 | |
| Minimum Charge | Job | | 275 | | 275 | |
| **Trim Set** | | | | | | |
| Demolish | Ea. | | 8.20 | | 8.20 | Cost includes material and labor to |
| Install | Opng. | 35 | 36 | | 71 | install 11/16" x 2-1/2" pine ranch |
| Demolish and Install | Opng. | 35 | 44.20 | | 79.20 | style window casing. |
| Reinstall | Opng. | | 28.60 | | 28.60 | |
| Clean | Opng. | .21 | 3.55 | | 3.76 | |
| Paint | Ea. | 5 | 7.75 | | 12.75 | |
| Minimum Charge | Job | | 232 | | 232 | |
| **Wood Picture Window** | | | | | | |
| **3' 6" x 4'** | | | | | | |
| Demolish | Ea. | | 22 | | 22 | Includes material and labor to install a |
| Install | Ea. | 490 | 77.50 | | 567.50 | wood framed picture window |
| Demolish and Install | Ea. | 490 | 99.50 | | 589.50 | including exterior trim. |
| Clean | Ea. | .04 | 19.05 | | 19.09 | |
| Paint | Ea. | 3.51 | 29.50 | | 33.01 | |
| Minimum Charge | Job | | 465 | | 465 | |
| **4' x 4' 6"** | | | | | | |
| Demolish | Ea. | | 22 | | 22 | Includes material and labor to install a |
| Install | Ea. | 585 | 84.50 | | 669.50 | wood framed picture window |
| Demolish and Install | Ea. | 585 | 106.50 | | 691.50 | including exterior trim. |
| Clean | Ea. | .04 | 19.05 | | 19.09 | |
| Paint | Ea. | 3.51 | 29.50 | | 33.01 | |
| Minimum Charge | Job | | 465 | | 465 | |
| **5' x 4'** | | | | | | |
| Demolish | Ea. | | 22 | | 22 | Includes material and labor to install a |
| Install | Ea. | 665 | 84.50 | | 749.50 | wood framed picture window |
| Demolish and Install | Ea. | 665 | 106.50 | | 771.50 | including exterior trim. |
| Clean | Ea. | .04 | 19.05 | | 19.09 | |
| Paint | Ea. | 7.50 | 48.50 | | 56 | |
| Minimum Charge | Job | | 465 | | 465 | |
| **6' x 4' 6"** | | | | | | |
| Demolish | Ea. | | 22 | | 22 | Includes material and labor to install a |
| Install | Ea. | 705 | 93 | | 798 | wood framed picture window |
| Demolish and Install | Ea. | 705 | 115 | | 820 | including exterior trim. |
| Clean | Ea. | .04 | 19.05 | | 19.09 | |
| Paint | Ea. | 7.50 | 48.50 | | 56 | |
| Minimum Charge | Job | | 465 | | 465 | |

## Window Screens

| Window Screens | Unit | Material | Labor | Equip. | Total | Specification |
|---|---|---|---|---|---|---|
| **Small** | | | | | | |
| Demolish | Ea. | | 4.11 | | 4.11 | Includes material and labor to install |
| Install | Ea. | 22 | 14.10 | | 36.10 | small window screen, 4 S.F. |
| Demolish and Install | Ea. | 22 | 18.21 | | 40.21 | |
| Reinstall | Ea. | | 11.27 | | 11.27 | |
| Clean | Ea. | .14 | 7.40 | | 7.54 | |
| Minimum Charge | Job | | 116 | | 116 | |

# Exterior Doors and Windows

| Window Screens | | Unit | Material | Labor | Equip. | Total | Specification |
|---|---|---|---|---|---|---|---|
| **Standard Size** | | | | | | | |
| | Demolish | Ea. | | 4.11 | | 4.11 | Includes material and labor to install |
| | Install | Ea. | 53.50 | 14.10 | | 67.60 | medium window screen, 12 S.F. |
| | Demolish and Install | Ea. | 53.50 | 18.21 | | 71.71 | |
| | Reinstall | Ea. | | 11.27 | | 11.27 | |
| | Clean | Ea. | .14 | 7.40 | | 7.54 | |
| | Minimum Charge | Job | | 116 | | 116 | |
| **Large** | | | | | | | |
| | Demolish | Ea. | | 4.11 | | 4.11 | Includes material and labor to install |
| | Install | Ea. | 71 | 14.10 | | 85.10 | large window screen, 16 S.F. |
| | Demolish and Install | Ea. | 71 | 18.21 | | 89.21 | |
| | Reinstall | Ea. | | 11.27 | | 11.27 | |
| | Clean | Ea. | .14 | 7.40 | | 7.54 | |
| | Minimum Charge | Job | | 116 | | 116 | |

| Window Glass | | Unit | Material | Labor | Equip. | Total | Specification |
|---|---|---|---|---|---|---|---|
| **Plate, Clear, 1/4"** | | | | | | | |
| | Demolish | S.F. | | 1.97 | | 1.97 | Includes material and labor to install |
| | Install | S.F. | 7.50 | 7.50 | | 15 | clear glass. |
| | Demolish and Install | S.F. | 7.50 | 9.47 | | 16.97 | |
| | Reinstall | S.F. | | 5.99 | | 5.99 | |
| | Clean | S.F. | .01 | .21 | | .22 | |
| | Minimum Charge | Job | | 225 | | 225 | |
| **Insulated** | | | | | | | |
| | Demolish | S.F. | | 1.97 | | 1.97 | Includes material and labor to install |
| | Install | S.F. | 16 | 12.85 | | 28.85 | insulated glass. |
| | Demolish and Install | S.F. | 16 | 14.82 | | 30.82 | |
| | Reinstall | S.F. | | 10.27 | | 10.27 | |
| | Clean | S.F. | .01 | .21 | | .22 | |
| | Minimum Charge | Job | | 225 | | 225 | |
| **Tinted** | | | | | | | |
| | Demolish | S.F. | | 1.97 | | 1.97 | Includes material and labor to install |
| | Install | S.F. | 7.50 | 6.90 | | 14.40 | tinted glass. |
| | Demolish and Install | S.F. | 7.50 | 8.87 | | 16.37 | |
| | Reinstall | S.F. | | 5.53 | | 5.53 | |
| | Clean | S.F. | .01 | .21 | | .22 | |
| | Minimum Charge | Job | | 225 | | 225 | |
| **Obscure / Florentine** | | | | | | | |
| | Demolish | S.F. | | 1.97 | | 1.97 | Includes material and labor to install |
| | Install | S.F. | 13.55 | 6.40 | | 19.95 | obscure glass. |
| | Demolish and Install | S.F. | 13.55 | 8.37 | | 21.92 | |
| | Reinstall | S.F. | | 5.13 | | 5.13 | |
| | Clean | S.F. | .01 | .21 | | .22 | |
| | Minimum Charge | Job | | 225 | | 225 | |
| **Laminated Safety** | | | | | | | |
| | Demolish | S.F. | | 1.97 | | 1.97 | Includes material and labor to install |
| | Install | S.F. | 14.05 | 10 | | 24.05 | clear laminated glass. |
| | Demolish and Install | S.F. | 14.05 | 11.97 | | 26.02 | |
| | Reinstall | S.F. | | 7.99 | | 7.99 | |
| | Clean | S.F. | .01 | .21 | | .22 | |
| | Minimum Charge | Job | | 225 | | 225 | |

# Exterior Doors and Windows

| Window Glass | | Unit | Material | Labor | Equip. | Total | Specification |
|---|---|---|---|---|---|---|---|
| **Wired** | | | | | | | |
| | Demolish | S.F. | | 1.97 | | 1.97 | Includes material and labor to install |
| | Install | S.F. | 28 | 6.65 | | 34.65 | rough obscure wire glass. |
| | Demolish and Install | S.F. | 28 | 8.62 | | 36.62 | |
| | Reinstall | S.F. | | 5.32 | | 5.32 | |
| | Clean | S.F. | .01 | .21 | | .22 | |
| | Minimum Charge | Job | | 225 | | 225 | |
| **Tempered** | | | | | | | |
| | Demolish | S.F. | | 1.97 | | 1.97 | Includes material and labor to install |
| | Install | S.F. | 7.50 | 7.50 | | 15 | clear glass. |
| | Demolish and Install | S.F. | 7.50 | 9.47 | | 16.97 | |
| | Reinstall | S.F. | | 5.99 | | 5.99 | |
| | Clean | S.F. | .01 | .21 | | .22 | |
| | Minimum Charge | Job | | 225 | | 225 | |
| **Stained** | | | | | | | |
| | Demolish | S.F. | | 1.97 | | 1.97 | Includes material and labor to install |
| | Install | S.F. | 69 | 7.50 | | 76.50 | stained glass. |
| | Demolish and Install | S.F. | 69 | 9.47 | | 78.47 | |
| | Reinstall | S.F. | | 5.99 | | 5.99 | |
| | Clean | S.F. | .01 | .21 | | .22 | |
| | Minimum Charge | Job | | 225 | | 225 | |
| **Etched** | | | | | | | |
| | Demolish | S.F. | | 1.97 | | 1.97 | Includes material and labor to install |
| | Install | S.F. | 56.50 | 7.50 | | 64 | etched glass. |
| | Demolish and Install | S.F. | 56.50 | 9.47 | | 65.97 | |
| | Reinstall | S.F. | | 5.99 | | 5.99 | |
| | Clean | S.F. | .01 | .21 | | .22 | |
| | Minimum Charge | Job | | 225 | | 225 | |
| **Wall Mirror** | | | | | | | |
| | Demolish | S.F. | | 1.97 | | 1.97 | Includes material and labor to install |
| | Install | S.F. | 10.80 | 7.20 | | 18 | distortion-free float glass mirror, 1/4" |
| | Demolish and Install | S.F. | 10.80 | 9.17 | | 19.97 | thick, cut and polished edges. |
| | Reinstall | S.F. | | 5.75 | | 5.75 | |
| | Clean | S.F. | .01 | .21 | | .22 | |
| | Minimum Charge | Job | | 225 | | 225 | |
| **Bullet Resistant** | | | | | | | |
| | Demolish | S.F. | | 1.97 | | 1.97 | Includes material and labor to install |
| | Install | S.F. | 140 | 90 | | 230 | bullet resistant clear laminated glass. |
| | Demolish and Install | S.F. | 140 | 91.97 | | 231.97 | |
| | Reinstall | S.F. | | 71.87 | | 71.87 | |
| | Clean | S.F. | .01 | .21 | | .22 | |
| | Minimum Charge | Job | | 225 | | 225 | |

| Storefront System | | Unit | Material | Labor | Equip. | Total | Specification |
|---|---|---|---|---|---|---|---|
| **Clear Glass** | | | | | | | |
| **Stub Wall to 8'** | | | | | | | |
| | Demolish | S.F. | | 3.08 | | 3.08 | Includes material and labor to install |
| | Install | S.F. | 27 | 12.55 | | 39.55 | commercial grade 4-1/2" section, |
| | Demolish and Install | S.F. | 27 | 15.63 | | 42.63 | anodized aluminum with clear glass in |
| | Reinstall | S.F. | | 10.03 | | 10.03 | upper section and safety glass in |
| | Clean | S.F. | | .44 | | .44 | lower section. |
| | Minimum Charge | Job | | 275 | | 275 | |

# Exterior Doors and Windows

| Storefront System | Unit | Material | Labor | Equip. | Total | Specification |
|---|---|---|---|---|---|---|
| **Floor to 10'** | | | | | | |
| Demolish | S.F. | | 3.08 | | 3.08 | Includes material and labor to install |
| Install | S.F. | 27.50 | 12.55 | | 40.05 | commercial grade 4-1/2" section, |
| Demolish and Install | S.F. | 27.50 | 15.63 | | 43.13 | anodized aluminum with clear glass in |
| Reinstall | S.F. | | 10.03 | | 10.03 | upper section and safety glass in |
| Clean | S.F. | | .44 | | .44 | lower section. |
| Minimum Charge | Job | | 275 | | 275 | |
| **Polished Plate Glass** | | | | | | |
| **Stub Wall to 9'** | | | | | | |
| Demolish | S.F. | | 3.08 | | 3.08 | Cost includes material and labor to |
| Install | S.F. | 17.80 | 12.55 | | 30.35 | install 4' x 9' polished plate glass |
| Demolish and Install | S.F. | 17.80 | 15.63 | | 33.43 | window built on 8" stub wall. |
| Reinstall | S.F. | | 10.03 | | 10.03 | |
| Clean | S.F. | | .44 | | .44 | |
| Minimum Charge | Job | | 275 | | 275 | |
| **Stub Wall to 13'** | | | | | | |
| Demolish | S.F. | | 3.08 | | 3.08 | Cost includes material and labor to |
| Install | S.F. | 19.20 | 12.55 | | 31.75 | install 4' x 13' polished plate glass |
| Demolish and Install | S.F. | 19.20 | 15.63 | | 34.83 | window built on 8" stub wall. |
| Reinstall | S.F. | | 10.03 | | 10.03 | |
| Clean | S.F. | | .44 | | .44 | |
| Minimum Charge | Job | | 275 | | 275 | |
| **Floor to 9'** | | | | | | |
| Demolish | S.F. | | 3.08 | | 3.08 | Cost includes material and labor to |
| Install | S.F. | 19.85 | 12.55 | | 32.40 | install 4' x 9' polished plate glass |
| Demolish and Install | S.F. | 19.85 | 15.63 | | 35.48 | window built on floor. |
| Reinstall | S.F. | | 10.03 | | 10.03 | |
| Clean | S.F. | | .44 | | .44 | |
| Minimum Charge | Job | | 275 | | 275 | |
| **Floor to 13'** | | | | | | |
| Demolish | S.F. | | 3.08 | | 3.08 | Cost includes material and labor to |
| Install | S.F. | 21.50 | 12.55 | | 34.05 | install 4' x 13' polished plate glass |
| Demolish and Install | S.F. | 21.50 | 15.63 | | 37.13 | window built on floor. |
| Reinstall | S.F. | | 10.03 | | 10.03 | |
| Clean | S.F. | | .44 | | .44 | |
| Minimum Charge | Job | | 275 | | 275 | |
| **Clear Plate Glass** | | | | | | |
| **Floor to 9'** | | | | | | |
| Demolish | S.F. | | 3.08 | | 3.08 | Cost includes material and labor to |
| Install | S.F. | 28 | 12.55 | | 40.55 | install 4' x 9' clear plate glass window |
| Demolish and Install | S.F. | 28 | 15.63 | | 43.63 | built on floor. |
| Reinstall | S.F. | | 10.03 | | 10.03 | |
| Clean | S.F. | | .44 | | .44 | |
| Minimum Charge | Job | | 275 | | 275 | |
| **Floor to 11'** | | | | | | |
| Demolish | S.F. | | 3.08 | | 3.08 | Cost includes material and labor to |
| Install | S.F. | 27.50 | 12.55 | | 40.05 | install 4' x 11' clear plate glass |
| Demolish and Install | S.F. | 27.50 | 15.63 | | 43.13 | window built on floor. |
| Reinstall | S.F. | | 10.03 | | 10.03 | |
| Clean | S.F. | | .44 | | .44 | |
| Minimum Charge | Job | | 275 | | 275 | |
| **Floor to 13'** | | | | | | |
| Demolish | S.F. | | 3.08 | | 3.08 | Cost includes material and labor to |
| Install | S.F. | 27.50 | 12.55 | | 40.05 | install 4' x 13' clear plate glass |
| Demolish and Install | S.F. | 27.50 | 15.63 | | 43.13 | window built on floor. |
| Reinstall | S.F. | | 10.03 | | 10.03 | |
| Clean | S.F. | | .44 | | .44 | |
| Minimum Charge | Job | | 275 | | 275 | |

# Exterior Doors and Windows

| Storefront System | | Unit | Material | Labor | Equip. | Total | Specification |
|---|---|---|---|---|---|---|---|
| **Entrance** | | | | | | | |
| **Narrow Stile 3' x 7'** | | | | | | | |
| | Demolish | Ea. | | 26 | | 26 | Cost includes material and labor to |
| | Install | Ea. | 1100 | 310 | | 1410 | install 3' x 7' narrow stile door, center |
| | Demolish and Install | Ea. | 1100 | 336 | | 1436 | pivot concealed closer. |
| | Reinstall | Ea. | | 247.89 | | 247.89 | |
| | Clean | Ea. | 1.84 | 44.50 | | 46.34 | |
| | Minimum Charge | Job | | 275 | | 275 | |
| **Tempered Glass** | | | | | | | |
| | Demolish | Ea. | | 26 | | 26 | Cost includes material and labor to |
| | Install | Ea. | 3375 | 595 | | 3970 | install 3' x 7' tempered glass door, |
| | Demolish and Install | Ea. | 3375 | 621 | | 3996 | center pivot concealed closer. |
| | Reinstall | Ea. | | 476.37 | | 476.37 | |
| | Clean | Ea. | 1.84 | 44.50 | | 46.34 | |
| | Minimum Charge | Job | | 275 | | 275 | |
| **Heavy Section** | | | | | | | |
| | Demolish | Ea. | | 26 | | 26 | Includes material and labor to install |
| | Install | Ea. | 1100 | 550 | | 1650 | wide stile door with 3' 6" x 7' |
| | Demolish and Install | Ea. | 1100 | 576 | | 1676 | opening. |
| | Reinstall | Ea. | | 440.64 | | 440.64 | |
| | Clean | Ea. | .36 | 14.80 | | 15.16 | |
| | Minimum Charge | Job | | 900 | | 900 | |

| Curtain Wall System | | Unit | Material | Labor | Equip. | Total | Specification |
|---|---|---|---|---|---|---|---|
| **Regular Weight** | | | | | | | |
| **Bronze Anodized** | | | | | | | |
| | Demolish | S.F. | | 3.08 | | 3.08 | Includes material and labor to install |
| | Install | S.F. | 44 | 12.55 | | 56.55 | structural curtain wall system with |
| | Demolish and Install | S.F. | 44 | 15.63 | | 59.63 | bronze anodized frame of regular |
| | Clean | S.F. | | .44 | | .44 | weight, float glass. |
| | Minimum Charge | Job | | 232 | | 232 | |
| **Black Anodized** | | | | | | | |
| | Demolish | S.F. | | 3.08 | | 3.08 | Includes material and labor to install |
| | Install | S.F. | 48 | 12.55 | | 60.55 | structural curtain wall system with |
| | Demolish and Install | S.F. | 48 | 15.63 | | 63.63 | black anodized frame of regular |
| | Clean | S.F. | | .44 | | .44 | weight, float glass. |
| | Minimum Charge | Job | | 232 | | 232 | |
| **Heavy Weight** | | | | | | | |
| **Bronze Anodized** | | | | | | | |
| | Demolish | S.F. | | 3.08 | | 3.08 | Includes material and labor to install |
| | Install | S.F. | 44 | 12.55 | | 56.55 | structural curtain wall system with |
| | Demolish and Install | S.F. | 44 | 15.63 | | 59.63 | bronze anodized frame of regular |
| | Clean | S.F. | | .44 | | .44 | weight, float glass. |
| | Minimum Charge | Job | | 232 | | 232 | |
| **Black Anodized** | | | | | | | |
| | Demolish | S.F. | | 3.08 | | 3.08 | Includes material and labor to install |
| | Install | S.F. | 45.50 | 12.55 | | 58.05 | structural curtain wall system with |
| | Demolish and Install | S.F. | 45.50 | 15.63 | | 61.13 | black anodized frame of heavy |
| | Clean | S.F. | | .44 | | .44 | weight, float glass. |
| | Minimum Charge | Job | | 232 | | 232 | |
| **Glass** | | | | | | | |
| **1" Insulated** | | | | | | | |
| | Install | S.F. | 19.10 | 12 | | 31.10 | Cost includes labor and material to |
| | Minimum Charge | Job | | 225 | | 225 | install 1" insulated glass. |

122

| Skylight | Unit | Material | Labor | Equip. | Total | Specification |
|---|---|---|---|---|---|---|
| **Dome** | | | | | | |
| **22″ x 22″** | | | | | | |
| Demolish | Ea. | | 91.50 | | 91.50 | Cost includes material and labor to |
| Install | Ea. | 238 | 144 | | 382 | install 22″ x 22″ fixed dome skylight. |
| Demolish and Install | Ea. | 238 | 235.50 | | 473.50 | |
| Reinstall | Ea. | | 115.31 | | 115.31 | |
| Clean | Ea. | .17 | 12.70 | | 12.87 | |
| Minimum Charge | Job | | 865 | | 865 | |
| **22″ x 46″** | | | | | | |
| Demolish | Ea. | | 91.50 | | 91.50 | Cost includes material and labor to |
| Install | Ea. | 296 | 173 | | 469 | install 22″ x 46″ fixed dome skylight. |
| Demolish and Install | Ea. | 296 | 264.50 | | 560.50 | |
| Reinstall | Ea. | | 138.37 | | 138.37 | |
| Clean | Ea. | | 19.75 | | 19.75 | |
| Minimum Charge | Job | | 865 | | 865 | |
| **30″ x 30″** | | | | | | |
| Demolish | Ea. | | 91.50 | | 91.50 | Cost includes material and labor to |
| Install | Ea. | 315 | 144 | | 459 | install 30″ x 30″ fixed dome skylight. |
| Demolish and Install | Ea. | 315 | 235.50 | | 550.50 | |
| Reinstall | Ea. | | 115.31 | | 115.31 | |
| Clean | Ea. | | 19.75 | | 19.75 | |
| Minimum Charge | Job | | 865 | | 865 | |
| **30″ x 46″** | | | | | | |
| Demolish | Ea. | | 91.50 | | 91.50 | Cost includes material and labor to |
| Install | Ea. | 420 | 173 | | 593 | install 30″ x 46″ fixed dome skylight. |
| Demolish and Install | Ea. | 420 | 264.50 | | 684.50 | |
| Reinstall | Ea. | | 138.37 | | 138.37 | |
| Clean | Ea. | | 19.75 | | 19.75 | |
| Minimum Charge | Job | | 865 | | 865 | |
| **Fixed** | | | | | | |
| **22″ x 27″** | | | | | | |
| Demolish | Ea. | | 91.50 | | 91.50 | Includes material and labor to install |
| Install | Ea. | 305 | 144 | | 449 | double glazed skylight, 22″ x 27″, |
| Demolish and Install | Ea. | 305 | 235.50 | | 540.50 | fixed. |
| Reinstall | Ea. | | 115.31 | | 115.31 | |
| Clean | Ea. | .17 | 12.70 | | 12.87 | |
| Minimum Charge | Job | | 865 | | 865 | |
| **22″ x 46″** | | | | | | |
| Demolish | Ea. | | 91.50 | | 91.50 | Includes material and labor to install |
| Install | Ea. | 320 | 173 | | 493 | double glazed skylight, 22″ x 46″, |
| Demolish and Install | Ea. | 320 | 264.50 | | 584.50 | fixed. |
| Reinstall | Ea. | | 138.37 | | 138.37 | |
| Clean | Ea. | | 19.75 | | 19.75 | |
| Minimum Charge | Job | | 865 | | 865 | |
| **44″ x 46″** | | | | | | |
| Demolish | Ea. | | 91.50 | | 91.50 | Includes material and labor to install |
| Install | Ea. | 475 | 173 | | 648 | double glazed skylight, 44″ x 46″, |
| Demolish and Install | Ea. | 475 | 264.50 | | 739.50 | fixed. |
| Reinstall | Ea. | | 138.37 | | 138.37 | |
| Clean | Ea. | | 22 | | 22 | |
| Minimum Charge | Job | | 865 | | 865 | |
| **Operable** | | | | | | |
| **22″ x 27″** | | | | | | |
| Demolish | Ea. | | 91.50 | | 91.50 | Includes material and labor to install |
| Install | Ea. | 445 | 144 | | 589 | double glazed skylight, 22″ x 27″, |
| Demolish and Install | Ea. | 445 | 235.50 | | 680.50 | operable. |
| Reinstall | Ea. | | 115.31 | | 115.31 | |
| Clean | Ea. | .17 | 12.70 | | 12.87 | |
| Minimum Charge | Job | | 865 | | 865 | |

# Exterior Doors and Windows

## Skylight

| | Unit | Material | Labor | Equip. | Total | Specification |
|---|---|---|---|---|---|---|
| **22" x 46"** | | | | | | |
| Demolish | Ea. | | 91.50 | | 91.50 | Includes material and labor to install |
| Install | Ea. | 510 | 173 | | 683 | double glazed skylight, 22" x 46", |
| Demolish and Install | Ea. | 510 | 264.50 | | 774.50 | operable. |
| Reinstall | Ea. | | 138.37 | | 138.37 | |
| Clean | Ea. | | 19.75 | | 19.75 | |
| Minimum Charge | Job | | 865 | | 865 | |
| **44" x 46"** | | | | | | |
| Demolish | Ea. | | 91.50 | | 91.50 | Includes material and labor to install |
| Install | Ea. | 1025 | 173 | | 1198 | double glazed skylight, 44" x 46", |
| Demolish and Install | Ea. | 1025 | 264.50 | | 1289.50 | operable. |
| Reinstall | Ea. | | 138.37 | | 138.37 | |
| Clean | Ea. | | 22 | | 22 | |
| Minimum Charge | Job | | 865 | | 865 | |

## Industrial Steel Window

| | Unit | Material | Labor | Equip. | Total | Specification |
|---|---|---|---|---|---|---|
| **100% Fixed** | | | | | | |
| Demolish | S.F. | | 2.53 | | 2.53 | Includes material and labor to install |
| Install | S.F. | 32.50 | 5.50 | | 38 | commercial grade steel windows, |
| Demolish and Install | S.F. | 32.50 | 8.03 | | 40.53 | including glazing. |
| Reinstall | S.F. | | 4.41 | | 4.41 | |
| Clean | Ea. | .07 | 1.39 | | 1.46 | |
| Paint | S.F. | .18 | .86 | | 1.04 | |
| Minimum Charge | Job | | 275 | | 275 | |
| **50% Vented** | | | | | | |
| Demolish | S.F. | | 2.53 | | 2.53 | Includes material and labor to install |
| Install | S.F. | 61 | 5.50 | | 66.50 | commercial grade steel windows, |
| Demolish and Install | S.F. | 61 | 8.03 | | 69.03 | including glazing. |
| Reinstall | S.F. | | 4.41 | | 4.41 | |
| Clean | Ea. | .07 | 1.39 | | 1.46 | |
| Paint | S.F. | .18 | .86 | | 1.04 | |
| Minimum Charge | Job | | 275 | | 275 | |

**Brick Veneer**

**Concrete Masonry Units**

| Masonry Wall | | Unit | Material | Labor | Equip. | Total | Specification |
|---|---|---|---|---|---|---|---|
| **Unreinforced CMU** | | | | | | | |
| **4" x 8" x 16"** | | | | | | | |
| | Demolish | S.F. | | .14 | .30 | .44 | Includes material and labor to install |
| | Install | S.F. | 2.11 | 4.69 | | 6.80 | normal weight 4" x 8" x 16" block. |
| | Demolish and Install | S.F. | 2.11 | 4.83 | .30 | 7.24 | |
| | Clean | S.F. | .09 | .55 | | .64 | |
| | Paint | S.F. | .18 | .88 | | 1.06 | |
| | Minimum Charge | Job | | 232 | | 232 | |
| **6" x 8" x 16"** | | | | | | | |
| | Demolish | S.F. | | .18 | .40 | .58 | Includes material and labor to install |
| | Install | S.F. | 2.75 | 4.90 | | 7.65 | normal weight 6" x 8" x 16" block. |
| | Demolish and Install | S.F. | 2.75 | 5.08 | .40 | 8.23 | |
| | Clean | S.F. | .09 | .55 | | .64 | |
| | Paint | S.F. | .18 | .88 | | 1.06 | |
| | Minimum Charge | Job | | 232 | | 232 | |
| **8" x 8" x 16"** | | | | | | | |
| | Demolish | S.F. | | .36 | .80 | 1.16 | Includes material and labor to install |
| | Install | S.F. | 2.93 | 5.40 | | 8.33 | normal weight 8" x 8" x 16" block. |
| | Demolish and Install | S.F. | 2.93 | 5.76 | .80 | 9.49 | |
| | Clean | S.F. | .09 | .55 | | .64 | |
| | Paint | S.F. | .18 | .88 | | 1.06 | |
| | Minimum Charge | Job | | 232 | | 232 | |
| **12" x 8" x 16"** | | | | | | | |
| | Demolish | S.F. | | .36 | .80 | 1.16 | Includes material and labor to install |
| | Install | S.F. | 4.80 | 8.20 | | 13 | normal weight 12" x 8" x 16" block. |
| | Demolish and Install | S.F. | 4.80 | 8.56 | .80 | 14.16 | |
| | Clean | S.F. | .09 | .55 | | .64 | |
| | Paint | S.F. | .18 | .88 | | 1.06 | |
| | Minimum Charge | Job | | 232 | | 232 | |
| **Reinforced CMU** | | | | | | | |
| **4" x 8" x 16"** | | | | | | | |
| | Demolish | S.F. | | 2.19 | .70 | 2.89 | Includes material and labor to install |
| | Install | S.F. | 2.31 | 4.79 | | 7.10 | normal weight 4" x 8" x 16" block, |
| | Demolish and Install | S.F. | 2.31 | 6.98 | .70 | 9.99 | mortar and 1/2" re-bar installed |
| | Clean | S.F. | .09 | .55 | | .64 | vertically every 24" O.C. |
| | Paint | S.F. | .18 | .88 | | 1.06 | |
| | Minimum Charge | Job | | 425 | | 425 | |
| **6" x 8" x 16"** | | | | | | | |
| | Demolish | S.F. | | 2.31 | .51 | 2.82 | Includes material and labor to install |
| | Install | S.F. | 2.95 | 5 | | 7.95 | normal weight 6" x 8" x 16" block, |
| | Demolish and Install | S.F. | 2.95 | 7.31 | .51 | 10.77 | mortar and 1/2" re-bar installed |
| | Clean | S.F. | .09 | .55 | | .64 | vertically every 24" O.C. |
| | Paint | S.F. | .18 | .88 | | 1.06 | |
| | Minimum Charge | Job | | 425 | | 425 | |

125

# Masonry

| Masonry Wall | | Unit | Material | Labor | Equip. | Total | Specification |
|---|---|---|---|---|---|---|---|
| **8" x 8" x 16"** | | | | | | | |
| | Demolish | S.F. | | 2.44 | .54 | 2.98 | Includes material and labor to install |
| | Install | S.F. | 3.17 | 5.45 | | 8.62 | normal weight 8" x 8" x 16" block, |
| | Demolish and Install | S.F. | 3.17 | 7.89 | .54 | 11.60 | mortar and 1/2" re-bar installed |
| | Clean | S.F. | .09 | .55 | | .64 | vertically every 24" O.C. |
| | Paint | S.F. | .18 | .88 | | 1.06 | |
| | Minimum Charge | Job | | 425 | | 425 | |
| **12" x 8" x 16"** | | | | | | | |
| | Demolish | S.F. | | 2.51 | .55 | 3.06 | Includes material and labor to install |
| | Install | S.F. | 4.99 | 8.45 | | 13.44 | normal weight 12" x 8" x 16" block, |
| | Demolish and Install | S.F. | 4.99 | 10.96 | .55 | 16.50 | mortar and 1/2" re-bar installed |
| | Clean | S.F. | .09 | .55 | | .64 | vertically every 24" O.C. |
| | Paint | S.F. | .18 | .88 | | 1.06 | |
| | Minimum Charge | Job | | 425 | | 425 | |

## Structural Brick

| | | Unit | Material | Labor | Equip. | Total | Specification |
|---|---|---|---|---|---|---|---|
| **4" Single Row** | | | | | | | |
| | Demolish | S.F. | | 3.16 | | 3.16 | Cost includes material and labor to |
| | Install | S.F. | 4.39 | 10.05 | | 14.44 | install 4" red brick wall, running bond |
| | Demolish and Install | S.F. | 4.39 | 13.21 | | 17.60 | with 3/8" concave joints, mortar and |
| | Clean | S.F. | .09 | .47 | | .56 | ties. |
| | Paint | S.F. | .18 | .88 | | 1.06 | |
| | Minimum Charge | Job | | 425 | | 425 | |
| **8" Double Row** | | | | | | | |
| | Demolish | S.F. | | 3.16 | | 3.16 | Cost includes material and labor to |
| | Install | S.F. | 10.20 | 16 | | 26.20 | install 8" red brick wall, running bond, |
| | Demolish and Install | S.F. | 10.20 | 19.16 | | 29.36 | with 3/8" concave joints, mortar and |
| | Clean | S.F. | .09 | .47 | | .56 | wall ties. |
| | Paint | S.F. | .18 | .88 | | 1.06 | |
| | Minimum Charge | Job | | 425 | | 425 | |
| **12" Triple Row** | | | | | | | |
| | Demolish | S.F. | | 4.39 | | 4.39 | Cost includes material and labor to |
| | Install | S.F. | 15.30 | 22.50 | | 37.80 | install 12" red brick wall, running |
| | Demolish and Install | S.F. | 15.30 | 26.89 | | 42.19 | bond, with 3/8" concave joints, |
| | Clean | S.F. | .09 | .47 | | .56 | mortar and wall ties. |
| | Paint | S.F. | .18 | .88 | | 1.06 | |
| | Minimum Charge | Job | | 425 | | 425 | |

## Brick Veneer

| | | Unit | Material | Labor | Equip. | Total | Specification |
|---|---|---|---|---|---|---|---|
| **4" Red Common** | | | | | | | |
| | Demolish | S.F. | | 3.16 | | 3.16 | Cost includes material and labor to |
| | Install | S.F. | 4.83 | 9.40 | | 14.23 | install 4" common red face brick, |
| | Demolish and Install | S.F. | 4.83 | 12.56 | | 17.39 | running bond with 3/8" concave |
| | Clean | S.F. | .09 | .47 | | .56 | joints, mortar and wall ties. |
| | Paint | S.F. | .18 | .88 | | 1.06 | |
| | Minimum Charge | Job | | 232 | | 232 | |
| **4" Used** | | | | | | | |
| | Demolish | S.F. | | 3.16 | | 3.16 | Cost includes material and labor to |
| | Install | S.F. | 4.39 | 10.05 | | 14.44 | install 4" red brick wall, running bond |
| | Demolish and Install | S.F. | 4.39 | 13.21 | | 17.60 | with 3/8" concave joints, mortar and |
| | Clean | S.F. | .09 | .47 | | .56 | ties. |
| | Paint | S.F. | .18 | .88 | | 1.06 | |
| | Minimum Charge | Job | | 232 | | 232 | |

# Masonry

| Masonry Wall | | Unit | Material | Labor | Equip. | Total | Specification |
|---|---|---|---|---|---|---|---|
| **6" Jumbo** | | | | | | | |
| | Demolish | S.F. | | 3.16 | | 3.16 | Cost includes material and labor to |
| | Install | S.F. | 6 | 4.96 | | 10.96 | install 6" red jumbo brick veneer, |
| | Demolish and Install | S.F. | 6 | 8.12 | | 14.12 | running bond with 3/8" concave |
| | Clean | S.F. | .09 | .47 | | .56 | joints, mortar and ties. |
| | Paint | S.F. | .18 | .88 | | 1.06 | |
| | Minimum Charge | Job | | 232 | | 232 | |
| **Norman (11-1/2")** | | | | | | | |
| | Demolish | S.F. | | 3.16 | | 3.16 | Cost includes material and labor to |
| | Install | S.F. | 7.05 | 6.75 | | 13.80 | install 4" norman red brick wall, |
| | Demolish and Install | S.F. | 7.05 | 9.91 | | 16.96 | running bond with 3/8" concave |
| | Clean | S.F. | .09 | .47 | | .56 | joints, mortar and ties. |
| | Paint | S.F. | .18 | .88 | | 1.06 | |
| | Minimum Charge | Job | | 232 | | 232 | |
| **Glazed** | | | | | | | |
| | Demolish | S.F. | | 3.16 | | 3.16 | Includes material, labor and |
| | Install | S.F. | 21 | 10.25 | | 31.25 | equipment to install glazed brick. |
| | Demolish and Install | S.F. | 21 | 13.41 | | 34.41 | |
| | Clean | S.F. | .09 | .47 | | .56 | |
| | Paint | S.F. | .18 | .88 | | 1.06 | |
| | Minimum Charge | Job | | 232 | | 232 | |
| **Slumpstone** | | | | | | | |
| **6" x 4"** | | | | | | | |
| | Demolish | S.F. | | .14 | .30 | .44 | Cost includes material and labor to |
| | Install | S.F. | 6.90 | 5.30 | | 12.20 | install 6" x 4" x 16" concrete slump |
| | Demolish and Install | S.F. | 6.90 | 5.44 | .30 | 12.64 | block. |
| | Clean | S.F. | .09 | .55 | | .64 | |
| | Paint | S.F. | .18 | .88 | | 1.06 | |
| | Minimum Charge | Job | | 232 | | 232 | |
| **6" x 6"** | | | | | | | |
| | Demolish | S.F. | | .18 | .40 | .58 | Cost includes material and labor to |
| | Install | S.F. | 6.40 | 5.45 | | 11.85 | install 6" x 6" x 16" concrete slump |
| | Demolish and Install | S.F. | 6.40 | 5.63 | .40 | 12.43 | block. |
| | Clean | S.F. | .09 | .55 | | .64 | |
| | Paint | S.F. | .18 | .88 | | 1.06 | |
| | Minimum Charge | Job | | 232 | | 232 | |
| **8" x 6"** | | | | | | | |
| | Demolish | S.F. | | .18 | .40 | .58 | Cost includes material and labor to |
| | Install | S.F. | 9.65 | 5.65 | | 15.30 | install 8" x 6" x 16" concrete slump |
| | Demolish and Install | S.F. | 9.65 | 5.83 | .40 | 15.88 | block. |
| | Clean | S.F. | .09 | .55 | | .64 | |
| | Paint | S.F. | .18 | .88 | | 1.06 | |
| | Minimum Charge | Job | | 232 | | 232 | |
| **12" x 4"** | | | | | | | |
| | Demolish | S.F. | | .14 | .30 | .44 | Cost includes material and labor to |
| | Install | S.F. | 14.40 | 6.50 | | 20.90 | install 12" x 4" x 16" concrete slump |
| | Demolish and Install | S.F. | 14.40 | 6.64 | .30 | 21.34 | block. |
| | Clean | S.F. | .09 | .55 | | .64 | |
| | Paint | S.F. | .18 | .88 | | 1.06 | |
| | Minimum Charge | Job | | 232 | | 232 | |
| **12" x 6"** | | | | | | | |
| | Demolish | S.F. | | .18 | .40 | .58 | Cost includes material and labor to |
| | Install | S.F. | 15.35 | 7.05 | | 22.40 | install 12" x 6" x 16" concrete slump |
| | Demolish and Install | S.F. | 15.35 | 7.23 | .40 | 22.98 | block. |
| | Clean | S.F. | .09 | .55 | | .64 | |
| | Paint | S.F. | .18 | .88 | | 1.06 | |
| | Minimum Charge | Job | | 232 | | 232 | |

# Masonry

| Masonry Wall | Unit | Material | Labor | Equip. | Total | Specification |
|---|---|---|---|---|---|---|
| **Stucco** | | | | | | |
| **Three Coat** | | | | | | |
| Demolish | S.Y. | | 2.72 | | 2.72 | Includes material and labor to install |
| Install | S.Y. | 4.80 | 10.50 | .73 | 16.03 | stucco on exterior masonry walls. |
| Demolish and Install | S.Y. | 4.80 | 13.22 | .73 | 18.75 | |
| Clean | S.Y. | | 4.93 | | 4.93 | |
| Paint | S.F. | .25 | .64 | | .89 | |
| Minimum Charge | Job | | 222 | | 222 | |
| **Three Coat w / Wire Mesh** | | | | | | |
| Demolish | S.Y. | | 4.93 | | 4.93 | Includes material and labor to install |
| Install | S.Y. | 9.20 | 40.50 | 2.32 | 52.02 | stucco on wire mesh over wood |
| Demolish and Install | S.Y. | 9.20 | 45.43 | 2.32 | 56.95 | framing. |
| Clean | S.Y. | | 4.93 | | 4.93 | |
| Paint | S.F. | .25 | .64 | | .89 | |
| Minimum Charge | Job | | 222 | | 222 | |
| **Fill Cracks** | | | | | | |
| Install | S.F. | .46 | 3.56 | | 4.02 | Includes labor and material to fill in |
| Minimum Charge | Job | | 222 | | 222 | hairline cracks up to 1/8" with filler, including sand and fill per S.F. |
| **Furring** | | | | | | |
| **1" x 2"** | | | | | | |
| Demolish | L.F. | | .16 | | .16 | Cost includes material and labor to |
| Install | L.F. | .32 | .94 | | 1.26 | install 1" x 2" furring on masonry wall. |
| Demolish and Install | L.F. | .32 | 1.10 | | 1.42 | |
| Minimum Charge | Job | | 232 | | 232 | |
| **2" x 2"** | | | | | | |
| Demolish | L.F. | | .17 | | .17 | Cost includes material and labor to |
| Install | L.F. | .23 | 1.03 | | 1.26 | install 2" x 2" furring on masonry wall. |
| Demolish and Install | L.F. | .23 | 1.20 | | 1.43 | |
| Minimum Charge | Job | | 232 | | 232 | |
| **2" x 4"** | | | | | | |
| Demolish | L.F. | | .19 | | .19 | Cost includes material and labor to |
| Install | L.F. | .56 | 1.18 | | 1.74 | install 2" x 4" furring on masonry wall. |
| Demolish and Install | L.F. | .56 | 1.37 | | 1.93 | |
| Reinstall | L.F. | | .94 | | .94 | |
| Minimum Charge | Job | | 232 | | 232 | |
| **Maintenance / Cleaning** | | | | | | |
| **Repoint** | | | | | | |
| Install | S.F. | .62 | 5.80 | | 6.42 | Includes labor and material to cut and |
| Minimum Charge | Hr. | | 58 | | 58 | repoint brick, hard mortar, running bond. |
| **Acid Etch** | | | | | | |
| Install | S.F. | .06 | .83 | | .89 | Includes labor and material to acid |
| Minimum Charge | Job | | 178 | 39 | 217 | wash smooth brick. |
| **Pressure Wash** | | | | | | |
| Clean | S.F. | | .95 | .19 | 1.14 | Includes labor and material to high |
| Minimum Charge | Job | | 178 | 39 | 217 | pressure water wash masonry, average. |
| **Waterproof** | | | | | | |
| Install | S.F. | .41 | .87 | | 1.28 | Includes labor and material to install |
| Minimum Charge | Job | | 218 | | 218 | masonry waterproofing. |

# Masonry

## Chimney

| Chimney | | Unit | Material | Labor | Equip. | Total | Specification |
|---|---|---|---|---|---|---|---|
| **Demo & Haul Exterior** | Demolish | C.F. | | 6.45 | | 6.45 | Includes minimum labor and |
| | Minimum Charge | Job | | 197 | | 197 | equipment to remove masonry and haul debris to truck or dumpster. |
| **Demo & Haul Interior** | Demolish | C.F. | | 8.90 | | 8.90 | Includes minimum labor and |
| | Minimum Charge | Job | | 197 | | 197 | equipment to remove masonry and haul debris to truck or dumpster. |
| **Install New (w / flue)** **Brick** | Install | V.L.F. | 44 | 53 | | 97 | Includes material and labor to install |
| | Minimum Charge | Job | | 232 | | 232 | brick chimney with flue. Foundation not included. |
| **Natural Stone** | Install | L.F. | 76.50 | 84.50 | | 161 | Includes labor and material to install |
| | Minimum Charge | Job | | 232 | | 232 | natural stone chimney. |
| **Install From Roof Up** **Brick** | Install | V.L.F. | 44 | 53 | | 97 | Includes material and labor to install |
| | Minimum Charge | Job | | 232 | | 232 | brick chimney with flue. Foundation not included. |
| **Natural Stone** | Install | L.F. | 76.50 | 70.50 | | 147 | Includes labor and material to install |
| | Minimum Charge | Job | | 232 | | 232 | natural stone chimney. |
| **Install Base Pad** | Install | Ea. | 297 | 258 | 1.17 | 556.17 | Includes labor and material to replace |
| | Minimum Charge | Job | | 1125 | 28 | 1153 | spread footing, 5' square x 16" deep. |
| **Metal Cap** | Install | Ea. | 61.50 | 29 | | 90.50 | Includes labor and materials to install |
| | Minimum Charge | Job | | 232 | | 232 | bird and squirrel screens in a chimney flue. |

## Fireplace

| Fireplace | | Unit | Material | Labor | Equip. | Total | Specification |
|---|---|---|---|---|---|---|---|
| **Common Brick** | | | | | | | |
| **30"** | Demolish | Ea. | | 1400 | | 1400 | Includes material and labor to install a |
| | Install | Ea. | 2175 | 4225 | | 6400 | complete unit which includes common |
| | Demolish and Install | Ea. | 2175 | 5625 | | 7800 | brick, foundation, damper, flue lining |
| | Minimum Charge | Job | | 232 | | 232 | stack to 15' high and a 30" fire box. |
| **36"** | Demolish | Ea. | | 1400 | | 1400 | Includes material and labor to install a |
| | Install | Ea. | 2275 | 4700 | | 6975 | complete unit which includes common |
| | Demolish and Install | Ea. | 2275 | 6100 | | 8375 | brick, foundation, damper, flue lining |
| | Minimum Charge | Job | | 232 | | 232 | stack to 15' high and a 36" fire box. |
| **42"** | Demolish | Ea. | | 1400 | | 1400 | Includes material and labor to install a |
| | Install | Ea. | 2475 | 5300 | | 7775 | complete unit which includes common |
| | Demolish and Install | Ea. | 2475 | 6700 | | 9175 | brick, foundation, damper, flue lining |
| | Minimum Charge | Job | | 232 | | 232 | stack to 15' high and a 42" fire box. |

129

| Fireplace | | Unit | Material | Labor | Equip. | Total | Specification |
|---|---|---|---|---|---|---|---|
| **48"** | | | | | | | Includes material and labor to install a complete unit which includes common brick, foundation, damper, flue lining stack to 15' high and a 48" fire box. |
| | Demolish | Ea. | | 1400 | | 1400 | |
| | Install | Ea. | 2625 | 6025 | | 8650 | |
| | Demolish and Install | Ea. | 2625 | 7425 | | 10050 | |
| | Minimum Charge | Job | | 232 | | 232 | |
| **Raised Hearth** | | | | | | | |
| **Brick Hearth** | | | | | | | Includes material and labor to install a brick fireplace hearth based on 30" x 29" opening. |
| | Demolish | S.F. | | 6.15 | | 6.15 | |
| | Install | S.F. | 19.45 | 35.50 | | 54.95 | |
| | Demolish and Install | S.F. | 19.45 | 41.65 | | 61.10 | |
| | Clean | S.F. | .09 | .47 | | .56 | |
| | Minimum Charge | Job | | 232 | | 232 | |
| **Marble Hearth / Facing** | | | | | | | Includes material and labor to install marble fireplace hearth. |
| | Demolish | Ea. | | 39.50 | | 39.50 | |
| | Install | Ea. | 385 | 142 | | 527 | |
| | Demolish and Install | Ea. | 385 | 181.50 | | 566.50 | |
| | Reinstall | Ea. | | 113.69 | | 113.69 | |
| | Clean | Ea. | 47.50 | 7.10 | | 54.60 | |
| | Minimum Charge | Job | | 232 | | 232 | |
| **Breast (face) Brick** | | | | | | | Includes material and labor to install standard red brick, running bond. |
| | Demolish | S.F. | | 6.15 | | 6.15 | |
| | Install | S.F. | 4.47 | 9.80 | | 14.27 | |
| | Demolish and Install | S.F. | 4.47 | 15.95 | | 20.42 | |
| | Clean | S.F. | | .95 | .19 | 1.14 | |
| | Minimum Charge | Hr. | | 58 | | 58 | |
| **Fire Box** | | | | | | | Includes material and labor to install fireplace box only (110 brick). |
| | Demolish | Ea. | | 315 | | 315 | |
| | Install | Ea. | 178 | 425 | | 603 | |
| | Demolish and Install | Ea. | 178 | 740 | | 918 | |
| | Clean | Ea. | .06 | 16.70 | | 16.76 | |
| | Minimum Charge | Hr. | | 58 | | 58 | |
| **Natural Stone** | | | | | | | Includes material and labor to install high priced stone, split or rock face. |
| | Demolish | S.F. | | 6.15 | | 6.15 | |
| | Install | S.F. | 19.30 | 18 | | 37.30 | |
| | Demolish and Install | S.F. | 19.30 | 24.15 | | 43.45 | |
| | Clean | S.F. | .10 | 1.06 | | 1.16 | |
| | Minimum Charge | Job | | 425 | | 425 | |
| **Slumpstone Veneer** | | | | | | | Cost includes material and labor to install 8" x 6" x 16" concrete slump block. |
| | Demolish | S.F. | | .18 | .40 | .58 | |
| | Install | S.F. | 9.65 | 5.65 | | 15.30 | |
| | Demolish and Install | S.F. | 9.65 | 5.83 | .40 | 15.88 | |
| | Clean | S.F. | .09 | .55 | | .64 | |
| | Paint | S.F. | .18 | .88 | | 1.06 | |
| | Minimum Charge | Job | | 232 | | 232 | |
| **Prefabricated** | | | | | | | |
| **Firebox** | | | | | | | Includes material and labor to install up to 43" radiant heat, zero clearance, prefabricated fireplace. Flues, doors or blowers not included. |
| | Demolish | Ea. | | 56.50 | | 56.50 | |
| | Install | Ea. | 2575 | 425 | | 3000 | |
| | Demolish and Install | Ea. | 2575 | 481.50 | | 3056.50 | |
| | Reinstall | Ea. | | 338.04 | | 338.04 | |
| | Clean | Ea. | .74 | 22 | | 22.74 | |
| | Minimum Charge | Job | | 232 | | 232 | |

# Masonry

## Fireplace

| | Unit | Material | Labor | Equip. | Total | Specification |
|---|---|---|---|---|---|---|
| **Flue** | | | | | | |
| Demolish | V.L.F. | | 6.60 | | 6.60 | Includes material and labor to install |
| Install | V.L.F. | 25 | 17.65 | | 42.65 | prefabricated chimney vent, 8" |
| Demolish and Install | V.L.F. | 25 | 24.25 | | 49.25 | exposed metal flue. |
| Reinstall | V.L.F. | | 14.12 | | 14.12 | |
| Clean | L.F. | .30 | 4.44 | | 4.74 | |
| Minimum Charge | Job | | 335 | | 335 | |
| **Blower** | | | | | | |
| Demolish | Ea. | | 5.25 | | 5.25 | Includes material and labor to install |
| Install | Ea. | 155 | 32 | | 187 | blower for metal fireplace flue. |
| Demolish and Install | Ea. | 155 | 37.25 | | 192.25 | |
| Reinstall | Ea. | | 25.48 | | 25.48 | |
| Clean | Ea. | 1.47 | 44.50 | | 45.97 | |
| Minimum Charge | Job | | 335 | | 335 | |
| **Accessories** | | | | | | |
| **Wood Mantle** | | | | | | |
| Demolish | L.F. | | 6.60 | | 6.60 | Includes material and labor to install |
| Install | L.F. | 11.90 | 13.30 | | 25.20 | wood fireplace mantel. |
| Demolish and Install | L.F. | 11.90 | 19.90 | | 31.80 | |
| Reinstall | L.F. | | 13.28 | | 13.28 | |
| Clean | S.F. | | .44 | | .44 | |
| Paint | L.F. | .12 | 10.50 | | 10.62 | |
| Minimum Charge | Job | | 232 | | 232 | |
| **Colonial Style** | | | | | | |
| Demolish | Ea. | | 57.50 | | 57.50 | Includes material and labor to install |
| Install | Opng. | 2000 | 232 | | 2232 | wood fireplace mantel. |
| Demolish and Install | Opng. | 2000 | 289.50 | | 2289.50 | |
| Reinstall | Opng. | | 232.40 | | 232.40 | |
| Clean | Ea. | | .74 | | .74 | |
| Paint | Ea. | 21.50 | 145 | | 166.50 | |
| Minimum Charge | Job | | 232 | | 232 | |
| **Modern Design** | | | | | | |
| Demolish | Ea. | | 57.50 | | 57.50 | Includes material and labor to install |
| Install | Opng. | 660 | 93 | | 753 | wood fireplace mantel. |
| Demolish and Install | Opng. | 660 | 150.50 | | 810.50 | |
| Reinstall | Opng. | | 92.96 | | 92.96 | |
| Clean | Ea. | | .74 | | .74 | |
| Paint | Ea. | 21.50 | 145 | | 166.50 | |
| Minimum Charge | Job | | 232 | | 232 | |
| **Glass Door Unit** | | | | | | |
| Demolish | Ea. | | 8.20 | | 8.20 | Includes material and labor to install a |
| Install | Ea. | 330 | 9.65 | | 339.65 | fireplace glass door unit. |
| Demolish and Install | Ea. | 330 | 17.85 | | 347.85 | |
| Reinstall | Ea. | | 9.67 | | 9.67 | |
| Clean | Ea. | .18 | 13.65 | | 13.83 | |
| Minimum Charge | Job | | 232 | | 232 | |
| **Fire Screen** | | | | | | |
| Demolish | Ea. | | 4.11 | | 4.11 | Includes material and labor to install a |
| Install | Ea. | 133 | 4.83 | | 137.83 | fireplace fire screen. |
| Demolish and Install | Ea. | 133 | 8.94 | | 141.94 | |
| Reinstall | Ea. | | 4.83 | | 4.83 | |
| Clean | Ea. | .09 | 11.85 | | 11.94 | |
| Minimum Charge | Job | | 232 | | 232 | |

**Shingles**

**Board & Batten**

**Texture 1-11**

| Exterior Siding | Unit | Material | Labor | Equip. | Total | Specification |
|---|---|---|---|---|---|---|
| **Trim** | | | | | | |
| **1″ x 2″** | | | | | | |
| Demolish | L.F. | | .33 | | .33 | Cost includes material and labor to |
| Install | L.F. | .31 | 1.41 | | 1.72 | install 1″ x 2″ pine trim. |
| Demolish and Install | L.F. | .31 | 1.74 | | 2.05 | |
| Clean | L.F. | | .24 | | .24 | |
| Paint | L.F. | .02 | .60 | | .62 | |
| Minimum Charge | Job | | 232 | | 232 | |
| **1″ x 3″** | | | | | | |
| Demolish | L.F. | | .33 | | .33 | Cost includes material and labor to |
| Install | L.F. | .33 | 1.60 | | 1.93 | install 1″ x 3″ pine trim. |
| Demolish and Install | L.F. | .33 | 1.93 | | 2.26 | |
| Clean | L.F. | | .24 | | .24 | |
| Paint | L.F. | .02 | .60 | | .62 | |
| Minimum Charge | Job | | 232 | | 232 | |
| **1″ x 4″** | | | | | | |
| Demolish | L.F. | | .33 | | .33 | Cost includes material and labor to |
| Install | L.F. | .59 | 1.86 | | 2.45 | install 1″ x 4″ pine trim. |
| Demolish and Install | L.F. | .59 | 2.19 | | 2.78 | |
| Clean | L.F. | | .24 | | .24 | |
| Paint | L.F. | .02 | .60 | | .62 | |
| Minimum Charge | Job | | 232 | | 232 | |
| **1″ x 6″** | | | | | | |
| Demolish | L.F. | | .33 | | .33 | Cost includes material and labor to |
| Install | L.F. | .90 | 1.86 | | 2.76 | install 1″ x 6″ pine trim. |
| Demolish and Install | L.F. | .90 | 2.19 | | 3.09 | |
| Clean | L.F. | .01 | .24 | | .25 | |
| Paint | L.F. | .09 | .60 | | .69 | |
| Minimum Charge | Job | | 232 | | 232 | |
| **1″ x 8″** | | | | | | |
| Demolish | L.F. | | .36 | | .36 | Cost includes material and labor to |
| Install | L.F. | 1.46 | 2.32 | | 3.78 | install 1″ x 8″ pine trim. |
| Demolish and Install | L.F. | 1.46 | 2.68 | | 4.14 | |
| Clean | L.F. | .01 | .25 | | .26 | |
| Paint | L.F. | .09 | .60 | | .69 | |
| Minimum Charge | Job | | 232 | | 232 | |
| **Aluminum** | | | | | | |
| **8″ Non-Insulated** | | | | | | |
| Demolish | S.F. | | .89 | | .89 | Includes material and labor to install |
| Install | S.F. | 2.75 | 1.88 | | 4.63 | horizontal, colored clapboard, 8″ to |
| Demolish and Install | S.F. | 2.75 | 2.77 | | 5.52 | 10″ wide, plain. |
| Reinstall | S.F. | | 1.88 | | 1.88 | |
| Clean | S.F. | | .36 | | .36 | |
| Minimum Charge | Job | | 232 | | 232 | |

# Siding

| Exterior Siding | Unit | Material | Labor | Equip. | Total | Specification |
|---|---|---|---|---|---|---|
| **8" Insulated** | | | | | | |
| Demolish | S.F. | | .89 | | .89 | Includes material and labor to install |
| Install | S.F. | 3.17 | 1.88 | | 5.05 | horizontal, colored insulated |
| Demolish and Install | S.F. | 3.17 | 2.77 | | 5.94 | clapboard, 8" wide, plain. |
| Reinstall | S.F. | | 1.88 | | 1.88 | |
| Clean | S.F. | | .36 | | .36 | |
| Minimum Charge | Job | | 232 | | 232 | |
| **12" Non-Insulated** | | | | | | |
| Demolish | S.F. | | .89 | | .89 | Includes material and labor to install |
| Install | S.F. | 2.97 | 1.61 | | 4.58 | horizontal, colored clapboard, plain. |
| Demolish and Install | S.F. | 2.97 | 2.50 | | 5.47 | |
| Reinstall | S.F. | | 1.61 | | 1.61 | |
| Clean | S.F. | | .36 | | .36 | |
| Minimum Charge | Job | | 232 | | 232 | |
| **12" Insulated** | | | | | | |
| Demolish | S.F. | | .89 | | .89 | Includes material and labor to install |
| Install | S.F. | 3.22 | 1.61 | | 4.83 | horizontal, colored insulated |
| Demolish and Install | S.F. | 3.22 | 2.50 | | 5.72 | clapboard, plain. |
| Reinstall | S.F. | | 1.61 | | 1.61 | |
| Clean | S.F. | | .36 | | .36 | |
| Minimum Charge | Job | | 232 | | 232 | |
| **Clapboard 8" to 10"** | | | | | | |
| Demolish | S.F. | | .89 | | .89 | Includes material and labor to install |
| Install | S.F. | 2.75 | 1.88 | | 4.63 | horizontal, colored clapboard, 8" to |
| Demolish and Install | S.F. | 2.75 | 2.77 | | 5.52 | 10" wide, plain. |
| Reinstall | S.F. | | 1.88 | | 1.88 | |
| Clean | S.F. | | .36 | | .36 | |
| Minimum Charge | Job | | 232 | | 232 | |
| **Corner Strips** | | | | | | |
| Demolish | L.F. | | .46 | | .46 | Includes material and labor to install |
| Install | V.L.F. | 4 | 1.88 | | 5.88 | corners for horizontal lapped |
| Demolish and Install | V.L.F. | 4 | 2.34 | | 6.34 | aluminum siding. |
| Reinstall | V.L.F. | | 1.88 | | 1.88 | |
| Clean | L.F. | | .24 | | .24 | |
| Minimum Charge | Job | | 232 | | 232 | |
| **Vinyl** | | | | | | |
| **Non-Insulated** | | | | | | |
| Demolish | S.F. | | .70 | | .70 | Cost includes material and labor to |
| Install | S.F. | 1.82 | 1.69 | | 3.51 | install 10" non-insulated solid vinyl |
| Demolish and Install | S.F. | 1.82 | 2.39 | | 4.21 | siding. |
| Reinstall | S.F. | | 1.35 | | 1.35 | |
| Clean | S.F. | | .36 | | .36 | |
| Minimum Charge | Job | | 232 | | 232 | |
| **Insulated** | | | | | | |
| Demolish | S.F. | | .70 | | .70 | Cost includes material and labor to |
| Install | S.F. | 3.56 | 2.32 | | 5.88 | install insulated solid vinyl siding |
| Demolish and Install | S.F. | 3.56 | 3.02 | | 6.58 | panels. |
| Reinstall | S.F. | | 1.86 | | 1.86 | |
| Clean | S.F. | | .36 | | .36 | |
| Minimum Charge | Job | | 232 | | 232 | |
| **Corner Strips** | | | | | | |
| Demolish | L.F. | | .44 | | .44 | Cost includes material and labor to |
| Install | L.F. | 2.34 | 1.33 | | 3.67 | install corner posts for vinyl siding |
| Demolish and Install | L.F. | 2.34 | 1.77 | | 4.11 | panels. |
| Reinstall | L.F. | | 1.06 | | 1.06 | |
| Clean | L.F. | | .24 | | .24 | |
| Minimum Charge | Job | | 232 | | 232 | |

# Siding

| Exterior Siding | Unit | Material | Labor | Equip. | Total | Specification |
|---|---|---|---|---|---|---|
| **Fiber Cement Siding** | | | | | | |
| **4' x 8' Panel** | | | | | | |
| Demolish | S.F. | | .70 | | .70 | Cost includes material and labor to |
| Install | S.F. | 1.19 | 1.02 | | 2.21 | install 4' x 8' fiber cement 7/16" |
| Demolish and Install | S.F. | 1.19 | 1.72 | | 2.91 | thick. |
| Clean | S.F. | | .36 | | .36 | |
| Paint | S.F. | .34 | 1.32 | | 1.66 | |
| Minimum Charge | Job | | 232 | | 232 | |
| **Lap** | | | | | | |
| Demolish | S.F. | | .69 | | .69 | Cost includes material and labor to |
| Install | S.F. | 1.17 | 1.84 | | 3.01 | install 8" or 12" fiber cement lap |
| Demolish and Install | S.F. | 1.17 | 2.53 | | 3.70 | siding 7/16" thick with wood grain |
| Clean | S.F. | | .36 | | .36 | finish. |
| Paint | S.F. | .34 | 1.32 | | 1.66 | |
| Minimum Charge | Job | | 232 | | 232 | |
| **Plywood Panel** | | | | | | |
| **Douglas Fir (T-1-11)** | | | | | | |
| Demolish | S.F. | | .54 | | .54 | Includes material and labor to install |
| Install | S.F. | 1.51 | 1.43 | | 2.94 | Douglas fir 4' x 8' plywood exterior |
| Demolish and Install | S.F. | 1.51 | 1.97 | | 3.48 | siding plain or patterns. |
| Reinstall | S.F. | | 1.43 | | 1.43 | |
| Clean | S.F. | | .36 | | .36 | |
| Paint | S.F. | .34 | 1.32 | | 1.66 | |
| Minimum Charge | Job | | 232 | | 232 | |
| **Redwood** | | | | | | |
| Demolish | S.F. | | .54 | | .54 | Includes material and labor to install |
| Install | S.F. | 2.21 | 1.43 | | 3.64 | redwood 4' x 8' plywood exterior |
| Demolish and Install | S.F. | 2.21 | 1.97 | | 4.18 | siding plain or patterns. |
| Reinstall | S.F. | | 1.43 | | 1.43 | |
| Clean | S.F. | | .36 | | .36 | |
| Paint | S.F. | .34 | 1.32 | | 1.66 | |
| Minimum Charge | Job | | 232 | | 232 | |
| **Cedar** | | | | | | |
| Demolish | S.F. | | .54 | | .54 | Cost includes material and labor to |
| Install | S.F. | 2.86 | 1.43 | | 4.29 | install 4' x 8' cedar plywood exterior |
| Demolish and Install | S.F. | 2.86 | 1.97 | | 4.83 | siding plain or patterns. |
| Reinstall | S.F. | | 1.43 | | 1.43 | |
| Clean | S.F. | | .36 | | .36 | |
| Paint | S.F. | .34 | 1.32 | | 1.66 | |
| Minimum Charge | Job | | 232 | | 232 | |
| **Tongue and Groove** | | | | | | |
| **Fir, Tongue and Groove** | | | | | | |
| Demolish | S.F. | | .99 | | .99 | Includes material and labor to install |
| Install | S.F. | 6.25 | 1.41 | | 7.66 | fir tongue and groove exterior siding |
| Demolish and Install | S.F. | 6.25 | 2.40 | | 8.65 | 1" x 8". |
| Reinstall | S.F. | | 1.13 | | 1.13 | |
| Clean | S.F. | | .36 | | .36 | |
| Paint | S.F. | .34 | 1.32 | | 1.66 | |
| Minimum Charge | Job | | 232 | | 232 | |
| **Pine, Tongue and Groove** | | | | | | |
| Demolish | S.F. | | .99 | | .99 | Includes labor and material to install |
| Install | S.F. | 2.76 | 1.41 | | 4.17 | Pine tongue and groove exterior |
| Demolish and Install | S.F. | 2.76 | 2.40 | | 5.16 | siding 1" x 8". |
| Reinstall | S.F. | | 1.41 | | 1.41 | |
| Clean | S.F. | | .36 | | .36 | |
| Paint | S.F. | .34 | 1.32 | | 1.66 | |
| Minimum Charge | Job | | 232 | | 232 | |

# Siding

| Exterior Siding | Unit | Material | Labor | Equip. | Total | Specification |
|---|---|---|---|---|---|---|
| **Cedar, Tongue and Groove** | | | | | | |
| Demolish | S.F. | | .99 | | .99 | Includes labor and material to install |
| Install | S.F. | 8.05 | 2.11 | | 10.16 | cedar tongue and groove exterior |
| Demolish and Install | S.F. | 8.05 | 3.10 | | 11.15 | siding 1″ x 8″. |
| Reinstall | S.F. | | 1.69 | | 1.69 | |
| Clean | S.F. | | .36 | | .36 | |
| Paint | S.F. | .34 | 1.32 | | 1.66 | |
| Minimum Charge | Job | | 232 | | 232 | |
| **Redwood, Tongue and Groove** | | | | | | |
| Demolish | S.F. | | .99 | | .99 | Includes labor and material to install |
| Install | S.F. | 8.25 | 2.11 | | 10.36 | redwood tongue and groove exterior |
| Demolish and Install | S.F. | 8.25 | 3.10 | | 11.35 | siding 1″ x 4″. |
| Reinstall | S.F. | | 2.11 | | 2.11 | |
| Clean | S.F. | | .36 | | .36 | |
| Paint | S.F. | .34 | 1.32 | | 1.66 | |
| Minimum Charge | Job | | 232 | | 232 | |
| **Board and Batten** | | | | | | |
| **Fir** | | | | | | |
| Demolish | S.F. | | .99 | | .99 | Cost includes material and labor to |
| Install | S.F. | 6.40 | 1.41 | | 7.81 | install 1″ x 8″ fir boards with 1″ x 3″ |
| Demolish and Install | S.F. | 6.40 | 2.40 | | 8.80 | battens. |
| Reinstall | S.F. | | 1.13 | | 1.13 | |
| Clean | S.F. | | .36 | | .36 | |
| Paint | S.F. | .34 | 1.32 | | 1.66 | |
| Minimum Charge | Job | | 232 | | 232 | |
| **Pine** | | | | | | |
| Demolish | S.F. | | .99 | | .99 | Cost includes material and labor to |
| Install | S.F. | 4.05 | 1.24 | | 5.29 | install 1″ x 10″ pine boards with 1″ x |
| Demolish and Install | S.F. | 4.05 | 2.23 | | 6.28 | 3″ battens. |
| Reinstall | S.F. | | .99 | | .99 | |
| Clean | S.F. | | .36 | | .36 | |
| Paint | S.F. | .34 | 1.32 | | 1.66 | |
| Minimum Charge | Job | | 232 | | 232 | |
| **Cedar** | | | | | | |
| Demolish | S.F. | | .99 | | .99 | Cost includes material and labor to |
| Install | S.F. | 7.25 | 1.24 | | 8.49 | install 1″ x 10″ cedar boards with 1″ x |
| Demolish and Install | S.F. | 7.25 | 2.23 | | 9.48 | 3″ battens. |
| Reinstall | S.F. | | .99 | | .99 | |
| Clean | S.F. | | .36 | | .36 | |
| Paint | S.F. | .34 | 1.32 | | 1.66 | |
| Minimum Charge | Job | | 232 | | 232 | |
| **Redwood** | | | | | | |
| Demolish | S.F. | | .99 | | .99 | Cost includes material and labor to |
| Install | S.F. | 6 | 1.24 | | 7.24 | install 1″ x 10″ redwood boards with |
| Demolish and Install | S.F. | 6 | 2.23 | | 8.23 | 1″ x 3″ battens. |
| Reinstall | S.F. | | .99 | | .99 | |
| Clean | S.F. | | .36 | | .36 | |
| Paint | S.F. | .34 | 1.32 | | 1.66 | |
| Minimum Charge | Job | | 232 | | 232 | |
| **Wood Shingle** | | | | | | |
| Demolish | Sq. | | 91 | | 91 | Includes material and labor to install |
| Install | Sq. | 205 | 207 | | 412 | wood shingles/shakes. |
| Demolish and Install | Sq. | 205 | 298 | | 503 | |
| Clean | Sq. | | 37.50 | | 37.50 | |
| Paint | Sq. | 22.50 | 117 | | 139.50 | |
| Minimum Charge | Job | | 232 | | 232 | |

# Siding

| Exterior Siding | Unit | Material | Labor | Equip. | Total | Specification |
|---|---|---|---|---|---|---|
| **Synthetic Stucco System** | | | | | | |
| Demolish | S.F. | | 3.29 | | 3.29 | Cost includes material and labor to |
| Install | S.F. | 2.85 | 7.80 | .55 | 11.20 | install 2 coats of adhesive mixture, |
| Demolish and Install | S.F. | 2.85 | 11.09 | .55 | 14.49 | glass fiber mesh, 1″ insulation board, |
| Clean | S.F. | | .32 | | .32 | and textured finish coat. |
| Paint | S.F. | .25 | .64 | | .89 | |
| Minimum Charge | Job | | 1050 | 73 | 1123 | |

**For customer support on your Contractor's Pricing Guide: Residential Repair & Remodeling, call 800.448.8182.**

**Front-end Loader**

**Bulldozer**

| Grading | | Unit | Material | Labor | Equip. | Total | Specification |
|---|---|---|---|---|---|---|---|
| **Hand (fine)** | | | | | | | |
| | Install | S.F. | | .71 | | .71 | Includes labor and material to install fine grade for slab on grade, hand grade. |
| | Minimum Charge | Job | | 178 | | 178 | |
| **Machine (fine)** | | | | | | | |
| | Install | S.Y. | | .62 | .52 | 1.14 | Includes labor and material to install fine grade for roadway, base or leveling course. |
| | Minimum Charge | Job | | 420 | 355 | 775 | |

| Asphalt Paving | | Unit | Material | Labor | Equip. | Total | Specification |
|---|---|---|---|---|---|---|---|
| **Seal Coat** | | | | | | | |
| | Install | S.F. | .10 | .13 | | .23 | Includes labor and material to install sealcoat for a small area. |
| | Minimum Charge | Job | | 178 | | 178 | |
| **Wearing Course** | | | | | | | |
| | Install | S.F. | .19 | .38 | .27 | .84 | Cost includes labor and material to install 1" thick wearing course. |
| | Minimum Charge | Job | | 178 | | 178 | |

| Masonry Pavers | | Unit | Material | Labor | Equip. | Total | Specification |
|---|---|---|---|---|---|---|---|
| **Natural Concrete** | | | | | | | |
| **Sand Base** | | | | | | | |
| | Demolish | S.F. | | 3.95 | | 3.95 | Includes material and labor to install concrete pavers on sand base. |
| | Install | S.F. | 3.78 | 7.75 | | 11.53 | |
| | Demolish and Install | S.F. | 3.78 | 11.70 | | 15.48 | |
| | Reinstall | S.F. | | 6.19 | | 6.19 | |
| | Clean | S.F. | | 1.42 | | 1.42 | |
| | Minimum Charge | Job | | 232 | | 232 | |
| **Mortar Base** | | | | | | | |
| | Demolish | S.F. | | 6.55 | 1.91 | 8.46 | Includes material and labor to install concrete pavers on mortar base. |
| | Install | S.F. | 3.72 | 6.45 | | 10.17 | |
| | Demolish and Install | S.F. | 3.72 | 13 | 1.91 | 18.63 | |
| | Reinstall | S.F. | | 5.16 | | 5.16 | |
| | Clean | S.F. | | 1.29 | | 1.29 | |
| | Minimum Charge | Job | | 232 | | 232 | |

## Masonry Pavers

| | Unit | Material | Labor | Equip. | Total | Specification |
|---|---|---|---|---|---|---|
| **Adobe** | | | | | | |
| **Sand Base** | | | | | | |
| Demolish | S.F. | | 3.95 | | 3.95 | Includes material and labor to install |
| Install | S.F. | 2.06 | 5.15 | | 7.21 | adobe pavers on sand base. |
| Demolish and Install | S.F. | 2.06 | 9.10 | | 11.16 | |
| Reinstall | S.F. | | 4.12 | | 4.12 | |
| Clean | S.F. | | 1.42 | | 1.42 | |
| Minimum Charge | Job | | 232 | | 232 | |
| **Mortar Base** | | | | | | |
| Demolish | S.F. | | 6.55 | 1.91 | 8.46 | Includes material and labor to install |
| Install | S.F. | 2.84 | 6.45 | | 9.29 | adobe pavers on mortar base. |
| Demolish and Install | S.F. | 2.84 | 13 | 1.91 | 17.75 | |
| Reinstall | S.F. | | 5.16 | | 5.16 | |
| Clean | S.F. | | 1.29 | | 1.29 | |
| Minimum Charge | Job | | 232 | | 232 | |
| **Standard Brick** | | | | | | |
| **Sand Base** | | | | | | |
| Demolish | S.F. | | 3.95 | | 3.95 | Includes material and labor to install |
| Install | S.F. | 3.14 | 8.45 | | 11.59 | standard brick pavers on sand base. |
| Demolish and Install | S.F. | 3.14 | 12.40 | | 15.54 | |
| Reinstall | S.F. | | 6.78 | | 6.78 | |
| Clean | S.F. | | 1.42 | | 1.42 | |
| Minimum Charge | Job | | 232 | | 232 | |
| **Mortar Base** | | | | | | |
| Demolish | S.F. | | 6.55 | 1.91 | 8.46 | Includes material and labor to install |
| Install | S.F. | 3.22 | 3.09 | | 6.31 | standard brick pavers on mortar base. |
| Demolish and Install | S.F. | 3.22 | 9.64 | 1.91 | 14.77 | |
| Reinstall | S.F. | | 2.47 | | 2.47 | |
| Clean | S.F. | | 1.29 | | 1.29 | |
| Minimum Charge | Job | | 232 | | 232 | |
| **Deluxe Brick** | | | | | | |
| **Sand Base** | | | | | | |
| Demolish | S.F. | | 3.95 | | 3.95 | Includes material and labor to install |
| Install | S.F. | 3.76 | 6.45 | | 10.21 | deluxe brick pavers on sand base. |
| Demolish and Install | S.F. | 3.76 | 10.40 | | 14.16 | |
| Reinstall | S.F. | | 5.16 | | 5.16 | |
| Clean | S.F. | | 1.42 | | 1.42 | |
| Minimum Charge | Job | | 232 | | 232 | |
| **Mortar Base** | | | | | | |
| Demolish | S.F. | | 6.55 | 1.91 | 8.46 | Includes material and labor to install |
| Install | S.F. | 4.31 | 11.60 | | 15.91 | deluxe brick pavers on mortar base. |
| Demolish and Install | S.F. | 4.31 | 18.15 | 1.91 | 24.37 | |
| Reinstall | S.F. | | 9.28 | | 9.28 | |
| Clean | S.F. | | 1.29 | | 1.29 | |
| Minimum Charge | Job | | 232 | | 232 | |

## Fencing

| | Unit | Material | Labor | Equip. | Total | Specification |
|---|---|---|---|---|---|---|
| **Block w / Footing** | | | | | | |
| **4"** | | | | | | |
| Demolish | S.F. | | .14 | .30 | .44 | Cost includes material and labor to |
| Install | S.F. | 2 | 5 | | 7 | install 4" non-reinforced concrete |
| Demolish and Install | S.F. | 2 | 5.14 | .30 | 7.44 | block. |
| Clean | S.F. | .29 | 1.18 | .23 | 1.70 | |
| Paint | S.F. | .33 | .74 | | 1.07 | |
| Minimum Charge | Job | | 232 | | 232 | |

## Fencing

| | Unit | Material | Labor | Equip. | Total | Specification |
|---|---|---|---|---|---|---|
| **4" Reinforced** | | | | | | |
| Demolish | S.F. | | 2.19 | .70 | 2.89 | Cost includes material and labor to |
| Install | S.F. | 2.20 | 5.10 | | 7.30 | install 4" reinforced concrete block. |
| Demolish and Install | S.F. | 2.20 | 7.29 | .70 | 10.19 | |
| Clean | S.F. | .29 | 1.18 | .23 | 1.70 | |
| Paint | S.F. | .33 | .74 | | 1.07 | |
| Minimum Charge | Job | | 232 | | 232 | |
| **6"** | | | | | | |
| Demolish | S.F. | | .18 | .40 | .58 | Cost includes material and labor to |
| Install | S.F. | 2.64 | 5.40 | | 8.04 | install 6" non-reinforced concrete |
| Demolish and Install | S.F. | 2.64 | 5.58 | .40 | 8.62 | block. |
| Clean | S.F. | .29 | 1.18 | .23 | 1.70 | |
| Paint | S.F. | .33 | .74 | | 1.07 | |
| Minimum Charge | Job | | 232 | | 232 | |
| **6" Reinforced** | | | | | | |
| Demolish | S.F. | | 2.31 | .51 | 2.82 | Cost includes material and labor to |
| Install | S.F. | 2.84 | 5.45 | | 8.29 | install 6" reinforced concrete block. |
| Demolish and Install | S.F. | 2.84 | 7.76 | .51 | 11.11 | |
| Clean | S.F. | .29 | 1.18 | .23 | 1.70 | |
| Paint | S.F. | .33 | .74 | | 1.07 | |
| Minimum Charge | Job | | 232 | | 232 | |
| **Cap for Block Wall** | | | | | | |
| Demolish | L.F. | | 2.47 | | 2.47 | Includes labor and equipment to install |
| Install | L.F. | 25 | 11.30 | | 36.30 | precast concrete coping. |
| Demolish and Install | L.F. | 25 | 13.77 | | 38.77 | |
| Clean | L.F. | .02 | 3.70 | | 3.72 | |
| Paint | S.F. | .33 | .74 | | 1.07 | |
| Minimum Charge | Job | | 232 | | 232 | |
| **Block Pilaster** | | | | | | |
| Demolish | Ea. | | 169 | | 169 | Includes material and labor to install |
| Install | Ea. | 54.50 | 169 | | 223.50 | block pilaster for fence. |
| Demolish and Install | Ea. | 54.50 | 338 | | 392.50 | |
| Reinstall | Ea. | | 169.38 | | 169.38 | |
| Paint | Ea. | 5.10 | 17.15 | | 22.25 | |
| Minimum Charge | Job | | 232 | | 232 | |
| **Block w / o Footing** | | | | | | |
| **4"** | | | | | | |
| Demolish | S.F. | | .14 | .30 | .44 | Cost includes material and labor to |
| Install | S.F. | 2.15 | 3.85 | | 6 | install 4" non-reinforced hollow |
| Demolish and Install | S.F. | 2.15 | 3.99 | .30 | 6.44 | lightweight block. |
| Clean | S.F. | .29 | 1.18 | .23 | 1.70 | |
| Paint | S.F. | .33 | .74 | | 1.07 | |
| Minimum Charge | Job | | 232 | | 232 | |
| **4" Reinforced** | | | | | | |
| Demolish | S.F. | | 2.19 | .70 | 2.89 | Cost includes material and labor to |
| Install | S.F. | 2.33 | 4.96 | | 7.29 | install 4" reinforced hollow lightweight |
| Demolish and Install | S.F. | 2.33 | 7.15 | .70 | 10.18 | block. |
| Clean | S.F. | .29 | 1.18 | .23 | 1.70 | |
| Paint | S.F. | .33 | .74 | | 1.07 | |
| Minimum Charge | Job | | 232 | | 232 | |
| **6"** | | | | | | |
| Demolish | S.F. | | .18 | .40 | .58 | Cost includes material and labor to |
| Install | S.F. | 3.03 | 4.23 | | 7.26 | install 6" non-reinforced hollow |
| Demolish and Install | S.F. | 3.03 | 4.41 | .40 | 7.84 | lightweight block. |
| Clean | S.F. | .29 | 1.18 | .23 | 1.70 | |
| Paint | S.F. | .33 | .74 | | 1.07 | |
| Minimum Charge | Job | | 232 | | 232 | |

| Fencing | Unit | Material | Labor | Equip. | Total | Specification |
|---|---|---|---|---|---|---|
| **6" Reinforced** | | | | | | |
| Demolish | S.F. | | 2.31 | .51 | 2.82 | Cost includes material and labor to |
| Install | S.F. | 3.19 | 5.35 | | 8.54 | install 6" reinforced hollow lightweight |
| Demolish and Install | S.F. | 3.19 | 7.66 | .51 | 11.36 | block. |
| Clean | S.F. | .29 | 1.18 | .23 | 1.70 | |
| Paint | S.F. | .33 | .74 | | 1.07 | |
| Minimum Charge | Job | | 232 | | 232 | |
| **Cap for Block Wall** | | | | | | |
| Demolish | L.F. | | 2.47 | | 2.47 | Includes labor and equipment to install |
| Install | L.F. | 25 | 11.30 | | 36.30 | precast concrete coping. |
| Demolish and Install | L.F. | 25 | 13.77 | | 38.77 | |
| Clean | L.F. | .02 | 3.70 | | 3.72 | |
| Paint | S.F. | .33 | .74 | | 1.07 | |
| Minimum Charge | Job | | 232 | | 232 | |
| **Block Pilaster** | | | | | | |
| Demolish | Ea. | | 169 | | 169 | Includes material and labor to install |
| Install | Ea. | 54.50 | 169 | | 223.50 | block pilaster for fence. |
| Demolish and Install | Ea. | 54.50 | 338 | | 392.50 | |
| Reinstall | Ea. | | 169.38 | | 169.38 | |
| Paint | Ea. | 5.10 | 17.15 | | 22.25 | |
| Minimum Charge | Job | | 232 | | 232 | |
| **Chain Link** | | | | | | |
| **3' High w / o Top Rail** | | | | | | |
| Demolish | L.F. | | 3.08 | .90 | 3.98 | Includes material and labor to install |
| Install | L.F. | 5.85 | 2.15 | .41 | 8.41 | residential galvanized fence 3' high |
| Demolish and Install | L.F. | 5.85 | 5.23 | 1.31 | 12.39 | with 1-3/8" steel post 10' O.C. set |
| Reinstall | L.F. | | 1.72 | .33 | 2.05 | without concrete. |
| Clean | L.F. | .04 | 1.29 | | 1.33 | |
| Minimum Charge | Job | | 545 | | 545 | |
| **4' High w / o Top Rail** | | | | | | |
| Demolish | L.F. | | 3.08 | .90 | 3.98 | Includes material and labor to install |
| Install | L.F. | 6.10 | 2.65 | .51 | 9.26 | residential galvanized fence 4' high |
| Demolish and Install | L.F. | 6.10 | 5.73 | 1.41 | 13.24 | with 1-3/8" steel post 10' O.C. set |
| Reinstall | L.F. | | 2.12 | .41 | 2.53 | without concrete. |
| Clean | L.F. | .06 | 1.42 | | 1.48 | |
| Minimum Charge | Job | | 545 | | 545 | |
| **5' High w / o Top Rail** | | | | | | |
| Demolish | L.F. | | 3.08 | .90 | 3.98 | Includes material and labor to install |
| Install | L.F. | 7.65 | 3.47 | .66 | 11.78 | residential galvanized fence 5' high |
| Demolish and Install | L.F. | 7.65 | 6.55 | 1.56 | 15.76 | with 1-3/8" steel post 10' O.C. set |
| Reinstall | L.F. | | 2.77 | .53 | 3.30 | without concrete. |
| Clean | L.F. | .08 | 1.58 | | 1.66 | |
| Minimum Charge | Job | | 545 | | 545 | |
| **3' High w / Top Rail** | | | | | | |
| Demolish | L.F. | | 3.27 | .95 | 4.22 | Includes material and labor to install |
| Install | L.F. | 2.20 | 2.25 | .43 | 4.88 | residential galvanized fence 3' high |
| Demolish and Install | L.F. | 2.20 | 5.52 | 1.38 | 9.10 | with 1-3/8" steel post 10' O.C. set |
| Reinstall | L.F. | | 1.80 | .34 | 2.14 | without concrete. |
| Clean | L.F. | .04 | 1.32 | | 1.36 | |
| Minimum Charge | Job | | 545 | | 545 | |
| **4' High w / Top Rail** | | | | | | |
| Demolish | L.F. | | 3.27 | .95 | 4.22 | Includes material and labor to install |
| Install | L.F. | 7.75 | 2.82 | .54 | 11.11 | residential galvanized fence 4' high |
| Demolish and Install | L.F. | 7.75 | 6.09 | 1.49 | 15.33 | with 1-5/8" steel post 10' O.C. set |
| Reinstall | L.F. | | 2.25 | .43 | 2.68 | without concrete and 1-5/8" top rail |
| Clean | L.F. | .06 | 1.45 | | 1.51 | with sleeves. |
| Minimum Charge | Job | | 545 | | 545 | |

| Fencing | Unit | Material | Labor | Equip. | Total | Specification |
|---|---|---|---|---|---|---|
| **5' High w / Top Rail** | | | | | | Includes material and labor to install |
| Demolish | L.F. | | 3.27 | .95 | 4.22 | residential galvanized fence 5' high |
| Install | L.F. | 8.30 | 3.75 | .72 | 12.77 | with 1-5/8" steel post 10' O.C. set |
| Demolish and Install | L.F. | 8.30 | 7.02 | 1.67 | 16.99 | without concrete and 1-5/8" top rail |
| Reinstall | L.F. | | 3 | .57 | 3.57 | with sleeves. |
| Clean | L.F. | .08 | 1.61 | | 1.69 | |
| Minimum Charge | Job | | 545 | | 545 | |
| **6' High w / Top Rail** | | | | | | Includes material and labor to install |
| Demolish | L.F. | | 3.27 | .95 | 4.22 | residential galvanized fence 6' high |
| Install | L.F. | 9.85 | 5.65 | 1.08 | 16.58 | with 2" steel post 10' O.C. set without |
| Demolish and Install | L.F. | 9.85 | 8.92 | 2.03 | 20.80 | concrete and 1-3/8" top rail with |
| Reinstall | L.F. | | 4.51 | .86 | 5.37 | sleeves. |
| Clean | L.F. | .09 | 1.82 | | 1.91 | |
| Minimum Charge | Job | | 545 | | 545 | |
| **Barbed Wire** | | | | | | Cost includes material and labor to |
| Demolish | L.F. | | 1.84 | | 1.84 | install 3 strand galvanized barbed |
| Install | L.F. | .43 | .71 | .17 | 1.31 | wire fence. |
| Demolish and Install | L.F. | .43 | 2.55 | .17 | 3.15 | |
| Reinstall | L.F. | | .57 | .14 | .71 | |
| Minimum Charge | Job | | 545 | 335 | 880 | |
| **6' Redwood** | | | | | | Cost includes material and labor to |
| Demolish | L.F. | | 12.15 | | 12.15 | install 6' high redwood fencing. |
| Install | L.F. | 29.50 | 9 | 1.72 | 40.22 | |
| Demolish and Install | L.F. | 29.50 | 21.15 | 1.72 | 52.37 | |
| Reinstall | L.F. | | 7.21 | 1.38 | 8.59 | |
| Clean | L.F. | .09 | 1.65 | | 1.74 | |
| Paint | L.F. | .91 | 4.07 | | 4.98 | |
| Minimum Charge | Job | | 545 | | 545 | |
| **6' Red Cedar** | | | | | | Cost includes material and labor to |
| Demolish | L.F. | | 12.15 | | 12.15 | install 6' cedar fence. |
| Install | L.F. | 23.50 | 9 | 1.72 | 34.22 | |
| Demolish and Install | L.F. | 23.50 | 21.15 | 1.72 | 46.37 | |
| Reinstall | L.F. | | 7.21 | 1.38 | 8.59 | |
| Clean | L.F. | .09 | 1.65 | | 1.74 | |
| Paint | L.F. | .91 | 4.07 | | 4.98 | |
| Minimum Charge | Job | | 545 | | 545 | |
| **6' Shadowbox** | | | | | | Cost includes material and labor to |
| Demolish | L.F. | | 12.15 | | 12.15 | install 6' pressure treated pine fence. |
| Install | L.F. | 26.50 | 7.50 | 1.44 | 35.44 | |
| Demolish and Install | L.F. | 26.50 | 19.65 | 1.44 | 47.59 | |
| Reinstall | L.F. | | 6.01 | 1.15 | 7.16 | |
| Clean | L.F. | .09 | 1.65 | | 1.74 | |
| Paint | L.F. | .91 | 4.07 | | 4.98 | |
| Minimum Charge | Job | | 545 | | 545 | |
| **3' Picket** | | | | | | Includes material, labor and |
| Demolish | L.F. | | 12.15 | | 12.15 | equipment to install pressure treated |
| Install | L.F. | 8.60 | 8.05 | 1.54 | 18.19 | pine picket fence. |
| Demolish and Install | L.F. | 8.60 | 20.20 | 1.54 | 30.34 | |
| Reinstall | L.F. | | 6.44 | 1.23 | 7.67 | |
| Clean | L.F. | .04 | 1.42 | | 1.46 | |
| Paint | L.F. | .46 | 2.21 | | 2.67 | |
| Minimum Charge | Job | | 545 | | 545 | |

| Fencing | Unit | Material | Labor | Equip. | Total | Specification |
|---|---|---|---|---|---|---|
| **Wrought Iron** | | | | | | |
| Demolish | L.F. | | 26.50 | | 26.50 | Cost includes material and labor to |
| Install | L.F. | 40 | 14.55 | | 54.55 | install 5' high wrought iron fence with |
| Demolish and Install | L.F. | 40 | 41.05 | | 81.05 | pickets at 4-1/2" O.C. |
| Reinstall | L.F. | | 11.65 | | 11.65 | |
| Clean | L.F. | | 1.42 | | 1.42 | |
| Paint | L.F. | 4.15 | 4.55 | | 8.70 | |
| Minimum Charge | Job | | 545 | | 545 | |
| **Post** | | | | | | |
| **Dirt Based Wood** | | | | | | |
| Demolish | Ea. | | 30.50 | | 30.50 | Includes material and labor to install |
| Install | Ea. | 32 | 137 | | 169 | wood post set in earth. |
| Demolish and Install | Ea. | 32 | 167.50 | | 199.50 | |
| Reinstall | Ea. | | 109.25 | | 109.25 | |
| Clean | Ea. | | 4.44 | | 4.44 | |
| Paint | Ea. | 1.01 | 1.55 | | 2.56 | |
| Minimum Charge | Job | | 545 | | 545 | |
| **Concrete Based Wood** | | | | | | |
| Demolish | Ea. | | 43.50 | 12.75 | 56.25 | Includes material and labor to install |
| Install | Ea. | 20 | 182 | | 202 | wood post set in concrete. |
| Demolish and Install | Ea. | 20 | 225.50 | 12.75 | 258.25 | |
| Clean | Ea. | | 4.44 | | 4.44 | |
| Paint | Ea. | 1.01 | 1.55 | | 2.56 | |
| Minimum Charge | Job | | 545 | | 545 | |
| **Block Pilaster** | | | | | | |
| Demolish | Ea. | | 169 | | 169 | Includes material and labor to install |
| Install | Ea. | 54.50 | 169 | | 223.50 | block pilaster for fence. |
| Demolish and Install | Ea. | 54.50 | 338 | | 392.50 | |
| Reinstall | Ea. | | 169.38 | | 169.38 | |
| Paint | Ea. | 5.10 | 17.15 | | 22.25 | |
| Minimum Charge | Job | | 232 | | 232 | |
| **Gate** | | | | | | |
| **Chain Link** | | | | | | |
| Demolish | Ea. | | 40.50 | | 40.50 | Cost includes material and labor to |
| Install | Ea. | 108 | 60.50 | | 168.50 | install 3' wide chain link pass gate |
| Demolish and Install | Ea. | 108 | 101 | | 209 | and hardware. |
| Reinstall | Ea. | | 48.55 | | 48.55 | |
| Clean | Ea. | | 4.93 | | 4.93 | |
| Minimum Charge | Job | | 545 | | 545 | |
| **Redwood** | | | | | | |
| Demolish | Ea. | | 34.50 | | 34.50 | Includes material, labor and |
| Install | Ea. | 262 | 60.50 | | 322.50 | equipment to install redwood pass |
| Demolish and Install | Ea. | 262 | 95 | | 357 | gate and hardware. |
| Reinstall | Ea. | | 48.55 | | 48.55 | |
| Clean | Ea. | | 4.93 | | 4.93 | |
| Paint | Ea. | 1.52 | 7.75 | | 9.27 | |
| Minimum Charge | Job | | 545 | | 545 | |
| **Red Cedar** | | | | | | |
| Demolish | Ea. | | 34.50 | | 34.50 | Cost includes material and labor to |
| Install | Ea. | 99 | 60.50 | | 159.50 | install 3' wide red cedar pass gate |
| Demolish and Install | Ea. | 99 | 95 | | 194 | and hardware. |
| Reinstall | Ea. | | 48.55 | | 48.55 | |
| Clean | Ea. | | 4.93 | | 4.93 | |
| Paint | Ea. | 1.52 | 7.75 | | 9.27 | |
| Minimum Charge | Job | | 545 | | 545 | |

## Fencing

| | Unit | Material | Labor | Equip. | Total | Specification |
|---|---|---|---|---|---|---|
| **Wrought Iron** | | | | | | |
| Demolish | Ea. | | 26.50 | | 26.50 | Includes material, labor and |
| Install | Ea. | 285 | 137 | | 422 | equipment to install wrought iron gate |
| Demolish and Install | Ea. | 285 | 163.50 | | 448.50 | with hardware. |
| Reinstall | Ea. | | 109.25 | | 109.25 | |
| Clean | Ea. | | 4.93 | | 4.93 | |
| Paint | Ea. | 8.30 | 7.75 | | 16.05 | |
| Minimum Charge | Job | | 545 | | 545 | |

145

Fire-extinguishing System | Space Heater | Hot-air Furnace

| Plumbing | Unit | Material | Labor | Equip. | Total | Specification |
|---|---|---|---|---|---|---|
| **Components** | | | | | | |
| **3/4" Supply** | | | | | | |
| Demolish | L.F. | | 2.91 | | 2.91 | Includes material and labor to install |
| Install | L.F. | 4.94 | 6.90 | | 11.84 | type L copper tubing including |
| Demolish and Install | L.F. | 4.94 | 9.81 | | 14.75 | couplings and hangers at 10' O.C. |
| Reinstall | L.F. | | 6.89 | | 6.89 | |
| Minimum Charge | Job | | 262 | | 262 | |
| **1/2" Fixture Supply** | | | | | | |
| Install | Ea. | 48 | 175 | | 223 | Includes labor and materials to install supply pipe only to a plumbing fixture. |
| **2" Single Vent Stack** | | | | | | |
| Demolish | Ea. | | 48.50 | | 48.50 | Includes material and labor to install |
| Install | Ea. | 27 | 262 | | 289 | PVC vent stack pipe with couplings at |
| Demolish and Install | Ea. | 27 | 310.50 | | 337.50 | 10' O.C. |
| Reinstall | Ea. | | 262 | | 262 | |
| Minimum Charge | Job | | 262 | | 262 | |
| **4" Drain** | | | | | | |
| Demolish | L.F. | | 3.88 | | 3.88 | Includes material and labor to install |
| Install | L.F. | 6.40 | 8.60 | | 15 | DVW type PVC pipe with couplings at |
| Demolish and Install | L.F. | 6.40 | 12.48 | | 18.88 | 10' O.C. and 3 strap hangers per 10' |
| Reinstall | L.F. | | 8.58 | | 8.58 | run. |
| Minimum Charge | Job | | 262 | | 262 | |
| **1-1/2" Drain** | | | | | | |
| Demolish | L.F. | | 2.91 | | 2.91 | Includes material and labor to install |
| Install | L.F. | 1.61 | 5.55 | | 7.16 | DVW type PVC pipe with couplings at |
| Demolish and Install | L.F. | 1.61 | 8.46 | | 10.07 | 10' O.C. and 3 strap hangers per 10' |
| Reinstall | L.F. | | 5.57 | | 5.57 | run. |
| Minimum Charge | Job | | 262 | | 262 | |
| **Maintenance** | | | | | | |
| **Flush Out / Unclog** | | | | | | |
| Clean | Job | | 262 | | 262 | Includes 4 hour minimum labor charge |
| Minimum Charge | Job | | 262 | | 262 | for a plumber. |
| **Vent Cap** | | | | | | |
| **Large** | | | | | | |
| Demolish | Ea. | | 13.25 | | 13.25 | Includes labor and material to install a |
| Install | Ea. | 48 | 24 | | 72 | PVC plumbing pipe vent cap. |
| Demolish and Install | Ea. | 48 | 37.25 | | 85.25 | |
| Clean | Ea. | .21 | 8.05 | | 8.26 | |
| Paint | Ea. | .19 | 8.80 | | 8.99 | |
| Minimum Charge | Job | | 131 | | 131 | |

# Rough Mechanical

## Plumbing

| | Unit | Material | Labor | Equip. | Total | Specification |
|---|---|---|---|---|---|---|
| **Medium** | | | | | | |
| Demolish | Ea. | | 13.25 | | 13.25 | Includes labor and material to install a |
| Install | Ea. | 41.50 | 23 | | 64.50 | PVC plumbing pipe vent cap. |
| Demolish and Install | Ea. | 41.50 | 36.25 | | 77.75 | |
| Clean | Ea. | .21 | 8.05 | | 8.26 | |
| Paint | Ea. | .19 | 8.80 | | 8.99 | |
| Minimum Charge | Job | | 131 | | 131 | |
| **Small** | | | | | | |
| Demolish | Ea. | | 13.25 | | 13.25 | Includes material and labor to install a |
| Install | Ea. | 14.25 | 22 | | 36.25 | PVC plumbing pipe vent cap. |
| Demolish and Install | Ea. | 14.25 | 35.25 | | 49.50 | |
| Clean | Ea. | .21 | 8.05 | | 8.26 | |
| Paint | Ea. | .19 | 8.80 | | 8.99 | |
| Minimum Charge | Job | | 131 | | 131 | |

## Plumbing

| | Unit | Material | Labor | Equip. | Total | Specification |
|---|---|---|---|---|---|---|
| **Sump Pump** | | | | | | |
| **1/4 HP** | | | | | | |
| Demolish | Ea. | | 52.50 | | 52.50 | Includes material and labor to install a |
| Install | Ea. | 165 | 82 | | 247 | submersible sump pump. |
| Demolish and Install | Ea. | 165 | 134.50 | | 299.50 | |
| Reinstall | Ea. | | 65.50 | | 65.50 | |
| Clean | Ea. | .52 | 22 | | 22.52 | |
| Minimum Charge | Job | | 262 | | 262 | |
| **1/3 HP** | | | | | | |
| Demolish | Ea. | | 52.50 | | 52.50 | Includes material and labor to install a |
| Install | Ea. | 239 | 87.50 | | 326.50 | submersible sump pump. |
| Demolish and Install | Ea. | 239 | 140 | | 379 | |
| Reinstall | Ea. | | 69.87 | | 69.87 | |
| Clean | Ea. | .52 | 22 | | 22.52 | |
| Minimum Charge | Job | | 262 | | 262 | |
| **1/2 HP** | | | | | | |
| Demolish | Ea. | | 52.50 | | 52.50 | Includes material and labor to install a |
| Install | Ea. | 293 | 97 | | 390 | submersible sump pump. |
| Demolish and Install | Ea. | 293 | 149.50 | | 442.50 | |
| Reinstall | Ea. | | 77.63 | | 77.63 | |
| Clean | Ea. | .52 | 22 | | 22.52 | |
| Minimum Charge | Job | | 262 | | 262 | |
| **Clean and Service** | | | | | | |
| Clean | Ea. | | 33 | | 33 | Includes labor to clean and service a |
| Minimum Charge | Job | | 262 | | 262 | sump pump. |

## Water Heater

| | Unit | Material | Labor | Equip. | Total | Specification |
|---|---|---|---|---|---|---|
| **Electric** | | | | | | |
| **30 Gallon** | | | | | | |
| Demolish | Ea. | | 164 | | 164 | Includes material and labor to install a |
| Install | Ea. | 1100 | 238 | | 1338 | 30 gallon electric water heater. |
| Demolish and Install | Ea. | 1100 | 402 | | 1502 | |
| Reinstall | Ea. | | 190.55 | | 190.55 | |
| Clean | Ea. | .52 | 22 | | 22.52 | |
| Minimum Charge | Job | | 262 | | 262 | |

**For customer support on your Contractor's Pricing Guide: Residential Repair & Remodeling, call 800.448.8182.**

| Water Heater | Unit | Material | Labor | Equip. | Total | Specification |
|---|---|---|---|---|---|---|
| **40 Gallon** | | | | | | |
| Demolish | Ea. | | 164 | | 164 | Includes material and labor to install a |
| Install | Ea. | 1175 | 262 | | 1437 | 40 gallon electric water heater. |
| Demolish and Install | Ea. | 1175 | 426 | | 1601 | |
| Reinstall | Ea. | | 209.60 | | 209.60 | |
| Clean | Ea. | .52 | 22 | | 22.52 | |
| Minimum Charge | Job | | 262 | | 262 | |
| **52 Gallon** | | | | | | |
| Demolish | Ea. | | 164 | | 164 | Includes material and labor to install a |
| Install | Ea. | 1325 | 262 | | 1587 | 52 gallon electric water heater. |
| Demolish and Install | Ea. | 1325 | 426 | | 1751 | |
| Reinstall | Ea. | | 209.60 | | 209.60 | |
| Clean | Ea. | .52 | 22 | | 22.52 | |
| Minimum Charge | Job | | 262 | | 262 | |
| **82 Gallon** | | | | | | |
| Demolish | Ea. | | 164 | | 164 | Includes material and labor to install |
| Install | Ea. | 2025 | 330 | | 2355 | an 82 gallon electric water heater. |
| Demolish and Install | Ea. | 2025 | 494 | | 2519 | |
| Reinstall | Ea. | | 262 | | 262 | |
| Clean | Ea. | .52 | 22 | | 22.52 | |
| Minimum Charge | Job | | 262 | | 262 | |
| **Gas-fired** | | | | | | |
| **30 Gallon Gas-fired** | | | | | | |
| Demolish | Ea. | | 202 | | 202 | Includes material and labor to install a |
| Install | Ea. | 1950 | 262 | | 2212 | 30 gallon gas water heater. |
| Demolish and Install | Ea. | 1950 | 464 | | 2414 | |
| Reinstall | Ea. | | 209.60 | | 209.60 | |
| Clean | Ea. | .52 | 22 | | 22.52 | |
| Minimum Charge | Job | | 262 | | 262 | |
| **30 Gallon Gas-fired, Flue** | | | | | | |
| Install | Ea. | 65 | 75 | | 140 | Includes labor and materials to install |
| Minimum Charge | Job | | 262 | | 262 | a flue for a water heater (up to 50 gallon size). |
| **40 Gallon Gas-fired** | | | | | | |
| Demolish | Ea. | | 202 | | 202 | Includes material and labor to install a |
| Install | Ea. | 1950 | 276 | | 2226 | 40 gallon gas water heater. |
| Demolish and Install | Ea. | 1950 | 478 | | 2428 | |
| Reinstall | Ea. | | 220.63 | | 220.63 | |
| Clean | Ea. | .52 | 22 | | 22.52 | |
| Minimum Charge | Job | | 262 | | 262 | |
| **40 Gallon Gas-fired, Flue** | | | | | | |
| Install | Ea. | 65 | 75 | | 140 | Includes labor and materials to install |
| Minimum Charge | Job | | 262 | | 262 | a flue for a water heater (up to 50 gallon size). |
| **50 Gallon Gas-fired** | | | | | | |
| Demolish | Ea. | | 202 | | 202 | Includes material and labor to install a |
| Install | Ea. | 2125 | 291 | | 2416 | 50 gallon gas water heater. |
| Demolish and Install | Ea. | 2125 | 493 | | 2618 | |
| Reinstall | Ea. | | 232.89 | | 232.89 | |
| Clean | Ea. | .52 | 22 | | 22.52 | |
| Minimum Charge | Job | | 262 | | 262 | |
| **50 Gallon Gas-fired, Flue** | | | | | | |
| Install | Ea. | 65 | 75 | | 140 | Includes labor and materials to install |
| Minimum Charge | Job | | 262 | | 262 | a flue for a water heater (up to 50 gallon size). |

## Water Heater

| Water Heater | Unit | Material | Labor | Equip. | Total | Specification |
|---|---|---|---|---|---|---|
| **Solar Hot Water System** | | | | | | |
| **Water Heater** | | | | | | |
| Demolish | Ea. | | 131 | | 131 | Includes material and labor to install a |
| Install | Ea. | 1750 | 350 | | 2100 | solar hot water heat exchanger |
| Demolish and Install | Ea. | 1750 | 481 | | 2231 | storage tank, 100 gallon. |
| Reinstall | Ea. | | 279.47 | | 279.47 | |
| Minimum Charge | Job | | 270 | | 270 | |
| **Solar Panel** | | | | | | |
| Demolish | S.F. | | 1.92 | | 1.92 | Includes material and labor to install |
| Install | S.F. | 34.50 | 3.45 | | 37.95 | solar panel. |
| Demolish and Install | S.F. | 34.50 | 5.37 | | 39.87 | |
| Reinstall | S.F. | | 2.76 | | 2.76 | |
| Minimum Charge | Job | | 270 | | 270 | |
| **Repair Plumbing** | | | | | | |
| Install | Job | | 270 | | 270 | Includes 4 hour minimum labor charge |
| Minimum Charge | Job | | 270 | | 270 | for a steamfitter or pipefitter. |
| **Insulating Wrap** | | | | | | |
| Demolish | Ea. | | 6.05 | | 6.05 | Includes material and labor to install |
| Install | Ea. | 20 | 16.40 | | 36.40 | insulating wrap around a 100 gallon |
| Demolish and Install | Ea. | 20 | 22.45 | | 42.45 | solar hot water heat exchanger |
| Reinstall | Ea. | | 13.10 | | 13.10 | storage tank. |
| Minimum Charge | Job | | 262 | | 262 | |
| **Water Softener** | | | | | | |
| Demolish | Ea. | | 97 | | 97 | Includes material and labor to install a |
| Install | Ea. | 2550 | 236 | | 2786 | 14 GPM water softener. |
| Demolish and Install | Ea. | 2550 | 333 | | 2883 | |
| Reinstall | Ea. | | 188.74 | | 188.74 | |
| Clean | Ea. | .52 | 22 | | 22.52 | |
| Minimum Charge | Ea. | | 262 | | 262 | |

## Ductwork

| Ductwork | Unit | Material | Labor | Equip. | Total | Specification |
|---|---|---|---|---|---|---|
| **Galvanized, Square or Rectangular** | | | | | | |
| **6"** | | | | | | |
| Demolish | L.F. | | 2.36 | | 2.36 | Includes material and labor to install |
| Install | L.F. | 1.45 | 11.30 | | 12.75 | rectangular ductwork with a greatest |
| Demolish and Install | L.F. | 1.45 | 13.66 | | 15.11 | dimension of 6". |
| Reinstall | L.F. | | 9.06 | | 9.06 | |
| Clean | L.F. | | 2.72 | | 2.72 | |
| Minimum Charge | Job | | 255 | | 255 | |
| **8"** | | | | | | |
| Demolish | L.F. | | 2.36 | | 2.36 | Includes material and labor to install |
| Install | L.F. | 1.93 | 14.55 | | 16.48 | rectangular ductwork with a greatest |
| Demolish and Install | L.F. | 1.93 | 16.91 | | 18.84 | dimension of 8". |
| Reinstall | L.F. | | 11.65 | | 11.65 | |
| Clean | L.F. | | 3.62 | | 3.62 | |
| Minimum Charge | Job | | 255 | | 255 | |
| **10"** | | | | | | |
| Demolish | L.F. | | 2.36 | | 2.36 | Includes material and labor to install |
| Install | L.F. | 2.41 | 18.85 | | 21.26 | rectangular ductwork with a greatest |
| Demolish and Install | L.F. | 2.41 | 21.21 | | 23.62 | dimension of 10". |
| Reinstall | L.F. | | 15.10 | | 15.10 | |
| Clean | L.F. | | 4.53 | | 4.53 | |
| Minimum Charge | Job | | 255 | | 255 | |

# Rough Mechanical

| Ductwork | Unit | Material | Labor | Equip. | Total | Specification |
|---|---|---|---|---|---|---|
| **Galvanized, Spiral** | | | | | | |
| **4"** | | | | | | |
| Demolish | L.F. | | 2.83 | | 2.83 | Cost includes material and labor to |
| Install | L.F. | 2.05 | 2.55 | | 4.60 | install 4" diameter spiral pre-formed |
| Demolish and Install | L.F. | 2.05 | 5.38 | | 7.43 | steel galvanized ductwork. |
| Reinstall | L.F. | | 2.04 | | 2.04 | |
| Clean | L.F. | | 1.42 | | 1.42 | |
| Minimum Charge | Job | | 255 | | 255 | |
| **6"** | | | | | | |
| Demolish | L.F. | | 2.83 | | 2.83 | Cost includes material and labor to |
| Install | L.F. | 2.05 | 3.28 | | 5.33 | install 6" diameter spiral pre-formed |
| Demolish and Install | L.F. | 2.05 | 6.11 | | 8.16 | steel galvanized ductwork. |
| Reinstall | L.F. | | 2.62 | | 2.62 | |
| Clean | L.F. | | 2.13 | | 2.13 | |
| Minimum Charge | Job | | 255 | | 255 | |
| **8"** | | | | | | |
| Demolish | L.F. | | 2.83 | | 2.83 | Cost includes material and labor to |
| Install | L.F. | 2.73 | 4.59 | | 7.32 | install 8" diameter spiral pre-formed |
| Demolish and Install | L.F. | 2.73 | 7.42 | | 10.15 | steel galvanized ductwork. |
| Reinstall | L.F. | | 3.67 | | 3.67 | |
| Clean | L.F. | | 2.85 | | 2.85 | |
| Minimum Charge | Job | | 255 | | 255 | |
| **10"** | | | | | | |
| Demolish | L.F. | | 3.54 | | 3.54 | Cost includes material and labor to |
| Install | L.F. | 3.41 | 5.75 | | 9.16 | install 10" diameter spiral pre-formed |
| Demolish and Install | L.F. | 3.41 | 9.29 | | 12.70 | steel galvanized ductwork. |
| Reinstall | L.F. | | 4.59 | | 4.59 | |
| Clean | L.F. | | 3.56 | | 3.56 | |
| Minimum Charge | Job | | 255 | | 255 | |
| **12"** | | | | | | |
| Demolish | L.F. | | 4.72 | | 4.72 | Cost includes material and labor to |
| Install | L.F. | 4.09 | 7.65 | | 11.74 | install 12" diameter spiral pre-formed |
| Demolish and Install | L.F. | 4.09 | 12.37 | | 16.46 | steel galvanized ductwork. |
| Reinstall | L.F. | | 6.12 | | 6.12 | |
| Clean | L.F. | | 4.27 | | 4.27 | |
| Minimum Charge | Job | | 255 | | 255 | |
| **16"** | | | | | | |
| Demolish | L.F. | | 9.45 | | 9.45 | Cost includes material and labor to |
| Install | L.F. | 6.75 | 15.30 | | 22.05 | install 16" diameter spiral pre-formed |
| Demolish and Install | L.F. | 6.75 | 24.75 | | 31.50 | steel galvanized ductwork. |
| Reinstall | L.F. | | 12.23 | | 12.23 | |
| Clean | L.F. | | 5.70 | | 5.70 | |
| Minimum Charge | Job | | 255 | | 255 | |
| **Flexible Fiberglass** | | | | | | |
| **6" Non-insulated** | | | | | | |
| Demolish | L.F. | | 2.18 | | 2.18 | Cost includes material and labor to |
| Install | L.F. | 1.76 | 3.28 | | 5.04 | install 6" diameter non-insulated |
| Demolish and Install | L.F. | 1.76 | 5.46 | | 7.22 | flexible fiberglass fabric on corrosion |
| Reinstall | L.F. | | 2.62 | | 2.62 | resistant metal ductwork. |
| Clean | L.F. | | 2.13 | | 2.13 | |
| Minimum Charge | Job | | 255 | | 255 | |
| **8" Non-insulated** | | | | | | |
| Demolish | L.F. | | 3.15 | | 3.15 | Cost includes material and labor to |
| Install | L.F. | 2.22 | 4.59 | | 6.81 | install 8" diameter non-insulated |
| Demolish and Install | L.F. | 2.22 | 7.74 | | 9.96 | flexible fiberglass fabric on corrosion |
| Reinstall | L.F. | | 3.67 | | 3.67 | resistant metal ductwork. |
| Clean | L.F. | | 2.85 | | 2.85 | |
| Minimum Charge | Job | | 255 | | 255 | |

## Ductwork

| | Unit | Material | Labor | Equip. | Total | Specification |
|---|---|---|---|---|---|---|
| **10″ Non-insulated** | | | | | | |
| Demolish | L.F. | | 4.04 | | 4.04 | Cost includes material and labor to |
| Install | L.F. | 2.87 | 5.75 | | 8.62 | install 10″ diameter non-insulated |
| Demolish and Install | L.F. | 2.87 | 9.79 | | 12.66 | flexible fiberglass fabric on corrosion |
| Reinstall | L.F. | | 4.59 | | 4.59 | resistant metal ductwork. |
| Clean | L.F. | | 3.56 | | 3.56 | |
| Minimum Charge | Job | | 255 | | 255 | |
| **12″ Non-insulated** | | | | | | |
| Demolish | L.F. | | 5.65 | | 5.65 | Cost includes material and labor to |
| Install | L.F. | 3.41 | 7.65 | | 11.06 | install 12″ diameter non-insulated |
| Demolish and Install | L.F. | 3.41 | 13.30 | | 16.71 | flexible fiberglass fabric on corrosion |
| Reinstall | L.F. | | 6.12 | | 6.12 | resistant metal ductwork. |
| Clean | L.F. | | 4.27 | | 4.27 | |
| Minimum Charge | Job | | 255 | | 255 | |
| **4″ Insulated** | | | | | | |
| Demolish | L.F. | | 1.67 | | 1.67 | Cost includes material and labor to |
| Install | L.F. | 3.08 | 2.70 | | 5.78 | install 4″ diameter insulated flexible |
| Demolish and Install | L.F. | 3.08 | 4.37 | | 7.45 | fiberglass fabric on corrosion resistant |
| Reinstall | L.F. | | 2.16 | | 2.16 | metal ductwork. |
| Clean | L.F. | | 1.42 | | 1.42 | |
| Minimum Charge | Job | | 255 | | 255 | |
| **6″ Insulated** | | | | | | |
| Demolish | L.F. | | 2.18 | | 2.18 | Cost includes material and labor to |
| Install | L.F. | 3.42 | 3.53 | | 6.95 | install 6″ diameter insulated flexible |
| Demolish and Install | L.F. | 3.42 | 5.71 | | 9.13 | fiberglass fabric on corrosion resistant |
| Reinstall | L.F. | | 2.82 | | 2.82 | metal ductwork. |
| Clean | L.F. | | 2.13 | | 2.13 | |
| Minimum Charge | Job | | 255 | | 255 | |
| **8″ Insulated** | | | | | | |
| Demolish | L.F. | | 3.15 | | 3.15 | Cost includes material and labor to |
| Install | L.F. | 4.18 | 5.10 | | 9.28 | install 8″ diameter insulated flexible |
| Demolish and Install | L.F. | 4.18 | 8.25 | | 12.43 | fiberglass fabric on corrosion resistant |
| Reinstall | L.F. | | 4.08 | | 4.08 | metal ductwork. |
| Clean | L.F. | | 2.85 | | 2.85 | |
| Minimum Charge | Job | | 255 | | 255 | |
| **10″ Insulated** | | | | | | |
| Demolish | L.F. | | 4.04 | | 4.04 | Cost includes material and labor to |
| Install | L.F. | 5 | 6.55 | | 11.55 | install 10″ diameter insulated flexible |
| Demolish and Install | L.F. | 5 | 10.59 | | 15.59 | fiberglass fabric on corrosion resistant |
| Reinstall | L.F. | | 5.24 | | 5.24 | metal ductwork. |
| Clean | L.F. | | 3.56 | | 3.56 | |
| Minimum Charge | Job | | 255 | | 255 | |
| **12″ Insulated** | | | | | | |
| Demolish | L.F. | | 5.65 | | 5.65 | Cost includes material and labor to |
| Install | L.F. | 5.80 | 9.20 | | 15 | install 12″ diameter insulated flexible |
| Demolish and Install | L.F. | 5.80 | 14.85 | | 20.65 | fiberglass fabric on corrosion resistant |
| Reinstall | L.F. | | 7.34 | | 7.34 | metal ductwork. |
| Clean | L.F. | | 4.27 | | 4.27 | |
| Minimum Charge | Job | | 255 | | 255 | |

| Heating | Unit | Material | Labor | Equip. | Total | Specification |
|---|---|---|---|---|---|---|
| **Hot Water Baseboard** | | | | | | |
| **Baseboard** | | | | | | |
| Demolish | L.F. | | 9.20 | | 9.20 | Includes material and labor to install |
| Install | L.F. | 12.75 | 16.15 | | 28.90 | baseboard radiation. |
| Demolish and Install | L.F. | 12.75 | 25.35 | | 38.10 | |
| Reinstall | L.F. | | 12.94 | | 12.94 | |
| Clean | L.F. | .52 | 5.55 | | 6.07 | |
| Minimum Charge | Job | | 262 | | 262 | |
| **3'** | | | | | | |
| Demolish | Ea. | | 29 | | 29 | Includes material and labor to install |
| Install | Ea. | 26 | 54 | | 80 | fin tube baseboard radiation in 3' |
| Demolish and Install | Ea. | 26 | 83 | | 109 | sections. |
| Reinstall | Ea. | | 43.13 | | 43.13 | |
| Clean | Ea. | .14 | 16.90 | | 17.04 | |
| Minimum Charge | Job | | 262 | | 262 | |
| **4'** | | | | | | |
| Demolish | Ea. | | 39 | | 39 | Includes material and labor to install |
| Install | Ea. | 34.50 | 60.50 | | 95 | fin tube baseboard radiation in 4' |
| Demolish and Install | Ea. | 34.50 | 99.50 | | 134 | sections. |
| Reinstall | Ea. | | 48.52 | | 48.52 | |
| Clean | Ea. | .30 | 22 | | 22.30 | |
| Minimum Charge | Job | | 262 | | 262 | |
| **5'** | | | | | | |
| Demolish | Ea. | | 48.50 | | 48.50 | Includes material and labor to install |
| Install | Ea. | 43.50 | 69.50 | | 113 | fin tube baseboard radiation in 5' |
| Demolish and Install | Ea. | 43.50 | 118 | | 161.50 | sections. |
| Reinstall | Ea. | | 55.45 | | 55.45 | |
| Clean | Ea. | .30 | 27.50 | | 27.80 | |
| Minimum Charge | Job | | 262 | | 262 | |
| **6'** | | | | | | |
| Demolish | Ea. | | 58 | | 58 | Includes material and labor to install |
| Install | Ea. | 51.50 | 81 | | 132.50 | fin tube baseboard radiation in 6' |
| Demolish and Install | Ea. | 51.50 | 139 | | 190.50 | sections. |
| Reinstall | Ea. | | 64.69 | | 64.69 | |
| Clean | Ea. | .30 | 32.50 | | 32.80 | |
| Minimum Charge | Job | | 262 | | 262 | |
| **7'** | | | | | | |
| Demolish | Ea. | | 64.50 | | 64.50 | Includes material and labor to install |
| Install | Ea. | 60.50 | 105 | | 165.50 | fin tube baseboard radiation in 7' |
| Demolish and Install | Ea. | 60.50 | 169.50 | | 230 | sections. |
| Reinstall | Ea. | | 83.84 | | 83.84 | |
| Clean | Ea. | .36 | 39.50 | | 39.86 | |
| Minimum Charge | Job | | 262 | | 262 | |
| **9'** | | | | | | |
| Demolish | Ea. | | 97 | | 97 | Includes material and labor to install |
| Install | Ea. | 77.50 | 150 | | 227.50 | fin tube baseboard radiation in 9' |
| Demolish and Install | Ea. | 77.50 | 247 | | 324.50 | sections. |
| Reinstall | Ea. | | 119.77 | | 119.77 | |
| Clean | Ea. | .59 | 50.50 | | 51.09 | |
| Minimum Charge | Job | | 262 | | 262 | |

# Rough Mechanical

| Heating | | Unit | Material | Labor | Equip. | Total | Specification |
|---|---|---|---|---|---|---|---|
| **Cabinet Unit Heater** | | | | | | | |
| **50 MBH** | | | | | | | |
| | Demolish | Ea. | | 48.50 | | 48.50 | Includes material and labor to install a |
| | Install | Ea. | 1975 | 139 | | 2114 | cabinet space heater with an output of |
| | Demolish and Install | Ea. | 1975 | 187.50 | | 2162.50 | 50 MBH. |
| | Reinstall | Ea. | | 110.90 | | 110.90 | |
| | Clean | Ea. | .36 | 44.50 | | 44.86 | |
| | Minimum Charge | Job | | 270 | | 270 | |
| **75 MBH** | | | | | | | |
| | Demolish | Ea. | | 48.50 | | 48.50 | Includes material and labor to install a |
| | Install | Ea. | 2175 | 162 | | 2337 | cabinet space heater with an output of |
| | Demolish and Install | Ea. | 2175 | 210.50 | | 2385.50 | 75 MBH. |
| | Reinstall | Ea. | | 129.39 | | 129.39 | |
| | Clean | Ea. | .36 | 44.50 | | 44.86 | |
| | Minimum Charge | Job | | 270 | | 270 | |
| **125 MBH** | | | | | | | |
| | Demolish | Ea. | | 58 | | 58 | Includes material and labor to install a |
| | Install | Ea. | 2575 | 194 | | 2769 | cabinet space heater with an output of |
| | Demolish and Install | Ea. | 2575 | 252 | | 2827 | 125 MBH. |
| | Reinstall | Ea. | | 155.26 | | 155.26 | |
| | Clean | Ea. | .36 | 44.50 | | 44.86 | |
| | Minimum Charge | Job | | 270 | | 270 | |
| **175 MBH** | | | | | | | |
| | Demolish | Ea. | | 73 | | 73 | Includes material and labor to install a |
| | Install | Ea. | 3475 | 277 | | 3752 | cabinet space heater with an output of |
| | Demolish and Install | Ea. | 3475 | 350 | | 3825 | 175 MBH. |
| | Reinstall | Ea. | | 221.81 | | 221.81 | |
| | Clean | Ea. | .36 | 44.50 | | 44.86 | |
| | Minimum Charge | Job | | 270 | | 270 | |

| Furnace Unit | | Unit | Material | Labor | Equip. | Total | Specification |
|---|---|---|---|---|---|---|---|
| **Wall (gas-fired)** | | | | | | | |
| **25 MBH w / Thermostat** | | | | | | | |
| | Demolish | Ea. | | 115 | | 115 | Includes material and labor to install a |
| | Install | Ea. | 870 | 184 | | 1054 | gas fired hot air furnace rated at 25 |
| | Demolish and Install | Ea. | 870 | 299 | | 1169 | MBH. |
| | Reinstall | Ea. | | 146.82 | | 146.82 | |
| | Clean | Ea. | .74 | 98.50 | | 99.24 | |
| | Minimum Charge | Job | | 270 | | 270 | |
| **35 MBH w / Thermostat** | | | | | | | |
| | Demolish | Ea. | | 115 | | 115 | Includes material and labor to install a |
| | Install | Ea. | 790 | 204 | | 994 | gas fired hot air furnace rated at 35 |
| | Demolish and Install | Ea. | 790 | 319 | | 1109 | MBH. |
| | Reinstall | Ea. | | 163.13 | | 163.13 | |
| | Clean | Ea. | .74 | 98.50 | | 99.24 | |
| | Minimum Charge | Job | | 270 | | 270 | |
| **50 MBH w / Thermostat** | | | | | | | |
| | Demolish | Ea. | | 115 | | 115 | Includes material and labor to install a |
| | Install | Ea. | 1600 | 229 | | 1829 | gas fired hot air furnace rated at 50 |
| | Demolish and Install | Ea. | 1600 | 344 | | 1944 | MBH. |
| | Reinstall | Ea. | | 183.52 | | 183.52 | |
| | Clean | Ea. | .74 | 98.50 | | 99.24 | |
| | Minimum Charge | Job | | 270 | | 270 | |

| Furnace Unit | Unit | Material | Labor | Equip. | Total | Specification |
|---|---|---|---|---|---|---|
| **60 MBH w / Thermostat** | | | | | | |
| Demolish | Ea. | | 115 | | 115 | Includes material and labor to install a |
| Install | Ea. | 1425 | 255 | | 1680 | gas fired hot air furnace rated at 60 |
| Demolish and Install | Ea. | 1425 | 370 | | 1795 | MBH. |
| Reinstall | Ea. | | 203.91 | | 203.91 | |
| Clean | Ea. | .74 | 98.50 | | 99.24 | |
| Minimum Charge | Job | | 270 | | 270 | |
| **Floor (gas-fired)** | | | | | | |
| **30 MBH w / Thermostat** | | | | | | |
| Demolish | Ea. | | 115 | | 115 | Includes material and labor to install a |
| Install | Ea. | 1350 | 184 | | 1534 | gas fired hot air floor furnace rated at |
| Demolish and Install | Ea. | 1350 | 299 | | 1649 | 30 MBH. |
| Reinstall | Ea. | | 146.82 | | 146.82 | |
| Clean | Ea. | .74 | 98.50 | | 99.24 | |
| Minimum Charge | Job | | 270 | | 270 | |
| **50 MBH w / Thermostat** | | | | | | |
| Demolish | Ea. | | 115 | | 115 | Includes material and labor to install a |
| Install | Ea. | 1550 | 229 | | 1779 | gas fired hot air floor furnace rated at |
| Demolish and Install | Ea. | 1550 | 344 | | 1894 | 50 MBH. |
| Reinstall | Ea. | | 183.52 | | 183.52 | |
| Clean | Ea. | .74 | 98.50 | | 99.24 | |
| Minimum Charge | Job | | 270 | | 270 | |
| **Gas Forced Air** | | | | | | |
| **50 MBH** | | | | | | |
| Demolish | Ea. | | 115 | | 115 | Includes material and labor to install |
| Install | Ea. | 1250 | 229 | | 1479 | an up flow gas fired hot air furnace |
| Demolish and Install | Ea. | 1250 | 344 | | 1594 | rated at 50 MBH. |
| Reinstall | Ea. | | 183.52 | | 183.52 | |
| Clean | Ea. | .74 | 98.50 | | 99.24 | |
| Minimum Charge | Job | | 270 | | 270 | |
| **80 MBH** | | | | | | |
| Demolish | Ea. | | 115 | | 115 | Includes material and labor to install |
| Install | Ea. | 1400 | 255 | | 1655 | an up flow gas fired hot air furnace |
| Demolish and Install | Ea. | 1400 | 370 | | 1770 | rated at 80 MBH. |
| Reinstall | Ea. | | 203.91 | | 203.91 | |
| Clean | Ea. | .74 | 98.50 | | 99.24 | |
| Minimum Charge | Job | | 270 | | 270 | |
| **100 MBH** | | | | | | |
| Demolish | Ea. | | 115 | | 115 | Includes material and labor to install |
| Install | Ea. | 1525 | 287 | | 1812 | an up flow gas fired hot air furnace |
| Demolish and Install | Ea. | 1525 | 402 | | 1927 | rated at 100 MBH. |
| Reinstall | Ea. | | 229.40 | | 229.40 | |
| Clean | Ea. | .74 | 98.50 | | 99.24 | |
| Minimum Charge | Job | | 270 | | 270 | |
| **120 MBH** | | | | | | |
| Demolish | Ea. | | 115 | | 115 | Includes material and labor to install |
| Install | Ea. | 1550 | 305 | | 1855 | an up flow gas fired hot air furnace |
| Demolish and Install | Ea. | 1550 | 420 | | 1970 | rated at 120 MBH. |
| Reinstall | Ea. | | 244.69 | | 244.69 | |
| Clean | Ea. | .74 | 98.50 | | 99.24 | |
| Minimum Charge | Job | | 270 | | 270 | |

**For customer support on your Contractor's Pricing Guide: Residential Repair & Remodeling, call 800.448.8182.**

| Furnace Unit | Unit | Material | Labor | Equip. | Total | Specification |
|---|---|---|---|---|---|---|
| **Electric Forced Air** | | | | | | |
| **31.4 MBH** | | | | | | |
| Demolish | Ea. | | 115 | | 115 | Includes material and labor to install |
| Install | Ea. | 2575 | 262 | | 2837 | an electric hot air furnace rated at |
| Demolish and Install | Ea. | 2575 | 377 | | 2952 | 31.4 MBH. |
| Reinstall | Ea. | | 209.81 | | 209.81 | |
| Clean | Ea. | .74 | 98.50 | | 99.24 | |
| Minimum Charge | Job | | 270 | | 270 | |
| **47.1 MBH** | | | | | | |
| Demolish | Ea. | | 115 | | 115 | Includes material and labor to install |
| Install | Ea. | 3025 | 274 | | 3299 | an electric hot air furnace rated at |
| Demolish and Install | Ea. | 3025 | 389 | | 3414 | 47.1 MBH. |
| Reinstall | Ea. | | 219.57 | | 219.57 | |
| Clean | Ea. | .74 | 98.50 | | 99.24 | |
| Minimum Charge | Job | | 270 | | 270 | |
| **62.9 MBH** | | | | | | |
| Demolish | Ea. | | 115 | | 115 | Includes material and labor to install |
| Install | Ea. | 3200 | 295 | | 3495 | an electric hot air furnace rated at |
| Demolish and Install | Ea. | 3200 | 410 | | 3610 | 62.9 MBH. |
| Reinstall | Ea. | | 236.04 | | 236.04 | |
| Clean | Ea. | .74 | 98.50 | | 99.24 | |
| Minimum Charge | Job | | 270 | | 270 | |
| **78.5 MBH** | | | | | | |
| Demolish | Ea. | | 115 | | 115 | Includes material and labor to install |
| Install | Ea. | 3850 | 305 | | 4155 | an electric hot air furnace rated at |
| Demolish and Install | Ea. | 3850 | 420 | | 4270 | 78.5 MBH. |
| Reinstall | Ea. | | 242.09 | | 242.09 | |
| Clean | Ea. | .74 | 98.50 | | 99.24 | |
| Minimum Charge | Job | | 270 | | 270 | |
| **110 MBH** | | | | | | |
| Demolish | Ea. | | 115 | | 115 | Includes material and labor to install |
| Install | Ea. | 4875 | 320 | | 5195 | an electric hot air furnace rated at |
| Demolish and Install | Ea. | 4875 | 435 | | 5310 | 110 MBH. |
| Reinstall | Ea. | | 255.18 | | 255.18 | |
| Clean | Ea. | .74 | 98.50 | | 99.24 | |
| Minimum Charge | Job | | 270 | | 270 | |
| **A/C Junction Box** | | | | | | |
| Install | Ea. | 57 | 131 | | 188 | Includes labor and materials to install |
| Minimum Charge | Job | | 270 | | 270 | a heating system emergency shut-off switch. Cost includes up to 40 feet of non-metallic cable. |

| Boiler | Unit | Material | Labor | Equip. | Total | Specification |
|---|---|---|---|---|---|---|
| **Gas-Fired** | | | | | | |
| **60 MBH** | | | | | | |
| Demolish | Ea. | | 980 | | 980 | Includes material and labor to install a |
| Install | Ea. | 2750 | 1775 | | 4525 | gas fired cast iron boiler rated at 60 |
| Demolish and Install | Ea. | 2750 | 2755 | | 5505 | MBH. |
| Reinstall | Ea. | | 1411.49 | | 1411.49 | |
| Clean | Ea. | 1.47 | 59 | | 60.47 | |
| Minimum Charge | Job | | 270 | | 270 | |

## Boiler

| Boiler | Unit | Material | Labor | Equip. | Total | Specification |
|---|---|---|---|---|---|---|
| **Gas-Fired** | | | | | | |
| **125 MBH** | | | | | | |
| Demolish | Ea. | | 2700 | | 2700 | Includes material and labor to install a |
| Install | Ea. | 4600 | 2150 | | 6750 | gas fired cast iron boiler rated at 125 |
| Demolish and Install | Ea. | 4600 | 4850 | | 9450 | MBH. |
| Reinstall | Ea. | | 1725.16 | | 1725.16 | |
| Clean | Ea. | 1.47 | 59 | | 60.47 | |
| Minimum Charge | Job | | 270 | | 270 | |
| **Gas-Fired** | | | | | | |
| **400 MBH** | | | | | | |
| Demolish | Ea. | | 2700 | | 2700 | Includes material and labor to install a |
| Install | Ea. | 5925 | 3025 | | 8950 | gas fired cast iron boiler rated at 400 |
| Demolish and Install | Ea. | 5925 | 5725 | | 11650 | MBH. |
| Reinstall | Ea. | | 3032.50 | | 3032.50 | |
| Clean | Ea. | 1.47 | 59 | | 60.47 | |
| Minimum Charge | Job | | 270 | | 270 | |

## Heat Pump

| Heat Pump | Unit | Material | Labor | Equip. | Total | Specification |
|---|---|---|---|---|---|---|
| **2 Ton** | | | | | | |
| Demolish | Ea. | | 270 | | 270 | Includes material and labor to install a |
| Install | Ea. | 1825 | 485 | | 2310 | 2 ton heat pump. |
| Demolish and Install | Ea. | 1825 | 755 | | 2580 | |
| Reinstall | Ea. | | 388.16 | | 388.16 | |
| Clean | Ea. | 1.47 | 44.50 | | 45.97 | |
| Minimum Charge | Job | | 270 | | 270 | |
| **3 Ton** | | | | | | |
| Demolish | Ea. | | 450 | | 450 | Includes material and labor to install a |
| Install | Ea. | 2300 | 810 | | 3110 | 3 ton heat pump. |
| Demolish and Install | Ea. | 2300 | 1260 | | 3560 | |
| Reinstall | Ea. | | 646.93 | | 646.93 | |
| Clean | Ea. | 1.47 | 44.50 | | 45.97 | |
| Minimum Charge | Job | | 270 | | 270 | |
| **4 Ton** | | | | | | |
| Demolish | Ea. | | 675 | | 675 | Includes material and labor to install a |
| Install | Ea. | 2700 | 1225 | | 3925 | 4 ton heat pump. |
| Demolish and Install | Ea. | 2700 | 1900 | | 4600 | |
| Reinstall | Ea. | | 970.40 | | 970.40 | |
| Clean | Ea. | 1.47 | 44.50 | | 45.97 | |
| Minimum Charge | Job | | 270 | | 270 | |
| **5 Ton** | | | | | | |
| Demolish | Ea. | | 1075 | | 1075 | Includes material and labor to install a |
| Install | Ea. | 2950 | 1950 | | 4900 | 5 ton heat pump. |
| Demolish and Install | Ea. | 2950 | 3025 | | 5975 | |
| Reinstall | Ea. | | 1552.64 | | 1552.64 | |
| Clean | Ea. | 1.47 | 44.50 | | 45.97 | |
| Minimum Charge | Job | | 270 | | 270 | |
| **Thermostat** | | | | | | |
| Demolish | Ea. | | 15 | | 15 | Includes material and labor to install |
| Install | Ea. | 251 | 67.50 | | 318.50 | an electric thermostat, 2 set back. |
| Demolish and Install | Ea. | 251 | 82.50 | | 333.50 | |
| Reinstall | Ea. | | 53.92 | | 53.92 | |
| Clean | Ea. | .03 | 11.10 | | 11.13 | |
| Minimum Charge | Job | | 270 | | 270 | |

# Rough Mechanical

| Heat Pump | | Unit | Material | Labor | Equip. | Total | Specification |
|---|---|---|---|---|---|---|---|
| **A/C Junction Box** | | | | | | | |
| | Install | Ea. | 525 | 405 | | 930 | Includes labor and materials to install |
| | Minimum Charge | Job | | 263 | | 263 | a heat pump circuit including up to 40 |
| | | | | | | | feet of non-metallic cable. |

| Heating | | Unit | Material | Labor | Equip. | Total | Specification |
|---|---|---|---|---|---|---|---|
| **Electric Baseboard** | | | | | | | |
| | Demolish | L.F. | | 7.30 | | 7.30 | Includes material and labor to install |
| | Install | L.F. | 10.20 | 17.50 | | 27.70 | electric baseboard heating elements. |
| | Demolish and Install | L.F. | 10.20 | 24.80 | | 35 | |
| | Reinstall | L.F. | | 14.01 | | 14.01 | |
| | Clean | L.F. | .52 | 5.55 | | 6.07 | |
| | Minimum Charge | Job | | 263 | | 263 | |
| **2' 6"** | | | | | | | |
| | Demolish | Ea. | | 24.50 | | 24.50 | Includes material and labor to install a |
| | Install | Ea. | 30.50 | 65.50 | | 96 | 2' 6" long section of electric |
| | Demolish and Install | Ea. | 30.50 | 90 | | 120.50 | baseboard heating elements. |
| | Reinstall | Ea. | | 52.52 | | 52.52 | |
| | Clean | Ea. | .14 | 13.65 | | 13.79 | |
| | Minimum Charge | Job | | 263 | | 263 | |
| **3'** | | | | | | | |
| | Demolish | Ea. | | 26.50 | | 26.50 | Includes material and labor to install a |
| | Install | Ea. | 34.50 | 65.50 | | 100 | 3' long section of electric baseboard |
| | Demolish and Install | Ea. | 34.50 | 92 | | 126.50 | heating elements. |
| | Reinstall | Ea. | | 52.52 | | 52.52 | |
| | Clean | Ea. | .14 | 16.90 | | 17.04 | |
| | Minimum Charge | Job | | 263 | | 263 | |
| **4'** | | | | | | | |
| | Demolish | Ea. | | 29 | | 29 | Includes material and labor to install a |
| | Install | Ea. | 40.50 | 78.50 | | 119 | 4' long section of electric baseboard |
| | Demolish and Install | Ea. | 40.50 | 107.50 | | 148 | heating elements. |
| | Reinstall | Ea. | | 62.71 | | 62.71 | |
| | Clean | Ea. | .06 | 3.70 | | 3.76 | |
| | Minimum Charge | Job | | 263 | | 263 | |
| **5'** | | | | | | | |
| | Demolish | Ea. | | 32.50 | | 32.50 | Includes material and labor to install a |
| | Install | Ea. | 51 | 92 | | 143 | 5' long section of electric baseboard |
| | Demolish and Install | Ea. | 51 | 124.50 | | 175.50 | heating elements. |
| | Reinstall | Ea. | | 73.71 | | 73.71 | |
| | Clean | Ea. | .30 | 27.50 | | 27.80 | |
| | Minimum Charge | Job | | 263 | | 263 | |
| **6'** | | | | | | | |
| | Demolish | Ea. | | 36.50 | | 36.50 | Includes material and labor to install a |
| | Install | Ea. | 56.50 | 105 | | 161.50 | 6' long section of electric baseboard |
| | Demolish and Install | Ea. | 56.50 | 141.50 | | 198 | heating elements. |
| | Reinstall | Ea. | | 84.03 | | 84.03 | |
| | Clean | Ea. | .06 | 4.44 | | 4.50 | |
| | Minimum Charge | Job | | 263 | | 263 | |
| **8'** | | | | | | | |
| | Demolish | Ea. | | 48.50 | | 48.50 | Includes material and labor to install |
| | Install | Ea. | 69 | 131 | | 200 | an 8' long section of electric |
| | Demolish and Install | Ea. | 69 | 179.50 | | 248.50 | baseboard heating elements. |
| | Reinstall | Ea. | | 105.04 | | 105.04 | |
| | Clean | Ea. | .06 | 5.55 | | 5.61 | |
| | Minimum Charge | Job | | 263 | | 263 | |

# Rough Mechanical

## Heating

| | Unit | Material | Labor | Equip. | Total | Specification |
|---|---|---|---|---|---|---|
| **10'** | | | | | | |
| Demolish | Ea. | | 58.50 | | 58.50 | Includes material and labor to install a |
| Install | Ea. | 169 | 159 | | 328 | 10' long section of electric baseboard |
| Demolish and Install | Ea. | 169 | 217.50 | | 386.50 | heating elements. |
| Reinstall | Ea. | | 127.32 | | 127.32 | |
| Clean | Ea. | .59 | 55.50 | | 56.09 | |
| Minimum Charge | Job | | 263 | | 263 | |
| **Electric Radiant Wall Panel** | | | | | | |
| Demolish | Ea. | | 39 | | 39 | Includes material and labor to install |
| Install | Ea. | 565 | 105 | | 670 | an 18" x 23" electric radiant heat wall |
| Demolish and Install | Ea. | 565 | 144 | | 709 | panel. |
| Reinstall | Ea. | | 84.03 | | 84.03 | |
| Clean | Ea. | .08 | 11.10 | | 11.18 | |
| Minimum Charge | Job | | 263 | | 263 | |

## Heating Control

| | Unit | Material | Labor | Equip. | Total | Specification |
|---|---|---|---|---|---|---|
| **Thermostat** | | | | | | |
| Demolish | Ea. | | 15 | | 15 | Includes material and labor to install |
| Install | Ea. | 251 | 67.50 | | 318.50 | an electric thermostat, 2 set back. |
| Demolish and Install | Ea. | 251 | 82.50 | | 333.50 | |
| Reinstall | Ea. | | 53.92 | | 53.92 | |
| Clean | Ea. | .03 | 11.10 | | 11.13 | |
| Minimum Charge | Job | | 270 | | 270 | |

## Air Conditioning

| | Unit | Material | Labor | Equip. | Total | Specification |
|---|---|---|---|---|---|---|
| **Self-contained** | | | | | | |
| **2 Ton** | | | | | | |
| Demolish | Ea. | | 485 | | 485 | Includes material and labor to install a |
| Install | Ea. | 4350 | 835 | | 5185 | rooftop air conditioning unit, 2 ton. |
| Demolish and Install | Ea. | 4350 | 1320 | | 5670 | No hoisting is included. |
| Reinstall | Ea. | | 837.27 | | 837.27 | |
| Clean | Ea. | 1.47 | 44.50 | | 45.97 | |
| Minimum Charge | Job | | 270 | | 270 | |
| **3 Ton** | | | | | | |
| Demolish | Ea. | | 485 | | 485 | Includes material and labor to install a |
| Install | Ea. | 4450 | 1100 | | 5550 | rooftop air conditioning unit, 3 ton. |
| Demolish and Install | Ea. | 4450 | 1585 | | 6035 | No hoisting is included. |
| Reinstall | Ea. | | 1103.98 | | 1103.98 | |
| Clean | Ea. | 1.47 | 44.50 | | 45.97 | |
| Minimum Charge | Job | | 270 | | 270 | |
| **4 Ton** | | | | | | |
| Demolish | Ea. | | 485 | | 485 | Includes material and labor to install a |
| Install | Ea. | 5375 | 1275 | | 6650 | rooftop air conditioning unit, 4 ton. |
| Demolish and Install | Ea. | 5375 | 1760 | | 7135 | No hoisting is included. |
| Reinstall | Ea. | | 1280.21 | | 1280.21 | |
| Clean | Ea. | 1.47 | 44.50 | | 45.97 | |
| Minimum Charge | Job | | 270 | | 270 | |
| **5 Ton** | | | | | | |
| Demolish | Ea. | | 485 | | 485 | Includes material and labor to install a |
| Install | Ea. | 5500 | 1375 | | 6875 | rooftop air conditioning unit, 5 ton. |
| Demolish and Install | Ea. | 5500 | 1860 | | 7360 | No hoisting is included. |
| Reinstall | Ea. | | 1382.34 | | 1382.34 | |
| Clean | Ea. | 1.47 | 44.50 | | 45.97 | |
| Minimum Charge | Job | | 270 | | 270 | |

159

| Air Conditioning | Unit | Material | Labor | Equip. | Total | Specification |
|---|---|---|---|---|---|---|
| **Compressor** | | | | | | |
| **2 Ton** | | | | | | |
| Demolish | Ea. | | 450 | | 450 | Includes material and labor to install a |
| Install | Ea. | 310 | 270 | | 580 | refrigeration compressor. |
| Demolish and Install | Ea. | 310 | 720 | | 1030 | |
| Reinstall | Ea. | | 215.68 | | 215.68 | |
| Clean | Ea. | 1.47 | 44.50 | | 45.97 | |
| Minimum Charge | Job | | 270 | | 270 | |
| **2.5 Ton** | | | | | | |
| Demolish | Ea. | | 450 | | 450 | Includes material and labor to install a |
| Install | Ea. | 370 | 292 | | 662 | refrigeration compressor. |
| Demolish and Install | Ea. | 370 | 742 | | 1112 | |
| Reinstall | Ea. | | 233.67 | | 233.67 | |
| Clean | Ea. | 1.47 | 44.50 | | 45.97 | |
| Minimum Charge | Job | | 270 | | 270 | |
| **3 Ton** | | | | | | |
| Demolish | Ea. | | 450 | | 450 | Includes material and labor to install a |
| Install | Ea. | 400 | 305 | | 705 | refrigeration compressor. |
| Demolish and Install | Ea. | 400 | 755 | | 1155 | |
| Reinstall | Ea. | | 243.84 | | 243.84 | |
| Clean | Ea. | 1.47 | 44.50 | | 45.97 | |
| Minimum Charge | Job | | 270 | | 270 | |
| **4 Ton** | | | | | | |
| Demolish | Ea. | | 450 | | 450 | Includes material and labor to install a |
| Install | Ea. | 515 | 245 | | 760 | refrigeration compressor. |
| Demolish and Install | Ea. | 515 | 695 | | 1210 | |
| Reinstall | Ea. | | 196.07 | | 196.07 | |
| Clean | Ea. | 1.47 | 44.50 | | 45.97 | |
| Minimum Charge | Job | | 270 | | 270 | |
| **5 Ton** | | | | | | |
| Demolish | Ea. | | 450 | | 450 | Includes material and labor to install a |
| Install | Ea. | 565 | 335 | | 900 | refrigeration compressor. |
| Demolish and Install | Ea. | 565 | 785 | | 1350 | |
| Reinstall | Ea. | | 267.10 | | 267.10 | |
| Clean | Ea. | 1.47 | 44.50 | | 45.97 | |
| Minimum Charge | Job | | 270 | | 270 | |
| **Condensing Unit** | | | | | | |
| **2 Ton Air-cooled** | | | | | | |
| Demolish | Ea. | | 128 | | 128 | Includes material and labor to install a |
| Install | Ea. | 1150 | 460 | | 1610 | 2 ton condensing unit. |
| Demolish and Install | Ea. | 1150 | 588 | | 1738 | |
| Reinstall | Ea. | | 369.68 | | 369.68 | |
| Clean | Ea. | 1.47 | 44.50 | | 45.97 | |
| Minimum Charge | Job | | 270 | | 270 | |
| **2.5 Ton Air-cooled** | | | | | | |
| Demolish | Ea. | | 150 | | 150 | Includes material and labor to install a |
| Install | Ea. | 1475 | 570 | | 2045 | 2.5 ton condensing unit. |
| Demolish and Install | Ea. | 1475 | 720 | | 2195 | |
| Reinstall | Ea. | | 456.66 | | 456.66 | |
| Clean | Ea. | 1.47 | 44.50 | | 45.97 | |
| Minimum Charge | Job | | 270 | | 270 | |
| **3 Ton Air-cooled** | | | | | | |
| Demolish | Ea. | | 180 | | 180 | Includes material and labor to install a |
| Install | Ea. | 1475 | 745 | | 2220 | 3 ton condensing unit. |
| Demolish and Install | Ea. | 1475 | 925 | | 2400 | |
| Reinstall | Ea. | | 597.17 | | 597.17 | |
| Clean | Ea. | 1.47 | 44.50 | | 45.97 | |
| Minimum Charge | Job | | 270 | | 270 | |

## Air Conditioning

| | Unit | Material | Labor | Equip. | Total | Specification |
|---|---|---|---|---|---|---|
| **4 Ton Air-cooled** | | | | | | |
| Demolish | Ea. | | 270 | | 270 | Includes material and labor to install a |
| Install | Ea. | 1950 | 1075 | | 3025 | 4 ton condensing unit. |
| Demolish and Install | Ea. | 1950 | 1345 | | 3295 | |
| Reinstall | Ea. | | 862.58 | | 862.58 | |
| Clean | Ea. | 1.47 | 44.50 | | 45.97 | |
| Minimum Charge | Job | | 270 | | 270 | |
| **5 Ton Air-cooled** | | | | | | |
| Demolish | Ea. | | 405 | | 405 | Includes material and labor to install a |
| Install | Ea. | 2375 | 1625 | | 4000 | 5 ton condensing unit. |
| Demolish and Install | Ea. | 2375 | 2030 | | 4405 | |
| Reinstall | Ea. | | 1293.87 | | 1293.87 | |
| Clean | Ea. | 1.47 | 44.50 | | 45.97 | |
| Minimum Charge | Job | | 270 | | 270 | |
| **Air Handler** | | | | | | |
| **1,200 CFM** | | | | | | |
| Demolish | Ea. | | 325 | | 325 | Includes material and labor to install a |
| Install | Ea. | 4825 | 745 | | 5570 | 1200 CFM capacity air handling unit. |
| Demolish and Install | Ea. | 4825 | 1070 | | 5895 | |
| Reinstall | Ea. | | 597.17 | | 597.17 | |
| Clean | Ea. | 1.47 | 44.50 | | 45.97 | |
| Minimum Charge | Job | | 270 | | 270 | |
| **2,000 CFM** | | | | | | |
| Demolish | Ea. | | 325 | | 325 | Includes material and labor to install a |
| Install | Ea. | 7200 | 880 | | 8080 | 2000 CFM capacity air handling unit. |
| Demolish and Install | Ea. | 7200 | 1205 | | 8405 | |
| Reinstall | Ea. | | 705.75 | | 705.75 | |
| Clean | Ea. | 1.47 | 44.50 | | 45.97 | |
| Minimum Charge | Job | | 270 | | 270 | |
| **3,000 CFM** | | | | | | |
| Demolish | Ea. | | 360 | | 360 | Includes material and labor to install a |
| Install | Ea. | 10600 | 970 | | 11570 | 3000 CFM capacity air handling unit. |
| Demolish and Install | Ea. | 10600 | 1330 | | 11930 | |
| Reinstall | Ea. | | 776.32 | | 776.32 | |
| Clean | Ea. | 1.47 | 44.50 | | 45.97 | |
| Minimum Charge | Job | | 270 | | 270 | |
| **Evaporative Cooler** | | | | | | |
| **4,300 CFM, 1/3 HP** | | | | | | |
| Demolish | Ea. | | 129 | | 129 | Includes material and labor to install |
| Install | Ea. | 800 | 287 | | 1087 | an evaporative cooler. |
| Demolish and Install | Ea. | 800 | 416 | | 1216 | |
| Reinstall | Ea. | | 229.40 | | 229.40 | |
| Clean | Ea. | .59 | 44.50 | | 45.09 | |
| Minimum Charge | Job | | 270 | | 270 | |
| **4,700 CFM, 1/2 HP** | | | | | | |
| Demolish | Ea. | | 129 | | 129 | Includes material and labor to install |
| Install | Ea. | 800 | 296 | | 1096 | an evaporative cooler. |
| Demolish and Install | Ea. | 800 | 425 | | 1225 | |
| Reinstall | Ea. | | 236.80 | | 236.80 | |
| Clean | Ea. | .59 | 44.50 | | 45.09 | |
| Minimum Charge | Job | | 270 | | 270 | |
| **5,600 CFM, 3/4 HP** | | | | | | |
| Demolish | Ea. | | 129 | | 129 | Includes material and labor to install |
| Install | Ea. | 1400 | 305 | | 1705 | an evaporative cooler. |
| Demolish and Install | Ea. | 1400 | 434 | | 1834 | |
| Reinstall | Ea. | | 244.69 | | 244.69 | |
| Clean | Ea. | .59 | 44.50 | | 45.09 | |
| Minimum Charge | Job | | 270 | | 270 | |

## Air Conditioning

| | Unit | Material | Labor | Equip. | Total | Specification |
|---|---|---|---|---|---|---|
| **Master Cool Brand** | | | | | | |
| Demolish | Ea. | | 129 | | 129 | Includes material and labor to install |
| Install | Ea. | 800 | 315 | | 1115 | an evaporative cooler. |
| Demolish and Install | Ea. | 800 | 444 | | 1244 | |
| Reinstall | Ea. | | 253.13 | | 253.13 | |
| Clean | Ea. | .59 | 44.50 | | 45.09 | |
| Minimum Charge | Job | | 270 | | 270 | |
| **Motor** | | | | | | |
| Demolish | Ea. | | 73 | | 73 | Includes material and labor to install |
| Install | Ea. | 220 | 223 | | 443 | an evaporative cooler motor. |
| Demolish and Install | Ea. | 220 | 296 | | 516 | |
| Reinstall | Ea. | | 178.34 | | 178.34 | |
| Clean | Ea. | .59 | 44.50 | | 45.09 | |
| Minimum Charge | Job | | 270 | | 270 | |
| **Water Pump** | | | | | | |
| Demolish | Ea. | | 73 | | 73 | Includes material and labor to install |
| Install | Ea. | 61.50 | 127 | | 188.50 | an evaporative cooler water pump. |
| Demolish and Install | Ea. | 61.50 | 200 | | 261.50 | |
| Reinstall | Ea. | | 101.92 | | 101.92 | |
| Clean | Ea. | .59 | 44.50 | | 45.09 | |
| Minimum Charge | Job | | 270 | | 270 | |
| **Fan Belt** | | | | | | |
| Demolish | Ea. | | 14.55 | | 14.55 | Includes material and labor to install |
| Install | Ea. | 44.50 | 25.50 | | 70 | an evaporative cooler fan belt. |
| Demolish and Install | Ea. | 44.50 | 40.05 | | 84.55 | |
| Reinstall | Ea. | | 20.38 | | 20.38 | |
| Minimum Charge | Job | | 270 | | 270 | |
| **Pads (4 ea.)** | | | | | | |
| Demolish | Ea. | | 14.55 | | 14.55 | Includes material and labor to install |
| Install | Ea. | 22.50 | 63.50 | | 86 | pre-fabricated roof pads for an |
| Demolish and Install | Ea. | 22.50 | 78.05 | | 100.55 | evaporative cooler. |
| Reinstall | Ea. | | 50.96 | | 50.96 | |
| Clean | Ea. | .59 | 44.50 | | 45.09 | |
| Minimum Charge | Job | | 270 | | 270 | |
| **Service call** | | | | | | |
| Install | Job | | 255 | | 255 | Includes 4 hour minimum labor charge for a sheetmetal worker. |
| **Window Mounted** | | | | | | |
| **Low Output BTU** | | | | | | |
| Install | Ea. | 274 | 58 | | 332 | Includes material and labor to install a |
| Reinstall | Ea. | | 46.48 | | 46.48 | window air conditioning unit. |
| Clean | Ea. | .22 | 11.10 | | 11.32 | |
| Minimum Charge | Job | | 270 | | 270 | |
| **Mid-range Output BTU** | | | | | | |
| Demolish | Ea. | | 41 | | 41 | Includes material and labor to install a |
| Install | Ea. | 440 | 77.50 | | 517.50 | window air conditioning unit. |
| Demolish and Install | Ea. | 440 | 118.50 | | 558.50 | |
| Reinstall | Ea. | | 61.97 | | 61.97 | |
| Clean | Ea. | .22 | 11.10 | | 11.32 | |
| Minimum Charge | Job | | 270 | | 270 | |
| **High Output BTU** | | | | | | |
| Demolish | Ea. | | 51.50 | | 51.50 | Includes material and labor to install a |
| Install | Ea. | 750 | 77.50 | | 827.50 | window air conditioning unit. |
| Demolish and Install | Ea. | 750 | 129 | | 879 | |
| Reinstall | Ea. | | 61.97 | | 61.97 | |
| Clean | Ea. | .22 | 11.10 | | 11.32 | |
| Minimum Charge | Job | | 270 | | 270 | |

# Rough Mechanical

| Air Conditioning | | Unit | Material | Labor | Equip. | Total | Specification |
|---|---|---|---|---|---|---|---|
| **Hotel-type Hot / Cold** | | | | | | | Includes material and labor to install |
| | Demolish | Ea. | | 86 | | 86 | packaged air conditioning unit, |
| | Install | Ea. | 1550 | 325 | | 1875 | 15000 BTU cooling, 13900 BTU |
| | Demolish and Install | Ea. | 1550 | 411 | | 1961 | heating. |
| | Reinstall | Ea. | | 258.77 | | 258.77 | |
| | Clean | Ea. | .30 | 22 | | 22.30 | |
| | Minimum Charge | Job | | 270 | | 270 | |
| **Thermostat** | | | | | | | Includes material and labor to install |
| | Demolish | Ea. | | 15 | | 15 | an electric thermostat, 2 set back. |
| | Install | Ea. | 251 | 67.50 | | 318.50 | |
| | Demolish and Install | Ea. | 251 | 82.50 | | 333.50 | |
| | Reinstall | Ea. | | 53.92 | | 53.92 | |
| | Clean | Ea. | .03 | 11.10 | | 11.13 | |
| | Minimum Charge | Job | | 270 | | 270 | |
| **Concrete Pad** | | | | | | | Includes material and labor to install |
| | Demolish | S.F. | | 2.47 | | 2.47 | air conditioner pad. |
| | Install | S.F. | 2.94 | 5.65 | | 8.59 | |
| | Demolish and Install | S.F. | 2.94 | 8.12 | | 11.06 | |
| | Reinstall | S.F. | | 4.50 | | 4.50 | |
| | Clean | S.F. | .03 | .71 | | .74 | |
| | Paint | S.F. | .18 | .40 | | .58 | |
| | Minimum Charge | Job | | 270 | | 270 | |
| **A/C Junction Box** | | | | | | | Includes labor and materials to install |
| | Install | Ea. | 147 | 150 | | 297 | an air conditioning circuit including up |
| | Minimum Charge | Job | | 263 | | 263 | to 40 feet of non-metallic cable. |

| Sprinkler System | | Unit | Material | Labor | Equip. | Total | Specification |
|---|---|---|---|---|---|---|---|
| **Exposed** | | | | | | | |
| **Wet Type** | | | | | | | Includes material and labor to install |
| | Demolish | SF Flr. | | .23 | | .23 | an exposed wet pipe fire sprinkler |
| | Install | SF Flr. | 2.84 | 2.25 | | 5.09 | system. |
| | Demolish and Install | SF Flr. | 2.84 | 2.48 | | 5.32 | |
| | Reinstall | SF Flr. | | 2.25 | | 2.25 | |
| | Minimum Charge | Job | | 262 | | 262 | |
| **Dry Type** | | | | | | | Includes material and labor to install |
| | Demolish | SF Flr. | | .23 | | .23 | an exposed dry pipe fire sprinkler |
| | Install | SF Flr. | 3.27 | 2.25 | | 5.52 | system. |
| | Demolish and Install | SF Flr. | 3.27 | 2.48 | | 5.75 | |
| | Reinstall | SF Flr. | | 2.25 | | 2.25 | |
| | Minimum Charge | Job | | 262 | | 262 | |
| **Concealed** | | | | | | | |
| **Wet Type** | | | | | | | Includes material and labor to install a |
| | Demolish | SF Flr. | | .47 | | .47 | concealed wet pipe fire sprinkler |
| | Install | SF Flr. | 2.84 | 3.08 | | 5.92 | system. |
| | Demolish and Install | SF Flr. | 2.84 | 3.55 | | 6.39 | |
| | Reinstall | SF Flr. | | 3.08 | | 3.08 | |
| | Minimum Charge | Job | | 262 | | 262 | |
| **Dry Type** | | | | | | | Includes material and labor to install a |
| | Demolish | SF Flr. | | .47 | | .47 | concealed dry pipe fire sprinkler |
| | Install | SF Flr. | 3.27 | 3.08 | | 6.35 | system. |
| | Demolish and Install | SF Flr. | 3.27 | 3.55 | | 6.82 | |
| | Reinstall | SF Flr. | | 3.08 | | 3.08 | |
| | Minimum Charge | Job | | 262 | | 262 | |

| Junction Boxes | | | Cable | | | Switch Box | |

| Electrical Per S.F. | | Unit | Material | Labor | Equip. | Total | Specification |
|---|---|---|---|---|---|---|---|
| **Residential** | | | | | | | |
| **Wiring, Lighting and Fixtures** | | | | | | | |
| | Install | S.F. | 1.49 | 2.19 | | 3.68 | Includes labor and material to replace |
| | Minimum Charge | Job | | 263 | | 263 | house wiring, lighting fixtures, and |
| | | | | | | | built-in appliances. Apply S.F. cost to |
| | | | | | | | floor area including garages but not |
| | | | | | | | basements. |
| **Wiring Only** | | | | | | | |
| | Install | S.F. | .15 | .86 | | 1.01 | Includes labor and material to replace |
| | | | | | | | house wiring only. Apply S.F. cost to |
| | | | | | | | floor area including garages but not |
| | | | | | | | basements. |

| Main Service | | Unit | Material | Labor | Equip. | Total | Specification |
|---|---|---|---|---|---|---|---|
| **Complete Panel** | | | | | | | |
| **100 Amp.** | | | | | | | |
| | Demolish | Ea. | | 216 | | 216 | Cost includes material and labor to |
| | Install | Ea. | 360 | 440 | | 800 | install 100 Amp. residential service |
| | Demolish and Install | Ea. | 360 | 656 | | 1016 | panel including 12 breaker interior |
| | Reinstall | Ea. | | 353.08 | | 353.08 | panel, 10 single pole breakers and |
| | Clean | Ea. | .14 | 89 | | 89.14 | hook-up. Wiring is not included. |
| | Minimum Charge | Job | | 263 | | 263 | |
| **150 Amp.** | | | | | | | |
| | Demolish | Ea. | | 216 | | 216 | Cost includes material and labor to |
| | Install | Ea. | 935 | 510 | | 1445 | install 150 Amp. residential service, |
| | Demolish and Install | Ea. | 935 | 726 | | 1661 | incl. 20 breaker exterior panel, main |
| | Reinstall | Ea. | | 407.92 | | 407.92 | switch, 1 GFI and 14 single pole |
| | Clean | Ea. | .14 | 89 | | 89.14 | breakers, meter socket including |
| | Minimum Charge | Job | | 263 | | 263 | hookup. Wiring is not included. |
| **200 Amp.** | | | | | | | |
| | Demolish | Ea. | | 216 | | 216 | Cost includes material and labor to |
| | Install | Ea. | 1100 | 585 | | 1685 | install 200 Amp. residential service, |
| | Demolish and Install | Ea. | 1100 | 801 | | 1901 | incl. 20 breaker exterior panel, main |
| | Reinstall | Ea. | | 466.84 | | 466.84 | switch, 1 GFI and 18 single pole |
| | Clean | Ea. | .14 | 89 | | 89.14 | breakers, meter socket including |
| | Minimum Charge | Job | | 263 | | 263 | hookup. Wiring is not included. |

| Main Service | Unit | Material | Labor | Equip. | Total | Specification |
|---|---|---|---|---|---|---|
| **Masthead** | | | | | | |
| Demolish | Ea. | | 36.50 | | 36.50 | Includes material and labor to install |
| Install | Ea. | 51 | 219 | | 270 | galvanized steel conduit including |
| Demolish and Install | Ea. | 51 | 255.50 | | 306.50 | masthead. |
| Reinstall | Ea. | | 175.07 | | 175.07 | |
| Minimum Charge | Job | | 263 | | 263 | |
| **Allowance to Repair** | | | | | | |
| Install | Job | | 263 | | 263 | Minimum labor charge for residential |
| Minimum Charge | Job | | 263 | | 263 | wiring. |

**Wall Insulation**

| Floor Insulation | | Unit | Material | Labor | Equip. | Total | Specification |
|---|---|---|---|---|---|---|---|
| **Kraft Faced Batt** | | | | | | | |
| **3-1/2″ (R-13)** | | | | | | | |
| | Demolish | S.F. | | .16 | | .16 | Cost includes material and labor to |
| | Install | S.F. | .45 | .66 | | 1.11 | install 3-1/2″ (R-13) kraft paper or foil |
| | Demolish and Install | S.F. | .45 | .82 | | 1.27 | faced roll and batt fiberglass |
| | Minimum Charge | Job | | 232 | | 232 | insulation, joists 16″ O.C. |
| **6″ (R-19)** | | | | | | | |
| | Demolish | S.F. | | .16 | | .16 | Cost includes material and labor to |
| | Install | S.F. | .56 | .77 | | 1.33 | install 6″ (R-19) kraft paper or foil |
| | Demolish and Install | S.F. | .56 | .93 | | 1.49 | faced roll and batt fiberglass |
| | Minimum Charge | Job | | 232 | | 232 | insulation, joists 16″ O.C. |
| **10″ (R-30)** | | | | | | | |
| | Demolish | S.F. | | .16 | | .16 | Cost includes material and labor to |
| | Install | S.F. | .80 | .93 | | 1.73 | install 10″ (R-30) kraft paper-faced roll |
| | Demolish and Install | S.F. | .80 | 1.09 | | 1.89 | and batt fiberglass insulation, joists |
| | Minimum Charge | Job | | 232 | | 232 | 16″ O.C. |
| **12″ (R-38)** | | | | | | | |
| | Demolish | S.F. | | .16 | | .16 | Cost includes material and labor to |
| | Install | S.F. | 1.16 | .98 | | 2.14 | install 12″ (R-38) kraft paper-faced roll |
| | Demolish and Install | S.F. | 1.16 | 1.14 | | 2.30 | and batt fiberglass insulation, joists |
| | Minimum Charge | Job | | 232 | | 232 | 16″ O.C. |
| **Foil Faced Batt** | | | | | | | |
| **3-1/2″ (R-13)** | | | | | | | |
| | Demolish | S.F. | | .16 | | .16 | Cost includes material and labor to |
| | Install | S.F. | .45 | .66 | | 1.11 | install 3-1/2″ (R-13) kraft paper or foil |
| | Demolish and Install | S.F. | .45 | .82 | | 1.27 | faced roll and batt fiberglass |
| | Minimum Charge | Job | | 232 | | 232 | insulation, joists 16″ O.C. |
| **6″ (R-19)** | | | | | | | |
| | Demolish | S.F. | | .16 | | .16 | Cost includes material and labor to |
| | Install | S.F. | .56 | .77 | | 1.33 | install 6″ (R-19) kraft paper or foil |
| | Demolish and Install | S.F. | .56 | .93 | | 1.49 | faced roll and batt fiberglass |
| | Minimum Charge | Job | | 232 | | 232 | insulation, joists 16″ O.C. |
| **Unfaced Batt** | | | | | | | |
| **3-1/2″ (R-13)** | | | | | | | |
| | Demolish | S.F. | | .16 | | .16 | Cost includes material and labor to |
| | Install | S.F. | .36 | .77 | | 1.13 | install 3-1/2″ (R-13) unfaced roll and |
| | Demolish and Install | S.F. | .36 | .93 | | 1.29 | batt fiberglass insulation, joists 16″ |
| | Minimum Charge | Job | | 232 | | 232 | O.C. |

# Insulation

## Floor Insulation

| Floor Insulation | Unit | Material | Labor | Equip. | Total | Specification |
|---|---|---|---|---|---|---|
| **6" (R-19)** | | | | | | |
| Demolish | S.F. | | .16 | | .16 | Cost includes material and labor to |
| Install | S.F. | .43 | .93 | | 1.36 | install 6" (R-19) unfaced roll and batt |
| Demolish and Install | S.F. | .43 | 1.09 | | 1.52 | fiberglass insulation, joists 16" O.C. |
| Minimum Charge | Job | | 232 | | 232 | |
| **10" (R-30)** | | | | | | |
| Demolish | S.F. | | .16 | | .16 | Cost includes material and labor to |
| Install | S.F. | .66 | 1.03 | | 1.69 | install 10" (R-30) unfaced roll and batt |
| Demolish and Install | S.F. | .66 | 1.19 | | 1.85 | fiberglass insulation, joists 16" O.C. |
| Minimum Charge | Job | | 232 | | 232 | |
| **12" (R-38)** | | | | | | |
| Demolish | S.F. | | .16 | | .16 | Cost includes material and labor to |
| Install | S.F. | .83 | 1.09 | | 1.92 | install 12" (R-38) unfaced roll and batt |
| Demolish and Install | S.F. | .83 | 1.25 | | 2.08 | fiberglass insulation, joists 16" O.C. |
| Minimum Charge | Job | | 232 | | 232 | |

### Rigid Board

| | Unit | Material | Labor | Equip. | Total | Specification |
|---|---|---|---|---|---|---|
| **1/2"** | | | | | | |
| Demolish | S.F. | | .16 | | .16 | Includes material and labor to install |
| Install | S.F. | .34 | .58 | | .92 | rigid foam insulation board foil faced |
| Demolish and Install | S.F. | .34 | .74 | | 1.08 | two sides. |
| Minimum Charge | Job | | 232 | | 232 | |
| **3/4"** | | | | | | |
| Demolish | S.F. | | .16 | | .16 | Includes material and labor to install |
| Install | S.F. | .50 | .58 | | 1.08 | rigid foam insulation board foil faced |
| Demolish and Install | S.F. | .50 | .74 | | 1.24 | two sides. |
| Minimum Charge | Job | | 232 | | 232 | |
| **1-1/2"** | | | | | | |
| Demolish | S.F. | | .16 | | .16 | Includes material and labor to install |
| Install | S.F. | .74 | .62 | | 1.36 | rigid foam insulation board foil faced |
| Demolish and Install | S.F. | .74 | .78 | | 1.52 | two sides. |
| Minimum Charge | Job | | 232 | | 232 | |

### Support Wire

| | Unit | Material | Labor | Equip. | Total | Specification |
|---|---|---|---|---|---|---|
| Demolish | S.F. | | .06 | | .06 | Includes material and labor to install |
| Install | S.F. | .02 | .13 | | .15 | friction fit support wire. |
| Demolish and Install | S.F. | .02 | .19 | | .21 | |
| Minimum Charge | Job | | 232 | | 232 | |

### Vapor Barrier

| | Unit | Material | Labor | Equip. | Total | Specification |
|---|---|---|---|---|---|---|
| Demolish | S.F. | | .09 | | .09 | Cost includes material and labor to |
| Install | S.F. | .03 | .13 | | .16 | install 4 mil polyethylene vapor barrier |
| Demolish and Install | S.F. | .03 | .22 | | .25 | 10' wide sheets with 6" overlaps. |
| Minimum Charge | Job | | 232 | | 232 | |

## Wall Insulation

| Wall Insulation | Unit | Material | Labor | Equip. | Total | Specification |
|---|---|---|---|---|---|---|
| **Kraft Faced Batt** | | | | | | |
| **3-1/2" (R-13)** | | | | | | |
| Demolish | S.F. | | .16 | | .16 | Cost includes material and labor to |
| Install | S.F. | .37 | .34 | | .71 | install 3-1/2" (R-13) kraft paper-faced |
| Demolish and Install | S.F. | .37 | .50 | | .87 | roll and batt fiberglass insulation, joist |
| Minimum Charge | Job | | 232 | | 232 | or studs 16" O.C. |
| **6" (R-19)** | | | | | | |
| Demolish | S.F. | | .16 | | .16 | Cost includes material and labor to |
| Install | S.F. | .48 | .34 | | .82 | install 6" (R-19) kraft paper-faced roll |
| Demolish and Install | S.F. | .48 | .50 | | .98 | and batt fiberglass insulation, joist or |
| Minimum Charge | Job | | 232 | | 232 | studs 16" O.C. |

# Insulation

| Wall Insulation | | Unit | Material | Labor | Equip. | Total | Specification |
|---|---|---|---|---|---|---|---|
| **10" (R-30)** | | | | | | | |
| | Demolish | S.F. | | .16 | | .16 | Cost includes material and labor to |
| | Install | S.F. | .80 | .40 | | 1.20 | install 10" (R-30) kraft paper-faced roll |
| | Demolish and Install | S.F. | .80 | .56 | | 1.36 | and batt fiberglass insulation, joist or |
| | Minimum Charge | Job | | 232 | | 232 | studs 16" O.C. |
| **12" (R-38)** | | | | | | | |
| | Demolish | S.F. | | .16 | | .16 | Cost includes material and labor to |
| | Install | S.F. | 1.16 | .40 | | 1.56 | install 12" (R-38) kraft paper-faced roll |
| | Demolish and Install | S.F. | 1.16 | .56 | | 1.72 | and batt fiberglass insulation, joist or |
| | Minimum Charge | Job | | 232 | | 232 | studs 16" O.C. |
| **Foil Faced Batt** | | | | | | | |
| **3-1/2" (R-13)** | | | | | | | |
| | Demolish | S.F. | | .16 | | .16 | Cost includes material and labor to |
| | Install | S.F. | .52 | .34 | | .86 | install 3-1/2" (R-13) foil faced roll and |
| | Demolish and Install | S.F. | .52 | .50 | | 1.02 | batt fiberglass insulation, joist or studs |
| | Minimum Charge | Job | | 232 | | 232 | 16" O.C. |
| **6" (R-19)** | | | | | | | |
| | Demolish | S.F. | | .16 | | .16 | Cost includes material and labor to |
| | Install | S.F. | .68 | .34 | | 1.02 | install 6" (R-19) foil faced roll and batt |
| | Demolish and Install | S.F. | .68 | .50 | | 1.18 | fiberglass insulation, joist or studs 16" |
| | Minimum Charge | Job | | 232 | | 232 | O.C. |
| **Unfaced Batt** | | | | | | | |
| **3-1/2" (R-13)** | | | | | | | |
| | Demolish | S.F. | | .16 | | .16 | Cost includes material and labor to |
| | Install | S.F. | .36 | .34 | | .70 | install 3-1/2" (R-13) unfaced roll and |
| | Demolish and Install | S.F. | .36 | .50 | | .86 | batt fiberglass insulation, joist or studs |
| | Minimum Charge | Job | | 232 | | 232 | 16" O.C. |
| **6" (R-19)** | | | | | | | |
| | Demolish | S.F. | | .16 | | .16 | Cost includes material and labor to |
| | Install | S.F. | .43 | .40 | | .83 | install 6" (R-19) unfaced roll and batt |
| | Demolish and Install | S.F. | .43 | .56 | | .99 | fiberglass insulation, joist or studs 16" |
| | Minimum Charge | Job | | 232 | | 232 | O.C. |
| **10" (R-30)** | | | | | | | |
| | Demolish | S.F. | | .16 | | .16 | Cost includes material and labor to |
| | Install | S.F. | .66 | .40 | | 1.06 | install 10" (R-30) unfaced roll and batt |
| | Demolish and Install | S.F. | .66 | .56 | | 1.22 | fiberglass insulation, joist or studs 16" |
| | Minimum Charge | Job | | 232 | | 232 | O.C. |
| **12" (R-38)** | | | | | | | |
| | Demolish | S.F. | | .16 | | .16 | Cost includes material and labor to |
| | Install | S.F. | .83 | .46 | | 1.29 | install 12" (R-38) unfaced roll and batt |
| | Demolish and Install | S.F. | .83 | .62 | | 1.45 | fiberglass insulation, joist or studs 16" |
| | Minimum Charge | Job | | 232 | | 232 | O.C. |
| **Blown** | | | | | | | |
| **R-13** | | | | | | | |
| | Demolish | S.F. | | .16 | | .16 | Includes material and labor to install |
| | Install | S.F. | .31 | 1.58 | .49 | 2.38 | fiberglass blown-in insulation in wall |
| | Demolish and Install | S.F. | .31 | 1.74 | .49 | 2.54 | with wood siding. |
| | Minimum Charge | Job | | 545 | 168 | 713 | |
| **R-19** | | | | | | | |
| | Demolish | S.F. | | .16 | | .16 | Includes material and labor to install |
| | Install | S.F. | .48 | 1.82 | .56 | 2.86 | fiberglass blown-in insulation in wall |
| | Demolish and Install | S.F. | .48 | 1.98 | .56 | 3.02 | with wood siding. |
| | Minimum Charge | Job | | 545 | 168 | 713 | |

**For customer support on your Contractor's Pricing Guide: Residential Repair & Remodeling, call 800.448.8182.**

# Insulation

## Wall Insulation

| Wall Insulation | | Unit | Material | Labor | Equip. | Total | Specification |
|---|---|---|---|---|---|---|---|
| **Rigid Board** | | | | | | | |
| 1/2" | Demolish | S.F. | | .20 | | .20 | Includes material and labor to install |
| | Install | S.F. | .34 | .58 | | .92 | rigid foam insulation board, foil faced, |
| | Demolish and Install | S.F. | .34 | .78 | | 1.12 | two sides. |
| | Minimum Charge | Job | | 232 | | 232 | |
| 3/4" | Demolish | S.F. | | .20 | | .20 | Includes material and labor to install |
| | Install | S.F. | .50 | .58 | | 1.08 | rigid foam insulation board, foil faced, |
| | Demolish and Install | S.F. | .50 | .78 | | 1.28 | two sides. |
| | Minimum Charge | Job | | 232 | | 232 | |
| 1-1/2" | Demolish | S.F. | | .20 | | .20 | Includes material and labor to install |
| | Install | S.F. | .74 | .64 | | 1.38 | rigid foam insulation board, foil faced, |
| | Demolish and Install | S.F. | .74 | .84 | | 1.58 | two sides. |
| | Minimum Charge | Job | | 232 | | 232 | |
| **Vapor Barrier** | Demolish | S.F. | | .05 | | .05 | Cost includes material and labor to |
| | Install | S.F. | .03 | .13 | | .16 | install 4 mil polyethylene vapor barrier |
| | Demolish and Install | S.F. | .03 | .18 | | .21 | 10' wide sheets with 6" overlaps. |
| | Minimum Charge | Job | | 232 | | 232 | |

## Ceiling Insulation

| Ceiling Insulation | | Unit | Material | Labor | Equip. | Total | Specification |
|---|---|---|---|---|---|---|---|
| **Kraft Faced Batt** | | | | | | | |
| 3-1/2" (R-13) | Demolish | S.F. | | .16 | | .16 | Cost includes material and labor to |
| | Install | S.F. | .37 | .34 | | .71 | install 3-1/2" (R-13) kraft paper-faced |
| | Demolish and Install | S.F. | .37 | .50 | | .87 | roll and batt fiberglass insulation, joist |
| | Minimum Charge | Job | | 232 | | 232 | or studs 16" O.C. |
| 6" (R-19) | Demolish | S.F. | | .16 | | .16 | Cost includes material and labor to |
| | Install | S.F. | .48 | .34 | | .82 | install 6" (R-19) kraft paper-faced roll |
| | Demolish and Install | S.F. | .48 | .50 | | .98 | and batt fiberglass insulation, joist or |
| | Minimum Charge | Job | | 232 | | 232 | studs 16" O.C. |
| 10" (R-30) | Demolish | S.F. | | .16 | | .16 | Cost includes material and labor to |
| | Install | S.F. | .80 | .40 | | 1.20 | install 10" (R-30) kraft paper-faced roll |
| | Demolish and Install | S.F. | .80 | .56 | | 1.36 | and batt fiberglass insulation, joist or |
| | Minimum Charge | Job | | 232 | | 232 | studs 16" O.C. |
| 12" (R-38) | Demolish | S.F. | | .16 | | .16 | Cost includes material and labor to |
| | Install | S.F. | 1.16 | .40 | | 1.56 | install 12" (R-38) kraft paper-faced roll |
| | Demolish and Install | S.F. | 1.16 | .56 | | 1.72 | and batt fiberglass insulation, joist or |
| | Minimum Charge | Job | | 232 | | 232 | studs 16" O.C. |
| **Foil Faced Batt** | | | | | | | |
| 3-1/2" (R-13) | Demolish | S.F. | | .16 | | .16 | Cost includes material and labor to |
| | Install | S.F. | .52 | .34 | | .86 | install 3-1/2" (R-13) foil faced roll and |
| | Demolish and Install | S.F. | .52 | .50 | | 1.02 | batt fiberglass insulation, joist or studs |
| | Minimum Charge | Job | | 232 | | 232 | 16" O.C. |

| Ceiling Insulation | | Unit | Material | Labor | Equip. | Total | Specification |
|---|---|---|---|---|---|---|---|
| **6" (R-19)** | | | | | | | |
| | Demolish | S.F. | | .16 | | .16 | Cost includes material and labor to |
| | Install | S.F. | .68 | .34 | | 1.02 | install 6" (R-19) foil faced roll and batt |
| | Demolish and Install | S.F. | .68 | .50 | | 1.18 | fiberglass insulation, joist or studs 16" |
| | Minimum Charge | Job | | 232 | | 232 | O.C. |
| **Unfaced Batt** | | | | | | | |
| **3-1/2" (R-13)** | | | | | | | |
| | Demolish | S.F. | | .16 | | .16 | Cost includes material and labor to |
| | Install | S.F. | .36 | .34 | | .70 | install 3-1/2" (R-13) unfaced roll and |
| | Demolish and Install | S.F. | .36 | .50 | | .86 | batt fiberglass insulation, joist or studs |
| | Minimum Charge | Job | | 232 | | 232 | 16" O.C. |
| **6" (R-19)** | | | | | | | |
| | Demolish | S.F. | | .16 | | .16 | Cost includes material and labor to |
| | Install | S.F. | .43 | .40 | | .83 | install 6" (R-19) unfaced roll and batt |
| | Demolish and Install | S.F. | .43 | .56 | | .99 | fiberglass insulation, joist or studs 16" |
| | Minimum Charge | Job | | 232 | | 232 | O.C. |
| **10" (R-30)** | | | | | | | |
| | Demolish | S.F. | | .16 | | .16 | Cost includes material and labor to |
| | Install | S.F. | .66 | .40 | | 1.06 | install 10" (R-30) unfaced roll and batt |
| | Demolish and Install | S.F. | .66 | .56 | | 1.22 | fiberglass insulation, joist or studs 16" |
| | Minimum Charge | Job | | 232 | | 232 | O.C. |
| **12" (R-38)** | | | | | | | |
| | Demolish | S.F. | | .16 | | .16 | Cost includes material and labor to |
| | Install | S.F. | .83 | .46 | | 1.29 | install 12" (R-38) unfaced roll and batt |
| | Demolish and Install | S.F. | .83 | .62 | | 1.45 | fiberglass insulation, joist or studs 16" |
| | Minimum Charge | Job | | 232 | | 232 | O.C. |
| **Blown** | | | | | | | |
| **5"** | | | | | | | |
| | Demolish | S.F. | | .49 | | .49 | Includes material and labor to install |
| | Install | S.F. | .23 | .29 | .09 | .61 | fiberglass blown-in insulation. |
| | Demolish and Install | S.F. | .23 | .78 | .09 | 1.10 | |
| | Minimum Charge | Job | | 545 | 168 | 713 | |
| **6"** | | | | | | | |
| | Demolish | S.F. | | .49 | | .49 | Includes material and labor to install |
| | Install | S.F. | .33 | .36 | .11 | .80 | fiberglass blown-in insulation. |
| | Demolish and Install | S.F. | .33 | .85 | .11 | 1.29 | |
| | Minimum Charge | Job | | 545 | 168 | 713 | |
| **9"** | | | | | | | |
| | Demolish | S.F. | | .49 | | .49 | Includes material and labor to install |
| | Install | S.F. | .41 | .50 | .15 | 1.06 | fiberglass blown-in insulation. |
| | Demolish and Install | S.F. | .41 | .99 | .15 | 1.55 | |
| | Minimum Charge | Job | | 545 | 168 | 713 | |
| **12"** | | | | | | | |
| | Demolish | S.F. | | .49 | | .49 | Includes material and labor to install |
| | Install | S.F. | .57 | .73 | .22 | 1.52 | fiberglass blown-in insulation. |
| | Demolish and Install | S.F. | .57 | 1.22 | .22 | 2.01 | |
| | Minimum Charge | Job | | 545 | 168 | 713 | |
| **Rigid Board** | | | | | | | |
| **1/2"** | | | | | | | |
| | Demolish | S.F. | | .27 | | .27 | Includes material and labor to install |
| | Install | S.F. | .34 | .58 | | .92 | rigid foam insulation board, foil faced, |
| | Demolish and Install | S.F. | .34 | .85 | | 1.19 | two sides. |
| | Minimum Charge | Job | | 232 | | 232 | |

# Insulation

## Ceiling Insulation

| | | Unit | Material | Labor | Equip. | Total | Specification |
|---|---|---|---|---|---|---|---|
| **3/4"** | | | | | | | |
| | Demolish | S.F. | | .27 | | .27 | Includes material and labor to install |
| | Install | S.F. | .50 | .58 | | 1.08 | rigid foam insulation board, foil faced, |
| | Demolish and Install | S.F. | .50 | .85 | | 1.35 | two sides. |
| | Minimum Charge | Job | | 232 | | 232 | |
| **1-1/2"** | | | | | | | |
| | Demolish | S.F. | | .27 | | .27 | Includes material and labor to install |
| | Install | S.F. | .74 | .64 | | 1.38 | rigid foam insulation board, foil faced, |
| | Demolish and Install | S.F. | .74 | .91 | | 1.65 | two sides. |
| | Minimum Charge | Job | | 232 | | 232 | |

## Exterior Insulation

| | | Unit | Material | Labor | Equip. | Total | Specification |
|---|---|---|---|---|---|---|---|
| **Rigid Board** | | | | | | | |
| **1/2"** | | | | | | | |
| | Demolish | S.F. | | .15 | | .15 | Includes material and labor to install |
| | Install | S.F. | .34 | .58 | | .92 | rigid foam insulation board foil faced |
| | Demolish and Install | S.F. | .34 | .73 | | 1.07 | two sides. |
| | Minimum Charge | Job | | 232 | | 232 | |
| **3/4"** | | | | | | | |
| | Demolish | S.F. | | .15 | | .15 | Includes material and labor to install |
| | Install | S.F. | .50 | .58 | | 1.08 | rigid foam insulation board foil faced |
| | Demolish and Install | S.F. | .50 | .73 | | 1.23 | two sides. |
| | Minimum Charge | Job | | 232 | | 232 | |
| **1-1/2"** | | | | | | | |
| | Demolish | S.F. | | .15 | | .15 | Includes material and labor to install |
| | Install | S.F. | .74 | .62 | | 1.36 | rigid foam insulation board foil faced |
| | Demolish and Install | S.F. | .74 | .77 | | 1.51 | two sides. |
| | Minimum Charge | Job | | 232 | | 232 | |
| **Vapor Barrier** | | | | | | | |
| | Demolish | S.F. | | .09 | | .09 | Cost includes material and labor to |
| | Install | S.F. | .03 | .13 | | .16 | install 4 mil polyethylene vapor barrier |
| | Demolish and Install | S.F. | .03 | .22 | | .25 | 10' wide sheets with 6" overlaps. |
| | Minimum Charge | Job | | 232 | | 232 | |

## Roof Insulation Board

| | | Unit | Material | Labor | Equip. | Total | Specification |
|---|---|---|---|---|---|---|---|
| **Fiberglass** | | | | | | | |
| **3/4" Fiberglass** | | | | | | | |
| | Demolish | Sq. | | 51.50 | | 51.50 | Includes material and labor to install |
| | Install | Sq. | 68 | 26.50 | | 94.50 | fiberglass board insulation 3/4" thick, |
| | Demolish and Install | Sq. | 68 | 78 | | 146 | R-2.80 and C-0.36 values. |
| | Minimum Charge | Job | | 232 | | 232 | |
| **1" Fiberglass** | | | | | | | |
| | Demolish | Sq. | | 51.50 | | 51.50 | Includes material and labor to install |
| | Install | Sq. | 96.50 | 30 | | 126.50 | fiberglass board insulation 1" thick, |
| | Demolish and Install | Sq. | 96.50 | 81.50 | | 178 | R-4.20 and C-0.24 values. |
| | Minimum Charge | Job | | 232 | | 232 | |
| **1-3/8" Fiberglass** | | | | | | | |
| | Demolish | Sq. | | 51.50 | | 51.50 | Includes material and labor to install |
| | Install | Sq. | 156 | 32.50 | | 188.50 | fiberglass board insulation 1-3/8" |
| | Demolish and Install | Sq. | 156 | 84 | | 240 | thick, R-5.30 and C-0.19 values. |
| | Minimum Charge | Job | | 232 | | 232 | |

| Roof Insulation Board | | Unit | Material | Labor | Equip. | Total | Specification |
|---|---|---|---|---|---|---|---|
| **1-5/8″ Fiberglass** | | | | | | | |
| | Demolish | Sq. | | 51.50 | | 51.50 | Includes material and labor to install |
| | Install | Sq. | 158 | 34 | | 192 | fiberglass board insulation 1-5/8″ |
| | Demolish and Install | Sq. | 158 | 85.50 | | 243.50 | thick, R-6.70 and C-0.15 values. |
| | Minimum Charge | Job | | 232 | | 232 | |
| **2-1/4″ Fiberglass** | | | | | | | |
| | Demolish | Sq. | | 51.50 | | 51.50 | Includes material and labor to install |
| | Install | Sq. | 189 | 37.50 | | 226.50 | fiberglass board insulation 2-1/4″ |
| | Demolish and Install | Sq. | 189 | 89 | | 278 | thick, R-8.30 and C-0.12 values. |
| | Minimum Charge | Job | | 232 | | 232 | |
| **Perlite** | | | | | | | |
| **1″ Perlite** | | | | | | | |
| | Demolish | Sq. | | 51.50 | | 51.50 | Includes material and labor to install |
| | Install | Sq. | 59.50 | 30 | | 89.50 | perlite board insulation 1″ thick, |
| | Demolish and Install | Sq. | 59.50 | 81.50 | | 141 | R-2.80 and C-0.36 values. |
| | Minimum Charge | Job | | 232 | | 232 | |
| **4″ Perlite** | | | | | | | |
| | Demolish | Sq. | | 51.50 | | 51.50 | Includes material and labor to install |
| | Install | Sq. | 238 | 57 | | 295 | perlite board insulation 4″ thick, |
| | Demolish and Install | Sq. | 238 | 108.50 | | 346.50 | R-10.0 and C-0.10 values. |
| | Minimum Charge | Job | | 232 | | 232 | |
| **1-1/2″ Perlite** | | | | | | | |
| | Demolish | Sq. | | 51.50 | | 51.50 | Includes material and labor to install |
| | Install | Sq. | 88 | 32.50 | | 120.50 | perlite board insulation 1-1/2″ thick, |
| | Demolish and Install | Sq. | 88 | 84 | | 172 | R-4.20 and C-0.24 values. |
| | Minimum Charge | Job | | 232 | | 232 | |
| **2″ Perlite** | | | | | | | |
| | Demolish | Sq. | | 51.50 | | 51.50 | Includes material and labor to install |
| | Install | Sq. | 119 | 39.50 | | 158.50 | perlite board insulation 2″ thick, |
| | Demolish and Install | Sq. | 119 | 91 | | 210 | R-5.30 and C-0.19 values. |
| | Minimum Charge | Job | | 232 | | 232 | |
| **2-1/2″ Perlite** | | | | | | | |
| | Demolish | Sq. | | 51.50 | | 51.50 | Includes material and labor to install |
| | Install | Sq. | 160 | 44.50 | | 204.50 | perlite board insulation 2-1/2″ thick, |
| | Demolish and Install | Sq. | 160 | 96 | | 256 | R-6.70 and C-0.15 values. |
| | Minimum Charge | Job | | 232 | | 232 | |
| **3″ Perlite** | | | | | | | |
| | Demolish | Sq. | | 51.50 | | 51.50 | Includes material and labor to install |
| | Install | Sq. | 178 | 47.50 | | 225.50 | perlite board insulation 3″ thick, |
| | Demolish and Install | Sq. | 178 | 99 | | 277 | R-8.30 and C-0.12 values. |
| | Minimum Charge | Job | | 232 | | 232 | |
| **Polystyrene** | | | | | | | |
| **2″ Polystyrene** | | | | | | | |
| | Demolish | Sq. | | 51.50 | | 51.50 | Includes material and labor to install |
| | Install | Sq. | 59.50 | 37.50 | | 97 | extruded polystyrene board, R5.0/ 2″ |
| | Demolish and Install | Sq. | 59.50 | 89 | | 148.50 | in thickness @25PSI compressive |
| | Minimum Charge | Job | | 232 | | 232 | strength. |
| **1″ Polystyrene** | | | | | | | |
| | Demolish | Sq. | | 51.50 | | 51.50 | Includes material and labor to install |
| | Install | Sq. | 29.50 | 25.50 | | 55 | extruded polystyrene board, R5.0/ 1″ |
| | Demolish and Install | Sq. | 29.50 | 77 | | 106.50 | in thickness @25PSI compressive |
| | Minimum Charge | Job | | 232 | | 232 | strength. |
| **1-1/2″ Polystyrene** | | | | | | | |
| | Demolish | Sq. | | 51.50 | | 51.50 | Includes material and labor to install |
| | Install | Sq. | 44.50 | 30 | | 74.50 | extruded polystyrene board, R5.0/ |
| | Demolish and Install | Sq. | 44.50 | 81.50 | | 126 | 1-1/2″ in thickness @25PSI |
| | Minimum Charge | Job | | 232 | | 232 | compressive strength. |

For customer support on your Contractor's Pricing Guide: Residential Repair & Remodeling, call 800.448.8182.

173

| Roof Insulation Board | Unit | Material | Labor | Equip. | Total | Specification |
|---|---|---|---|---|---|---|
| **Urethane** | | | | | | |
| **3/4″ Urethane** | | | | | | |
| Demolish | Sq. | | 51.50 | | 51.50 | Includes material and labor to install |
| Install | Sq. | 52.50 | 22.50 | | 75 | urethane board insulation 3/4″ thick, |
| Demolish and Install | Sq. | 52.50 | 74 | | 126.50 | R-5.30 and C-0.19 values. |
| Minimum Charge | Job | | 232 | | 232 | |
| **1″ Urethane** | | | | | | |
| Demolish | Sq. | | 51.50 | | 51.50 | Includes material and labor to install |
| Install | Sq. | 94 | 24 | | 118 | urethane board insulation 1″ thick, |
| Demolish and Install | Sq. | 94 | 75.50 | | 169.50 | R-6.70 and C-0.15 values. |
| Minimum Charge | Job | | 232 | | 232 | |
| **1-1/2″ Urethane** | | | | | | |
| Demolish | Sq. | | 51.50 | | 51.50 | Includes material and labor to install |
| Install | Sq. | 77 | 27.50 | | 104.50 | urethane board insulation 1-1/2″ |
| Demolish and Install | Sq. | 77 | 79 | | 156 | thick, R-10.1 and C-0.09 values. |
| Minimum Charge | Job | | 232 | | 232 | |
| **2″ Urethane** | | | | | | |
| Demolish | Sq. | | 51.50 | | 51.50 | Includes material and labor to install |
| Install | Sq. | 93.50 | 32.50 | | 126 | urethane board insulation 2″ thick, |
| Demolish and Install | Sq. | 93.50 | 84 | | 177.50 | R-13.4 and C-0.06 values. |
| Minimum Charge | Job | | 232 | | 232 | |

**Gypsum Wallboard**

**Tile Ceiling**

**Gypsum Wallboard Ceiling**

| Wall Plaster | | Unit | Material | Labor | Equip. | Total | Specification |
|---|---|---|---|---|---|---|---|
| **Interior** | | | | | | | |
| **Cement** | | | | | | | |
| | Demolish | S.Y. | | 6.60 | | 6.60 | Includes material and labor to install |
| | Install | S.Y. | 14.85 | 28.50 | 1.98 | 45.33 | cement plaster on interior wood |
| | Demolish and Install | S.Y. | 14.85 | 35.10 | 1.98 | 51.93 | framing with metal lath. |
| | Clean | S.Y. | | 3.23 | | 3.23 | |
| | Paint | S.Y. | 1.88 | 5.15 | | 7.03 | |
| | Minimum Charge | Job | | 222 | | 222 | |
| **Acoustic** | | | | | | | |
| | Demolish | S.Y. | | 6.60 | | 6.60 | Includes material and labor to install |
| | Install | S.Y. | 11.60 | 34.50 | 1.98 | 48.08 | acoustical cement plaster on interior |
| | Demolish and Install | S.Y. | 11.60 | 41.10 | 1.98 | 54.68 | wood framing with metal lath. |
| | Clean | S.Y. | | 3.23 | | 3.23 | |
| | Paint | S.Y. | 1.88 | 5.15 | | 7.03 | |
| | Minimum Charge | Job | | 222 | | 222 | |
| **Gypsum** | | | | | | | |
| | Demolish | S.Y. | | 6.60 | | 6.60 | Includes material and labor to install |
| | Install | S.Y. | 10.50 | 29 | 1.68 | 41.18 | gypsum wall or ceiling plaster on |
| | Demolish and Install | S.Y. | 10.50 | 35.60 | 1.68 | 47.78 | interior wood framing with metal lath. |
| | Clean | S.Y. | | 3.23 | | 3.23 | |
| | Paint | S.Y. | 1.88 | 5.15 | | 7.03 | |
| | Minimum Charge | Job | | 222 | | 222 | |
| **Stucco** | | | | | | | |
| | Demolish | S.Y. | | 15.80 | | 15.80 | Includes material and labor to install |
| | Install | S.Y. | 9.20 | 40.50 | 2.32 | 52.02 | stucco on wire mesh over wood |
| | Demolish and Install | S.Y. | 9.20 | 56.30 | 2.32 | 67.82 | framing. |
| | Clean | S.Y. | | 3.23 | | 3.23 | |
| | Paint | S.Y. | 1.88 | 5.15 | | 7.03 | |
| | Minimum Charge | Job | | 222 | | 222 | |
| **Thin Coat** | | | | | | | |
| **Veneer** | | | | | | | |
| | Demolish | S.Y. | | 6.60 | | 6.60 | Includes materials and labor to install |
| | Install | S.Y. | 1 | 5.25 | .37 | 6.62 | thin coat veneer plaster. |
| | Demolish and Install | S.Y. | 1 | 11.85 | .37 | 13.22 | |
| | Clean | S.Y. | | 3.23 | | 3.23 | |
| | Paint | S.Y. | 1.88 | 5.15 | | 7.03 | |
| | Minimum Charge | Job | | 222 | | 222 | |

# Walls / Ceilings

## Wall Plaster

| Wall Plaster | | Unit | Material | Labor | Equip. | Total | Specification |
|---|---|---|---|---|---|---|---|
| **Lath** | | | | | | | |
| **Metal** | | | | | | | |
| | Demolish | S.Y. | | 1.55 | | 1.55 | Includes material and labor to install |
| | Install | S.Y. | 4.99 | 5.95 | | 10.94 | wire lath on steel framed walls. |
| | Demolish and Install | S.Y. | 4.99 | 7.50 | | 12.49 | |
| | Clean | S.Y. | | 3.23 | | 3.23 | |
| | Minimum Charge | Job | | 105 | | 105 | |
| **Gypsum** | | | | | | | |
| | Demolish | S.Y. | | 1.55 | | 1.55 | Includes material and labor to install |
| | Install | S.Y. | 2.77 | 5.55 | | 8.32 | gypsum lath on walls. |
| | Demolish and Install | S.Y. | 2.77 | 7.10 | | 9.87 | |
| | Clean | S.Y. | | 3.23 | | 3.23 | |
| | Minimum Charge | Job | | 222 | | 222 | |
| **Repairs** | | | | | | | |
| **Fill in Small Cracks** | | | | | | | |
| | Install | L.F. | .07 | 3.56 | | 3.63 | Includes labor and material to fill in |
| | Minimum Charge | Job | | 222 | | 222 | hairline cracks up to 1/8" with filler, sand and prep for paint. |
| **Repair up to 1" Stock** | | | | | | | |
| | Install | L.F. | .29 | 4.45 | | 4.74 | Includes labor and material to fill in |
| | Minimum Charge | Job | | 222 | | 222 | hairline cracks up to 1" with filler, sand and prep for paint. |
| **Patch Small Hole (2")** | | | | | | | |
| | Install | Ea. | .07 | 4.68 | | 4.75 | Includes labor and material to patch |
| | Minimum Charge | Job | | 222 | | 222 | up to a 2" diameter hole in plaster. |
| **Patch Large Hole** | | | | | | | |
| | Install | Ea. | 8.75 | 148 | | 156.75 | Includes labor and material to patch |
| | Minimum Charge | Job | | 222 | | 222 | up to a 20" diameter hole in plaster. |
| **Patch Plaster Section** | | | | | | | |
| | Install | Ea. | 2.51 | 4.45 | | 6.96 | Includes labor and material to repair |
| | Minimum Charge | Job | | 222 | | 222 | section of plaster wall, area less than 20 S.F. |
| **Repair Inside Corner** | | | | | | | |
| | Install | L.F. | 3.49 | 11.10 | | 14.59 | Includes material and labor to patch |
| | Minimum Charge | Job | | 222 | | 222 | plaster corners. |

## Ceiling Plaster

| Ceiling Plaster | | Unit | Material | Labor | Equip. | Total | Specification |
|---|---|---|---|---|---|---|---|
| **Interior** | | | | | | | |
| **Cement** | | | | | | | |
| | Demolish | S.Y. | | 6.60 | | 6.60 | Includes material and labor to install |
| | Install | S.Y. | 14.85 | 28.50 | 1.98 | 45.33 | cement plaster on interior wood |
| | Demolish and Install | S.Y. | 14.85 | 35.10 | 1.98 | 51.93 | framing with metal lath. |
| | Clean | S.Y. | | 3.38 | | 3.38 | |
| | Paint | S.Y. | 1.88 | 5.15 | | 7.03 | |
| | Minimum Charge | Job | | 222 | | 222 | |
| **Acoustic** | | | | | | | |
| | Demolish | S.Y. | | 6.60 | | 6.60 | Includes material and labor to install |
| | Install | S.Y. | 11.60 | 34.50 | 1.98 | 48.08 | acoustical cement plaster on interior |
| | Demolish and Install | S.Y. | 11.60 | 41.10 | 1.98 | 54.68 | wood framing with metal lath. |
| | Clean | S.Y. | | 3.38 | | 3.38 | |
| | Paint | S.Y. | 1.88 | 5.15 | | 7.03 | |
| | Minimum Charge | Job | | 222 | | 222 | |

# Walls / Ceilings

## Ceiling Plaster

| | Unit | Material | Labor | Equip. | Total | Specification |
|---|---|---|---|---|---|---|
| **Gypsum** | | | | | | |
| Demolish | S.Y. | | 6.60 | | 6.60 | Includes material and labor to install |
| Install | S.Y. | 10.50 | 29 | 1.68 | 41.18 | gypsum wall or ceiling plaster on |
| Demolish and Install | S.Y. | 10.50 | 35.60 | 1.68 | 47.78 | interior wood framing with metal lath. |
| Clean | S.Y. | | 3.38 | | 3.38 | |
| Paint | S.Y. | 1.88 | 5.15 | | 7.03 | |
| Minimum Charge | Job | | 222 | | 222 | |
| **Stucco** | | | | | | |
| Demolish | S.Y. | | 15.80 | | 15.80 | Includes material and labor to install |
| Install | S.Y. | 9.20 | 40.50 | 2.32 | 52.02 | stucco on wire mesh over wood |
| Demolish and Install | S.Y. | 9.20 | 56.30 | 2.32 | 67.82 | framing. |
| Clean | S.Y. | | 3.38 | | 3.38 | |
| Paint | S.Y. | 1.88 | 5.15 | | 7.03 | |
| Minimum Charge | Job | | 222 | | 222 | |

## Thin Coat

| | Unit | Material | Labor | Equip. | Total | Specification |
|---|---|---|---|---|---|---|
| **Veneer** | | | | | | |
| Demolish | S.Y. | | 6.60 | | 6.60 | Includes materials and labor to install |
| Install | S.Y. | 1 | 5.25 | .37 | 6.62 | thin coat veneer plaster. |
| Demolish and Install | S.Y. | 1 | 11.85 | .37 | 13.22 | |
| Clean | S.Y. | | 3.38 | | 3.38 | |
| Paint | S.Y. | 1.88 | 5.15 | | 7.03 | |
| Minimum Charge | Job | | 222 | | 222 | |

## Lath

| | Unit | Material | Labor | Equip. | Total | Specification |
|---|---|---|---|---|---|---|
| **Metal** | | | | | | |
| Demolish | S.Y. | | 1.55 | | 1.55 | Includes material and labor to install |
| Install | S.Y. | 4.99 | 7.40 | | 12.39 | wire lath on steel framed ceilings. |
| Demolish and Install | S.Y. | 4.99 | 8.95 | | 13.94 | |
| Minimum Charge | Job | | 222 | | 222 | |
| **Gypsum** | | | | | | |
| Demolish | S.Y. | | 1.55 | | 1.55 | Includes material and labor to install |
| Install | S.Y. | 2.77 | 4.64 | | 7.41 | gypsum lath on ceilings. |
| Demolish and Install | S.Y. | 2.77 | 6.19 | | 8.96 | |
| Minimum Charge | Job | | 222 | | 222 | |

## Repairs

| | Unit | Material | Labor | Equip. | Total | Specification |
|---|---|---|---|---|---|---|
| **Fill in Small Cracks** | | | | | | |
| Install | S.F. | .07 | .81 | | .88 | Includes labor and material to fill in |
| Minimum Charge | Job | | 445 | | 445 | hairline cracks up to 1/8" with filler, |
| | | | | | | sand and prep for paint. |

## Gypsum Wallboard

| | Unit | Material | Labor | Equip. | Total | Specification |
|---|---|---|---|---|---|---|
| **Finished** | | | | | | |
| **1/4"** | | | | | | |
| Demolish | S.F. | | .39 | | .39 | Cost includes material and labor to |
| Install | S.F. | .35 | 1.43 | | 1.78 | install 1/4" drywall, including joint |
| Demolish and Install | S.F. | .35 | 1.82 | | 2.17 | taping, premixed compound and |
| Clean | S.F. | | .32 | | .32 | screws. |
| Paint | S.F. | .28 | .34 | | .62 | |
| Minimum Charge | Job | | 232 | | 232 | |
| **3/8"** | | | | | | |
| Demolish | S.F. | | .39 | | .39 | Cost includes material and labor to |
| Install | S.F. | .34 | 1.45 | | 1.79 | install 3/8" drywall including joint |
| Demolish and Install | S.F. | .34 | 1.84 | | 2.18 | tape, premixed compound and |
| Clean | S.F. | | .32 | | .32 | screws. |
| Paint | S.F. | .28 | .34 | | .62 | |
| Minimum Charge | Job | | 232 | | 232 | |

For customer support on your Contractor's Pricing Guide: Residential Repair & Remodeling, call 800.448.8182.

| Gypsum Wallboard | | Unit | Material | Labor | Equip. | Total | Specification |
|---|---|---|---|---|---|---|---|
| **1/2″** | | | | | | | |
| | Demolish | S.F. | | .39 | | .39 | Cost includes material and labor to |
| | Install | S.F. | .32 | 1.46 | | 1.78 | install 1/2″ drywall including joint |
| | Demolish and Install | S.F. | .32 | 1.85 | | 2.17 | tape, premixed compound and |
| | Clean | S.F. | | .32 | | .32 | screws. |
| | Paint | S.F. | .28 | .34 | | .62 | |
| | Minimum Charge | Job | | 232 | | 232 | |
| **1/2″ Water Resistant** | | | | | | | |
| | Demolish | S.F. | | .39 | | .39 | Cost includes material and labor to |
| | Install | S.F. | .36 | 1.46 | | 1.82 | install 1/2″ WR drywall including joint |
| | Demolish and Install | S.F. | .36 | 1.85 | | 2.21 | tape, premixed compound and |
| | Clean | S.F. | | .32 | | .32 | screws. |
| | Paint | S.F. | .28 | .34 | | .62 | |
| | Minimum Charge | Job | | 232 | | 232 | |
| **1/2″ Fire-rated** | | | | | | | |
| | Demolish | S.F. | | .39 | | .39 | Cost includes material and labor to |
| | Install | S.F. | .35 | 1.46 | | 1.81 | install 1/2″ FR drywall including |
| | Demolish and Install | S.F. | .35 | 1.85 | | 2.20 | screws. Taping and finishing not |
| | Clean | S.F. | | .32 | | .32 | included. |
| | Paint | S.F. | .28 | .34 | | .62 | |
| | Minimum Charge | Job | | 232 | | 232 | |
| **5/8″** | | | | | | | |
| | Demolish | S.F. | | .39 | | .39 | Cost includes material and labor to |
| | Install | S.F. | .37 | 1.41 | | 1.78 | install 5/8″ drywall including joint |
| | Demolish and Install | S.F. | .37 | 1.80 | | 2.17 | tape, premixed compound and |
| | Clean | S.F. | | .32 | | .32 | screws. |
| | Paint | S.F. | .28 | .34 | | .62 | |
| | Minimum Charge | Job | | 232 | | 232 | |
| **5/8″ Fire-rated** | | | | | | | |
| | Demolish | S.F. | | .39 | | .39 | Cost includes material and labor to |
| | Install | S.F. | .37 | 1.66 | | 2.03 | install 5/8″ FR drywall, including joint |
| | Demolish and Install | S.F. | .37 | 2.05 | | 2.42 | taping, premixed compound and |
| | Clean | S.F. | | .32 | | .32 | screws. |
| | Paint | S.F. | .28 | .34 | | .62 | |
| | Minimum Charge | Job | | 232 | | 232 | |
| **2-hour Fire-rated** | | | | | | | |
| | Demolish | S.F. | | .99 | | .99 | Cost includes material and labor to |
| | Install | S.F. | .83 | 3.29 | | 4.12 | install 2 layers of 5/8″ type X drywall |
| | Demolish and Install | S.F. | .83 | 4.28 | | 5.11 | including joint taping, premixed |
| | Clean | S.F. | | .32 | | .32 | compound and screws. |
| | Paint | S.F. | .28 | .34 | | .62 | |
| | Minimum Charge | Job | | 232 | | 232 | |
| **Unfinished** | | | | | | | |
| **1/4″** | | | | | | | |
| | Demolish | S.F. | | .39 | | .39 | Cost includes material and labor to |
| | Install | S.F. | .31 | .50 | | .81 | install 1/4″ drywall including screws. |
| | Demolish and Install | S.F. | .31 | .89 | | 1.20 | Taping and finishing not included. |
| | Clean | S.F. | | .32 | | .32 | |
| | Minimum Charge | Job | | 232 | | 232 | |
| **3/8″** | | | | | | | |
| | Demolish | S.F. | | .39 | | .39 | Cost includes material and labor to |
| | Install | S.F. | .30 | .50 | | .80 | install 3/8″ drywall including screws. |
| | Demolish and Install | S.F. | .30 | .89 | | 1.19 | Tape, finish or texture is not included. |
| | Clean | S.F. | | .32 | | .32 | |
| | Minimum Charge | Job | | 232 | | 232 | |

## Gypsum Wallboard

| | Unit | Material | Labor | Equip. | Total | Specification |
|---|---|---|---|---|---|---|
| **1/2"** | | | | | | |
| Demolish | S.F. | | .39 | | .39 | Cost includes material and labor to |
| Install | S.F. | .28 | .50 | | .78 | install 1/2" drywall including screws. |
| Demolish and Install | S.F. | .28 | .89 | | 1.17 | Taping and finishing not included. |
| Clean | S.F. | | .32 | | .32 | |
| Minimum Charge | Job | | 232 | | 232 | |
| **1/2" Water Resistant** | | | | | | |
| Demolish | S.F. | | .39 | | .39 | Cost includes material and labor to |
| Install | S.F. | .32 | .50 | | .82 | install 1/2" WR drywall including |
| Demolish and Install | S.F. | .32 | .89 | | 1.21 | screws. taping and finishing not |
| Clean | S.F. | | .32 | | .32 | included. |
| Minimum Charge | Job | | 232 | | 232 | |
| **1/2" Fire-rated** | | | | | | |
| Demolish | S.F. | | .39 | | .39 | Cost includes labor and material to |
| Install | S.F. | .28 | .34 | | .62 | install 1/2" FR drywall including |
| Demolish and Install | S.F. | .28 | .73 | | 1.01 | screws. Tape, finish or texture is not |
| Clean | S.F. | | .32 | | .32 | included. |
| Minimum Charge | Job | | 232 | | 232 | |
| **5/8"** | | | | | | |
| Demolish | S.F. | | .39 | | .39 | Cost includes material and labor to |
| Install | S.F. | .32 | .56 | | .88 | install 5/8" drywall including screws. |
| Demolish and Install | S.F. | .32 | .95 | | 1.27 | Taping and finishing not included. |
| Clean | S.F. | | .32 | | .32 | |
| Minimum Charge | Job | | 232 | | 232 | |
| **5/8" Fire-rated** | | | | | | |
| Demolish | S.F. | | .39 | | .39 | Cost includes material and labor to |
| Install | S.F. | .33 | .56 | | .89 | install 5/8" FR drywall including |
| Demolish and Install | S.F. | .33 | .95 | | 1.28 | screws. Taping and finishing not |
| Clean | S.F. | | .32 | | .32 | included. |
| Minimum Charge | Job | | 232 | | 232 | |
| **2-hour Fire-rated** | | | | | | |
| Demolish | S.F. | | .99 | | .99 | Cost includes material and labor to |
| Install | S.F. | .83 | 3.29 | | 4.12 | install 2 layers of 5/8" type X drywall |
| Demolish and Install | S.F. | .83 | 4.28 | | 5.11 | including joint taping, premixed |
| Clean | S.F. | | .32 | | .32 | compound and screws. |
| Minimum Charge | Job | | 232 | | 232 | |
| **Accessories** | | | | | | |
| **Tape & Finish** | | | | | | |
| Install | S.F. | .06 | .46 | | .52 | Includes labor and material for taping, |
| Minimum Charge | Job | | 232 | | 232 | sanding and finish. |
| **Texture** | | | | | | |
| **Sprayed** | | | | | | |
| Install | S.F. | .04 | .56 | | .60 | Includes labor and material to install |
| Minimum Charge | Job | | 232 | | 232 | by spray, texture finish. |
| **Trowel** | | | | | | |
| Install | S.F. | .43 | .36 | | .79 | Includes labor and material to apply |
| Minimum Charge | Job | | 232 | | 232 | hand troweled texture. |
| **Patch Hole in Gypsum Wallboard** | | | | | | |
| Install | S.F. | .52 | 63.50 | | 64.02 | Includes labor and material to patch a |
| Minimum Charge | Job | | 232 | | 232 | one square foot area of damaged |
| | | | | | | drywall. |
| **Furring Strips** | | | | | | |
| Demolish | S.F. | | .10 | | .10 | Cost includes material and labor to |
| Install | S.F. | .22 | 1.03 | | 1.25 | install 1" x 2" furring strips, 16" O.C. |
| Demolish and Install | S.F. | .22 | 1.13 | | 1.35 | per S.F. of surface area to be covered. |
| Minimum Charge | Job | | 116 | | 116 | |

179

| Ceramic Wall Tile | Unit | Material | Labor | Equip. | Total | Specification |
|---|---|---|---|---|---|---|
| **Thin Set** | | | | | | |
| **2″ x 2″** | | | | | | |
| Demolish | S.F. | | 1.32 | | 1.32 | Cost includes material and labor to |
| Install | S.F. | 7.35 | 7.30 | | 14.65 | install 2″ x 2″ ceramic tile in organic |
| Demolish and Install | S.F. | 7.35 | 8.62 | | 15.97 | adhesive including cutting and grout. |
| Clean | S.F. | | .68 | | .68 | |
| Minimum Charge | Job | | 219 | | 219 | |
| **4-1/4″ x 4-1/4″** | | | | | | |
| Demolish | S.F. | | 1.32 | | 1.32 | Cost includes material and labor to |
| Install | S.F. | 2.87 | 4.08 | | 6.95 | install 4-1/4″ x 4-1/4″ ceramic tile in |
| Demolish and Install | S.F. | 2.87 | 5.40 | | 8.27 | organic adhesive including grout. |
| Clean | S.F. | | .68 | | .68 | |
| Minimum Charge | Job | | 219 | | 219 | |
| **6″ x 6″** | | | | | | |
| Demolish | S.F. | | 1.32 | | 1.32 | Cost includes material and labor to |
| Install | S.F. | 3.93 | 4.43 | | 8.36 | install 6″ x 6″ ceramic tile in organic |
| Demolish and Install | S.F. | 3.93 | 5.75 | | 9.68 | adhesive including cutting and grout. |
| Clean | S.F. | | .68 | | .68 | |
| Minimum Charge | Job | | 219 | | 219 | |
| **8″ x 8″** | | | | | | |
| Demolish | S.F. | | 1.32 | | 1.32 | Cost includes material and labor to |
| Install | S.F. | 5.85 | 4.56 | | 10.41 | install 8″ x 8″ ceramic tile in organic |
| Demolish and Install | S.F. | 5.85 | 5.88 | | 11.73 | adhesive including cutting and grout. |
| Clean | S.F. | | .68 | | .68 | |
| Minimum Charge | Job | | 219 | | 219 | |
| **Thick Set** | | | | | | |
| **2″ x 2″** | | | | | | |
| Demolish | S.F. | | 1.58 | | 1.58 | Cost includes material and labor to |
| Install | S.F. | 7.15 | 8.75 | | 15.90 | install 2″ x 2″ ceramic tile in mortar |
| Demolish and Install | S.F. | 7.15 | 10.33 | | 17.48 | bed including cutting and grout. |
| Clean | S.F. | | .68 | | .68 | |
| Minimum Charge | Job | | 219 | | 219 | |
| **4-1/4″ x 4-1/4″** | | | | | | |
| Demolish | S.F. | | 1.58 | | 1.58 | Cost includes material and labor to |
| Install | S.F. | 10.45 | 8.75 | | 19.20 | install 4-1/4″ x 4-1/4″ ceramic tile in |
| Demolish and Install | S.F. | 10.45 | 10.33 | | 20.78 | mortar bed including cutting and |
| Clean | S.F. | | .68 | | .68 | grout. |
| Minimum Charge | Job | | 219 | | 219 | |
| **6″ x 6″** | | | | | | |
| Demolish | S.F. | | 1.58 | | 1.58 | Cost includes material and labor to |
| Install | S.F. | 11.15 | 7.95 | | 19.10 | install 6″ x 6″ ceramic tile in mortar |
| Demolish and Install | S.F. | 11.15 | 9.53 | | 20.68 | bed including cutting and grout. |
| Clean | S.F. | | .68 | | .68 | |
| Minimum Charge | Job | | 219 | | 219 | |
| **8″ x 8″** | | | | | | |
| Demolish | S.F. | | 1.58 | | 1.58 | Cost includes material and labor to |
| Install | S.F. | 10.35 | 7.30 | | 17.65 | install 8″ x 8″ ceramic tile in mortar |
| Demolish and Install | S.F. | 10.35 | 8.88 | | 19.23 | bed including cutting and grout. |
| Clean | S.F. | | .68 | | .68 | |
| Minimum Charge | Job | | 219 | | 219 | |
| **Re-grout** | | | | | | |
| Install | S.F. | .17 | 4.38 | | 4.55 | Includes material and labor to regrout |
| Minimum Charge | Job | | 219 | | 219 | wall tiles. |

| Wall Tile | | Unit | Material | Labor | Equip. | Total | Specification |
|---|---|---|---|---|---|---|---|
| **Mirror** | | | | | | | |
| **Custom Cut Wall** | | | | | | | |
| | Demolish | S.F. | | .75 | | .75 | Includes material and labor to install |
| | Install | S.F. | 2.87 | 4.61 | | 7.48 | distortion-free float glass mirror, 1/8" |
| | Demolish and Install | S.F. | 2.87 | 5.36 | | 8.23 | to 3/16" thick with cut and polished |
| | Clean | S.F. | .01 | .21 | | .22 | edges. |
| | Minimum Charge | Job | | 225 | | 225 | |
| **Beveled Wall** | | | | | | | |
| | Demolish | S.F. | | .75 | | .75 | Includes material and labor to install |
| | Install | S.F. | 7.30 | 8.15 | | 15.45 | distortion-free float glass mirror, 1/8" |
| | Demolish and Install | S.F. | 7.30 | 8.90 | | 16.20 | to 3/16" thick with beveled edges. |
| | Clean | S.F. | .01 | .21 | | .22 | |
| | Minimum Charge | Job | | 225 | | 225 | |
| **Smoked Wall** | | | | | | | |
| | Demolish | S.F. | | .75 | | .75 | Includes material and labor to install |
| | Install | S.F. | 9.85 | 8.15 | | 18 | distortion-free smoked float glass |
| | Demolish and Install | S.F. | 9.85 | 8.90 | | 18.75 | mirror, 1/8" to 3/16" thick with cut |
| | Clean | S.F. | .01 | .21 | | .22 | and polished edges. |
| | Minimum Charge | Job | | 225 | | 225 | |
| **Marble** | | | | | | | |
| **Good Grade** | | | | | | | |
| | Demolish | S.F. | | 2.47 | | 2.47 | Cost includes material and labor to |
| | Install | S.F. | 9.90 | 19.05 | | 28.95 | install 3/8" x 12" x 12" marble tile in |
| | Demolish and Install | S.F. | 9.90 | 21.52 | | 31.42 | mortar bed with grout. |
| | Clean | S.F. | | .68 | | .68 | |
| | Minimum Charge | Job | | 219 | | 219 | |
| **Better Grade** | | | | | | | |
| | Demolish | S.F. | | 2.47 | | 2.47 | Cost includes material and labor to |
| | Install | S.F. | 11 | 19.05 | | 30.05 | install 3/8" x 12" x 12" marble, in |
| | Demolish and Install | S.F. | 11 | 21.52 | | 32.52 | mortar bed with grout. |
| | Clean | S.F. | | .68 | | .68 | |
| | Minimum Charge | Job | | 219 | | 219 | |
| **Premium Grade** | | | | | | | |
| | Demolish | S.F. | | 2.47 | | 2.47 | Cost includes material and labor to |
| | Install | S.F. | 21 | 19.05 | | 40.05 | install 3/8" x 12" x 12" marble, in |
| | Demolish and Install | S.F. | 21 | 21.52 | | 42.52 | mortar bed with grout. |
| | Clean | S.F. | | .68 | | .68 | |
| | Minimum Charge | Job | | 219 | | 219 | |
| **Glass Block** | | | | | | | |
| | Demolish | S.F. | | 1.97 | | 1.97 | Cost includes material and labor to |
| | Install | S.F. | 28.50 | 18.75 | | 47.25 | install 3-7/8" thick, 6" x 6" smooth |
| | Demolish and Install | S.F. | 28.50 | 20.72 | | 49.22 | face glass block including mortar and |
| | Clean | S.F. | | .68 | | .68 | cleaning. |
| | Minimum Charge | Job | | 425 | | 425 | |
| **Cork** | | | | | | | |
| | Demolish | S.F. | | .31 | | .31 | Cost includes material and labor to |
| | Install | S.F. | 4.69 | 1.61 | | 6.30 | install 3/16" thick cork tile and |
| | Demolish and Install | S.F. | 4.69 | 1.92 | | 6.61 | installation. |
| | Clean | S.F. | | .68 | | .68 | |
| | Minimum Charge | Job | | 193 | | 193 | |
| **Re-grout** | | | | | | | |
| | Install | S.F. | .17 | 4.38 | | 4.55 | Includes material and labor to regrout |
| | Minimum Charge | Job | | 219 | | 219 | wall tiles. |

| Wall Paneling | Unit | Material | Labor | Equip. | Total | Specification |
|---|---|---|---|---|---|---|
| **Plywood** | | | | | | |
| **Good Grade** | | | | | | |
| Demolish | S.F. | | .39 | | .39 | Cost includes material and labor to |
| Install | S.F. | 1.62 | 1.86 | | 3.48 | install 4' x 8' x 1/4" paneling, |
| Demolish and Install | S.F. | 1.62 | 2.25 | | 3.87 | including nails and adhesive. |
| Clean | S.F. | | .32 | | .32 | |
| Paint | S.F. | .33 | .74 | | 1.07 | |
| Minimum Charge | Job | | 232 | | 232 | |
| **Better Grade** | | | | | | |
| Demolish | S.F. | | .39 | | .39 | Cost includes material and labor to |
| Install | S.F. | 1.32 | 2.21 | | 3.53 | install 4' x 8' x 1/4" plywood wall |
| Demolish and Install | S.F. | 1.32 | 2.60 | | 3.92 | panel, including nails and adhesive. |
| Clean | S.F. | | .32 | | .32 | |
| Paint | S.F. | .33 | .74 | | 1.07 | |
| Minimum Charge | Job | | 232 | | 232 | |
| **Premium Grade** | | | | | | |
| Demolish | S.F. | | .39 | | .39 | Cost includes material and labor to |
| Install | S.F. | 1.21 | 2.66 | | 3.87 | install 4' x 8' x 1/4" plywood wall |
| Demolish and Install | S.F. | 1.21 | 3.05 | | 4.26 | panel, including nails and adhesive. |
| Clean | S.F. | | .32 | | .32 | |
| Paint | S.F. | .33 | .74 | | 1.07 | |
| Minimum Charge | Job | | 232 | | 232 | |
| **Hardboard** | | | | | | |
| **Embossed Paneling** | | | | | | |
| Demolish | S.F. | | .39 | | .39 | Includes material and labor to install |
| Install | S.F. | .68 | 1.86 | | 2.54 | hardboard embossed paneling |
| Demolish and Install | S.F. | .68 | 2.25 | | 2.93 | including adhesives. |
| Clean | S.F. | | .32 | | .32 | |
| Minimum Charge | Job | | 232 | | 232 | |
| **Plastic Molding** | | | | | | |
| Demolish | L.F. | | .39 | | .39 | Includes material and labor to install |
| Install | L.F. | .87 | 1.16 | | 2.03 | plastic trim for paneling. |
| Demolish and Install | L.F. | .87 | 1.55 | | 2.42 | |
| Clean | L.F. | .01 | .21 | | .22 | |
| Minimum Charge | Job | | 232 | | 232 | |
| **Hardboard** | | | | | | |
| **Embossed Brick** | | | | | | |
| Demolish | S.F. | | .39 | | .39 | Cost includes material and labor to |
| Install | S.F. | .89 | 2.07 | | 2.96 | install 4' x 8' non-ceramic mineral |
| Demolish and Install | S.F. | .89 | 2.46 | | 3.35 | brick-like veneer wall panels including |
| Clean | S.F. | | .32 | | .32 | adhesive. |
| Paint | S.F. | .21 | .58 | | .79 | |
| Minimum Charge | Job | | 232 | | 232 | |
| **Embossed Ceramic Tile** | | | | | | |
| Demolish | S.F. | | .39 | | .39 | Includes material and labor to install |
| Install | S.F. | .68 | 1.86 | | 2.54 | hardboard embossed paneling |
| Demolish and Install | S.F. | .68 | 2.25 | | 2.93 | including adhesives. |
| Clean | S.F. | | .32 | | .32 | |
| Paint | S.F. | .21 | .58 | | .79 | |
| Minimum Charge | Job | | 232 | | 232 | |
| **Pegboard** | | | | | | |
| Demolish | S.F. | | .39 | | .39 | Cost includes material and labor to |
| Install | S.F. | .53 | 1.86 | | 2.39 | install 4' x 8' x 1/4" pegboard panel, |
| Demolish and Install | S.F. | .53 | 2.25 | | 2.78 | including adhesive. |
| Clean | S.F. | | .32 | | .32 | |
| Paint | S.F. | .21 | .58 | | .79 | |
| Minimum Charge | Job | | 232 | | 232 | |

# Walls / Ceilings

| Wall Paneling | Unit | Material | Labor | Equip. | Total | Specification |
|---|---|---|---|---|---|---|
| **Furring Strips** | | | | | | |
| Demolish | S.F. | | .10 | | .10 | Cost includes material and labor to |
| Install | S.F. | .22 | 1.03 | | 1.25 | install 1" x 2" furring strips, 16" O.C. |
| Demolish and Install | S.F. | .22 | 1.13 | | 1.35 | per S.F. of surface area to be covered. |
| Minimum Charge | Job | | 232 | | 232 | |
| **Molding** | | | | | | |
| **Unfinished Base** | | | | | | |
| Demolish | L.F. | | .66 | | .66 | Cost includes material and labor to |
| Install | L.F. | .60 | 1.40 | | 2 | install 1/2" x 2-1/2" pine or birch |
| Demolish and Install | L.F. | .60 | 2.06 | | 2.66 | base molding. |
| Reinstall | L.F. | | 1.12 | | 1.12 | |
| Clean | L.F. | .01 | .21 | | .22 | |
| Paint | L.F. | .06 | .42 | | .48 | |
| Minimum Charge | Job | | 232 | | 232 | |
| **Unfinished Corner** | | | | | | |
| Demolish | L.F. | | .66 | | .66 | Includes material and labor to install |
| Install | L.F. | 1.78 | 1.94 | | 3.72 | unfinished pine corner molding. |
| Demolish and Install | L.F. | 1.78 | 2.60 | | 4.38 | |
| Reinstall | L.F. | | 1.55 | | 1.55 | |
| Clean | L.F. | .01 | .21 | | .22 | |
| Paint | L.F. | .17 | .62 | | .79 | |
| Minimum Charge | Job | | 232 | | 232 | |
| **Unfinished Cove** | | | | | | |
| Demolish | L.F. | | .66 | | .66 | Cost includes material and labor to |
| Install | L.F. | 2.06 | 1.75 | | 3.81 | install 11/16" x 2-3/4" pine cove |
| Demolish and Install | L.F. | 2.06 | 2.41 | | 4.47 | molding. |
| Reinstall | L.F. | | 1.40 | | 1.40 | |
| Clean | L.F. | .01 | .21 | | .22 | |
| Paint | L.F. | .17 | .62 | | .79 | |
| Minimum Charge | Job | | 232 | | 232 | |
| **Unfinished Crown** | | | | | | |
| Demolish | L.F. | | .79 | | .79 | Includes material and labor to install |
| Install | L.F. | 2.31 | 1.86 | | 4.17 | unfinished pine crown molding. |
| Demolish and Install | L.F. | 2.31 | 2.65 | | 4.96 | |
| Reinstall | L.F. | | 1.49 | | 1.49 | |
| Clean | L.F. | .01 | .27 | | .28 | |
| Paint | L.F. | .10 | .42 | | .52 | |
| Minimum Charge | Job | | 232 | | 232 | |
| **Prefinished Base** | | | | | | |
| Demolish | L.F. | | .66 | | .66 | Cost includes material and labor to |
| Install | L.F. | 1.32 | 1.49 | | 2.81 | install 9/16" x 3-5/16" prefinished |
| Demolish and Install | L.F. | 1.32 | 2.15 | | 3.47 | base molding. |
| Reinstall | L.F. | | 1.19 | | 1.19 | |
| Clean | L.F. | .01 | .21 | | .22 | |
| Paint | L.F. | .06 | .42 | | .48 | |
| Minimum Charge | Job | | 232 | | 232 | |
| **Prefinished Corner** | | | | | | |
| Demolish | L.F. | | .66 | | .66 | Includes material and labor to install |
| Install | L.F. | 2.28 | 1.94 | | 4.22 | inside or outside prefinished corner |
| Demolish and Install | L.F. | 2.28 | 2.60 | | 4.88 | molding. |
| Reinstall | L.F. | | 1.55 | | 1.55 | |
| Clean | L.F. | .01 | .21 | | .22 | |
| Paint | L.F. | .17 | .62 | | .79 | |
| Minimum Charge | Job | | 232 | | 232 | |

# Walls / Ceilings

## Wall Paneling

| Wall Paneling | Unit | Material | Labor | Equip. | Total | Specification |
|---|---|---|---|---|---|---|
| **Prefinished Crown** | | | | | | |
| Demolish | L.F. | | .79 | | .79 | Cost includes material and labor to |
| Install | L.F. | 1.96 | 1.86 | | 3.82 | install 3/4" x 3-13/16" prefinished |
| Demolish and Install | L.F. | 1.96 | 2.65 | | 4.61 | crown molding. |
| Reinstall | L.F. | | 1.49 | | 1.49 | |
| Clean | L.F. | .01 | .27 | | .28 | |
| Paint | L.F. | .10 | .42 | | .52 | |
| Minimum Charge | Job | | 232 | | 232 | |

## Wallcovering

| Wallcovering | Unit | Material | Labor | Equip. | Total | Specification |
|---|---|---|---|---|---|---|
| **Surface Prep** | | | | | | |
| **Patch Walls** | | | | | | |
| Install | S.F. | .12 | 4.45 | | 4.57 | Includes labor and material to repair |
| Minimum Charge | Job | | 193 | | 193 | section of plaster wall, area less than |
| | | | | | | 20 S.F. |
| **Sizing** | | | | | | |
| Install | S.F. | .09 | .19 | | .28 | Includes labor and materials to apply |
| Minimum Charge | Job | | 193 | | 193 | wall sizing. |
| **Good Grade** | | | | | | |
| **Walls** | | | | | | |
| Demolish | S.F. | | .72 | | .72 | Includes material and labor to install |
| Install | S.F. | 1 | .81 | | 1.81 | pre-pasted, vinyl-coated wallcovering. |
| Demolish and Install | S.F. | 1 | 1.53 | | 2.53 | |
| Clean | S.F. | | .08 | | .08 | |
| Minimum Charge | Job | | 193 | | 193 | |
| **Ceiling** | | | | | | |
| Demolish | S.F. | | .72 | | .72 | Includes material and labor to install |
| Install | S.F. | 1 | .81 | | 1.81 | pre-pasted, vinyl-coated wallcovering. |
| Demolish and Install | S.F. | 1 | 1.53 | | 2.53 | |
| Clean | S.F. | .01 | .21 | | .22 | |
| Minimum Charge | Job | | 193 | | 193 | |
| **Better Grade** | | | | | | |
| **Walls** | | | | | | |
| Demolish | S.F. | | .72 | | .72 | Includes material and labor to install |
| Install | S.F. | 1.12 | .81 | | 1.93 | pre-pasted, vinyl-coated wallcovering. |
| Demolish and Install | S.F. | 1.12 | 1.53 | | 2.65 | |
| Clean | S.F. | | .08 | | .08 | |
| Minimum Charge | Job | | 193 | | 193 | |
| **Ceiling** | | | | | | |
| Demolish | S.F. | | .72 | | .72 | Includes material and labor to install |
| Install | S.F. | 1.12 | .81 | | 1.93 | pre-pasted, vinyl-coated wallcovering. |
| Demolish and Install | S.F. | 1.12 | 1.53 | | 2.65 | |
| Clean | S.F. | .01 | .21 | | .22 | |
| Minimum Charge | Job | | 193 | | 193 | |
| **Premium Grade** | | | | | | |
| **Walls** | | | | | | |
| Demolish | S.F. | | .90 | | .90 | Includes material and labor to install |
| Install | S.F. | 1.21 | .81 | | 2.02 | pre-pasted, vinyl-coated wallcovering. |
| Demolish and Install | S.F. | 1.21 | 1.71 | | 2.92 | |
| Clean | S.F. | | .08 | | .08 | |
| Minimum Charge | Job | | 193 | | 193 | |
| **Ceiling** | | | | | | |
| Demolish | S.F. | | .90 | | .90 | Includes material and labor to install |
| Install | S.F. | 1.21 | .81 | | 2.02 | pre-pasted, vinyl-coated wallcovering. |
| Demolish and Install | S.F. | 1.21 | 1.71 | | 2.92 | |
| Clean | S.F. | .01 | .21 | | .22 | |
| Minimum Charge | Job | | 193 | | 193 | |

# Walls / Ceilings

| Wallcovering | | Unit | Material | Labor | Equip. | Total | Specification |
|---|---|---|---|---|---|---|---|
| **Mural** | | | | | | | |
| **Pre-printed** | | | | | | | |
| | Demolish | Ea. | | 123 | | 123 | Includes material and labor to install |
| | Install | Ea. | 273 | 291 | | 564 | wallcovering with a pre-printed mural. |
| | Demolish and Install | Ea. | 273 | 414 | | 687 | |
| | Clean | Ea. | | 8.90 | | 8.90 | |
| | Minimum Charge | Job | | 193 | | 193 | |
| **Hand-painted** | | | | | | | |
| | Demolish | Ea. | | 123 | | 123 | Includes material and labor to install |
| | Install | Ea. | 535 | 585 | | 1120 | wallcovering with a hand-painted |
| | Demolish and Install | Ea. | 535 | 708 | | 1243 | mural. |
| | Clean | Ea. | | 8.90 | | 8.90 | |
| | Minimum Charge | Job | | 193 | | 193 | |
| **Vinyl-coated** | | | | | | | |
| **Good Grade** | | | | | | | |
| | Demolish | Ea. | | 22 | | 22 | Includes material and labor to install |
| | Install | Ea. | 16.95 | 27 | | 43.95 | pre-pasted, vinyl-coated wallcovering. |
| | Demolish and Install | Ea. | 16.95 | 49 | | 65.95 | Based upon 30 S.F. rolls. |
| | Clean | Ea. | | 8.90 | | 8.90 | |
| | Minimum Charge | Job | | 193 | | 193 | |
| **Better Grade** | | | | | | | |
| | Demolish | Ea. | | 22 | | 22 | Includes material and labor to install |
| | Install | Ea. | 26.50 | 27 | | 53.50 | pre-pasted, vinyl-coated wallcovering. |
| | Demolish and Install | Ea. | 26.50 | 49 | | 75.50 | Based upon 30 S.F. rolls. |
| | Clean | Ea. | | 8.90 | | 8.90 | |
| | Minimum Charge | Job | | 193 | | 193 | |
| **Premium Grade** | | | | | | | |
| | Demolish | Ea. | | 22 | | 22 | Includes material and labor to install |
| | Install | Ea. | 31.50 | 27 | | 58.50 | pre-pasted, vinyl-coated wallcovering. |
| | Demolish and Install | Ea. | 31.50 | 49 | | 80.50 | Based upon 30 S.F. rolls. |
| | Clean | Ea. | | 8.90 | | 8.90 | |
| | Minimum Charge | Job | | 193 | | 193 | |
| **Blank Underliner** | | | | | | | |
| | Demolish | Ea. | | 22 | | 22 | Includes material and labor to install |
| | Install | Ea. | 9.55 | 27.50 | | 37.05 | blank stock liner paper or bridging |
| | Demolish and Install | Ea. | 9.55 | 49.50 | | 59.05 | paper. Based upon 30 S.F. rolls. |
| | Clean | Ea. | | 8.90 | | 8.90 | |
| | Minimum Charge | Job | | 193 | | 193 | |
| **Solid Vinyl** | | | | | | | |
| **Good Grade** | | | | | | | |
| | Demolish | Roll | | 24 | | 24 | Includes material and labor to install |
| | Install | Ea. | 24 | 27 | | 51 | pre-pasted, vinyl wallcovering. Based |
| | Demolish and Install | Ea. | 24 | 51 | | 75 | upon 30 S.F. rolls. |
| | Clean | Ea. | | 8.90 | | 8.90 | |
| | Minimum Charge | Job | | 193 | | 193 | |
| **Better Grade** | | | | | | | |
| | Demolish | Roll | | 24 | | 24 | Includes material and labor to install |
| | Install | Ea. | 31.50 | 27 | | 58.50 | pre-pasted, vinyl wallcovering. Based |
| | Demolish and Install | Ea. | 31.50 | 51 | | 82.50 | upon 30 S.F. rolls. |
| | Clean | Ea. | | 8.90 | | 8.90 | |
| | Minimum Charge | Job | | 193 | | 193 | |
| **Premium Grade** | | | | | | | |
| | Demolish | Roll | | 24 | | 24 | Includes material and labor to install |
| | Install | Ea. | 50.50 | 27 | | 77.50 | pre-pasted, vinyl wallcovering. Based |
| | Demolish and Install | Ea. | 50.50 | 51 | | 101.50 | upon 30 S.F. rolls. |
| | Clean | Ea. | | 8.90 | | 8.90 | |
| | Minimum Charge | Job | | 193 | | 193 | |

| Wallcovering | Unit | Material | Labor | Equip. | Total | Specification |
|---|---|---|---|---|---|---|
| **Blank Underliner** | | | | | | |
| Demolish | Ea. | | 22 | | 22 | Includes material and labor to install |
| Install | Ea. | 9.55 | 27.50 | | 37.05 | blank stock liner paper or bridging |
| Demolish and Install | Ea. | 9.55 | 49.50 | | 59.05 | paper. Based upon 30 S.F. rolls. |
| Clean | Ea. | | 8.90 | | 8.90 | |
| Minimum Charge | Job | | 193 | | 193 | |
| **Grasscloth / String** | | | | | | |
| **Good Grade** | | | | | | |
| Demolish | Ea. | | 22 | | 22 | Includes material and labor to install |
| Install | Ea. | 54 | 35 | | 89 | grasscloth or string wallcovering. |
| Demolish and Install | Ea. | 54 | 57 | | 111 | Based upon 30 S.F. rolls. |
| Clean | Ea. | | 11.85 | | 11.85 | |
| Minimum Charge | Job | | 193 | | 193 | |
| **Better Grade** | | | | | | |
| Demolish | Ea. | | 22 | | 22 | Includes material and labor to install |
| Install | Ea. | 60.50 | 35 | | 95.50 | grasscloth or string wallcovering. |
| Demolish and Install | Ea. | 60.50 | 57 | | 117.50 | Based upon 30 S.F. rolls. |
| Clean | Ea. | | 11.85 | | 11.85 | |
| Minimum Charge | Job | | 193 | | 193 | |
| **Premium Grade** | | | | | | |
| Demolish | Ea. | | 22 | | 22 | Includes material and labor to install |
| Install | Ea. | 67.50 | 35 | | 102.50 | grasscloth or string wallcovering. |
| Demolish and Install | Ea. | 67.50 | 57 | | 124.50 | Based upon 30 S.F. rolls. |
| Clean | Ea. | | 11.85 | | 11.85 | |
| Minimum Charge | Job | | 193 | | 193 | |
| **Blank Underliner** | | | | | | |
| Demolish | Ea. | | 22 | | 22 | Includes material and labor to install |
| Install | Ea. | 9.55 | 27.50 | | 37.05 | blank stock liner paper or bridging |
| Demolish and Install | Ea. | 9.55 | 49.50 | | 59.05 | paper. Based upon 30 S.F. rolls. |
| Clean | Ea. | | 8.90 | | 8.90 | |
| Minimum Charge | Job | | 193 | | 193 | |
| **Designer** | | | | | | |
| **Premium Grade** | | | | | | |
| Demolish | S.F. | | .90 | | .90 | Includes material and labor to install |
| Install | Ea. | 127 | 40.50 | | 167.50 | paperback vinyl, untrimmed and |
| Demolish and Install | Ea. | 127 | 41.40 | | 168.40 | hand-painted. Based upon 30 S.F. |
| Clean | Ea. | | 8.90 | | 8.90 | rolls. |
| Minimum Charge | Job | | 193 | | 193 | |
| **Blank Underliner** | | | | | | |
| Demolish | Ea. | | 22 | | 22 | Includes material and labor to install |
| Install | Ea. | 9.55 | 27.50 | | 37.05 | blank stock liner paper or bridging |
| Demolish and Install | Ea. | 9.55 | 49.50 | | 59.05 | paper. Based upon 30 S.F. rolls. |
| Clean | Ea. | | 8.90 | | 8.90 | |
| Minimum Charge | Job | | 193 | | 193 | |
| **Foil** | | | | | | |
| Demolish | Ea. | | 22 | | 22 | Includes material and labor to install |
| Install | Ea. | 29 | 35 | | 64 | foil coated wallcovering. Based upon |
| Demolish and Install | Ea. | 29 | 57 | | 86 | 30 S.F. rolls. |
| Clean | Ea. | | 8.90 | | 8.90 | |
| Minimum Charge | Job | | 193 | | 193 | |
| **Wallcovering Border** | | | | | | |
| **Good Grade** | | | | | | |
| Demolish | L.F. | | 1.44 | | 1.44 | Includes material and labor install |
| Install | L.F. | 1.65 | 1.49 | | 3.14 | vinyl-coated border 4" to 6" wide. |
| Demolish and Install | L.F. | 1.65 | 2.93 | | 4.58 | |
| Clean | L.F. | | .13 | | .13 | |
| Minimum Charge | Job | | 193 | | 193 | |

# Walls / Ceilings

## Wallcovering

| | Unit | Material | Labor | Equip. | Total | Specification |
|---|---|---|---|---|---|---|
| **Better Grade** | | | | | | |
| Demolish | L.F. | | 1.44 | | 1.44 | Includes material and labor install |
| Install | L.F. | 2.96 | 1.49 | | 4.45 | vinyl-coated border 4" to 6" wide. |
| Demolish and Install | L.F. | 2.96 | 2.93 | | 5.89 | |
| Clean | L.F. | | .13 | | .13 | |
| Minimum Charge | Job | | 193 | | 193 | |
| **Premium Grade** | | | | | | |
| Demolish | L.F. | | 1.44 | | 1.44 | Includes material and labor install |
| Install | L.F. | 3.69 | 1.49 | | 5.18 | vinyl-coated border 4" to 6" wide. |
| Demolish and Install | L.F. | 3.69 | 2.93 | | 6.62 | |
| Clean | L.F. | | .13 | | .13 | |
| Minimum Charge | Job | | 193 | | 193 | |

## Acoustical Ceiling

| | Unit | Material | Labor | Equip. | Total | Specification |
|---|---|---|---|---|---|---|
| **Complete System** | | | | | | |
| **2' x 2' Square Edge** | | | | | | |
| Demolish | S.F. | | .66 | | .66 | Includes material and labor to install |
| Install | S.F. | 2.09 | 1.43 | | 3.52 | T-bar grid acoustical ceiling system, |
| Demolish and Install | S.F. | 2.09 | 2.09 | | 4.18 | random pinhole 5/8" thick square |
| Reinstall | S.F. | | 1.14 | | 1.14 | edge tile. |
| Clean | S.F. | .01 | .27 | | .28 | |
| Minimum Charge | Job | | 232 | | 232 | |
| **2' x 2' Reveal Edge** | | | | | | |
| Demolish | S.F. | | .66 | | .66 | Includes material and labor to install |
| Install | S.F. | 2.84 | 1.86 | | 4.70 | T-bar grid acoustical ceiling system, |
| Demolish and Install | S.F. | 2.84 | 2.52 | | 5.36 | random pinhole 5/8" thick reveal |
| Reinstall | S.F. | | 1.49 | | 1.49 | edge tile. |
| Clean | S.F. | .01 | .27 | | .28 | |
| Minimum Charge | Job | | 232 | | 232 | |
| **2' x 2' Fire-rated** | | | | | | |
| Demolish | S.F. | | .66 | | .66 | Includes material and labor to install |
| Install | S.F. | 3.99 | 1.43 | | 5.42 | T-bar grid FR acoustical ceiling system, |
| Demolish and Install | S.F. | 3.99 | 2.09 | | 6.08 | random pinhole 5/8" thick square |
| Reinstall | S.F. | | 1.14 | | 1.14 | edge tile. |
| Clean | S.F. | .01 | .27 | | .28 | |
| Minimum Charge | Job | | 232 | | 232 | |
| **2' x 4' Square Edge** | | | | | | |
| Demolish | S.F. | | .66 | | .66 | Cost includes material and labor to |
| Install | S.F. | 2.43 | 1.22 | | 3.65 | install 2' x 4' T-bar grid acoustical |
| Demolish and Install | S.F. | 2.43 | 1.88 | | 4.31 | ceiling system, random pinhole 5/8" |
| Reinstall | S.F. | | .98 | | .98 | thick square edge tile. |
| Clean | S.F. | .01 | .27 | | .28 | |
| Minimum Charge | Job | | 232 | | 232 | |
| **2' x 4' Reveal Edge** | | | | | | |
| Demolish | S.F. | | .66 | | .66 | Includes material and labor to install |
| Install | S.F. | 2.53 | 1.69 | | 4.22 | T-bar grid acoustical ceiling system, |
| Demolish and Install | S.F. | 2.53 | 2.35 | | 4.88 | random pinhole 5/8" thick reveal |
| Reinstall | S.F. | | 1.35 | | 1.35 | edge tile. |
| Clean | S.F. | .01 | .27 | | .28 | |
| Minimum Charge | Job | | 232 | | 232 | |
| **2' x 4' Fire-rated** | | | | | | |
| Demolish | S.F. | | .66 | | .66 | Includes material and labor to install |
| Install | S.F. | 3.99 | 1.43 | | 5.42 | T-bar grid FR acoustical ceiling system, |
| Demolish and Install | S.F. | 3.99 | 2.09 | | 6.08 | random pinhole 5/8" thick square |
| Reinstall | S.F. | | 1.14 | | 1.14 | edge tile. |
| Clean | S.F. | .01 | .27 | | .28 | |
| Minimum Charge | Job | | 232 | | 232 | |

# Walls / Ceilings

| Acoustical Ceiling | Unit | Material | Labor | Equip. | Total | Specification |
|---|---|---|---|---|---|---|
| **T-bar Grid** | | | | | | |
| **2' x 2'** | | | | | | |
| Demolish | S.F. | | .77 | | .77 | Includes material and labor to install |
| Install | S.F. | 1.12 | .72 | | 1.84 | T-bar grid acoustical ceiling |
| Demolish and Install | S.F. | 1.12 | 1.49 | | 2.61 | suspension system. |
| Reinstall | S.F. | | .57 | | .57 | |
| Clean | S.F. | .01 | .27 | | .28 | |
| Minimum Charge | Job | | 222 | | 222 | |
| **2' x 4'** | | | | | | |
| Demolish | S.F. | | .77 | | .77 | Includes material and labor to install |
| Install | S.F. | .87 | .58 | | 1.45 | T-bar grid acoustical ceiling |
| Demolish and Install | S.F. | .87 | 1.35 | | 2.22 | suspension system. |
| Reinstall | S.F. | | .46 | | .46 | |
| Clean | S.F. | .01 | .27 | | .28 | |
| Minimum Charge | Job | | 222 | | 222 | |
| **Tile Panels** | | | | | | |
| **2' x 2' Square Edge** | | | | | | |
| Demolish | S.F. | | .22 | | .22 | Includes material and labor to install |
| Install | S.F. | .87 | .74 | | 1.61 | square edge panels 5/8" thick with |
| Demolish and Install | S.F. | .87 | .96 | | 1.83 | pinhole pattern. T-bar grid not |
| Reinstall | S.F. | | .59 | | .59 | included. |
| Clean | S.F. | .01 | .27 | | .28 | |
| Minimum Charge | Job | | 232 | | 232 | |
| **2' x 2' Reveal Edge** | | | | | | |
| Demolish | S.F. | | .22 | | .22 | Includes material and labor to install |
| Install | S.F. | 1.23 | .99 | | 2.22 | reveal edge panels 5/8" thick with |
| Demolish and Install | S.F. | 1.23 | 1.21 | | 2.44 | pinhole pattern. T-bar grid not |
| Reinstall | S.F. | | .79 | | .79 | included. |
| Clean | S.F. | .01 | .27 | | .28 | |
| Minimum Charge | Job | | 232 | | 232 | |
| **2' x 2' Fire-rated** | | | | | | |
| Demolish | S.F. | | .22 | | .22 | Includes material and labor to install |
| Install | S.F. | 1.41 | .69 | | 2.10 | FR square edge panels 5/8" thick with |
| Demolish and Install | S.F. | 1.41 | .91 | | 2.32 | pinhole pattern. T-bar grid not |
| Reinstall | S.F. | | .55 | | .55 | included. |
| Clean | S.F. | .01 | .27 | | .28 | |
| Minimum Charge | Job | | 232 | | 232 | |
| **2' x 4' Square Edge** | | | | | | |
| Demolish | S.F. | | .22 | | .22 | Includes material and labor to install |
| Install | S.F. | .87 | .74 | | 1.61 | square edge panels 5/8" thick with |
| Demolish and Install | S.F. | .87 | .96 | | 1.83 | pinhole pattern. T-bar grid not |
| Reinstall | S.F. | | .59 | | .59 | included. |
| Clean | S.F. | .01 | .27 | | .28 | |
| Minimum Charge | Job | | 232 | | 232 | |
| **2' x 4' Reveal Edge** | | | | | | |
| Demolish | S.F. | | .22 | | .22 | Includes material and labor to install |
| Install | S.F. | 1.23 | .99 | | 2.22 | reveal edge panels 5/8" thick with |
| Demolish and Install | S.F. | 1.23 | 1.21 | | 2.44 | pinhole pattern. T-bar grid not |
| Reinstall | S.F. | | .79 | | .79 | included. |
| Clean | S.F. | .01 | .27 | | .28 | |
| Minimum Charge | Job | | 232 | | 232 | |
| **2' x 4' Fire-rated** | | | | | | |
| Demolish | S.F. | | .22 | | .22 | Includes material and labor to install |
| Install | S.F. | 1.41 | .69 | | 2.10 | FR square edge panels 5/8" thick with |
| Demolish and Install | S.F. | 1.41 | .91 | | 2.32 | pinhole pattern. T-bar grid not |
| Reinstall | S.F. | | .55 | | .55 | included. |
| Clean | S.F. | .01 | .27 | | .28 | |
| Minimum Charge | Job | | 232 | | 232 | |

# Walls / Ceilings

## Acoustical Ceiling

| Bleach Out Stains | Unit | Material | Labor | Equip. | Total | Specification |
|---|---|---|---|---|---|---|
| Install | S.F. | .18 | .36 | | .54 | Includes labor and material to apply |
| Minimum Charge | Job | | 232 | | 232 | laundry type chlorine bleach to remove light staining of acoustic tile material. |

### Adhesive Tile
#### Good Grade

| | Unit | Material | Labor | Equip. | Total | Specification |
|---|---|---|---|---|---|---|
| Demolish | S.F. | | .88 | | .88 | Cost includes material and labor to |
| Install | S.F. | 1.24 | 1.55 | | 2.79 | install 12" x 12" tile with embossed |
| Demolish and Install | S.F. | 1.24 | 2.43 | | 3.67 | texture pattern. Furring strips not |
| Clean | S.F. | .01 | .21 | | .22 | included. |
| Minimum Charge | Job | | 232 | | 232 | |

#### Better Grade

| | Unit | Material | Labor | Equip. | Total | Specification |
|---|---|---|---|---|---|---|
| Demolish | S.F. | | .88 | | .88 | Cost includes material and labor to |
| Install | S.F. | 1.53 | 1.55 | | 3.08 | install 12" x 12" tile with embossed |
| Demolish and Install | S.F. | 1.53 | 2.43 | | 3.96 | texture pattern. Furring strips not |
| Clean | S.F. | .01 | .21 | | .22 | included. |
| Minimum Charge | Job | | 232 | | 232 | |

#### Premium Grade

| | Unit | Material | Labor | Equip. | Total | Specification |
|---|---|---|---|---|---|---|
| Demolish | S.F. | | .88 | | .88 | Cost includes material and labor to |
| Install | S.F. | 5.60 | 1.55 | | 7.15 | install 12" x 12" tile with embossed |
| Demolish and Install | S.F. | 5.60 | 2.43 | | 8.03 | texture pattern. Furring strips not |
| Clean | S.F. | .01 | .21 | | .22 | included. |
| Minimum Charge | Job | | 232 | | 232 | |

#### Bleach Out Stains

| | Unit | Material | Labor | Equip. | Total | Specification |
|---|---|---|---|---|---|---|
| Install | S.F. | .18 | .36 | | .54 | Includes labor and material to apply |
| Minimum Charge | Job | | 232 | | 232 | laundry type chlorine bleach to remove light staining of acoustic tile material. |

### Stapled Tile
#### Good Grade

| | Unit | Material | Labor | Equip. | Total | Specification |
|---|---|---|---|---|---|---|
| Demolish | S.F. | | .53 | | .53 | Cost includes material and labor to |
| Install | S.F. | 1.24 | 1.55 | | 2.79 | install 12" x 12" tile with embossed |
| Demolish and Install | S.F. | 1.24 | 2.08 | | 3.32 | texture pattern. Furring strips not |
| Clean | S.F. | .01 | .21 | | .22 | included. |
| Minimum Charge | Job | | 232 | | 232 | |

#### Better Grade

| | Unit | Material | Labor | Equip. | Total | Specification |
|---|---|---|---|---|---|---|
| Demolish | S.F. | | .53 | | .53 | Cost includes material and labor to |
| Install | S.F. | 1.53 | 1.55 | | 3.08 | install 12" x 12" tile with embossed |
| Demolish and Install | S.F. | 1.53 | 2.08 | | 3.61 | texture pattern. Furring strips not |
| Clean | S.F. | .01 | .21 | | .22 | included. |
| Minimum Charge | Job | | 232 | | 232 | |

#### Premium Grade

| | Unit | Material | Labor | Equip. | Total | Specification |
|---|---|---|---|---|---|---|
| Demolish | S.F. | | .53 | | .53 | Cost includes material and labor to |
| Install | S.F. | 5.60 | 1.55 | | 7.15 | install 12" x 12" tile with embossed |
| Demolish and Install | S.F. | 5.60 | 2.08 | | 7.68 | texture pattern. Furring strips not |
| Clean | S.F. | .01 | .21 | | .22 | included. |
| Minimum Charge | Job | | 232 | | 232 | |

#### Bleach Out Stains

| | Unit | Material | Labor | Equip. | Total | Specification |
|---|---|---|---|---|---|---|
| Install | S.F. | .18 | .36 | | .54 | Includes labor and material to apply |
| Minimum Charge | Job | | 232 | | 232 | laundry type chlorine bleach to remove light staining of acoustic tile material. |

For customer support on your Contractor's Pricing Guide: Residential Repair & Remodeling, call 800.448.8182.

## Acoustical Ceiling

| Acoustical Ceiling | Unit | Material | Labor | Equip. | Total | Specification |
|---|---|---|---|---|---|---|
| **Blown** | | | | | | |
| **Scrape Off Existing** | | | | | | |
| Install | S.F. | | .73 | | .73 | Includes labor and material to remove existing blown acoustical ceiling material. |
| Minimum Charge | Job | | 232 | | 232 | |
| **Seal Ceiling Gypsum Wallboard** | | | | | | |
| Install | S.F. | .07 | .34 | | .41 | Includes labor and material to seal existing drywall with shellac-based material. |
| Minimum Charge | Job | | 193 | | 193 | |
| **Reblow Existing** | | | | | | |
| Install | S.F. | .04 | .56 | | .60 | Includes labor and material to install by spray, texture finish. |
| Minimum Charge | Job | | 590 | 70.50 | 660.50 | |
| **Bleach Out Stains** | | | | | | |
| Install | S.F. | .18 | .36 | | .54 | Includes labor and material to apply laundry type chlorine bleach to remove light staining of acoustic tile material. |
| Minimum Charge | Job | | 232 | | 232 | |
| **New Blown** | | | | | | |
| Install | S.F. | .04 | .56 | | .60 | Includes labor and material to install by spray, texture finish. |
| Minimum Charge | Job | | 193 | | 193 | |
| **Apply Glitter To** | | | | | | |
| Install | S.F. | .23 | .18 | | .41 | Includes labor and material to apply glitter material to existing or new drywall. |
| Minimum Charge | Job | | 193 | | 193 | |
| **Furring Strips** | | | | | | |
| Demolish | S.F. | | .10 | | .10 | Cost includes material and labor to install 1" x 2" furring strips, 16" O.C. per S.F. of surface area to be covered. |
| Install | S.F. | .22 | 1.03 | | 1.25 | |
| Demolish and Install | S.F. | .22 | 1.13 | | 1.35 | |
| Minimum Charge | Job | | 116 | | 116 | |

**Finished Stair**

**Bi-Fold Door**

**Raised Panel Door**

| Hollow Core Door | Unit | Material | Labor | Equip. | Total | Specification |
|---|---|---|---|---|---|---|
| **Pre-hung Wood** | | | | | | |
| **1' 6" x 6' 8"** | | | | | | |
| Demolish | Ea. | | 43 | | 43 | Cost includes material and labor to |
| Install | Ea. | 166 | 26 | | 192 | install 1' 6" x 6' 8" pre-hung 1-3/8" |
| Demolish and Install | Ea. | 166 | 69 | | 235 | lauan door with split pine jamb |
| Reinstall | Ea. | | 25.82 | | 25.82 | including casing, stops, and hinges. |
| Clean | Ea. | 1.47 | 14.80 | | 16.27 | |
| Paint | Ea. | 10.60 | 55 | | 65.60 | |
| Minimum Charge | Job | | 232 | | 232 | |
| **2' x 6' 8"** | | | | | | |
| Demolish | Ea. | | 43 | | 43 | Cost includes material and labor to |
| Install | Ea. | 179 | 27.50 | | 206.50 | install 2' x 6' 8" pre-hung 1-3/8" |
| Demolish and Install | Ea. | 179 | 70.50 | | 249.50 | lauan door with split pine jamb |
| Reinstall | Ea. | | 27.34 | | 27.34 | including casing, stops, and hinges. |
| Clean | Ea. | 1.47 | 14.80 | | 16.27 | |
| Paint | Ea. | 10.60 | 55 | | 65.60 | |
| Minimum Charge | Job | | 232 | | 232 | |
| **2' 4" x 6' 8"** | | | | | | |
| Demolish | Ea. | | 43 | | 43 | Cost includes material and labor to |
| Install | Ea. | 184 | 27.50 | | 211.50 | install 2' 4" x 6' 8" pre-hung 1-3/8" |
| Demolish and Install | Ea. | 184 | 70.50 | | 254.50 | lauan door with split pine jamb |
| Reinstall | Ea. | | 27.34 | | 27.34 | including casing, stops, and hinges. |
| Clean | Ea. | 1.47 | 14.80 | | 16.27 | |
| Paint | Ea. | 10.60 | 55 | | 65.60 | |
| Minimum Charge | Job | | 232 | | 232 | |
| **2' 6" x 6' 8"** | | | | | | |
| Demolish | Ea. | | 43 | | 43 | Cost includes material and labor to |
| Install | Ea. | 188 | 27.50 | | 215.50 | install 2' 6" x 6' 8" pre-hung 1-3/8" |
| Demolish and Install | Ea. | 188 | 70.50 | | 258.50 | lauan door with split pine jamb |
| Reinstall | Ea. | | 27.34 | | 27.34 | including casing, stops, and hinges. |
| Clean | Ea. | 1.47 | 14.80 | | 16.27 | |
| Paint | Ea. | 10.60 | 55 | | 65.60 | |
| Minimum Charge | Job | | 232 | | 232 | |
| **2' 8" x 6' 8"** | | | | | | |
| Demolish | Ea. | | 43 | | 43 | Cost includes material and labor to |
| Install | Ea. | 177 | 27.50 | | 204.50 | install 2' 8" x 6' 8" pre-hung 1-3/8" |
| Demolish and Install | Ea. | 177 | 70.50 | | 247.50 | lauan door with split pine jamb, |
| Reinstall | Ea. | | 27.34 | | 27.34 | including casing, stop and hinges. |
| Clean | Ea. | 1.47 | 14.80 | | 16.27 | |
| Paint | Ea. | 10.60 | 55 | | 65.60 | |
| Minimum Charge | Job | | 232 | | 232 | |

191

# Finish Carpentry

## Hollow Core Door

| | Unit | Material | Labor | Equip. | Total | Specification |
|---|---|---|---|---|---|---|
| **3' x 6' 8"** | | | | | | |
| Demolish | Ea. | | 43 | | 43 | Cost includes material and labor to |
| Install | Ea. | 186 | 29 | | 215 | install 3' x 6' 8" pre-hung 1-3/8" |
| Demolish and Install | Ea. | 186 | 72 | | 258 | lauan door with split pine jamb |
| Reinstall | Ea. | | 29.05 | | 29.05 | including casing, stops, and hinges. |
| Clean | Ea. | 1.47 | 14.80 | | 16.27 | |
| Paint | Ea. | 10.60 | 55 | | 65.60 | |
| Minimum Charge | Job | | 232 | | 232 | |
| **Pocket Type** | | | | | | |
| Demolish | Ea. | | 64.50 | | 64.50 | Cost includes material and labor to |
| Install | Ea. | 77 | 51.50 | | 128.50 | install 2' 8" x 6' 8" lauan door. |
| Demolish and Install | Ea. | 77 | 116 | | 193 | |
| Reinstall | Ea. | | 51.64 | | 51.64 | |
| Clean | Ea. | 1.47 | 14.80 | | 16.27 | |
| Paint | Ea. | 10.60 | 55 | | 65.60 | |
| Minimum Charge | Job | | 232 | | 232 | |
| **Stain** | | | | | | |
| Install | Ea. | 8.55 | 32 | | 40.55 | Includes labor and material to stain |
| Minimum Charge | Job | | 193 | | 193 | single door and trim on both sides. |
| **Pre-hung Masonite** | | | | | | |
| **1' 6" x 6' 8"** | | | | | | |
| Demolish | Ea. | | 43 | | 43 | Cost includes material and labor to |
| Install | Ea. | 119 | 26 | | 145 | install 1' 6" x 6' 8" pre-hung 1-3/8" |
| Demolish and Install | Ea. | 119 | 69 | | 188 | masonite door with split pine jamb |
| Reinstall | Ea. | | 25.82 | | 25.82 | including casing, stops, and hinges. |
| Clean | Ea. | 1.47 | 14.80 | | 16.27 | |
| Paint | Ea. | 10.60 | 55 | | 65.60 | |
| Minimum Charge | Job | | 232 | | 232 | |
| **2' x 6' 8"** | | | | | | |
| Demolish | Ea. | | 43 | | 43 | Cost includes material and labor to |
| Install | Ea. | 240 | 27.50 | | 267.50 | install 2' x 6' 8" pre-hung 1-3/8" |
| Demolish and Install | Ea. | 240 | 70.50 | | 310.50 | masonite door with split pine jamb |
| Reinstall | Ea. | | 27.34 | | 27.34 | including casing, stops, and hinges. |
| Clean | Ea. | 1.47 | 14.80 | | 16.27 | |
| Paint | Ea. | 10.60 | 55 | | 65.60 | |
| Minimum Charge | Job | | 232 | | 232 | |
| **2' 4" x 6' 8"** | | | | | | |
| Demolish | Ea. | | 43 | | 43 | Cost includes material and labor to |
| Install | Ea. | 235 | 27.50 | | 262.50 | install 2' 4" x 6' 8" pre-hung 1-3/8" |
| Demolish and Install | Ea. | 235 | 70.50 | | 305.50 | masonite door with split pine jamb |
| Reinstall | Ea. | | 27.34 | | 27.34 | including casing, stops, and hinges. |
| Clean | Ea. | 1.47 | 14.80 | | 16.27 | |
| Paint | Ea. | 10.60 | 55 | | 65.60 | |
| Minimum Charge | Job | | 232 | | 232 | |
| **2' 8" x 6' 8"** | | | | | | |
| Demolish | Ea. | | 43 | | 43 | Cost includes material and labor to |
| Install | Ea. | 238 | 27.50 | | 265.50 | install 2' 8" x 6' 8" pre-hung 1-3/8" |
| Demolish and Install | Ea. | 238 | 70.50 | | 308.50 | masonite door with split pine jamb |
| Reinstall | Ea. | | 27.34 | | 27.34 | including casing, stops, and hinges. |
| Clean | Ea. | 1.47 | 14.80 | | 16.27 | |
| Paint | Ea. | 10.60 | 55 | | 65.60 | |
| Minimum Charge | Job | | 232 | | 232 | |
| **3' x 6' 8"** | | | | | | |
| Demolish | Ea. | | 43 | | 43 | Cost includes material and labor to |
| Install | Ea. | 252 | 29 | | 281 | install 3' x 6' 8" pre-hung 1-3/8" |
| Demolish and Install | Ea. | 252 | 72 | | 324 | masonite door with split pine jamb |
| Reinstall | Ea. | | 29.05 | | 29.05 | including casing, stops, and hinges. |
| Clean | Ea. | 1.47 | 14.80 | | 16.27 | |
| Paint | Ea. | 10.60 | 55 | | 65.60 | |
| Minimum Charge | Job | | 232 | | 232 | |

| Hollow Core Door | Unit | Material | Labor | Equip. | Total | Specification |
|---|---|---|---|---|---|---|
| **Pocket Type** | | | | | | |
| Demolish | Ea. | | 64.50 | | 64.50 | Cost includes material and labor to |
| Install | Ea. | 77 | 51.50 | | 128.50 | install 2' 8" x 6' 8" lauan door. |
| Demolish and Install | Ea. | 77 | 116 | | 193 | |
| Reinstall | Ea. | | 51.64 | | 51.64 | |
| Clean | Ea. | 1.47 | 14.80 | | 16.27 | |
| Paint | Ea. | 10.60 | 55 | | 65.60 | |
| Minimum Charge | Job | | 232 | | 232 | |
| **Stain** | | | | | | |
| Install | Ea. | 8.55 | 32 | | 40.55 | Includes labor and material to stain |
| Minimum Charge | Job | | 193 | | 193 | single door and trim on both sides. |
| **Hollow Core Door Only** | | | | | | |
| **1' 6" x 6' 8"** | | | | | | |
| Demolish | Ea. | | 9.20 | | 9.20 | Cost includes material and labor to |
| Install | Ea. | 75.50 | 24.50 | | 100 | install 1' 6" x 6' 8" lauan hollow core |
| Demolish and Install | Ea. | 75.50 | 33.70 | | 109.20 | door slab. |
| Reinstall | Ea. | | 24.46 | | 24.46 | |
| Clean | Ea. | .62 | 11.10 | | 11.72 | |
| Paint | Ea. | 4.92 | 48.50 | | 53.42 | |
| Minimum Charge | Job | | 232 | | 232 | |
| **2' x 6' 8"** | | | | | | |
| Demolish | Ea. | | 9.20 | | 9.20 | Cost includes material and labor to |
| Install | Ea. | 83 | 26 | | 109 | install 2' x 6' 8" lauan hollow core |
| Demolish and Install | Ea. | 83 | 35.20 | | 118.20 | door slab. |
| Reinstall | Ea. | | 25.82 | | 25.82 | |
| Clean | Ea. | .62 | 11.10 | | 11.72 | |
| Paint | Ea. | 4.92 | 48.50 | | 53.42 | |
| Minimum Charge | Job | | 232 | | 232 | |
| **2' 4" x 6' 8"** | | | | | | |
| Demolish | Ea. | | 9.20 | | 9.20 | Cost includes material and labor to |
| Install | Ea. | 92 | 26 | | 118 | install 2' 4" x 6' 8" lauan hollow core |
| Demolish and Install | Ea. | 92 | 35.20 | | 127.20 | door slab. |
| Reinstall | Ea. | | 25.82 | | 25.82 | |
| Clean | Ea. | .62 | 11.10 | | 11.72 | |
| Paint | Ea. | 4.92 | 48.50 | | 53.42 | |
| Minimum Charge | Job | | 232 | | 232 | |
| **2' 8" x 6' 8"** | | | | | | |
| Demolish | Ea. | | 9.20 | | 9.20 | Cost includes material and labor to |
| Install | Ea. | 94 | 26 | | 120 | install 2' 8" x 6' 8" lauan hollow core |
| Demolish and Install | Ea. | 94 | 35.20 | | 129.20 | door slab. |
| Reinstall | Ea. | | 25.82 | | 25.82 | |
| Clean | Ea. | .62 | 11.10 | | 11.72 | |
| Paint | Ea. | 4.92 | 48.50 | | 53.42 | |
| Minimum Charge | Job | | 232 | | 232 | |
| **3' x 6' 8"** | | | | | | |
| Demolish | Ea. | | 9.20 | | 9.20 | Cost includes material and labor to |
| Install | Ea. | 99.50 | 27.50 | | 127 | install 3' x 6' 8" lauan hollow core |
| Demolish and Install | Ea. | 99.50 | 36.70 | | 136.20 | door slab. |
| Reinstall | Ea. | | 27.34 | | 27.34 | |
| Clean | Ea. | .62 | 11.10 | | 11.72 | |
| Paint | Ea. | 4.92 | 48.50 | | 53.42 | |
| Minimum Charge | Job | | 232 | | 232 | |
| **Pocket Type** | | | | | | |
| Demolish | Ea. | | 9.20 | | 9.20 | Cost includes material and labor to |
| Install | Ea. | 77 | 51.50 | | 128.50 | install 2' 8" x 6' 8" lauan door. |
| Demolish and Install | Ea. | 77 | 60.70 | | 137.70 | |
| Reinstall | Ea. | | 51.64 | | 51.64 | |
| Clean | Ea. | .62 | 11.10 | | 11.72 | |
| Paint | Ea. | 4.92 | 48.50 | | 53.42 | |
| Minimum Charge | Job | | 232 | | 232 | |

193

| Hollow Core Door | | Unit | Material | Labor | Equip. | Total | Specification |
|---|---|---|---|---|---|---|---|
| **Stain** | | | | | | | |
| | Install | Ea. | 4.73 | 43 | | 47.73 | Includes labor and materials to stain a |
| | Minimum Charge | Job | | 193 | | 193 | flush style door on both sides by brush. |
| **Casing Trim** | | | | | | | |
| **Single Width** | | | | | | | |
| | Demolish | Ea. | | 8.20 | | 8.20 | Cost includes material and labor to |
| | Install | Opng. | 26.50 | 38.50 | | 65 | install 11/16" x 2-1/2" pine ranch |
| | Demolish and Install | Opng. | 26.50 | 46.70 | | 73.20 | style casing for one side of a standard |
| | Reinstall | Opng. | | 30.99 | | 30.99 | door opening. |
| | Clean | Opng. | .21 | 3.55 | | 3.76 | |
| | Paint | Ea. | 5 | 7.75 | | 12.75 | |
| | Minimum Charge | Job | | 232 | | 232 | |
| **Double Width** | | | | | | | |
| | Demolish | Ea. | | 9.75 | | 9.75 | Cost includes material and labor to |
| | Install | Opng. | 31 | 46.50 | | 77.50 | install 11/16" x 2-1/2" pine ranch |
| | Demolish and Install | Opng. | 31 | 56.25 | | 87.25 | style door casing for one side of a |
| | Reinstall | Opng. | | 37.18 | | 37.18 | double door opening. |
| | Clean | Opng. | .24 | 7.10 | | 7.34 | |
| | Paint | Ea. | 5 | 8.80 | | 13.80 | |
| | Minimum Charge | Job | | 232 | | 232 | |
| **Hardware** | | | | | | | |
| **Doorknob w / Lock** | | | | | | | |
| | Demolish | Ea. | | 18.45 | | 18.45 | Includes material and labor to install |
| | Install | Ea. | 40.50 | 29 | | 69.50 | privacy lockset. |
| | Demolish and Install | Ea. | 40.50 | 47.45 | | 87.95 | |
| | Reinstall | Ea. | | 23.24 | | 23.24 | |
| | Clean | Ea. | .14 | 11.10 | | 11.24 | |
| | Minimum Charge | Job | | 232 | | 232 | |
| **Doorknob** | | | | | | | |
| | Demolish | Ea. | | 18.45 | | 18.45 | Includes material and labor to install |
| | Install | Ea. | 42.50 | 29 | | 71.50 | residential passage lockset, keyless |
| | Demolish and Install | Ea. | 42.50 | 47.45 | | 89.95 | bored type. |
| | Reinstall | Ea. | | 23.24 | | 23.24 | |
| | Clean | Ea. | .14 | 11.10 | | 11.24 | |
| | Minimum Charge | Job | | 232 | | 232 | |
| **Deadbolt** | | | | | | | |
| | Demolish | Ea. | | 16.15 | | 16.15 | Includes material and labor to install a |
| | Install | Ea. | 69.50 | 33 | | 102.50 | deadbolt. |
| | Demolish and Install | Ea. | 69.50 | 49.15 | | 118.65 | |
| | Reinstall | Ea. | | 26.56 | | 26.56 | |
| | Clean | Ea. | .14 | 11.10 | | 11.24 | |
| | Minimum Charge | Job | | 232 | | 232 | |
| **Lever Handle** | | | | | | | |
| | Demolish | Ea. | | 18.45 | | 18.45 | Includes material and labor to install |
| | Install | Ea. | 51 | 46.50 | | 97.50 | residential passage lockset, keyless |
| | Demolish and Install | Ea. | 51 | 64.95 | | 115.95 | bored type with lever handle. |
| | Reinstall | Ea. | | 37.18 | | 37.18 | |
| | Clean | Ea. | .14 | 11.10 | | 11.24 | |
| | Minimum Charge | Job | | 232 | | 232 | |
| **Closer** | | | | | | | |
| | Demolish | Ea. | | 9.70 | | 9.70 | Includes material and labor to install |
| | Install | Ea. | 183 | 71.50 | | 254.50 | pneumatic light duty closer for interior |
| | Demolish and Install | Ea. | 183 | 81.20 | | 264.20 | type doors. |
| | Reinstall | Ea. | | 57.21 | | 57.21 | |
| | Clean | Ea. | .08 | 14.80 | | 14.88 | |
| | Minimum Charge | Job | | 232 | | 232 | |

# Finish Carpentry

## Hollow Core Door

| Hollow Core Door | | Unit | Material | Labor | Equip. | Total | Specification |
|---|---|---|---|---|---|---|---|
| **Push Plate** | | | | | | | |
| | Demolish | Ea. | | 14.35 | | 14.35 | Includes material and labor to install |
| | Install | Ea. | 28.50 | 38.50 | | 67 | bronze push-pull plate. |
| | Demolish and Install | Ea. | 28.50 | 52.85 | | 81.35 | |
| | Reinstall | Ea. | | 30.99 | | 30.99 | |
| | Clean | Ea. | .14 | 11.10 | | 11.24 | |
| | Minimum Charge | Job | | 232 | | 232 | |
| **Kickplate** | | | | | | | |
| | Demolish | Ea. | | 14.35 | | 14.35 | Includes material and labor to install |
| | Install | Ea. | 26 | 31 | | 57 | aluminum kickplate, 10" x 28". |
| | Demolish and Install | Ea. | 26 | 45.35 | | 71.35 | |
| | Reinstall | Ea. | | 24.79 | | 24.79 | |
| | Clean | Ea. | .14 | 11.10 | | 11.24 | |
| | Minimum Charge | Job | | 232 | | 232 | |
| **Exit Sign** | | | | | | | |
| | Demolish | Ea. | | 26.50 | | 26.50 | Includes material and labor to install a |
| | Install | Ea. | 104 | 65.50 | | 169.50 | wall mounted interior electric exit sign. |
| | Demolish and Install | Ea. | 104 | 92 | | 196 | |
| | Reinstall | Ea. | | 52.52 | | 52.52 | |
| | Clean | Ea. | .01 | 8.35 | | 8.36 | |
| | Minimum Charge | Job | | 232 | | 232 | |
| **Jamb** | | | | | | | |
| | Demolish | Ea. | | 32.50 | | 32.50 | Includes material and labor to install |
| | Install | Ea. | 116 | 12.55 | | 128.55 | flat pine jamb with square cut heads |
| | Demolish and Install | Ea. | 116 | 45.05 | | 161.05 | and rabbeted sides for 6' 8" high and |
| | Clean | Ea. | .04 | 8.90 | | 8.94 | 3-9/16" door including trim sets for |
| | Paint | Ea. | 7 | 26 | | 33 | both sides. |
| | Minimum Charge | Job | | 232 | | 232 | |
| **Shave & Refit** | | | | | | | |
| | Install | Ea. | | 38.50 | | 38.50 | Includes labor to shave and rework |
| | Minimum Charge | Job | | 232 | | 232 | door to fit opening at the job site. |

## Solid Core Door

| Solid Core Door | | Unit | Material | Labor | Equip. | Total | Specification |
|---|---|---|---|---|---|---|---|
| **Pre-hung Wood** | | | | | | | |
| **2' 4" x 6' 8"** | | | | | | | |
| | Demolish | Ea. | | 43 | | 43 | Cost includes material and labor to |
| | Install | Ea. | 261 | 27.50 | | 288.50 | install 2' 4" x 6' 8" pre-hung 1-3/4" |
| | Demolish and Install | Ea. | 261 | 70.50 | | 331.50 | door, 4-1/2" split jamb, including |
| | Reinstall | Ea. | | 27.34 | | 27.34 | casing and stop, hinges, aluminum sill, |
| | Clean | Ea. | 1.47 | 14.80 | | 16.27 | weatherstripping. |
| | Paint | Ea. | 11.20 | 55 | | 66.20 | |
| | Minimum Charge | Job | | 232 | | 232 | |
| **2' 6" x 6' 8"** | | | | | | | |
| | Demolish | Ea. | | 43 | | 43 | Cost includes material and labor to |
| | Install | Ea. | 252 | 27.50 | | 279.50 | install 2' 6" x 6' 8" pre-hung 1-3/4" |
| | Demolish and Install | Ea. | 252 | 70.50 | | 322.50 | door, 4-1/2" split jamb, including |
| | Reinstall | Ea. | | 27.34 | | 27.34 | casing and stop, hinges, aluminum sill, |
| | Clean | Ea. | 1.47 | 14.80 | | 16.27 | weatherstripping. |
| | Paint | Ea. | 11.20 | 55 | | 66.20 | |
| | Minimum Charge | Job | | 232 | | 232 | |

# Finish Carpentry

## Solid Core Door

| Solid Core Door | Unit | Material | Labor | Equip. | Total | Specification |
|---|---|---|---|---|---|---|
| **2' 8" x 6' 8"** | | | | | | |
| Demolish | Ea. | | 43 | | 43 | Cost includes material and labor to |
| Install | Ea. | 260 | 27.50 | | 287.50 | install 2' 8" x 6' 8" pre-hung 1-3/4" |
| Demolish and Install | Ea. | 260 | 70.50 | | 330.50 | door, 4-1/2" split jamb, including |
| Reinstall | Ea. | | 27.34 | | 27.34 | casing and stop, hinges, aluminum sill, |
| Clean | Ea. | 1.47 | 14.80 | | 16.27 | weatherstripping. |
| Paint | Ea. | 11.20 | 55 | | 66.20 | |
| Minimum Charge | Job | | 232 | | 232 | |
| **3' x 6' 8"** | | | | | | |
| Demolish | Ea. | | 43 | | 43 | Cost includes material and labor to |
| Install | Ea. | 264 | 29 | | 293 | install 3' x 6' 8" pre-hung 1-3/4" |
| Demolish and Install | Ea. | 264 | 72 | | 336 | door, including casing and stop, |
| Reinstall | Ea. | | 29.05 | | 29.05 | hinges, jamb, aluminum sill, |
| Clean | Ea. | 1.47 | 14.80 | | 16.27 | weatherstripped. |
| Paint | Ea. | 11.20 | 55 | | 66.20 | |
| Minimum Charge | Job | | 232 | | 232 | |
| **3' 6" x 6' 8"** | | | | | | |
| Demolish | Ea. | | 43 | | 43 | Cost includes material and labor to |
| Install | Ea. | 415 | 31 | | 446 | install 3' 6" x 6' 8" pre-hung 1-3/4" |
| Demolish and Install | Ea. | 415 | 74 | | 489 | door, 4-1/2" split jamb, including |
| Reinstall | Ea. | | 30.99 | | 30.99 | casing and stop, hinges, aluminum sill, |
| Clean | Ea. | 1.47 | 14.80 | | 16.27 | weatherstripping. |
| Paint | Ea. | 11.20 | 55 | | 66.20 | |
| Minimum Charge | Job | | 232 | | 232 | |
| **Raised Panel** | | | | | | |
| Demolish | Ea. | | 43 | | 43 | Cost includes material and labor to |
| Install | Ea. | 505 | 26 | | 531 | install 3' x 6' 8" pre-hung 1-3/4" |
| Demolish and Install | Ea. | 505 | 69 | | 574 | door, jamb, including casing and |
| Reinstall | Ea. | | 25.82 | | 25.82 | stop, hinges, aluminum sill, |
| Clean | Ea. | 1.47 | 14.80 | | 16.27 | weatherstripping. |
| Paint | Ea. | 11.20 | 129 | | 140.20 | |
| Minimum Charge | Job | | 232 | | 232 | |
| **Stain** | | | | | | |
| Install | Ea. | 8.55 | 32 | | 40.55 | Includes labor and material to stain |
| Minimum Charge | Job | | 193 | | 193 | single door and trim on both sides. |

## Bifold Door

| Bifold Door | Unit | Material | Labor | Equip. | Total | Specification |
|---|---|---|---|---|---|---|
| **Wood, Single** | | | | | | |
| **3'** | | | | | | |
| Demolish | Ea. | | 32.50 | | 32.50 | Cost includes material and labor to |
| Install | Ea. | 82 | 71.50 | | 153.50 | install 3' wide solid wood bi-fold door |
| Demolish and Install | Ea. | 82 | 104 | | 186 | not including jamb, trim, track, and |
| Reinstall | Ea. | | 71.51 | | 71.51 | hardware. |
| Clean | Ea. | 1.47 | 14.80 | | 16.27 | |
| Paint | Ea. | 10.60 | 55 | | 65.60 | |
| Minimum Charge | Job | | 232 | | 232 | |
| **5'** | | | | | | |
| Demolish | Ea. | | 32.50 | | 32.50 | Includes material and labor to install |
| Install | Ea. | 125 | 84.50 | | 209.50 | pair of 2' 6" wide solid wood bi-fold |
| Demolish and Install | Ea. | 125 | 117 | | 242 | doors not including jamb, trim, track, |
| Reinstall | Ea. | | 84.51 | | 84.51 | and hardware. |
| Clean | Ea. | .65 | 25.50 | | 26.15 | |
| Paint | Ea. | 34 | 96.50 | | 130.50 | |
| Minimum Charge | Job | | 232 | | 232 | |

## Bifold Door

| Bifold Door | | Unit | Material | Labor | Equip. | Total | Specification |
|---|---|---|---|---|---|---|---|
| **6'** | | | | | | | |
| | Demolish | Ea. | | 32.50 | | 32.50 | Includes material and labor to install |
| | Install | Ea. | 149 | 93 | | 242 | pair of 3' wide solid wood bi-fold |
| | Demolish and Install | Ea. | 149 | 125.50 | | 274.50 | doors not including jamb, trim, track, |
| | Reinstall | Ea. | | 92.96 | | 92.96 | and hardware. |
| | Clean | Ea. | .74 | 29.50 | | 30.24 | |
| | Paint | Ea. | 34 | 96.50 | | 130.50 | |
| | Minimum Charge | Job | | 232 | | 232 | |
| **8'** | | | | | | | |
| | Demolish | Ea. | | 32.50 | | 32.50 | Includes material and labor to install |
| | Install | Ea. | 223 | 103 | | 326 | pair of 4' wide solid wood bi-fold |
| | Demolish and Install | Ea. | 223 | 135.50 | | 358.50 | doors not including jamb, trim, track, |
| | Reinstall | Ea. | | 103.29 | | 103.29 | and hardware. |
| | Clean | Ea. | 1.01 | 35.50 | | 36.51 | |
| | Paint | Ea. | 17.75 | 96.50 | | 114.25 | |
| | Minimum Charge | Job | | 232 | | 232 | |

## Mirrored

| Mirrored | | Unit | Material | Labor | Equip. | Total | Specification |
|---|---|---|---|---|---|---|---|
| **6'** | | | | | | | |
| | Demolish | Ea. | | 32.50 | | 32.50 | Includes material and labor to install |
| | Install | Ea. | 445 | 51.50 | | 496.50 | pair of 3' wide mirrored bi-fold doors |
| | Demolish and Install | Ea. | 445 | 84 | | 529 | including jamb, trim, track, and |
| | Reinstall | Ea. | | 51.64 | | 51.64 | hardware. |
| | Clean | Ea. | .74 | 14.80 | | 15.54 | |
| | Minimum Charge | Job | | 232 | | 232 | |
| **8'** | | | | | | | |
| | Demolish | Ea. | | 32.50 | | 32.50 | Includes material and labor to install |
| | Install | Ea. | 730 | 77.50 | | 807.50 | pair of 4' wide mirrored bi-fold doors |
| | Demolish and Install | Ea. | 730 | 110 | | 840 | including jamb, trim, track, and |
| | Reinstall | Ea. | | 77.47 | | 77.47 | hardware. |
| | Clean | Ea. | 1.01 | 17.75 | | 18.76 | |
| | Minimum Charge | Job | | 232 | | 232 | |

## Louvered

| Louvered | | Unit | Material | Labor | Equip. | Total | Specification |
|---|---|---|---|---|---|---|---|
| **3'** | | | | | | | |
| | Demolish | Ea. | | 32.50 | | 32.50 | Cost includes material and labor to |
| | Install | Ea. | 263 | 71.50 | | 334.50 | install 3' wide louvered bi-fold door |
| | Demolish and Install | Ea. | 263 | 104 | | 367 | not including jamb, trim, track, and |
| | Reinstall | Ea. | | 71.51 | | 71.51 | hardware. |
| | Clean | Ea. | 1.47 | 14.80 | | 16.27 | |
| | Paint | Ea. | 10.60 | 55 | | 65.60 | |
| | Minimum Charge | Job | | 232 | | 232 | |
| **5'** | | | | | | | |
| | Demolish | Ea. | | 32.50 | | 32.50 | Includes material and labor to install |
| | Install | Ea. | 305 | 84.50 | | 389.50 | pair of 2' 6" wide louvered wood |
| | Demolish and Install | Ea. | 305 | 117 | | 422 | bi-fold doors not including jamb, trim, |
| | Reinstall | Ea. | | 84.51 | | 84.51 | track, and hardware. |
| | Clean | Ea. | .65 | 25.50 | | 26.15 | |
| | Paint | Ea. | 34 | 96.50 | | 130.50 | |
| | Minimum Charge | Job | | 232 | | 232 | |
| **6'** | | | | | | | |
| | Demolish | Ea. | | 32.50 | | 32.50 | Includes material and labor to install |
| | Install | Ea. | 340 | 93 | | 433 | pair of 3' wide louvered wood bi-fold |
| | Demolish and Install | Ea. | 340 | 125.50 | | 465.50 | doors not including jamb, trim, track, |
| | Reinstall | Ea. | | 92.96 | | 92.96 | and hardware. |
| | Clean | Ea. | .74 | 29.50 | | 30.24 | |
| | Paint | Ea. | 34 | 96.50 | | 130.50 | |
| | Minimum Charge | Job | | 232 | | 232 | |

| Bifold Door | Unit | Material | Labor | Equip. | Total | Specification |
|---|---|---|---|---|---|---|
| **8'** | | | | | | |
| Demolish | Ea. | | 32.50 | | 32.50 | Includes material and labor to install |
| Install | Ea. | 515 | 93 | | 608 | pair of 4' wide louvered wood bi-fold |
| Demolish and Install | Ea. | 515 | 125.50 | | 640.50 | doors not including jamb, trim, track, |
| Reinstall | Ea. | | 92.96 | | 92.96 | and hardware. |
| Clean | Ea. | 3.50 | 25.50 | | 29 | |
| Paint | Ea. | 17.75 | 96.50 | | 114.25 | |
| Minimum Charge | Job | | 232 | | 232 | |
| **Accordion Type** | | | | | | |
| Demolish | S.F. | | 1.16 | | 1.16 | Includes material and labor to install |
| Install | Ea. | 43 | 93 | | 136 | vinyl folding accordion closet doors, |
| Demolish and Install | Ea. | 43 | 94.16 | | 137.16 | including track and frame. |
| Reinstall | Ea. | | 92.96 | | 92.96 | |
| Clean | S.F. | | .44 | | .44 | |
| Minimum Charge | Job | | 232 | | 232 | |
| **Casing Trim** | | | | | | |
| **Single Width** | | | | | | |
| Demolish | Ea. | | 8.20 | | 8.20 | Cost includes material and labor to |
| Install | Opng. | 26.50 | 38.50 | | 65 | install 11/16" x 2-1/2" pine ranch |
| Demolish and Install | Opng. | 26.50 | 46.70 | | 73.20 | style casing for one side of a standard |
| Reinstall | Opng. | | 30.99 | | 30.99 | door opening. |
| Clean | Opng. | .21 | 3.55 | | 3.76 | |
| Paint | Ea. | 5 | 7.75 | | 12.75 | |
| Minimum Charge | Job | | 232 | | 232 | |
| **Double Width** | | | | | | |
| Demolish | Ea. | | 9.75 | | 9.75 | Cost includes material and labor to |
| Install | Opng. | 31 | 46.50 | | 77.50 | install 11/16" x 2-1/2" pine ranch |
| Demolish and Install | Opng. | 31 | 56.25 | | 87.25 | style door casing for one side of a |
| Reinstall | Opng. | | 37.18 | | 37.18 | double door opening. |
| Clean | Opng. | .24 | 7.10 | | 7.34 | |
| Paint | Ea. | 5 | 8.80 | | 13.80 | |
| Minimum Charge | Job | | 232 | | 232 | |
| **Jamb** | | | | | | |
| Demolish | Ea. | | 32.50 | | 32.50 | Includes material and labor to install |
| Install | Ea. | 116 | 12.55 | | 128.55 | flat pine jamb with square cut heads |
| Demolish and Install | Ea. | 116 | 45.05 | | 161.05 | and rabbeted sides for 6' 8" high and |
| Clean | Ea. | .04 | 8.90 | | 8.94 | 3-9/16" door including trim sets for |
| Paint | Ea. | 7 | 26 | | 33 | both sides. |
| Minimum Charge | Job | | 116 | | 116 | |
| **Stain** | | | | | | |
| Install | Ea. | 8.55 | 32 | | 40.55 | Includes labor and material to stain |
| Minimum Charge | Job | | 193 | | 193 | single door and trim on both sides. |
| **Shave & Refit** | | | | | | |
| Install | Ea. | | 38.50 | | 38.50 | Includes labor to shave and rework |
| Minimum Charge | Job | | 116 | | 116 | door to fit opening at the job site. |

| Louver Door | Unit | Material | Labor | Equip. | Total | Specification |
|---|---|---|---|---|---|---|
| **Full** | | | | | | |
| **2' x 6' 8"** | | | | | | |
| Demolish | Ea. | | 9.20 | | 9.20 | Cost includes material and labor to |
| Install | Ea. | 295 | 46.50 | | 341.50 | install 2' x 6' 8" pine full louver door, |
| Demolish and Install | Ea. | 295 | 55.70 | | 350.70 | jamb, hinges. |
| Reinstall | Ea. | | 46.48 | | 46.48 | |
| Clean | Ea. | 2.73 | 44.50 | | 47.23 | |
| Paint | Ea. | 10.60 | 55 | | 65.60 | |
| Minimum Charge | Job | | 232 | | 232 | |
| **2' 8" x 6' 8"** | | | | | | |
| Demolish | Ea. | | 9.20 | | 9.20 | Cost includes material and labor to |
| Install | Ea. | 345 | 46.50 | | 391.50 | install 2' 8" x 6' 8" pine full louver |
| Demolish and Install | Ea. | 345 | 55.70 | | 400.70 | door, jamb, hinges. |
| Reinstall | Ea. | | 46.48 | | 46.48 | |
| Clean | Ea. | 2.73 | 44.50 | | 47.23 | |
| Paint | Ea. | 10.60 | 55 | | 65.60 | |
| Minimum Charge | Job | | 232 | | 232 | |
| **3' x 6' 8"** | | | | | | |
| Demolish | Ea. | | 9.20 | | 9.20 | Cost includes material and labor to |
| Install | Ea. | 360 | 49 | | 409 | install 3' x 6' 8" pine full louver door, |
| Demolish and Install | Ea. | 360 | 58.20 | | 418.20 | jamb, hinges. |
| Reinstall | Ea. | | 48.93 | | 48.93 | |
| Clean | Ea. | 2.73 | 44.50 | | 47.23 | |
| Paint | Ea. | 10.60 | 55 | | 65.60 | |
| Minimum Charge | Job | | 232 | | 232 | |
| **Half** | | | | | | |
| **2' x 6' 8"** | | | | | | |
| Demolish | Ea. | | 9.20 | | 9.20 | Cost includes material and labor to |
| Install | Ea. | 395 | 46.50 | | 441.50 | install 2' x 6' 8" pine half louver door, |
| Demolish and Install | Ea. | 395 | 55.70 | | 450.70 | jamb, hinges. |
| Reinstall | Ea. | | 46.48 | | 46.48 | |
| Clean | Ea. | 2.73 | 44.50 | | 47.23 | |
| Paint | Ea. | 10.60 | 55 | | 65.60 | |
| Minimum Charge | Job | | 232 | | 232 | |
| **2' 8" x 6' 8"** | | | | | | |
| Demolish | Ea. | | 9.20 | | 9.20 | Cost includes material and labor to |
| Install | Ea. | 405 | 46.50 | | 451.50 | install 2' 8" x 6' 8" pine half louver |
| Demolish and Install | Ea. | 405 | 55.70 | | 460.70 | door, jamb, hinges. |
| Reinstall | Ea. | | 46.48 | | 46.48 | |
| Clean | Ea. | 2.73 | 44.50 | | 47.23 | |
| Paint | Ea. | 10.60 | 55 | | 65.60 | |
| Minimum Charge | Job | | 232 | | 232 | |
| **3' x 6' 8"** | | | | | | |
| Demolish | Ea. | | 9.20 | | 9.20 | Cost includes material and labor to |
| Install | Ea. | 430 | 49 | | 479 | install 3' x 6' 8" pine half louver door, |
| Demolish and Install | Ea. | 430 | 58.20 | | 488.20 | jamb, hinges. |
| Reinstall | Ea. | | 48.93 | | 48.93 | |
| Clean | Ea. | 2.73 | 44.50 | | 47.23 | |
| Paint | Ea. | 10.60 | 55 | | 65.60 | |
| Minimum Charge | Job | | 232 | | 232 | |
| **Jamb** | | | | | | |
| Demolish | Ea. | | 32.50 | | 32.50 | Includes material and labor to install |
| Install | Ea. | 116 | 12.55 | | 128.55 | flat pine jamb with square cut heads |
| Demolish and Install | Ea. | 116 | 45.05 | | 161.05 | and rabbeted sides for 6' 8" high and |
| Clean | Ea. | .04 | 8.90 | | 8.94 | 3-9/16" door including trim sets for |
| Paint | Ea. | 7 | 26 | | 33 | both sides. |
| Minimum Charge | Job | | 116 | | 116 | |

199

# Finish Carpentry

## Louver Door

| Louver Door | | Unit | Material | Labor | Equip. | Total | Specification |
|---|---|---|---|---|---|---|---|
| **Stain** | | | | | | | |
| | Install | Ea. | 8.55 | 32 | | 40.55 | Includes labor and material to stain |
| | Minimum Charge | Job | | 193 | | 193 | single door and trim on both sides. |
| **Shave & Refit** | | | | | | | |
| | Install | Ea. | | 38.50 | | 38.50 | Includes labor to shave and rework |
| | Minimum Charge | Job | | 193 | | 193 | door to fit opening at the job site. |

## Bypass Sliding Door

| Bypass Sliding Door | | Unit | Material | Labor | Equip. | Total | Specification |
|---|---|---|---|---|---|---|---|
| **Wood** | | | | | | | |
| **5'** | | | | | | | |
| | Demolish | Ea. | | 43 | | 43 | Includes material and labor to install |
| | Install | Opng. | 201 | 84.50 | | 285.50 | pair of 2' 6" wide lauan hollow core |
| | Demolish and Install | Opng. | 201 | 127.50 | | 328.50 | wood bi-pass doors with jamb, track |
| | Reinstall | Opng. | | 84.51 | | 84.51 | hardware. |
| | Clean | Ea. | .65 | 25.50 | | 26.15 | |
| | Paint | Ea. | 34 | 96.50 | | 130.50 | |
| | Minimum Charge | Job | | 232 | | 232 | |
| **6'** | | | | | | | |
| | Demolish | Ea. | | 43 | | 43 | Includes material and labor to install |
| | Install | Opng. | 186 | 93 | | 279 | pair of 3' wide lauan hollow core |
| | Demolish and Install | Opng. | 186 | 136 | | 322 | wood bi-pass doors with jamb, track |
| | Reinstall | Opng. | | 92.96 | | 92.96 | and hardware. |
| | Clean | Ea. | .74 | 29.50 | | 30.24 | |
| | Paint | Ea. | 34 | 96.50 | | 130.50 | |
| | Minimum Charge | Job | | 232 | | 232 | |
| **8'** | | | | | | | |
| | Demolish | Ea. | | 43 | | 43 | Includes material and labor to install |
| | Install | Opng. | 370 | 116 | | 486 | pair of 4' wide lauan hollow core |
| | Demolish and Install | Opng. | 370 | 159 | | 529 | wood bi-pass doors with jamb, track |
| | Reinstall | Opng. | | 116.20 | | 116.20 | and hardware. |
| | Clean | Ea. | 1.01 | 35.50 | | 36.51 | |
| | Paint | Ea. | 17.75 | 96.50 | | 114.25 | |
| | Minimum Charge | Job | | 232 | | 232 | |
| **Mirrored** | | | | | | | |
| **6'** | | | | | | | |
| | Demolish | Ea. | | 43 | | 43 | Includes material and labor to install |
| | Install | Opng. | 325 | 93 | | 418 | pair of 3' wide mirrored bi-pass doors |
| | Demolish and Install | Opng. | 325 | 136 | | 461 | with jamb, trim, track hardware. |
| | Reinstall | Opng. | | 92.96 | | 92.96 | |
| | Clean | Ea. | .74 | 14.80 | | 15.54 | |
| | Minimum Charge | Job | | 232 | | 232 | |
| **8'** | | | | | | | |
| | Demolish | Ea. | | 43 | | 43 | Includes material and labor to install |
| | Install | Opng. | 505 | 103 | | 608 | pair of 4' wide mirrored bi-pass doors |
| | Demolish and Install | Opng. | 505 | 146 | | 651 | with jamb, trim, track hardware. |
| | Reinstall | Opng. | | 103.29 | | 103.29 | |
| | Clean | Ea. | 1.01 | 17.75 | | 18.76 | |
| | Minimum Charge | Job | | 232 | | 232 | |

# Finish Carpentry

## Bypass Sliding Door

| Bypass Sliding Door | Unit | Material | Labor | Equip. | Total | Specification |
|---|---|---|---|---|---|---|
| **Casing Trim** | | | | | | |
| **Single Width** | | | | | | |
| Demolish | Ea. | | 8.20 | | 8.20 | Cost includes material and labor to |
| Install | Opng. | 26.50 | 38.50 | | 65 | install 11/16" x 2-1/2" pine ranch |
| Demolish and Install | Opng. | 26.50 | 46.70 | | 73.20 | style casing for one side of a standard |
| Reinstall | Opng. | | 30.99 | | 30.99 | door opening. |
| Clean | Opng. | .21 | 3.55 | | 3.76 | |
| Paint | Ea. | 5 | 7.75 | | 12.75 | |
| Minimum Charge | Job | | 232 | | 232 | |
| **Double Width** | | | | | | |
| Demolish | Ea. | | 9.75 | | 9.75 | Cost includes material and labor to |
| Install | Opng. | 31 | 46.50 | | 77.50 | install 11/16" x 2-1/2" pine ranch |
| Demolish and Install | Opng. | 31 | 56.25 | | 87.25 | style door casing for one side of a |
| Reinstall | Opng. | | 37.18 | | 37.18 | double door opening. |
| Clean | Opng. | .24 | 7.10 | | 7.34 | |
| Paint | Ea. | 5 | 8.80 | | 13.80 | |
| Minimum Charge | Job | | 232 | | 232 | |
| **Jamb** | | | | | | |
| Demolish | Ea. | | 32.50 | | 32.50 | Includes material and labor to install |
| Install | Ea. | 116 | 12.55 | | 128.55 | flat pine jamb with square cut heads |
| Demolish and Install | Ea. | 116 | 45.05 | | 161.05 | and rabbeted sides for 6' 8" high and |
| Clean | Ea. | .04 | 8.90 | | 8.94 | 3-9/16" door including trim sets for |
| Paint | Ea. | 7 | 26 | | 33 | both sides. |
| Minimum Charge | Job | | 116 | | 116 | |
| **Stain** | | | | | | |
| Install | Ea. | 8.55 | 32 | | 40.55 | Includes labor and material to stain |
| Minimum Charge | Job | | 193 | | 193 | single door and trim on both sides. |
| **Shave & Refit** | | | | | | |
| Install | Ea. | | 38.50 | | 38.50 | Includes labor to shave and rework |
| Minimum Charge | Job | | 116 | | 116 | door to fit opening at the job site. |

## Base Molding

| Base Molding | Unit | Material | Labor | Equip. | Total | Specification |
|---|---|---|---|---|---|---|
| **One Piece** | | | | | | |
| **1-5/8"** | | | | | | |
| Demolish | L.F. | | .66 | | .66 | Includes material and labor to install |
| Install | L.F. | .39 | 1.38 | | 1.77 | pine or birch base molding. |
| Demolish and Install | L.F. | .39 | 2.04 | | 2.43 | |
| Reinstall | L.F. | | 1.10 | | 1.10 | |
| Clean | L.F. | .01 | .21 | | .22 | |
| Paint | L.F. | .06 | .42 | | .48 | |
| Minimum Charge | Job | | 232 | | 232 | |
| **2-1/2"** | | | | | | |
| Demolish | L.F. | | .66 | | .66 | Cost includes material and labor to |
| Install | L.F. | .60 | 1.40 | | 2 | install 1/2" x 2-1/2" pine or birch |
| Demolish and Install | L.F. | .60 | 2.06 | | 2.66 | base molding. |
| Reinstall | L.F. | | 1.12 | | 1.12 | |
| Clean | L.F. | .01 | .21 | | .22 | |
| Paint | L.F. | .06 | .42 | | .48 | |
| Minimum Charge | Job | | 232 | | 232 | |

# Finish Carpentry

| Base Molding | | Unit | Material | Labor | Equip. | Total | Specification |
|---|---|---|---|---|---|---|---|
| **3-1/2"** | | | | | | | |
| | Demolish | L.F. | | .66 | | .66 | Cost includes material and labor to |
| | Install | L.F. | .90 | 1.49 | | 2.39 | install 9/16" x 3-1/2" pine or birch |
| | Demolish and Install | L.F. | .90 | 2.15 | | 3.05 | base molding. |
| | Reinstall | L.F. | | 1.19 | | 1.19 | |
| | Clean | L.F. | .01 | .21 | | .22 | |
| | Paint | L.F. | .08 | .46 | | .54 | |
| | Minimum Charge | Job | | 232 | | 232 | |
| **6"** | | | | | | | |
| | Demolish | L.F. | | .66 | | .66 | Cost includes material and labor to |
| | Install | L.F. | .55 | 1.49 | | 2.04 | install 1" x 6" pine or birch base |
| | Demolish and Install | L.F. | .55 | 2.15 | | 2.70 | molding. |
| | Reinstall | L.F. | | 1.19 | | 1.19 | |
| | Clean | L.F. | .01 | .30 | | .31 | |
| | Paint | L.F. | .10 | .58 | | .68 | |
| | Minimum Charge | Job | | 232 | | 232 | |
| **8"** | | | | | | | |
| | Demolish | L.F. | | .79 | | .79 | Cost includes material and labor to |
| | Install | L.F. | .72 | 1.49 | | 2.21 | install 1" x 8" pine or birch base |
| | Demolish and Install | L.F. | .72 | 2.28 | | 3 | molding. |
| | Reinstall | L.F. | | 1.19 | | 1.19 | |
| | Clean | L.F. | .02 | .36 | | .38 | |
| | Paint | L.F. | .11 | .64 | | .75 | |
| | Minimum Charge | Job | | 232 | | 232 | |
| **Premium / Custom Grade** | | | | | | | |
| | Demolish | L.F. | | .79 | | .79 | Includes material and labor to install |
| | Install | L.F. | 4.58 | 4.47 | | 9.05 | custom 2 or 3 piece oak or other |
| | Demolish and Install | L.F. | 4.58 | 5.26 | | 9.84 | hardwood, stain grade trim. |
| | Reinstall | L.F. | | 3.58 | | 3.58 | |
| | Clean | L.F. | .02 | .36 | | .38 | |
| | Paint | L.F. | .12 | .64 | | .76 | |
| | Minimum Charge | Job | | 232 | | 232 | |
| **One Piece Oak** | | | | | | | |
| **1-5/8"** | | | | | | | |
| | Demolish | L.F. | | .66 | | .66 | Cost includes material and labor to |
| | Install | L.F. | .95 | 1.38 | | 2.33 | install 1/2" x 1-5/8" oak base |
| | Demolish and Install | L.F. | .95 | 2.04 | | 2.99 | molding. |
| | Reinstall | L.F. | | 1.10 | | 1.10 | |
| | Clean | L.F. | .01 | .21 | | .22 | |
| | Paint | L.F. | .13 | .57 | | .70 | |
| | Minimum Charge | Job | | 232 | | 232 | |
| **2-1/2"** | | | | | | | |
| | Demolish | L.F. | | .66 | | .66 | Cost includes material and labor to |
| | Install | L.F. | 1.81 | 1.40 | | 3.21 | install 1/2" x 2-1/2" oak base |
| | Demolish and Install | L.F. | 1.81 | 2.06 | | 3.87 | molding. |
| | Reinstall | L.F. | | 1.12 | | 1.12 | |
| | Clean | L.F. | .01 | .21 | | .22 | |
| | Paint | L.F. | .13 | .59 | | .72 | |
| | Minimum Charge | Job | | 232 | | 232 | |
| **3-1/2"** | | | | | | | |
| | Demolish | L.F. | | .66 | | .66 | Cost includes material and labor to |
| | Install | L.F. | 2.14 | 1.43 | | 3.57 | install 1/2" x 3-1/2" oak base |
| | Demolish and Install | L.F. | 2.14 | 2.09 | | 4.23 | molding. |
| | Reinstall | L.F. | | 1.14 | | 1.14 | |
| | Clean | L.F. | .01 | .21 | | .22 | |
| | Paint | L.F. | .14 | .59 | | .73 | |
| | Minimum Charge | Job | | 232 | | 232 | |

**For customer support on your Contractor's Pricing Guide: Residential Repair & Remodeling, call 800.448.8182.**

| Base Molding | Unit | Material | Labor | Equip. | Total | Specification |
|---|---|---|---|---|---|---|
| **6"** | | | | | | |
| Demolish | L.F. | | .66 | | .66 | Cost includes material and labor to |
| Install | L.F. | 3.19 | 1.49 | | 4.68 | install 1" x 6" oak base molding. |
| Demolish and Install | L.F. | 3.19 | 2.15 | | 5.34 | |
| Reinstall | L.F. | | 1.19 | | 1.19 | |
| Clean | L.F. | .02 | .36 | | .38 | |
| Paint | L.F. | .14 | .62 | | .76 | |
| Minimum Charge | Job | | 232 | | 232 | |
| **8"** | | | | | | |
| Demolish | L.F. | | .79 | | .79 | Cost includes material and labor to |
| Install | L.F. | 3.97 | 1.49 | | 5.46 | install 1" x 8" oak base molding. |
| Demolish and Install | L.F. | 3.97 | 2.28 | | 6.25 | |
| Reinstall | L.F. | | 1.19 | | 1.19 | |
| Clean | L.F. | .02 | .36 | | .38 | |
| Paint | L.F. | .15 | .64 | | .79 | |
| Minimum Charge | Job | | 232 | | 232 | |
| **Vinyl Base Molding** | | | | | | |
| **2-1/2"** | | | | | | |
| Demolish | L.F. | | .66 | | .66 | Cost includes material and labor to |
| Install | L.F. | 1.41 | 1.39 | | 2.80 | install 2-1/2" vinyl or rubber cove |
| Demolish and Install | L.F. | 1.41 | 2.05 | | 3.46 | base including adhesive. |
| Reinstall | L.F. | | 1.39 | | 1.39 | |
| Clean | L.F. | | .32 | | .32 | |
| Minimum Charge | Job | | 223 | | 223 | |
| **4"** | | | | | | |
| Demolish | L.F. | | .66 | | .66 | Cost includes material and labor to |
| Install | Ea. | 2.71 | 1.39 | | 4.10 | install 4" vinyl or rubber cove base |
| Demolish and Install | Ea. | 2.71 | 2.05 | | 4.76 | including adhesive. |
| Reinstall | Ea. | | 1.39 | | 1.39 | |
| Clean | L.F. | | .32 | | .32 | |
| Minimum Charge | Job | | 223 | | 223 | |
| **6"** | | | | | | |
| Demolish | L.F. | | .66 | | .66 | Cost includes material and labor to |
| Install | L.F. | 1.31 | 1.39 | | 2.70 | install 6" vinyl or rubber cove base |
| Demolish and Install | L.F. | 1.31 | 2.05 | | 3.36 | including adhesive. |
| Reinstall | L.F. | | 1.39 | | 1.39 | |
| Clean | L.F. | | .32 | | .32 | |
| Minimum Charge | Job | | 223 | | 223 | |
| **Prefinished** | | | | | | |
| Demolish | L.F. | | .66 | | .66 | Cost includes material and labor to |
| Install | L.F. | 1.32 | 1.49 | | 2.81 | install 9/16" x 3-5/16" prefinished |
| Demolish and Install | L.F. | 1.32 | 2.15 | | 3.47 | base molding. |
| Reinstall | L.F. | | 1.19 | | 1.19 | |
| Clean | L.F. | .01 | .21 | | .22 | |
| Paint | L.F. | .08 | .46 | | .54 | |
| Minimum Charge | Job | | 232 | | 232 | |
| **Shoe** | | | | | | |
| Demolish | L.F. | | .66 | | .66 | Includes material and labor to install |
| Install | L.F. | .39 | 1.38 | | 1.77 | pine or birch base molding. |
| Demolish and Install | L.F. | .39 | 2.04 | | 2.43 | |
| Reinstall | L.F. | | 1.10 | | 1.10 | |
| Clean | L.F. | .01 | .21 | | .22 | |
| Paint | L.F. | .17 | .62 | | .79 | |
| Minimum Charge | Job | | 232 | | 232 | |

# Finish Carpentry

## Base Molding

| Base Molding | Unit | Material | Labor | Equip. | Total | Specification |
|---|---|---|---|---|---|---|
| **Carpeted** | | | | | | |
| Demolish | L.F. | | .26 | | .26 | Includes material and labor to install |
| Install | L.F. | .99 | 2.65 | | 3.64 | carpet baseboard with adhesive. |
| Demolish and Install | L.F. | .99 | 2.91 | | 3.90 | |
| Reinstall | L.F. | | 2.12 | | 2.12 | |
| Clean | L.F. | | .13 | | .13 | |
| Minimum Charge | Job | | 219 | | 219 | |
| **Ceramic Tile** | | | | | | |
| Demolish | L.F. | | 1.44 | | 1.44 | Includes material and labor to install |
| Install | L.F. | 4.61 | 6.05 | | 10.66 | thin set ceramic tile cove base |
| Demolish and Install | L.F. | 4.61 | 7.49 | | 12.10 | including grout. |
| Clean | L.F. | | .68 | | .68 | |
| Minimum Charge | Job | | 219 | | 219 | |
| **Quarry Tile** | | | | | | |
| Demolish | L.F. | | 1.44 | | 1.44 | Includes material and labor to install |
| Install | L.F. | 7.05 | 7.05 | | 14.10 | quarry tile base including grout. |
| Demolish and Install | L.F. | 7.05 | 8.49 | | 15.54 | |
| Clean | L.F. | | .68 | | .68 | |
| Minimum Charge | Job | | 219 | | 219 | |
| **Caulk and Renail** | | | | | | |
| Install | L.F. | .01 | .73 | | .74 | Includes labor and material to renail |
| Minimum Charge | Job | | 232 | | 232 | base and caulk where necessary. |

## Molding

| Molding | Unit | Material | Labor | Equip. | Total | Specification |
|---|---|---|---|---|---|---|
| **Oak Base** | | | | | | |
| Demolish | L.F. | | .66 | | .66 | Cost includes material and labor to |
| Install | L.F. | 1.81 | 1.40 | | 3.21 | install 1/2" x 2-1/2" oak base |
| Demolish and Install | L.F. | 1.81 | 2.06 | | 3.87 | molding. |
| Reinstall | L.F. | | 1.12 | | 1.12 | |
| Clean | L.F. | .01 | .21 | | .22 | |
| Paint | L.F. | .13 | .59 | | .72 | |
| Minimum Charge | Job | | 232 | | 232 | |
| **Shoe** | | | | | | |
| Demolish | L.F. | | .66 | | .66 | Includes material and labor to install |
| Install | L.F. | .39 | 1.38 | | 1.77 | pine or birch base molding. |
| Demolish and Install | L.F. | .39 | 2.04 | | 2.43 | |
| Reinstall | L.F. | | 1.10 | | 1.10 | |
| Clean | L.F. | .01 | .21 | | .22 | |
| Paint | L.F. | .17 | .62 | | .79 | |
| Minimum Charge | Job | | 232 | | 232 | |
| **Quarter Round** | | | | | | |
| Demolish | L.F. | | .66 | | .66 | Includes material and labor to install |
| Install | L.F. | .57 | 1.82 | | 2.39 | pine base quarter round molding. |
| Demolish and Install | L.F. | .57 | 2.48 | | 3.05 | |
| Reinstall | L.F. | | 1.46 | | 1.46 | |
| Clean | L.F. | .01 | .21 | | .22 | |
| Paint | L.F. | .12 | .97 | | 1.09 | |
| Minimum Charge | Job | | 232 | | 232 | |
| **Chair Rail** | | | | | | |
| Demolish | L.F. | | .66 | | .66 | Cost includes material and labor to |
| Install | L.F. | 1.69 | 1.72 | | 3.41 | install 5/8" x 2-1/2" oak chair rail |
| Demolish and Install | L.F. | 1.69 | 2.38 | | 4.07 | molding. |
| Reinstall | L.F. | | 1.38 | | 1.38 | |
| Clean | L.F. | .01 | .21 | | .22 | |
| Paint | L.F. | .12 | .97 | | 1.09 | |
| Minimum Charge | Job | | 232 | | 232 | |

# Finish Carpentry

## Molding

| | Unit | Material | Labor | Equip. | Total | Specification |
|---|---|---|---|---|---|---|
| **Crown** | | | | | | |
| Demolish | L.F. | | .66 | | .66 | Includes material and labor to install |
| Install | L.F. | 2.31 | 1.86 | | 4.17 | unfinished pine crown molding. |
| Demolish and Install | L.F. | 2.31 | 2.52 | | 4.83 | |
| Reinstall | L.F. | | 1.49 | | 1.49 | |
| Clean | L.F. | .01 | .27 | | .28 | |
| Paint | L.F. | .12 | .97 | | 1.09 | |
| Minimum Charge | Job | | 232 | | 232 | |
| **Cove** | | | | | | |
| Demolish | L.F. | | .66 | | .66 | Cost includes material and labor to |
| Install | L.F. | 2.06 | 1.75 | | 3.81 | install 11/16" x 2-3/4" pine cove |
| Demolish and Install | L.F. | 2.06 | 2.41 | | 4.47 | molding. |
| Reinstall | L.F. | | 1.40 | | 1.40 | |
| Clean | L.F. | .01 | .21 | | .22 | |
| Paint | L.F. | .12 | .97 | | 1.09 | |
| Minimum Charge | Job | | 232 | | 232 | |
| **Corner** | | | | | | |
| Demolish | L.F. | | .66 | | .66 | Includes material and labor to install |
| Install | L.F. | 1.80 | .73 | | 2.53 | inside or outside pine corner molding. |
| Demolish and Install | L.F. | 1.80 | 1.39 | | 3.19 | |
| Reinstall | L.F. | | .58 | | .58 | |
| Clean | L.F. | .01 | .21 | | .22 | |
| Paint | L.F. | .12 | .97 | | 1.09 | |
| Minimum Charge | Job | | 232 | | 232 | |
| **Picture** | | | | | | |
| Demolish | L.F. | | .66 | | .66 | Cost includes material and labor to |
| Install | L.F. | 1.55 | 1.94 | | 3.49 | install 9/16" x 2-1/2" pine casing. |
| Demolish and Install | L.F. | 1.55 | 2.60 | | 4.15 | |
| Reinstall | L.F. | | 1.55 | | 1.55 | |
| Clean | L.F. | .01 | .21 | | .22 | |
| Paint | L.F. | .12 | .97 | | 1.09 | |
| Minimum Charge | Job | | 232 | | 232 | |
| **Single Cased Opening** | | | | | | |
| Demolish | Ea. | | 32.50 | | 32.50 | Includes material and labor to install |
| Install | Ea. | 26 | 29 | | 55 | flat pine jamb with square cut heads |
| Demolish and Install | Ea. | 26 | 61.50 | | 87.50 | and rabbeted sides for 6' 8" high |
| Clean | Ea. | .10 | 7.40 | | 7.50 | opening including trim sets for both |
| Paint | Ea. | 1.55 | 9.65 | | 11.20 | sides. |
| Minimum Charge | Job | | 232 | | 232 | |
| **Double Cased Opening** | | | | | | |
| Demolish | Ea. | | 32.50 | | 32.50 | Includes material and labor to install |
| Install | Ea. | 51 | 33 | | 84 | flat pine jamb with square cut heads |
| Demolish and Install | Ea. | 51 | 65.50 | | 116.50 | and rabbeted sides for 6' 8" high |
| Clean | Ea. | .12 | 8.65 | | 8.77 | opening including trim sets for both |
| Paint | Ea. | 1.80 | 11.35 | | 13.15 | sides. |
| Minimum Charge | Job | | 232 | | 232 | |

## Window Trim Set

| | Unit | Material | Labor | Equip. | Total | Specification |
|---|---|---|---|---|---|---|
| **Single** | | | | | | |
| Demolish | Ea. | | 8.75 | | 8.75 | Cost includes material and labor to |
| Install | Opng. | 42 | 46.50 | | 88.50 | install 11/16" x 2-1/2" pine ranch |
| Demolish and Install | Opng. | 42 | 55.25 | | 97.25 | style window casing. |
| Reinstall | Opng. | | 37.18 | | 37.18 | |
| Clean | Ea. | .14 | 5.05 | | 5.19 | |
| Paint | Ea. | 1.45 | 9.20 | | 10.65 | |
| Minimum Charge | Job | | 232 | | 232 | |

For customer support on your Contractor's Pricing Guide: Residential Repair & Remodeling, call 800.448.8182.

# Finish Carpentry

## Window Trim Set

| | Unit | Material | Labor | Equip. | Total | Specification |
|---|---|---|---|---|---|---|
| **Double** | | | | | | |
| Demolish | Ea. | | 9.85 | | 9.85 | Cost includes material and labor to |
| Install | Ea. | 56 | 58 | | 114 | install 11/16" x 2-1/2" pine ranch |
| Demolish and Install | Ea. | 56 | 67.85 | | 123.85 | style trim for casing one side of a |
| Reinstall | Ea. | | 46.48 | | 46.48 | double window opening. |
| Clean | Ea. | .26 | 5.90 | | 6.16 | |
| Paint | Ea. | 2.44 | 15.45 | | 17.89 | |
| Minimum Charge | Job | | 232 | | 232 | |
| **Triple** | | | | | | |
| Demolish | Ea. | | 11.30 | | 11.30 | Cost includes material and labor to |
| Install | Opng. | 84.50 | 77.50 | | 162 | install 11/16" x 2-1/2" pine ranch |
| Demolish and Install | Opng. | 84.50 | 88.80 | | 173.30 | style trim for casing one side of a |
| Reinstall | Opng. | | 61.97 | | 61.97 | triple window opening. |
| Clean | Ea. | .36 | 7.10 | | 7.46 | |
| Paint | Ea. | 3.42 | 21.50 | | 24.92 | |
| Minimum Charge | Job | | 232 | | 232 | |
| **Window Casing Per L.F.** | | | | | | |
| Demolish | L.F. | | .66 | | .66 | Cost includes material and labor to |
| Install | L.F. | 1.56 | 1.86 | | 3.42 | install 11/16" x 2-1/2" pine ranch |
| Demolish and Install | L.F. | 1.56 | 2.52 | | 4.08 | style trim for casing. |
| Reinstall | L.F. | | 1.49 | | 1.49 | |
| Clean | L.F. | .01 | .21 | | .22 | |
| Paint | L.F. | .12 | .97 | | 1.09 | |
| Minimum Charge | Job | | 232 | | 232 | |

## Window Sill

| | Unit | Material | Labor | Equip. | Total | Specification |
|---|---|---|---|---|---|---|
| **Wood** | | | | | | |
| Demolish | L.F. | | .66 | | .66 | Includes material and labor to install |
| Install | L.F. | 2.39 | 2.32 | | 4.71 | flat, wood window stool. |
| Demolish and Install | L.F. | 2.39 | 2.98 | | 5.37 | |
| Reinstall | L.F. | | 1.86 | | 1.86 | |
| Clean | L.F. | .01 | .21 | | .22 | |
| Paint | L.F. | .12 | .97 | | 1.09 | |
| Minimum Charge | Job | | 232 | | 232 | |
| **Marble** | | | | | | |
| Demolish | L.F. | | .99 | | .99 | Includes material and labor to install |
| Install | L.F. | 12.20 | 9.95 | | 22.15 | marble window sills, 6" x 3/4" thick. |
| Demolish and Install | L.F. | 12.20 | 10.94 | | 23.14 | |
| Reinstall | L.F. | | 7.97 | | 7.97 | |
| Clean | L.F. | .01 | .21 | | .22 | |
| Minimum Charge | Job | | 219 | | 219 | |

## Shelving

| | Unit | Material | Labor | Equip. | Total | Specification |
|---|---|---|---|---|---|---|
| **Pine** | | | | | | |
| **18"** | | | | | | |
| Demolish | L.F. | | 1.13 | | 1.13 | Cost includes material and labor to |
| Install | L.F. | 8.70 | 4.89 | | 13.59 | install 1" thick custom shelving board |
| Demolish and Install | L.F. | 8.70 | 6.02 | | 14.72 | 18" wide. |
| Reinstall | L.F. | | 3.91 | | 3.91 | |
| Clean | L.F. | .02 | .59 | | .61 | |
| Paint | L.F. | 1.51 | .61 | | 2.12 | |
| Minimum Charge | Job | | 232 | | 232 | |

# Finish Carpentry

## Shelving

| Shelving | | Unit | Material | Labor | Equip. | Total | Specification |
|---|---|---|---|---|---|---|---|
| **24"** | | | | | | | |
| | Demolish | L.F. | | 1.13 | | 1.13 | Cost includes material and labor to |
| | Install | L.F. | 11.60 | 5.45 | | 17.05 | install 1" thick custom shelving board. |
| | Demolish and Install | L.F. | 11.60 | 6.58 | | 18.18 | |
| | Reinstall | L.F. | | 4.37 | | 4.37 | |
| | Clean | L.F. | .03 | .89 | | .92 | |
| | Paint | L.F. | 2.05 | .81 | | 2.86 | |
| | Minimum Charge | Job | | 232 | | 232 | |
| **Particle Board** | | | | | | | |
| | Demolish | L.F. | | 1.04 | | 1.04 | Cost includes material and labor to |
| | Install | L.F. | 1.41 | 2.45 | | 3.86 | install 3/4" thick particle shelving |
| | Demolish and Install | L.F. | 1.41 | 3.49 | | 4.90 | board 18" wide. |
| | Reinstall | L.F. | | 1.96 | | 1.96 | |
| | Clean | L.F. | .02 | .59 | | .61 | |
| | Paint | S.F. | .14 | .90 | | 1.04 | |
| | Minimum Charge | Job | | 232 | | 232 | |
| **Plywood** | | | | | | | |
| | Demolish | L.F. | | 1.04 | | 1.04 | Cost includes material and labor to |
| | Install | L.F. | 1.06 | 7.15 | | 8.21 | install 3/4" thick plywood shelving |
| | Demolish and Install | L.F. | 1.06 | 8.19 | | 9.25 | board. |
| | Reinstall | L.F. | | 5.72 | | 5.72 | |
| | Clean | L.F. | .02 | .59 | | .61 | |
| | Paint | S.F. | .14 | .90 | | 1.04 | |
| | Minimum Charge | Job | | 232 | | 232 | |
| **Hardwood (oak)** | | | | | | | |
| | Demolish | L.F. | | 1.04 | | 1.04 | Includes material and labor to install |
| | Install | L.F. | 6.95 | 3.21 | | 10.16 | oak shelving, 1" x 12", incl. cleats and |
| | Demolish and Install | L.F. | 6.95 | 4.25 | | 11.20 | bracing. |
| | Reinstall | L.F. | | 2.56 | | 2.56 | |
| | Clean | L.F. | .02 | .59 | | .61 | |
| | Paint | S.F. | .14 | .90 | | 1.04 | |
| | Minimum Charge | Job | | 232 | | 232 | |
| **Custom Bookcase** | | | | | | | |
| | Demolish | L.F. | | 13.15 | | 13.15 | Includes material and labor to install |
| | Install | L.F. | 44.50 | 46.50 | | 91 | custom modular bookcase unit with |
| | Demolish and Install | L.F. | 44.50 | 59.65 | | 104.15 | clear pine faced frames, shelves 12" |
| | Reinstall | L.F. | | 37.18 | | 37.18 | O.C., 7' high, 8" deep. |
| | Clean | L.F. | .59 | 12.70 | | 13.29 | |
| | Paint | L.F. | 1.73 | 29.50 | | 31.23 | |
| | Minimum Charge | Job | | 232 | | 232 | |
| **Glass** | | | | | | | |
| | Demolish | L.F. | | 1.04 | | 1.04 | Includes material and labor to install |
| | Install | L.F. | 27 | 3.21 | | 30.21 | glass shelving, 1" x 12", incl. cleats |
| | Demolish and Install | L.F. | 27 | 4.25 | | 31.25 | and bracing. |
| | Reinstall | L.F. | | 2.56 | | 2.56 | |
| | Clean | L.F. | .02 | .55 | | .57 | |
| | Minimum Charge | Job | | 225 | | 225 | |

## Closet Shelving

| Closet Shelving | | Unit | Material | Labor | Equip. | Total | Specification |
|---|---|---|---|---|---|---|---|
| **Closet Shelf and Rod** | | | | | | | |
| | Demolish | L.F. | | 1.04 | | 1.04 | Cost includes material and labor to |
| | Install | L.F. | 3.31 | 7.75 | | 11.06 | install 1" thick custom shelving board |
| | Demolish and Install | L.F. | 3.31 | 8.79 | | 12.10 | 18" wide and 1" diameter clothes pole |
| | Reinstall | L.F. | | 6.20 | | 6.20 | with brackets 3' O.C. |
| | Clean | L.F. | .01 | .44 | | .45 | |
| | Paint | L.F. | .23 | .97 | | 1.20 | |
| | Minimum Charge | Job | | 232 | | 232 | |

207

# Finish Carpentry

## Closet Shelving

| | Unit | Material | Labor | Equip. | Total | Specification |
|---|---|---|---|---|---|---|
| **Clothing Rod** | | | | | | |
| Demolish | L.F. | | .69 | | .69 | Includes material and labor to install |
| Install | L.F. | 2.26 | 2.32 | | 4.58 | fir closet pole, 1-5/8" diameter. |
| Demolish and Install | L.F. | 2.26 | 3.01 | | 5.27 | |
| Reinstall | L.F. | | 1.86 | | 1.86 | |
| Clean | L.F. | .01 | .60 | | .61 | |
| Paint | L.F. | .22 | 1.07 | | 1.29 | |
| Minimum Charge | Job | | 232 | | 232 | |

## Stair Assemblies

| | Unit | Material | Labor | Equip. | Total | Specification |
|---|---|---|---|---|---|---|
| **Straight Hardwood** | | | | | | |
| Demolish | Ea. | | 455 | | 455 | Includes material and labor to install |
| Install | Flight | 1325 | 310 | | 1635 | factory cut and assembled straight |
| Demolish and Install | Flight | 1325 | 765 | | 2090 | closed box stairs with oak treads and |
| Clean | Flight | 1.47 | 30 | | 31.47 | prefinished stair rail with balusters. |
| Paint | Flight | 21 | 51.50 | | 72.50 | |
| Minimum Charge | Job | | 232 | | 232 | |
| **Clear Oak Tread** | | | | | | |
| Demolish | Ea. | | 11.75 | | 11.75 | Includes material and labor to install |
| Install | Ea. | 106 | 26 | | 132 | 1-1/4" thick clear oak treads per riser. |
| Demolish and Install | Ea. | 106 | 37.75 | | 143.75 | |
| Clean | Ea. | .04 | 2.99 | | 3.03 | |
| Paint | Ea. | .70 | 2.86 | | 3.56 | |
| Minimum Charge | Job | | 232 | | 232 | |
| **Landing** | | | | | | |
| Demolish | S.F. | | 4.11 | | 4.11 | Includes material and labor to install |
| Install | S.F. | 12.80 | 26 | | 38.80 | clear oak landing. |
| Demolish and Install | S.F. | 12.80 | 30.11 | | 42.91 | |
| Clean | S.F. | | .44 | | .44 | |
| Paint | S.F. | .48 | .95 | | 1.43 | |
| Minimum Charge | Job | | 232 | | 232 | |
| **Refinish / Stain** | | | | | | |
| Install | S.F. | .23 | 2.73 | | 2.96 | Includes labor and material to refinish |
| Minimum Charge | Job | | 193 | | 193 | wood floor. |
| **Spiral Hardwood** | | | | | | |
| Demolish | Ea. | | 455 | | 455 | Cost includes material and labor to |
| Install | Flight | 3025 | 620 | | 3645 | install 4' - 6' diameter spiral stairs with |
| Demolish and Install | Flight | 3025 | 1075 | | 4100 | oak treads, factory cut and assembled |
| Clean | Flight | 1.47 | 30 | | 31.47 | with double handrails. |
| Paint | Flight | 21 | 51.50 | | 72.50 | |
| Minimum Charge | Job | | 232 | | 232 | |

## Stairs

| | Unit | Material | Labor | Equip. | Total | Specification |
|---|---|---|---|---|---|---|
| **Disappearing** | | | | | | |
| Demolish | Ea. | | 24.50 | | 24.50 | Includes material and labor to install a |
| Install | Ea. | 267 | 133 | | 400 | folding staircase. |
| Demolish and Install | Ea. | 267 | 157.50 | | 424.50 | |
| Reinstall | Ea. | | 106.24 | | 106.24 | |
| Clean | Ea. | .22 | 22 | | 22.22 | |
| Minimum Charge | Job | | 232 | | 232 | |

# Finish Carpentry

## Stair Components

| | Unit | Material | Labor | Equip. | Total | Specification |
|---|---|---|---|---|---|---|
| **Handrail w / Balusters** | | | | | | |
| Demolish | L.F. | | 3.29 | | 3.29 | Includes material and labor to install |
| Install | L.F. | 51 | 9.70 | | 60.70 | prefinished assembled stair rail with |
| Demolish and Install | L.F. | 51 | 12.99 | | 63.99 | brackets, turned balusters and newel. |
| Clean | L.F. | .30 | 2.45 | | 2.75 | |
| Paint | L.F. | 2.46 | 4.55 | | 7.01 | |
| Minimum Charge | Job | | 232 | | 232 | |
| **Banister** | | | | | | |
| Demolish | L.F. | | 3.29 | | 3.29 | Includes material and labor to install |
| Install | L.F. | 42 | 7.75 | | 49.75 | built-up oak railings. |
| Demolish and Install | L.F. | 42 | 11.04 | | 53.04 | |
| Clean | L.F. | .30 | 2.22 | | 2.52 | |
| Paint | L.F. | 2.46 | 3.51 | | 5.97 | |
| Minimum Charge | Job | | 232 | | 232 | |

For customer support on your Contractor's Pricing Guide: Residential Repair & Remodeling, call 800.448.8182.

**Wall and Base Cabinets**　　　　　　　　　　　　　**Countertop**

| Wall Cabinets | Unit | Material | Labor | Equip. | Total | Specification |
|---|---|---|---|---|---|---|
| **Good Grade Laminated** | | | | | | |
| **12" Wide, 1 Door** | | | | | | |
| Demolish | Ea. | | 9.85 | | 9.85 | Includes material and labor to install |
| Install | Ea. | 216 | 42.50 | | 258.50 | good modular unit with melamine |
| Demolish and Install | Ea. | 216 | 52.35 | | 268.35 | laminated to particle board, in |
| Reinstall | Ea. | | 33.80 | | 33.80 | textured colors or wood grain print |
| Clean | Ea. | .22 | 6.85 | | 7.07 | finish, hinges and pulls. |
| Paint | Ea. | .74 | 13.80 | | 14.54 | |
| Minimum Charge | Job | | 232 | | 232 | |
| **15" Wide, 1 Door** | | | | | | |
| Demolish | Ea. | | 12.35 | | 12.35 | Includes material and labor to install |
| Install | Ea. | 226 | 43.50 | | 269.50 | good modular unit with melamine |
| Demolish and Install | Ea. | 226 | 55.85 | | 281.85 | laminated to particle board, in |
| Reinstall | Ea. | | 34.75 | | 34.75 | textured colors or wood grain print |
| Clean | Ea. | .28 | 8.45 | | 8.73 | finish, hinges and pulls. |
| Paint | Ea. | .92 | 17.55 | | 18.47 | |
| Minimum Charge | Job | | 232 | | 232 | |
| **18" Wide, 1 Door** | | | | | | |
| Demolish | Ea. | | 14.90 | | 14.90 | Includes material and labor to install |
| Install | Ea. | 249 | 44.50 | | 293.50 | good modular unit with melamine |
| Demolish and Install | Ea. | 249 | 59.40 | | 308.40 | laminated to particle board, in |
| Reinstall | Ea. | | 35.58 | | 35.58 | textured colors or wood grain print |
| Clean | Ea. | .33 | 10.15 | | 10.48 | finish, hinges and pulls. |
| Paint | Ea. | 1.10 | 20.50 | | 21.60 | |
| Minimum Charge | Job | | 232 | | 232 | |
| **21" Wide, 1 Door** | | | | | | |
| Demolish | Ea. | | 17.15 | | 17.15 | Includes material and labor to install |
| Install | Ea. | 261 | 45 | | 306 | good modular unit with melamine |
| Demolish and Install | Ea. | 261 | 62.15 | | 323.15 | laminated to particle board, in |
| Reinstall | Ea. | | 36.10 | | 36.10 | textured colors or wood grain print |
| Clean | Ea. | .39 | 11.85 | | 12.24 | finish, hinges and pulls. |
| Paint | Ea. | 1.30 | 24 | | 25.30 | |
| Minimum Charge | Job | | 232 | | 232 | |
| **24" w / Blind Corner** | | | | | | |
| Demolish | Ea. | | 19.75 | | 19.75 | Includes material and labor to install |
| Install | Ea. | 290 | 46 | | 336 | good modular unit with melamine |
| Demolish and Install | Ea. | 290 | 65.75 | | 355.75 | laminated to particle board, in |
| Reinstall | Ea. | | 36.63 | | 36.63 | textured colors or wood grain print |
| Clean | Ea. | .44 | 13.65 | | 14.09 | finish, hinges and pulls. |
| Paint | Ea. | 1.47 | 27.50 | | 28.97 | |
| Minimum Charge | Job | | 232 | | 232 | |

211

# Cabinets and Countertops

| Wall Cabinets | Unit | Material | Labor | Equip. | Total | Specification |
|---|---|---|---|---|---|---|
| **27" Wide, 2 Door** | | | | | | |
| Demolish | Ea. | | 22 | | 22 | Includes material and labor to install |
| Install | Ea. | 76 | 47 | | 123 | good modular unit with melamine |
| Demolish and Install | Ea. | 76 | 69 | | 145 | laminated to particle board, in |
| Reinstall | Ea. | | 37.56 | | 37.56 | textured colors or wood grain print |
| Clean | Ea. | .50 | 15.45 | | 15.95 | finish, hinges and pulls. |
| Paint | Ea. | 1.67 | 32 | | 33.67 | |
| Minimum Charge | Job | | 232 | | 232 | |
| **30" Wide, 2 Door** | | | | | | |
| Demolish | Ea. | | 24.50 | | 24.50 | Includes material and labor to install |
| Install | Ea. | 335 | 48 | | 383 | good modular unit with melamine |
| Demolish and Install | Ea. | 335 | 72.50 | | 407.50 | laminated to particle board, in |
| Reinstall | Ea. | | 38.53 | | 38.53 | textured colors or wood grain print |
| Clean | Ea. | .55 | 16.90 | | 17.45 | finish, hinges and pulls. |
| Paint | Ea. | 1.85 | 35 | | 36.85 | |
| Minimum Charge | Job | | 232 | | 232 | |
| **36" Wide, 2 Door** | | | | | | |
| Demolish | Ea. | | 29 | | 29 | Includes material and labor to install |
| Install | Ea. | 385 | 49.50 | | 434.50 | good modular unit with melamine |
| Demolish and Install | Ea. | 385 | 78.50 | | 463.50 | laminated to particle board, in |
| Reinstall | Ea. | | 39.56 | | 39.56 | textured colors or wood grain print |
| Clean | Ea. | .66 | 21 | | 21.66 | finish, hinges and pulls. |
| Paint | Ea. | 2.23 | 43 | | 45.23 | |
| Minimum Charge | Job | | 232 | | 232 | |
| **48" Wide, 2 Door** | | | | | | |
| Demolish | Ea. | | 39.50 | | 39.50 | Includes material and labor to install |
| Install | Ea. | 470 | 50.50 | | 520.50 | good modular unit with melamine |
| Demolish and Install | Ea. | 470 | 90 | | 560 | laminated to particle board, in |
| Reinstall | Ea. | | 40.42 | | 40.42 | textured colors or wood grain print |
| Clean | Ea. | .88 | 27.50 | | 28.38 | finish, hinges and pulls. |
| Paint | Ea. | 2.97 | 55 | | 57.97 | |
| Minimum Charge | Job | | 232 | | 232 | |
| **Above-Appliance** | | | | | | |
| Demolish | Ea. | | 24.50 | | 24.50 | Includes material and labor to install |
| Install | Ea. | 224 | 37.50 | | 261.50 | good modular unit with melamine |
| Demolish and Install | Ea. | 224 | 62 | | 286 | laminated to particle board, in |
| Reinstall | Ea. | | 29.99 | | 29.99 | textured colors or wood grain print |
| Clean | Ea. | .55 | 16.90 | | 17.45 | finish, hinges and pulls. |
| Paint | Ea. | 1.85 | 35 | | 36.85 | |
| Minimum Charge | Job | | 232 | | 232 | |
| **Better Grade Wood** | | | | | | |
| **12" Wide, 1 Door** | | | | | | |
| Demolish | Ea. | | 9.85 | | 9.85 | Includes material and labor to install |
| Install | Ea. | 287 | 42.50 | | 329.50 | custom modular unit with solid |
| Demolish and Install | Ea. | 287 | 52.35 | | 339.35 | hardwood faced frames, hardwood |
| Reinstall | Ea. | | 33.80 | | 33.80 | door frames and drawer fronts, |
| Clean | Ea. | .22 | 6.85 | | 7.07 | hardwood veneer on raised door |
| Paint | Ea. | .74 | 14.85 | | 15.59 | panels, hinges and pulls. |
| Minimum Charge | Job | | 232 | | 232 | |
| **15" Wide, 1 Door** | | | | | | |
| Demolish | Ea. | | 12.35 | | 12.35 | Includes material and labor to install |
| Install | Ea. | 300 | 43.50 | | 343.50 | custom modular unit with solid |
| Demolish and Install | Ea. | 300 | 55.85 | | 355.85 | hardwood faced frames, hardwood |
| Reinstall | Ea. | | 34.75 | | 34.75 | door frames and drawer fronts, |
| Clean | Ea. | .28 | 8.45 | | 8.73 | hardwood veneer on raised door |
| Paint | Ea. | .92 | 18.40 | | 19.32 | panels, hinges and pulls. |
| Minimum Charge | Job | | 232 | | 232 | |

| Wall Cabinets | Unit | Material | Labor | Equip. | Total | Specification |
|---|---|---|---|---|---|---|
| **18″ Wide, 1 Door** | | | | | | |
| Demolish | Ea. | | 14.90 | | 14.90 | Includes material and labor to install |
| Install | Ea. | 330 | 44.50 | | 374.50 | custom modular unit with solid |
| Demolish and Install | Ea. | 330 | 59.40 | | 389.40 | hardwood faced frames, hardwood |
| Reinstall | Ea. | | 35.58 | | 35.58 | door frames and drawer fronts, |
| Clean | Ea. | .33 | 10.15 | | 10.48 | hardwood veneer on raised door |
| Paint | Ea. | 1.10 | 22.50 | | 23.60 | panels, hinges and pulls. |
| Minimum Charge | Job | | 232 | | 232 | |
| **21″ Wide, 1 Door** | | | | | | |
| Demolish | Ea. | | 17.15 | | 17.15 | Includes material and labor to install |
| Install | Ea. | 345 | 45 | | 390 | custom modular unit with solid |
| Demolish and Install | Ea. | 345 | 62.15 | | 407.15 | hardwood faced frames, hardwood |
| Reinstall | Ea. | | 36.10 | | 36.10 | door frames and drawer fronts, |
| Clean | Ea. | .39 | 11.85 | | 12.24 | hardwood veneer on raised door |
| Paint | Ea. | 1.30 | 26 | | 27.30 | panels, hinges and pulls. |
| Minimum Charge | Job | | 232 | | 232 | |
| **24″ w / Blind Corner** | | | | | | |
| Demolish | Ea. | | 19.75 | | 19.75 | Includes material and labor to install |
| Install | Ea. | 385 | 46 | | 431 | custom modular unit with solid |
| Demolish and Install | Ea. | 385 | 65.75 | | 450.75 | hardwood faced frames, hardwood |
| Reinstall | Ea. | | 36.63 | | 36.63 | door frames and drawer fronts, |
| Clean | Ea. | .44 | 13.65 | | 14.09 | hardwood veneer on raised door |
| Paint | Ea. | 1.47 | 29.50 | | 30.97 | panels, hinges and pulls. |
| Minimum Charge | Job | | 232 | | 232 | |
| **27″ Wide, 2 Door** | | | | | | |
| Demolish | Ea. | | 22 | | 22 | Includes material and labor to install |
| Install | Ea. | 102 | 47 | | 149 | custom modular unit with solid |
| Demolish and Install | Ea. | 102 | 69 | | 171 | hardwood faced frames, hardwood |
| Reinstall | Ea. | | 37.56 | | 37.56 | door frames and drawer fronts, |
| Clean | Ea. | .50 | 15.45 | | 15.95 | hardwood veneer on raised door |
| Paint | Ea. | 1.67 | 32 | | 33.67 | panels, hinges and pulls. |
| Minimum Charge | Job | | 232 | | 232 | |
| **30″ Wide, 2 Door** | | | | | | |
| Demolish | Ea. | | 24.50 | | 24.50 | Includes material and labor to install |
| Install | Ea. | 450 | 48 | | 498 | custom modular unit with solid |
| Demolish and Install | Ea. | 450 | 72.50 | | 522.50 | hardwood faced frames, hardwood |
| Reinstall | Ea. | | 38.53 | | 38.53 | door frames and drawer fronts, |
| Clean | Ea. | .55 | 16.90 | | 17.45 | hardwood veneer on raised door |
| Paint | Ea. | 1.85 | 38.50 | | 40.35 | panels, hinges and pulls. |
| Minimum Charge | Job | | 232 | | 232 | |
| **36″ Wide, 2 Door** | | | | | | |
| Demolish | Ea. | | 29 | | 29 | Includes material and labor to install |
| Install | Ea. | 510 | 49.50 | | 559.50 | custom modular unit with solid |
| Demolish and Install | Ea. | 510 | 78.50 | | 588.50 | hardwood faced frames, hardwood |
| Reinstall | Ea. | | 39.56 | | 39.56 | door frames and drawer fronts, |
| Clean | Ea. | .66 | 21 | | 21.66 | hardwood veneer on raised door |
| Paint | Ea. | 2.23 | 43 | | 45.23 | panels, hinges and pulls. |
| Minimum Charge | Job | | 232 | | 232 | |
| **48″ Wide, 2 Door** | | | | | | |
| Demolish | Ea. | | 39.50 | | 39.50 | Includes material and labor to install |
| Install | Ea. | 625 | 50.50 | | 675.50 | custom modular unit with solid |
| Demolish and Install | Ea. | 625 | 90 | | 715 | hardwood faced frames, hardwood |
| Reinstall | Ea. | | 40.42 | | 40.42 | door frames and drawer fronts, |
| Clean | Ea. | .88 | 27.50 | | 28.38 | hardwood veneer on raised door |
| Paint | Ea. | 2.97 | 55 | | 57.97 | panels, hinges and pulls. |
| Minimum Charge | Job | | 232 | | 232 | |

# Cabinets and Countertops

| Wall Cabinets | Unit | Material | Labor | Equip. | Total | Specification |
|---|---|---|---|---|---|---|
| **Above-appliance** | | | | | | |
| Demolish | Ea. | | 24.50 | | 24.50 | Includes material and labor to install |
| Install | Ea. | 299 | 37.50 | | 336.50 | custom modular unit with solid |
| Demolish and Install | Ea. | 299 | 62 | | 361 | hardwood faced frames, hardwood |
| Reinstall | Ea. | | 29.99 | | 29.99 | door frames and drawer fronts, |
| Clean | Ea. | .55 | 16.90 | | 17.45 | hardwood veneer on raised door |
| Paint | Ea. | 1.85 | 38.50 | | 40.35 | panels, hinges and pulls. |
| Minimum Charge | Job | | 232 | | 232 | |
| **Premium Grade Wood** | | | | | | |
| **12″ Wide, 1 Door** | | | | | | |
| Demolish | Ea. | | 9.85 | | 9.85 | Includes material and labor to install |
| Install | Ea. | 345 | 42.50 | | 387.50 | premium modular unit with solid |
| Demolish and Install | Ea. | 345 | 52.35 | | 397.35 | hardwood faced frames, hardwood |
| Reinstall | Ea. | | 33.80 | | 33.80 | door frames and drawer fronts, |
| Clean | Ea. | .22 | 6.85 | | 7.07 | hardwood veneer on raised door |
| Paint | Ea. | .74 | 16.10 | | 16.84 | panels, hinges and pulls. |
| Minimum Charge | Job | | 232 | | 232 | |
| **15″ Wide, 1 Door** | | | | | | |
| Demolish | Ea. | | 12.35 | | 12.35 | Includes material and labor to install |
| Install | Ea. | 400 | 43.50 | | 443.50 | premium modular unit with solid |
| Demolish and Install | Ea. | 400 | 55.85 | | 455.85 | hardwood faced frames, hardwood |
| Reinstall | Ea. | | 34.75 | | 34.75 | door frames and drawer fronts, |
| Clean | Ea. | .28 | 8.45 | | 8.73 | hardwood veneer on raised door |
| Paint | Ea. | .92 | 20.50 | | 21.42 | panels, hinges and pulls. |
| Minimum Charge | Job | | 232 | | 232 | |
| **18″ Wide, 1 Door** | | | | | | |
| Demolish | Ea. | | 14.90 | | 14.90 | Includes material and labor to install |
| Install | Ea. | 395 | 44.50 | | 439.50 | premium modular unit with solid |
| Demolish and Install | Ea. | 395 | 59.40 | | 454.40 | hardwood faced frames, hardwood |
| Reinstall | Ea. | | 35.58 | | 35.58 | door frames and drawer fronts, |
| Clean | Ea. | .33 | 10.15 | | 10.48 | hardwood veneer on raised door |
| Paint | Ea. | 1.10 | 24 | | 25.10 | panels, hinges and pulls. |
| Minimum Charge | Job | | 232 | | 232 | |
| **21″ Wide, 1 Door** | | | | | | |
| Demolish | Ea. | | 17.15 | | 17.15 | Includes material and labor to install |
| Install | Ea. | 420 | 45 | | 465 | premium modular unit with solid |
| Demolish and Install | Ea. | 420 | 62.15 | | 482.15 | hardwood faced frames, hardwood |
| Reinstall | Ea. | | 36.10 | | 36.10 | door frames and drawer fronts, |
| Clean | Ea. | .39 | 11.85 | | 12.24 | hardwood veneer on raised door |
| Paint | Ea. | 1.30 | 27.50 | | 28.80 | panels, hinges and pulls. |
| Minimum Charge | Job | | 232 | | 232 | |
| **24″ w / Blind Corner** | | | | | | |
| Demolish | Ea. | | 19.75 | | 19.75 | Includes material and labor to install |
| Install | Ea. | 460 | 46 | | 506 | premium modular unit with solid |
| Demolish and Install | Ea. | 460 | 65.75 | | 525.75 | hardwood faced frames, hardwood |
| Reinstall | Ea. | | 36.63 | | 36.63 | door frames and drawer fronts, |
| Clean | Ea. | .44 | 13.65 | | 14.09 | hardwood veneer on raised door |
| Paint | Ea. | 1.47 | 32 | | 33.47 | panels, hinges and pulls. |
| Minimum Charge | Job | | 232 | | 232 | |
| **27″ Wide, 2 Door** | | | | | | |
| Demolish | Ea. | | 22 | | 22 | Includes material and labor to install |
| Install | Ea. | 430 | 47 | | 477 | premium modular unit with solid |
| Demolish and Install | Ea. | 430 | 69 | | 499 | hardwood faced frames, hardwood |
| Reinstall | Ea. | | 37.56 | | 37.56 | door frames and drawer fronts, |
| Clean | Ea. | .50 | 15.45 | | 15.95 | hardwood veneer on raised door |
| Paint | Ea. | 1.67 | 35 | | 36.67 | panels, hinges and pulls. |
| Minimum Charge | Job | | 232 | | 232 | |

# Cabinets and Countertops

| Wall Cabinets | Unit | Material | Labor | Equip. | Total | Specification |
|---|---|---|---|---|---|---|
| **30" Wide, 2 Door** | | | | | | |
| Demolish | Ea. | | 24.50 | | 24.50 | Includes material and labor to install |
| Install | Ea. | 540 | 48 | | 588 | premium modular unit with solid |
| Demolish and Install | Ea. | 540 | 72.50 | | 612.50 | hardwood faced frames, hardwood |
| Reinstall | Ea. | | 38.53 | | 38.53 | door frames and drawer fronts, |
| Clean | Ea. | .55 | 16.90 | | 17.45 | hardwood veneer on raised door |
| Paint | Ea. | 1.85 | 38.50 | | 40.35 | panels, hinges and pulls. |
| Minimum Charge | Job | | 232 | | 232 | |
| **36" Wide, 2 Door** | | | | | | |
| Demolish | Ea. | | 29 | | 29 | Includes material and labor to install |
| Install | Ea. | 615 | 49.50 | | 664.50 | premium modular unit with solid |
| Demolish and Install | Ea. | 615 | 78.50 | | 693.50 | hardwood faced frames, hardwood |
| Reinstall | Ea. | | 39.56 | | 39.56 | door frames and drawer fronts, |
| Clean | Ea. | .66 | 21 | | 21.66 | hardwood veneer on raised door |
| Paint | Ea. | 2.23 | 48.50 | | 50.73 | panels, hinges and pulls. |
| Minimum Charge | Job | | 232 | | 232 | |
| **48" Wide, 2 Door** | | | | | | |
| Demolish | Ea. | | 39.50 | | 39.50 | Includes material and labor to install |
| Install | Ea. | 755 | 50.50 | | 805.50 | premium modular unit with solid |
| Demolish and Install | Ea. | 755 | 90 | | 845 | hardwood faced frames, hardwood |
| Reinstall | Ea. | | 40.42 | | 40.42 | door frames and drawer fronts, |
| Clean | Ea. | .88 | 27.50 | | 28.38 | hardwood veneer on raised door |
| Paint | Ea. | 2.97 | 64.50 | | 67.47 | panels, hinges and pulls. |
| Minimum Charge | Job | | 232 | | 232 | |
| **Above-appliance** | | | | | | |
| Demolish | Ea. | | 24.50 | | 24.50 | Includes material and labor to install |
| Install | Ea. | 360 | 37.50 | | 397.50 | premium modular unit with solid |
| Demolish and Install | Ea. | 360 | 62 | | 422 | hardwood faced frames, hardwood |
| Reinstall | Ea. | | 29.99 | | 29.99 | door frames and drawer fronts, |
| Clean | Ea. | .55 | 16.90 | | 17.45 | hardwood veneer on raised door |
| Paint | Ea. | 1.85 | 38.50 | | 40.35 | panels, hinges, pulls. |
| Minimum Charge | Job | | 232 | | 232 | |
| **Premium Hardwood** | | | | | | |
| **12" Wide, 1 Door** | | | | | | |
| Demolish | Ea. | | 9.85 | | 9.85 | Includes material and labor to install |
| Install | Ea. | 375 | 42.50 | | 417.50 | premium hardwood modular unit with |
| Demolish and Install | Ea. | 375 | 52.35 | | 427.35 | solid hardwood faced frames, drawer |
| Reinstall | Ea. | | 33.80 | | 33.80 | fronts, and door panels, steel drawer |
| Clean | Ea. | .22 | 6.85 | | 7.07 | guides, hinges, pulls, laminated |
| Paint | Ea. | .74 | 17.55 | | 18.29 | interior. |
| Minimum Charge | Job | | 232 | | 232 | |
| **15" Wide, 1 Door** | | | | | | |
| Demolish | Ea. | | 12.35 | | 12.35 | Includes material and labor to install |
| Install | Ea. | 530 | 43.50 | | 573.50 | premium hardwood modular unit with |
| Demolish and Install | Ea. | 530 | 55.85 | | 585.85 | solid hardwood faced frames, drawer |
| Reinstall | Ea. | | 34.75 | | 34.75 | fronts, and door panels, steel drawer |
| Clean | Ea. | .28 | 8.45 | | 8.73 | guides, hinges, pulls, laminated |
| Paint | Ea. | .92 | 21.50 | | 22.42 | interior. |
| Minimum Charge | Job | | 232 | | 232 | |
| **18" Wide, 1 Door** | | | | | | |
| Demolish | Ea. | | 14.90 | | 14.90 | Includes material and labor to install |
| Install | Ea. | 435 | 44.50 | | 479.50 | premium hardwood modular unit with |
| Demolish and Install | Ea. | 435 | 59.40 | | 494.40 | solid hardwood faced frames, drawer |
| Reinstall | Ea. | | 35.58 | | 35.58 | fronts, and door panels, steel drawer |
| Clean | Ea. | .33 | 10.15 | | 10.48 | guides, hinges, pulls, laminated |
| Paint | Ea. | 1.10 | 26 | | 27.10 | interior. |
| Minimum Charge | Job | | 232 | | 232 | |

# Cabinets and Countertops

| Wall Cabinets | Unit | Material | Labor | Equip. | Total | Specification |
|---|---|---|---|---|---|---|
| **21" Wide, 1 Door** | | | | | | |
| Demolish | Ea. | | 17.15 | | 17.15 | Includes material and labor to install |
| Install | Ea. | 455 | 45 | | 500 | premium hardwood modular unit with |
| Demolish and Install | Ea. | 455 | 62.15 | | 517.15 | solid hardwood faced frames, drawer |
| Reinstall | Ea. | | 36.10 | | 36.10 | fronts, and door panels, steel drawer |
| Clean | Ea. | .39 | 11.85 | | 12.24 | guides, hinges, pulls, laminated |
| Paint | Ea. | 1.30 | 29.50 | | 30.80 | interior. |
| Minimum Charge | Job | | 232 | | 232 | |
| **24" w / Blind Corner** | | | | | | |
| Demolish | Ea. | | 19.75 | | 19.75 | Includes material and labor to install |
| Install | Ea. | 505 | 46 | | 551 | premium hardwood modular unit with |
| Demolish and Install | Ea. | 505 | 65.75 | | 570.75 | solid hardwood faced frames, drawer |
| Reinstall | Ea. | | 36.63 | | 36.63 | fronts, and door panels, steel drawer |
| Clean | Ea. | .44 | 13.65 | | 14.09 | guides, hinges, pulls, laminated |
| Paint | Ea. | 1.47 | 35 | | 36.47 | interior. |
| Minimum Charge | Job | | 232 | | 232 | |
| **27" Wide, 2 Door** | | | | | | |
| Demolish | Ea. | | 22 | | 22 | Includes material and labor to install |
| Install | Ea. | 133 | 47 | | 180 | premium hardwood modular unit with |
| Demolish and Install | Ea. | 133 | 69 | | 202 | solid hardwood faced frames, drawer |
| Reinstall | Ea. | | 37.56 | | 37.56 | fronts, and door panels, steel drawer |
| Clean | Ea. | .50 | 15.45 | | 15.95 | guides, hinges, pulls, laminated |
| Paint | Ea. | 1.67 | 38.50 | | 40.17 | interior. |
| Minimum Charge | Job | | 232 | | 232 | |
| **30" Wide, 2 Door** | | | | | | |
| Demolish | Ea. | | 24.50 | | 24.50 | Includes material and labor to install |
| Install | Ea. | 590 | 48 | | 638 | premium hardwood modular unit with |
| Demolish and Install | Ea. | 590 | 72.50 | | 662.50 | solid hardwood faced frames, drawer |
| Reinstall | Ea. | | 38.53 | | 38.53 | fronts, and door panels, steel drawer |
| Clean | Ea. | .55 | 16.90 | | 17.45 | guides, hinges, pulls, laminated |
| Paint | Ea. | 1.85 | 43 | | 44.85 | interior. |
| Minimum Charge | Job | | 232 | | 232 | |
| **36" Wide, 2 Door** | | | | | | |
| Demolish | Ea. | | 29 | | 29 | Includes material and labor to install |
| Install | Ea. | 670 | 49.50 | | 719.50 | premium hardwood modular unit with |
| Demolish and Install | Ea. | 670 | 78.50 | | 748.50 | solid hardwood faced frames, drawer |
| Reinstall | Ea. | | 39.56 | | 39.56 | fronts, and door panels, steel drawer |
| Clean | Ea. | .66 | 21 | | 21.66 | guides, hinges, pulls, laminated |
| Paint | Ea. | 2.23 | 55 | | 57.23 | interior. |
| Minimum Charge | Job | | 232 | | 232 | |
| **48" Wide, 2 Door** | | | | | | |
| Demolish | Ea. | | 39.50 | | 39.50 | Includes material and labor to install |
| Install | Ea. | 820 | 50.50 | | 870.50 | premium hardwood modular unit with |
| Demolish and Install | Ea. | 820 | 90 | | 910 | solid hardwood faced frames, drawer |
| Reinstall | Ea. | | 40.42 | | 40.42 | fronts, and door panels, steel drawer |
| Clean | Ea. | .88 | 27.50 | | 28.38 | guides, hinges, pulls, laminated |
| Paint | Ea. | 2.97 | 64.50 | | 67.47 | interior. |
| Minimum Charge | Job | | 232 | | 232 | |
| **Above-appliance** | | | | | | |
| Demolish | Ea. | | 24.50 | | 24.50 | Includes material and labor to install |
| Install | Ea. | 390 | 37.50 | | 427.50 | premium hardwood modular unit with |
| Demolish and Install | Ea. | 390 | 62 | | 452 | solid hardwood faced frames, drawer |
| Reinstall | Ea. | | 29.99 | | 29.99 | fronts, and door panels, steel drawer |
| Clean | Ea. | .55 | 16.90 | | 17.45 | guides, hinges, pulls, laminated |
| Paint | Ea. | 1.85 | 43 | | 44.85 | interior. |
| Minimum Charge | Job | | 232 | | 232 | |

# Cabinets and Countertops

| Wall Cabinets | Unit | Material | Labor | Equip. | Total | Specification |
|---|---|---|---|---|---|---|
| **Medicine Cabinet** | | | | | | |
| **Good Grade** | | | | | | |
| Demolish | Ea. | | 16.45 | | 16.45 | Cost includes material and labor to |
| Install | Ea. | 165 | 33 | | 198 | install 14" x 18" recessed medicine |
| Demolish and Install | Ea. | 165 | 49.45 | | 214.45 | cabinet with stainless steel frame, |
| Reinstall | Ea. | | 26.56 | | 26.56 | mirror. Light not included. |
| Clean | Ea. | .01 | 16.70 | | 16.71 | |
| Paint | Ea. | 2.05 | 24 | | 26.05 | |
| Minimum Charge | Job | | 232 | | 232 | |
| **Good Grade Lighted** | | | | | | |
| Demolish | Ea. | | 16.45 | | 16.45 | Cost includes material and labor to |
| Install | Ea. | 430 | 38.50 | | 468.50 | install 14" x 18" recessed medicine |
| Demolish and Install | Ea. | 430 | 54.95 | | 484.95 | cabinet with stainless steel frame, |
| Reinstall | Ea. | | 30.99 | | 30.99 | mirror and light. |
| Clean | Ea. | .01 | 16.70 | | 16.71 | |
| Paint | Ea. | 2.05 | 24 | | 26.05 | |
| Minimum Charge | Job | | 232 | | 232 | |
| **Better Grade** | | | | | | |
| Demolish | Ea. | | 16.45 | | 16.45 | Includes material and labor to install |
| Install | Ea. | 286 | 33 | | 319 | recessed medicine cabinet with |
| Demolish and Install | Ea. | 286 | 49.45 | | 335.45 | polished chrome frame and mirror. |
| Reinstall | Ea. | | 26.56 | | 26.56 | Light not included. |
| Clean | Ea. | .01 | 16.70 | | 16.71 | |
| Paint | Ea. | 2.05 | 24 | | 26.05 | |
| Minimum Charge | Job | | 232 | | 232 | |
| **Better Grade Lighted** | | | | | | |
| Demolish | Ea. | | 16.45 | | 16.45 | Includes material and labor to install |
| Install | Ea. | 550 | 38.50 | | 588.50 | recessed medicine cabinet with |
| Demolish and Install | Ea. | 550 | 54.95 | | 604.95 | polished chrome frame, mirror and |
| Reinstall | Ea. | | 30.99 | | 30.99 | light. |
| Clean | Ea. | .01 | 16.70 | | 16.71 | |
| Paint | Ea. | 2.05 | 24 | | 26.05 | |
| Minimum Charge | Job | | 232 | | 232 | |
| **Premium Grade** | | | | | | |
| Demolish | Ea. | | 16.45 | | 16.45 | Cost includes material and labor to |
| Install | Ea. | 340 | 33 | | 373 | install 14" x 24" recessed medicine |
| Demolish and Install | Ea. | 340 | 49.45 | | 389.45 | cabinet with beveled mirror, frameless |
| Reinstall | Ea. | | 26.56 | | 26.56 | swing door. Light not included. |
| Clean | Ea. | .01 | 16.70 | | 16.71 | |
| Paint | Ea. | 2.05 | 24 | | 26.05 | |
| Minimum Charge | Job | | 232 | | 232 | |
| **Premium Grade Lighted** | | | | | | |
| Demolish | Ea. | | 16.45 | | 16.45 | Cost includes material and labor to |
| Install | Ea. | 605 | 38.50 | | 643.50 | install 36" x 30" surface mounted |
| Demolish and Install | Ea. | 605 | 54.95 | | 659.95 | medicine cabinet with beveled tri-view |
| Reinstall | Ea. | | 30.99 | | 30.99 | mirror, frameless swing door and light. |
| Clean | Ea. | .01 | 16.70 | | 16.71 | |
| Paint | Ea. | 2.05 | 24 | | 26.05 | |
| Minimum Charge | Job | | 232 | | 232 | |
| **Bath Mirror** | | | | | | |
| Demolish | Ea. | | 12.35 | | 12.35 | Cost includes material and labor to |
| Install | Ea. | 51.50 | 23 | | 74.50 | install 18" x 24" surface mount mirror |
| Demolish and Install | Ea. | 51.50 | 35.35 | | 86.85 | with stainless steel frame. |
| Reinstall | Ea. | | 18.59 | | 18.59 | |
| Clean | Ea. | .01 | 1.11 | | 1.12 | |
| Minimum Charge | Job | | 232 | | 232 | |

# Cabinets and Countertops

## Wall Cabinets

| | Unit | Material | Labor | Equip. | Total | Specification |
|---|---|---|---|---|---|---|
| **Garage** | | | | | | |
| Demolish | L.F. | | 9.85 | | 9.85 | Includes material and labor to install |
| Install | L.F. | 109 | 23 | | 132 | good modular unit with melamine |
| Demolish and Install | L.F. | 109 | 32.85 | | 141.85 | laminated to particle board, in |
| Reinstall | L.F. | | 18.59 | | 18.59 | textured colors or wood grain print |
| Clean | L.F. | .22 | 6.85 | | 7.07 | finish, hinges and pulls. |
| Paint | L.F. | .74 | 13.80 | | 14.54 | |
| Minimum Charge | Job | | 232 | | 232 | |
| **Strip and Refinish** | | | | | | |
| Install | L.F. | 5.35 | 48.50 | | 53.85 | Includes labor and material to strip, |
| Minimum Charge | Job | | 193 | | 193 | prep and refinish exterior of cabinets. |

## Tall Cabinets

| | Unit | Material | Labor | Equip. | Total | Specification |
|---|---|---|---|---|---|---|
| **Good Grade Laminated** | | | | | | |
| Demolish | L.F. | | 17.55 | | 17.55 | Includes material and labor to install |
| Install | L.F. | 271 | 51.50 | | 322.50 | custom modular unit with solid |
| Demolish and Install | L.F. | 271 | 69.05 | | 340.05 | hardwood faced frames, hardwood |
| Reinstall | L.F. | | 41.32 | | 41.32 | door frames and drawer fronts, |
| Clean | L.F. | .59 | 12.70 | | 13.29 | hardwood veneer on raised door |
| Paint | L.F. | 1.73 | 27.50 | | 29.23 | panels, hinges and pulls. |
| Minimum Charge | Job | | 232 | | 232 | |
| **Better Grade Wood** | | | | | | |
| Demolish | L.F. | | 17.55 | | 17.55 | Includes material and labor to install |
| Install | L.F. | 390 | 51.50 | | 441.50 | custom modular unit with solid |
| Demolish and Install | L.F. | 390 | 69.05 | | 459.05 | hardwood faced frames, hardwood |
| Reinstall | L.F. | | 41.32 | | 41.32 | door frames and drawer fronts, |
| Clean | L.F. | .59 | 12.70 | | 13.29 | hardwood veneer on raised door |
| Paint | L.F. | 1.73 | 29.50 | | 31.23 | panels, hinges and pulls. |
| Minimum Charge | Job | | 232 | | 232 | |
| **Premium Grade Wood** | | | | | | |
| Demolish | L.F. | | 17.55 | | 17.55 | Includes material and labor to install |
| Install | L.F. | 555 | 51.50 | | 606.50 | premium modular unit with solid |
| Demolish and Install | L.F. | 555 | 69.05 | | 624.05 | hardwood faced frames, hardwood |
| Reinstall | L.F. | | 41.32 | | 41.32 | door frames and drawer fronts, |
| Clean | L.F. | .59 | 12.70 | | 13.29 | hardwood veneer on raised door |
| Paint | L.F. | 1.73 | 32 | | 33.73 | panels, hinges and pulls. |
| Minimum Charge | Job | | 232 | | 232 | |
| **Premium Hardwood** | | | | | | |
| Demolish | L.F. | | 17.55 | | 17.55 | Includes material and labor to install |
| Install | L.F. | 775 | 51.50 | | 826.50 | premium hardwood modular unit with |
| Demolish and Install | L.F. | 775 | 69.05 | | 844.05 | solid hardwood faced frames, drawer |
| Reinstall | L.F. | | 41.32 | | 41.32 | fronts, and door panels, steel drawer |
| Clean | L.F. | .59 | 12.70 | | 13.29 | guides, hinges and pulls, laminated |
| Paint | L.F. | 1.73 | 35 | | 36.73 | interior. |
| Minimum Charge | Job | | 232 | | 232 | |

| Tall Cabinets | Unit | Material | Labor | Equip. | Total | Specification |
|---|---|---|---|---|---|---|
| **Single Oven, 27" Wide** | | | | | | |
| **Good Grade Laminated** | | | | | | Includes material and labor to install |
| Demolish | Ea. | | 39.50 | | 39.50 | good modular unit with melamine |
| Install | Ea. | 935 | 116 | | 1051 | laminated to particle board, in |
| Demolish and Install | Ea. | 935 | 155.50 | | 1090.50 | textured colors or wood grain print |
| Reinstall | Ea. | | 92.96 | | 92.96 | finish, hinges and pulls. |
| Clean | Ea. | 1.84 | 22 | | 23.84 | |
| Paint | Ea. | 2.88 | 62 | | 64.88 | |
| Minimum Charge | Job | | 232 | | 232 | |
| **Better Grade Veneer** | | | | | | Includes material and labor to install |
| Demolish | Ea. | | 39.50 | | 39.50 | custom modular unit with solid |
| Install | Ea. | 1650 | 116 | | 1766 | hardwood faced frames, hardwood |
| Demolish and Install | Ea. | 1650 | 155.50 | | 1805.50 | door frames and drawer fronts, |
| Reinstall | Ea. | | 92.96 | | 92.96 | hardwood veneer on raised door |
| Clean | Ea. | 1.84 | 22 | | 23.84 | panels, hinges and pulls. |
| Paint | Ea. | 2.88 | 66.50 | | 69.38 | |
| Minimum Charge | Job | | 232 | | 232 | |
| **Premium Grade Veneer** | | | | | | Includes material and labor to install |
| Demolish | Ea. | | 39.50 | | 39.50 | premium modular unit with solid |
| Install | Ea. | 1475 | 116 | | 1591 | hardwood faced frames, hardwood |
| Demolish and Install | Ea. | 1475 | 155.50 | | 1630.50 | door frames and drawer fronts, |
| Reinstall | Ea. | | 92.96 | | 92.96 | hardwood veneer on raised door |
| Clean | Ea. | 1.84 | 22 | | 23.84 | panels, hinges and pulls. |
| Paint | Ea. | 2.88 | 73 | | 75.88 | |
| Minimum Charge | Job | | 232 | | 232 | |
| **Premium Hardwood** | | | | | | Includes material and labor to install |
| Demolish | Ea. | | 39.50 | | 39.50 | premium hardwood modular unit with |
| Install | Ea. | 1625 | 116 | | 1741 | solid hardwood faced frames, drawer |
| Demolish and Install | Ea. | 1625 | 155.50 | | 1780.50 | fronts, and door panels, steel drawer |
| Reinstall | Ea. | | 92.96 | | 92.96 | guides, hinges and pulls, laminated |
| Clean | Ea. | 1.84 | 22 | | 23.84 | interior. |
| Paint | Ea. | 2.88 | 79 | | 81.88 | |
| Minimum Charge | Job | | 232 | | 232 | |
| **Single Oven, 30" Wide** | | | | | | |
| **Good Grade Laminated** | | | | | | Includes material and labor to install |
| Demolish | Ea. | | 39.50 | | 39.50 | good modular unit with melamine |
| Install | Ea. | 1100 | 116 | | 1216 | laminated to particle board, in |
| Demolish and Install | Ea. | 1100 | 155.50 | | 1255.50 | textured colors or wood grain print |
| Reinstall | Ea. | | 92.96 | | 92.96 | finish, hinges and pulls. |
| Clean | Ea. | 1.84 | 22 | | 23.84 | |
| Paint | Ea. | 2.88 | 62 | | 64.88 | |
| Minimum Charge | Job | | 232 | | 232 | |
| **Better Grade Veneer** | | | | | | Includes material and labor to install |
| Demolish | Ea. | | 39.50 | | 39.50 | custom modular unit with solid |
| Install | Ea. | 1250 | 116 | | 1366 | hardwood faced frames, hardwood |
| Demolish and Install | Ea. | 1250 | 155.50 | | 1405.50 | door frames and drawer fronts, |
| Reinstall | Ea. | | 92.96 | | 92.96 | hardwood veneer on raised door |
| Clean | Ea. | 1.84 | 22 | | 23.84 | panels, hinges and pulls. |
| Paint | Ea. | 2.88 | 62 | | 64.88 | |
| Minimum Charge | Job | | 232 | | 232 | |

## Tall Cabinets

| Tall Cabinets | Unit | Material | Labor | Equip. | Total | Specification |
|---|---|---|---|---|---|---|
| **Premium Grade Veneer** | | | | | | Includes material and labor to install |
| Demolish | Ea. | | 39.50 | | 39.50 | premium modular unit with solid |
| Install | Ea. | 1375 | 116 | | 1491 | hardwood faced frames, hardwood |
| Demolish and Install | Ea. | 1375 | 155.50 | | 1530.50 | door frames and drawer fronts, |
| Reinstall | Ea. | | 92.96 | | 92.96 | hardwood veneer on raised door |
| Clean | Ea. | 1.84 | 22 | | 23.84 | panels, hinges and pulls. |
| Paint | Ea. | 2.88 | 62 | | 64.88 | |
| Minimum Charge | Job | | 232 | | 232 | |
| **Premium Hardwood** | | | | | | Includes material and labor to install |
| Demolish | Ea. | | 39.50 | | 39.50 | premium hardwood modular unit with |
| Install | Ea. | 1525 | 116 | | 1641 | solid hardwood faced frames, drawer |
| Demolish and Install | Ea. | 1525 | 155.50 | | 1680.50 | fronts, and door panels, steel drawer |
| Reinstall | Ea. | | 92.96 | | 92.96 | guides, hinges and pulls, laminated |
| Clean | Ea. | 1.84 | 22 | | 23.84 | interior. |
| Paint | Ea. | 2.88 | 62 | | 64.88 | |
| Minimum Charge | Job | | 232 | | 232 | |

## Utility 18" W x 12" D

| | Unit | Material | Labor | Equip. | Total | Specification |
|---|---|---|---|---|---|---|
| **Good Grade Laminated** | | | | | | Includes material and labor to install |
| Demolish | Ea. | | 39.50 | | 39.50 | good modular unit with melamine |
| Install | Ea. | 665 | 93 | | 758 | laminated to particle board, in |
| Demolish and Install | Ea. | 665 | 132.50 | | 797.50 | textured colors or wood grain print |
| Reinstall | Ea. | | 74.37 | | 74.37 | finish, hinges and pulls. |
| Clean | Ea. | .43 | 10.75 | | 11.18 | |
| Paint | Ea. | 1.10 | 24 | | 25.10 | |
| Minimum Charge | Job | | 232 | | 232 | |
| **Better Grade Veneer** | | | | | | Includes material and labor to install |
| Demolish | Ea. | | 39.50 | | 39.50 | custom modular unit with solid |
| Install | Ea. | 885 | 93 | | 978 | hardwood faced frames, hardwood |
| Demolish and Install | Ea. | 885 | 132.50 | | 1017.50 | door frames and drawer fronts, |
| Reinstall | Ea. | | 74.37 | | 74.37 | hardwood veneer on raised door |
| Clean | Ea. | .43 | 10.75 | | 11.18 | panels, hinges and pulls. |
| Paint | Ea. | 1.10 | 26 | | 27.10 | |
| Minimum Charge | Job | | 232 | | 232 | |
| **Premium Grade Veneer** | | | | | | Includes material and labor to install |
| Demolish | Ea. | | 39.50 | | 39.50 | premium modular unit with solid |
| Install | Ea. | 1050 | 93 | | 1143 | hardwood faced frames, hardwood |
| Demolish and Install | Ea. | 1050 | 132.50 | | 1182.50 | door frames and drawer fronts, |
| Reinstall | Ea. | | 74.37 | | 74.37 | hardwood veneer on raised door |
| Clean | Ea. | .43 | 10.75 | | 11.18 | panels, hinges and pulls. |
| Paint | Ea. | 1.10 | 29.50 | | 30.60 | |
| Minimum Charge | Job | | 232 | | 232 | |
| **Premium Hardwood** | | | | | | Includes material and labor to install |
| Demolish | Ea. | | 39.50 | | 39.50 | premium hardwood modular unit with |
| Install | Ea. | 1150 | 93 | | 1243 | solid hardwood faced frames, drawer |
| Demolish and Install | Ea. | 1150 | 132.50 | | 1282.50 | fronts, and door panels, steel drawer |
| Reinstall | Ea. | | 74.37 | | 74.37 | guides, hinges and pulls, laminated |
| Clean | Ea. | .43 | 10.75 | | 11.18 | interior. |
| Paint | Ea. | 1.10 | 32 | | 33.10 | |
| Minimum Charge | Job | | 232 | | 232 | |

| Tall Cabinets | Unit | Material | Labor | Equip. | Total | Specification |
|---|---|---|---|---|---|---|
| **Utility 24″ W x 24″ D** | | | | | | |
| **Good Grade Laminated** | | | | | | |
| Demolish | Ea. | | 39.50 | | 39.50 | Includes material and labor to install |
| Install | Ea. | 780 | 116 | | 896 | good modular unit with melamine |
| Demolish and Install | Ea. | 780 | 155.50 | | 935.50 | laminated to particle board, in |
| Reinstall | Ea. | | 92.96 | | 92.96 | textured colors or wood grain print |
| Clean | Ea. | .56 | 14.20 | | 14.76 | finish, hinges and pulls. |
| Paint | Ea. | 1.47 | 32 | | 33.47 | |
| Minimum Charge | Job | | 232 | | 232 | |
| **Better Grade Veneer** | | | | | | |
| Demolish | Ea. | | 39.50 | | 39.50 | Includes material and labor to install |
| Install | Ea. | 1050 | 116 | | 1166 | custom modular unit with solid |
| Demolish and Install | Ea. | 1050 | 155.50 | | 1205.50 | hardwood faced frames, hardwood |
| Reinstall | Ea. | | 92.96 | | 92.96 | door frames and drawer fronts, |
| Clean | Ea. | .56 | 14.20 | | 14.76 | hardwood veneer on raised door |
| Paint | Ea. | 1.47 | 35 | | 36.47 | panels, hinges and pulls. |
| Minimum Charge | Job | | 232 | | 232 | |
| **Premium Grade Veneer** | | | | | | |
| Demolish | Ea. | | 39.50 | | 39.50 | Includes material and labor to install |
| Install | Ea. | 1275 | 116 | | 1391 | premium modular unit with solid |
| Demolish and Install | Ea. | 1275 | 155.50 | | 1430.50 | hardwood faced frames, hardwood |
| Reinstall | Ea. | | 92.96 | | 92.96 | door frames and drawer fronts, |
| Clean | Ea. | .56 | 14.20 | | 14.76 | hardwood veneer on raised door |
| Paint | Ea. | 1.47 | 38.50 | | 39.97 | panels, hinges and pulls. |
| Minimum Charge | Job | | 232 | | 232 | |
| **Premium Hardwood** | | | | | | |
| Demolish | Ea. | | 39.50 | | 39.50 | Includes material and labor to install |
| Install | Ea. | 1375 | 116 | | 1491 | premium hardwood modular unit with |
| Demolish and Install | Ea. | 1375 | 155.50 | | 1530.50 | solid hardwood faced frames, drawer |
| Reinstall | Ea. | | 92.96 | | 92.96 | fronts, and door panels, steel drawer |
| Clean | Ea. | .56 | 14.20 | | 14.76 | guides, hinges and pulls, laminated |
| Paint | Ea. | 1.47 | 43 | | 44.47 | interior. |
| Minimum Charge | Job | | 232 | | 232 | |
| **Plain Shelves** | | | | | | |
| Install | Ea. | 33.50 | 26 | | 59.50 | Includes labor and material to install |
| | | | | | | tall cabinet shelving, per set. |
| **Rotating Shelves** | | | | | | |
| Install | Ea. | 111 | 23 | | 134 | Includes labor and material to install |
| | | | | | | lazy susan shelving for wall cabinet. |
| **Strip and Refinish** | | | | | | |
| Install | L.F. | 10.75 | 77.50 | | 88.25 | Includes labor and material to strip, |
| Minimum Charge | Job | | 193 | | 193 | prep and refinish exterior of cabinets. |
| **Stain** | | | | | | |
| Install | L.F. | .33 | 3.51 | | 3.84 | Includes labor and material to replace |
| Minimum Charge | Job | | 193 | | 193 | normal prep and stain on doors and |
| | | | | | | exterior cabinets. |

| Base Cabinets | Unit | Material | Labor | Equip. | Total | Specification |
|---|---|---|---|---|---|---|
| **Good Grade Laminated** | | | | | | |
| **12″ w, 1 Door, 1 Drawer** | | | | | | |
| Demolish | Ea. | | 9.85 | | 9.85 | Includes material and labor to install |
| Install | Ea. | 249 | 37.50 | | 286.50 | good modular unit with melamine |
| Demolish and Install | Ea. | 249 | 47.35 | | 296.35 | laminate, textured colors or wood |
| Reinstall | Ea. | | 29.99 | | 29.99 | grain print finish including hinges and |
| Clean | Ea. | .30 | 7.10 | | 7.40 | pulls. |
| Paint | Ea. | .74 | 16.10 | | 16.84 | |
| Minimum Charge | Job | | 232 | | 232 | |
| **15″ w, 1 Door, 1 Drawer** | | | | | | |
| Demolish | Ea. | | 12.35 | | 12.35 | Includes material and labor to install |
| Install | Ea. | 260 | 38.50 | | 298.50 | good modular unit with melamine |
| Demolish and Install | Ea. | 260 | 50.85 | | 310.85 | laminate, textured colors or wood |
| Reinstall | Ea. | | 30.99 | | 30.99 | grain print finish including hinges and |
| Clean | Ea. | .36 | 8.90 | | 9.26 | pulls. |
| Paint | Ea. | .92 | 20.50 | | 21.42 | |
| Minimum Charge | Job | | 232 | | 232 | |
| **18″ w, 1 Door, 1 Drawer** | | | | | | |
| Demolish | Ea. | | 14.60 | | 14.60 | Includes material and labor to install |
| Install | Ea. | 282 | 40 | | 322 | good modular unit with melamine |
| Demolish and Install | Ea. | 282 | 54.60 | | 336.60 | laminate, textured colors or wood |
| Reinstall | Ea. | | 31.92 | | 31.92 | grain print finish including hinges and |
| Clean | Ea. | .43 | 10.75 | | 11.18 | pulls. |
| Paint | Ea. | 1.10 | 24 | | 25.10 | |
| Minimum Charge | Job | | 232 | | 232 | |
| **21″ w, 1 Door, 1 Drawer** | | | | | | |
| Demolish | Ea. | | 17.15 | | 17.15 | Includes material and labor to install |
| Install | Ea. | 290 | 41 | | 331 | good modular unit with melamine |
| Demolish and Install | Ea. | 290 | 58.15 | | 348.15 | laminate, textured colors or wood |
| Reinstall | Ea. | | 32.76 | | 32.76 | grain print finish including hinges and |
| Clean | Ea. | .53 | 12.70 | | 13.23 | pulls. |
| Paint | Ea. | 1.30 | 27.50 | | 28.80 | |
| Minimum Charge | Job | | 232 | | 232 | |
| **24″ w, 1 Door, 1 Drawer** | | | | | | |
| Demolish | Ea. | | 19.75 | | 19.75 | Includes material and labor to install |
| Install | Ea. | 340 | 41.50 | | 381.50 | good modular unit with melamine |
| Demolish and Install | Ea. | 340 | 61.25 | | 401.25 | laminate, textured colors or wood |
| Reinstall | Ea. | | 33.35 | | 33.35 | grain print finish including hinges and |
| Clean | Ea. | .56 | 14.20 | | 14.76 | pulls. |
| Paint | Ea. | 1.47 | 32 | | 33.47 | |
| Minimum Charge | Job | | 232 | | 232 | |
| **30″ Wide Sink Base** | | | | | | |
| Demolish | Ea. | | 24.50 | | 24.50 | Includes material and labor to install |
| Install | Ea. | 335 | 43.50 | | 378.50 | good modular unit with melamine |
| Demolish and Install | Ea. | 335 | 68 | | 403 | laminate, textured colors or wood |
| Reinstall | Ea. | | 34.75 | | 34.75 | grain print finish including hinges and |
| Clean | Ea. | .74 | 17.75 | | 18.49 | pulls. |
| Paint | Ea. | 1.85 | 38.50 | | 40.35 | |
| Minimum Charge | Job | | 232 | | 232 | |
| **36″ Wide Sink Base** | | | | | | |
| Demolish | Ea. | | 30.50 | | 30.50 | Includes material and labor to install |
| Install | Ea. | 375 | 46 | | 421 | good modular unit with melamine |
| Demolish and Install | Ea. | 375 | 76.50 | | 451.50 | laminate, textured colors or wood |
| Reinstall | Ea. | | 36.63 | | 36.63 | grain print finish including hinges and |
| Clean | Ea. | .81 | 21 | | 21.81 | pulls. |
| Paint | Ea. | 2.23 | 48.50 | | 50.73 | |
| Minimum Charge | Job | | 232 | | 232 | |

# Cabinets and Countertops

| Base Cabinets | Unit | Material | Labor | Equip. | Total | Specification |
|---|---|---|---|---|---|---|
| **36" Blind Corner** | | | | | | |
| Demolish | Ea. | | 30.50 | | 30.50 | Includes material and labor to install |
| Install | Ea. | 615 | 51.50 | | 666.50 | good modular unit with melamine |
| Demolish and Install | Ea. | 615 | 82 | | 697 | laminate, textured colors or wood |
| Reinstall | Ea. | | 41.32 | | 41.32 | grain print finish including hinges and |
| Clean | Ea. | .81 | 21 | | 21.81 | pulls. |
| Paint | Ea. | 2.23 | 48.50 | | 50.73 | |
| Minimum Charge | Job | | 232 | | 232 | |
| **42" w, 2 Door, 2 Drawer** | | | | | | |
| Demolish | Ea. | | 36 | | 36 | Includes material and labor to install |
| Install | Ea. | 455 | 47 | | 502 | good modular unit with melamine |
| Demolish and Install | Ea. | 455 | 83 | | 538 | laminate, textured colors or wood |
| Reinstall | Ea. | | 37.56 | | 37.56 | grain print finish including hinges and |
| Clean | Ea. | 1.06 | 25.50 | | 26.56 | pulls. |
| Paint | Ea. | 2.56 | 55 | | 57.56 | |
| Minimum Charge | Job | | 232 | | 232 | |
| **Clean Interior** | | | | | | |
| Clean | L.F. | .30 | 5.40 | | 5.70 | Includes labor and materials to clean |
| Minimum Charge | Job | | 178 | | 178 | the interior of a 27" cabinet. |
| **Better Grade Wood** | | | | | | |
| **12" w, 1 Door, 1 Drawer** | | | | | | |
| Demolish | Ea. | | 9.85 | | 9.85 | Includes material and labor to install |
| Install | Ea. | 330 | 37.50 | | 367.50 | custom modular unit including solid |
| Demolish and Install | Ea. | 330 | 47.35 | | 377.35 | hardwood faced frames, hardwood |
| Reinstall | Ea. | | 29.99 | | 29.99 | door frames and drawer fronts, |
| Clean | Ea. | .30 | 7.10 | | 7.40 | hardwood veneer on raised door |
| Paint | Ea. | .74 | 17.55 | | 18.29 | panels, hinges and pulls. |
| Minimum Charge | Job | | 232 | | 232 | |
| **15" w, 1 Door, 1 Drawer** | | | | | | |
| Demolish | Ea. | | 12.35 | | 12.35 | Includes material and labor to install |
| Install | Ea. | 345 | 38.50 | | 383.50 | custom modular unit including solid |
| Demolish and Install | Ea. | 345 | 50.85 | | 395.85 | hardwood faced frames, hardwood |
| Reinstall | Ea. | | 30.99 | | 30.99 | door frames and drawer fronts, |
| Clean | Ea. | .36 | 8.90 | | 9.26 | hardwood veneer on raised door |
| Paint | Ea. | .92 | 22.50 | | 23.42 | panels, hinges and pulls. |
| Minimum Charge | Job | | 232 | | 232 | |
| **18" w, 1 Door, 1 Drawer** | | | | | | |
| Demolish | Ea. | | 14.60 | | 14.60 | Includes material and labor to install |
| Install | Ea. | 375 | 40 | | 415 | custom modular unit including solid |
| Demolish and Install | Ea. | 375 | 54.60 | | 429.60 | hardwood faced frames, hardwood |
| Reinstall | Ea. | | 31.92 | | 31.92 | door frames and drawer fronts, |
| Clean | Ea. | .43 | 10.75 | | 11.18 | hardwood veneer on raised door |
| Paint | Ea. | 1.10 | 26 | | 27.10 | panels, hinges and pulls. |
| Minimum Charge | Job | | 232 | | 232 | |
| **21" w, 1 Door, 1 Drawer** | | | | | | |
| Demolish | Ea. | | 17.15 | | 17.15 | Includes material and labor to install |
| Install | Ea. | 385 | 41 | | 426 | custom modular unit including solid |
| Demolish and Install | Ea. | 385 | 58.15 | | 443.15 | hardwood faced frames, hardwood |
| Reinstall | Ea. | | 32.76 | | 32.76 | door frames and drawer fronts, |
| Clean | Ea. | .53 | 12.70 | | 13.23 | hardwood veneer on raised door |
| Paint | Ea. | 1.30 | 29.50 | | 30.80 | panels, hinges and pulls. |
| Minimum Charge | Job | | 232 | | 232 | |

# Cabinets and Countertops

| Base Cabinets | Unit | Material | Labor | Equip. | Total | Specification |
|---|---|---|---|---|---|---|
| **24" w, 1 Door, 1 Drawer** | | | | | | |
| Demolish | Ea. | | 19.75 | | 19.75 | Includes material and labor to install |
| Install | Ea. | 455 | 41.50 | | 496.50 | custom modular unit including solid |
| Demolish and Install | Ea. | 455 | 61.25 | | 516.25 | hardwood faced frames, hardwood |
| Reinstall | Ea. | | 33.35 | | 33.35 | door frames and drawer fronts, |
| Clean | Ea. | .56 | 14.20 | | 14.76 | hardwood veneer on raised door |
| Paint | Ea. | 1.47 | 35 | | 36.47 | panels, hinges and pulls. |
| Minimum Charge | Job | | 232 | | 232 | |
| **30" Wide Sink Base** | | | | | | |
| Demolish | Ea. | | 24.50 | | 24.50 | Includes material and labor to install |
| Install | Ea. | 445 | 43.50 | | 488.50 | custom modular unit including solid |
| Demolish and Install | Ea. | 445 | 68 | | 513 | hardwood faced frames, hardwood |
| Reinstall | Ea. | | 34.75 | | 34.75 | door frames and drawer fronts, |
| Clean | Ea. | .74 | 17.75 | | 18.49 | hardwood veneer on raised door |
| Paint | Ea. | 1.85 | 43 | | 44.85 | panels, hinges and pulls. |
| Minimum Charge | Job | | 232 | | 232 | |
| **36" Wide Sink Base** | | | | | | |
| Demolish | Ea. | | 30.50 | | 30.50 | Includes material and labor to install |
| Install | Ea. | 500 | 46 | | 546 | custom modular unit including solid |
| Demolish and Install | Ea. | 500 | 76.50 | | 576.50 | hardwood faced frames, hardwood |
| Reinstall | Ea. | | 36.63 | | 36.63 | door frames and drawer fronts, |
| Clean | Ea. | .81 | 21 | | 21.81 | hardwood veneer on raised door |
| Paint | Ea. | 2.23 | 55 | | 57.23 | panels, hinges and pulls. |
| Minimum Charge | Job | | 232 | | 232 | |
| **36" Blind Corner** | | | | | | |
| Demolish | Ea. | | 30.50 | | 30.50 | Includes material and labor to install |
| Install | Ea. | 820 | 51.50 | | 871.50 | custom modular unit including solid |
| Demolish and Install | Ea. | 820 | 82 | | 902 | hardwood faced frames, hardwood |
| Reinstall | Ea. | | 41.32 | | 41.32 | door frames and drawer fronts, |
| Clean | Ea. | .81 | 21 | | 21.81 | hardwood veneer on raised door |
| Paint | Ea. | 2.23 | 55 | | 57.23 | panels, hinges and pulls. |
| Minimum Charge | Job | | 232 | | 232 | |
| **42" w, 2 Door, 2 Drawer** | | | | | | |
| Demolish | Ea. | | 36 | | 36 | Includes material and labor to install |
| Install | Ea. | 605 | 47 | | 652 | modular unit including solid hardwood |
| Demolish and Install | Ea. | 605 | 83 | | 688 | faced frames, hardwood door frames |
| Reinstall | Ea. | | 37.56 | | 37.56 | and drawer fronts, hardwood veneer |
| Clean | Ea. | 1.06 | 25.50 | | 26.56 | on raised door panels, hinges and |
| Paint | Ea. | 2.56 | 64.50 | | 67.06 | pulls. |
| Minimum Charge | Job | | 232 | | 232 | |
| **Clean Interior** | | | | | | |
| Clean | L.F. | .30 | 5.40 | | 5.70 | Includes labor and materials to clean |
| Minimum Charge | Job | | 178 | | 178 | the interior of a 27" cabinet. |
| **Premium Grade Wood** | | | | | | |
| **12" w, 1 Door, 1 Drawer** | | | | | | |
| Demolish | Ea. | | 9.85 | | 9.85 | Includes material and labor to install |
| Install | Ea. | 395 | 18.75 | | 413.75 | premium hardwood modular unit with |
| Demolish and Install | Ea. | 395 | 28.60 | | 423.60 | solid hardwood faced frames, drawer |
| Reinstall | Ea. | | 14.99 | | 14.99 | fronts, and door panels, steel drawer |
| Clean | Ea. | .30 | 7.10 | | 7.40 | guides, hinges, pulls and laminated |
| Paint | Ea. | .74 | 19.30 | | 20.04 | interior. |
| Minimum Charge | Job | | 232 | | 232 | |

# Cabinets and Countertops

| Base Cabinets | Unit | Material | Labor | Equip. | Total | Specification |
|---|---|---|---|---|---|---|
| **15" w, 1 Door, 1 Drawer** | | | | | | |
| Demolish | Ea. | | 12.35 | | 12.35 | Includes material and labor to install |
| Install | Ea. | 415 | 38.50 | | 453.50 | premium modular unit including solid |
| Demolish and Install | Ea. | 415 | 50.85 | | 465.85 | hardwood faced frames, hardwood |
| Reinstall | Ea. | | 30.99 | | 30.99 | door frames and drawer fronts, |
| Clean | Ea. | .36 | 8.90 | | 9.26 | hardwood veneer on raised door |
| Paint | Ea. | .92 | 24 | | 24.92 | panels, hinges and pulls. |
| Minimum Charge | Job | | 232 | | 232 | |
| **18" w, 1 Door, 1 Drawer** | | | | | | |
| Demolish | Ea. | | 14.60 | | 14.60 | Includes material and labor to install |
| Install | Ea. | 450 | 40 | | 490 | premium modular unit including solid |
| Demolish and Install | Ea. | 450 | 54.60 | | 504.60 | hardwood faced frames, hardwood |
| Reinstall | Ea. | | 31.92 | | 31.92 | door frames and drawer fronts, |
| Clean | Ea. | .43 | 10.75 | | 11.18 | hardwood veneer on raised door |
| Paint | Ea. | 1.10 | 29.50 | | 30.60 | panels, hinges and pulls. |
| Minimum Charge | Job | | 232 | | 232 | |
| **21" w, 1 Door, 1 Drawer** | | | | | | |
| Demolish | Ea. | | 17.15 | | 17.15 | Includes material and labor to install |
| Install | Ea. | 470 | 41 | | 511 | premium modular unit including solid |
| Demolish and Install | Ea. | 470 | 58.15 | | 528.15 | hardwood faced frames, hardwood |
| Reinstall | Ea. | | 32.76 | | 32.76 | door frames and drawer fronts, |
| Clean | Ea. | .53 | 12.70 | | 13.23 | hardwood veneer on raised door |
| Paint | Ea. | 1.30 | 32 | | 33.30 | panels, hinges and pulls. |
| Minimum Charge | Job | | 232 | | 232 | |
| **24" w, 1 Door, 1 Drawer** | | | | | | |
| Demolish | Ea. | | 19.75 | | 19.75 | Includes material and labor to install |
| Install | Ea. | 545 | 41.50 | | 586.50 | premium modular unit including solid |
| Demolish and Install | Ea. | 545 | 61.25 | | 606.25 | hardwood faced frames, hardwood |
| Reinstall | Ea. | | 33.35 | | 33.35 | door frames and drawer fronts, |
| Clean | Ea. | .56 | 14.20 | | 14.76 | hardwood veneer on raised door |
| Paint | Ea. | 1.47 | 38.50 | | 39.97 | panels, hinges and pulls. |
| Minimum Charge | Job | | 232 | | 232 | |
| **30" Wide Sink Base** | | | | | | |
| Demolish | Ea. | | 24.50 | | 24.50 | Includes material and labor to install |
| Install | Ea. | 535 | 43.50 | | 578.50 | premium modular unit including solid |
| Demolish and Install | Ea. | 535 | 68 | | 603 | hardwood faced frames, hardwood |
| Reinstall | Ea. | | 34.75 | | 34.75 | door frames and drawer fronts, |
| Clean | Ea. | .74 | 17.75 | | 18.49 | hardwood veneer on raised door |
| Paint | Ea. | 1.85 | 48.50 | | 50.35 | panels, hinges and pulls. |
| Minimum Charge | Job | | 232 | | 232 | |
| **36" Wide Sink Base** | | | | | | |
| Demolish | Ea. | | 30.50 | | 30.50 | Includes material and labor to install |
| Install | Ea. | 600 | 46 | | 646 | premium modular unit including solid |
| Demolish and Install | Ea. | 600 | 76.50 | | 676.50 | hardwood faced frames, hardwood |
| Reinstall | Ea. | | 36.63 | | 36.63 | door frames and drawer fronts, |
| Clean | Ea. | .81 | 21 | | 21.81 | hardwood veneer on raised door |
| Paint | Ea. | 2.23 | 55 | | 57.23 | panels, hinges and pulls. |
| Minimum Charge | Job | | 232 | | 232 | |
| **36" Blind Corner** | | | | | | |
| Demolish | Ea. | | 30.50 | | 30.50 | Includes material and labor to install |
| Install | Ea. | 985 | 51.50 | | 1036.50 | premium modular unit including solid |
| Demolish and Install | Ea. | 985 | 82 | | 1067 | hardwood faced frames, hardwood |
| Reinstall | Ea. | | 41.32 | | 41.32 | door frames and drawer fronts, |
| Clean | Ea. | .81 | 21 | | 21.81 | hardwood veneer on raised door |
| Paint | Ea. | 2.23 | 55 | | 57.23 | panels, hinges and pulls. |
| Minimum Charge | Job | | 232 | | 232 | |

# Cabinets and Countertops

| Base Cabinets | Unit | Material | Labor | Equip. | Total | Specification |
|---|---|---|---|---|---|---|
| **42″ w, 2 Door, 2 Drawer** | | | | | | |
| Demolish | Ea. | | 36 | | 36 | Includes material and labor to install |
| Install | Ea. | 810 | 47 | | 857 | premium modular unit including solid |
| Demolish and Install | Ea. | 810 | 83 | | 893 | hardwood faced frames, hardwood |
| Reinstall | Ea. | | 37.56 | | 37.56 | door frames and drawer fronts, |
| Clean | Ea. | 1.06 | 25.50 | | 26.56 | hardwood veneer on raised door |
| Paint | Ea. | 2.56 | 64.50 | | 67.06 | panels, hinges and pulls. |
| Minimum Charge | Job | | 232 | | 232 | |
| **Clean Interior** | | | | | | |
| Clean | L.F. | .30 | 5.40 | | 5.70 | Includes labor and materials to clean |
| Minimum Charge | Job | | 178 | | 178 | the interior of a 27″ cabinet. |
| **Premium Hardwood** | | | | | | |
| **12″ w, 1 Door, 1 Drawer** | | | | | | |
| Demolish | Ea. | | 9.85 | | 9.85 | Includes material and labor to install |
| Install | Ea. | 435 | 37.50 | | 472.50 | premium hardwood modular unit with |
| Demolish and Install | Ea. | 435 | 47.35 | | 482.35 | solid hardwood faced frames, drawer |
| Reinstall | Ea. | | 29.99 | | 29.99 | fronts, and door panels, steel drawer |
| Clean | Ea. | .30 | 7.10 | | 7.40 | guides, hinges, pulls and laminated |
| Paint | Ea. | .74 | 21.50 | | 22.24 | interior. |
| Minimum Charge | Job | | 232 | | 232 | |
| **15″ w, 1 Door, 1 Drawer** | | | | | | |
| Demolish | Ea. | | 12.35 | | 12.35 | Includes material and labor to install |
| Install | Ea. | 450 | 38.50 | | 488.50 | premium hardwood modular unit with |
| Demolish and Install | Ea. | 450 | 50.85 | | 500.85 | solid hardwood faced frames, drawer |
| Reinstall | Ea. | | 30.99 | | 30.99 | fronts, and door panels, steel drawer |
| Clean | Ea. | .36 | 8.90 | | 9.26 | guides, hinges, pulls and laminated |
| Paint | Ea. | .92 | 27.50 | | 28.42 | interior. |
| Minimum Charge | Job | | 232 | | 232 | |
| **18″ w, 1 Door, 1 Drawer** | | | | | | |
| Demolish | Ea. | | 14.60 | | 14.60 | Includes material and labor to install |
| Install | Ea. | 490 | 40 | | 530 | premium hardwood modular unit with |
| Demolish and Install | Ea. | 490 | 54.60 | | 544.60 | solid hardwood faced frames, drawer |
| Reinstall | Ea. | | 31.92 | | 31.92 | fronts, and door panels, steel drawer |
| Clean | Ea. | .43 | 10.75 | | 11.18 | guides, hinges, pulls and laminated |
| Paint | Ea. | 1.10 | 32 | | 33.10 | interior. |
| Minimum Charge | Job | | 232 | | 232 | |
| **21″ w, 1 Door, 1 Drawer** | | | | | | |
| Demolish | Ea. | | 17.15 | | 17.15 | Includes material and labor to install |
| Install | Ea. | 505 | 41 | | 546 | premium hardwood modular unit with |
| Demolish and Install | Ea. | 505 | 58.15 | | 563.15 | solid hardwood faced frames, drawer |
| Reinstall | Ea. | | 32.76 | | 32.76 | fronts, and door panels, steel drawer |
| Clean | Ea. | .53 | 12.70 | | 13.23 | guides, hinges, pulls and laminated |
| Paint | Ea. | 1.30 | 35 | | 36.30 | interior. |
| Minimum Charge | Job | | 232 | | 232 | |
| **24″ w, 1 Door, 1 Drawer** | | | | | | |
| Demolish | Ea. | | 19.75 | | 19.75 | Includes material and labor to install |
| Install | Ea. | 595 | 41.50 | | 636.50 | premium hardwood modular unit with |
| Demolish and Install | Ea. | 595 | 61.25 | | 656.25 | solid hardwood faced frames, drawer |
| Reinstall | Ea. | | 33.35 | | 33.35 | fronts, and door panels, steel drawer |
| Clean | Ea. | .56 | 14.20 | | 14.76 | guides, hinges and pulls, laminated |
| Paint | Ea. | 1.47 | 43 | | 44.47 | interior. |
| Minimum Charge | Job | | 232 | | 232 | |

# Cabinets and Countertops

| Base Cabinets | | Unit | Material | Labor | Equip. | Total | Specification |
|---|---|---|---|---|---|---|---|
| **30" Wide Sink Base** | | | | | | | |
| | Demolish | Ea. | | 24.50 | | 24.50 | Includes material and labor to install |
| | Install | Ea. | 585 | 43.50 | | 628.50 | premium hardwood modular unit with |
| | Demolish and Install | Ea. | 585 | 68 | | 653 | solid hardwood faced frames, drawer |
| | Reinstall | Ea. | | 34.75 | | 34.75 | fronts, and door panels, steel drawer |
| | Clean | Ea. | .74 | 17.75 | | 18.49 | guides, hinges, pulls and laminated |
| | Paint | Ea. | 1.85 | 48.50 | | 50.35 | interior. |
| | Minimum Charge | Job | | 232 | | 232 | |
| **36" Wide Sink Base** | | | | | | | |
| | Demolish | Ea. | | 30.50 | | 30.50 | Includes material and labor to install |
| | Install | Ea. | 655 | 46 | | 701 | premium hardwood modular unit with |
| | Demolish and Install | Ea. | 655 | 76.50 | | 731.50 | solid hardwood faced frames, drawer |
| | Reinstall | Ea. | | 36.63 | | 36.63 | fronts, and door panels, steel drawer |
| | Clean | Ea. | .81 | 21 | | 21.81 | guides, hinges, pulls and laminated |
| | Paint | Ea. | 2.23 | 64.50 | | 66.73 | interior. |
| | Minimum Charge | Job | | 232 | | 232 | |
| **36" Blind Corner** | | | | | | | |
| | Demolish | Ea. | | 30.50 | | 30.50 | Includes material and labor to install |
| | Install | Ea. | 1075 | 51.50 | | 1126.50 | premium hardwood modular unit with |
| | Demolish and Install | Ea. | 1075 | 82 | | 1157 | solid hardwood faced frames, drawer |
| | Reinstall | Ea. | | 41.32 | | 41.32 | fronts, and door panels, steel drawer |
| | Clean | Ea. | .81 | 21 | | 21.81 | guides, hinges, pulls and laminated |
| | Paint | Ea. | 2.23 | 64.50 | | 66.73 | interior. |
| | Minimum Charge | Job | | 232 | | 232 | |
| **42" w, 2 Door, 2 Drawer** | | | | | | | |
| | Demolish | Ea. | | 36 | | 36 | Includes material and labor to install |
| | Install | Ea. | 1050 | 47 | | 1097 | premium hardwood modular unit with |
| | Demolish and Install | Ea. | 1050 | 83 | | 1133 | solid hardwood faced frames, drawer |
| | Reinstall | Ea. | | 37.56 | | 37.56 | fronts, and door panels, steel drawer |
| | Clean | Ea. | 1.06 | 25.50 | | 26.56 | guides, hinges, pulls and laminated |
| | Paint | Ea. | 2.56 | 77.50 | | 80.06 | interior. |
| | Minimum Charge | Job | | 232 | | 232 | |
| **Clean Interior** | | | | | | | |
| | Clean | L.F. | .30 | 5.40 | | 5.70 | Includes labor and materials to clean |
| | Minimum Charge | Job | | 178 | | 178 | the interior of a 27" cabinet. |
| **Island** | | | | | | | |
| **24" Wide** | | | | | | | |
| | Demolish | Ea. | | 19.75 | | 19.75 | Includes material and labor to install |
| | Install | Ea. | 700 | 38.50 | | 738.50 | custom modular unit with solid |
| | Demolish and Install | Ea. | 700 | 58.25 | | 758.25 | hardwood faced frames, hardwood |
| | Reinstall | Ea. | | 30.99 | | 30.99 | door frames and drawer fronts, |
| | Clean | Ea. | .81 | 21 | | 21.81 | hardwood veneer on raised door |
| | Paint | Ea. | 2.21 | 55 | | 57.21 | panels, hinges and pulls. |
| | Minimum Charge | Job | | 232 | | 232 | |
| **30" Wide** | | | | | | | |
| | Demolish | Ea. | | 24.50 | | 24.50 | Includes material and labor to install |
| | Install | Ea. | 780 | 48.50 | | 828.50 | custom modular unit with solid |
| | Demolish and Install | Ea. | 780 | 73 | | 853 | hardwood faced frames, hardwood |
| | Reinstall | Ea. | | 38.73 | | 38.73 | door frames and drawer fronts, |
| | Clean | Ea. | 1.06 | 25.50 | | 26.56 | hardwood veneer on raised door |
| | Paint | Ea. | 2.77 | 64.50 | | 67.27 | panels, hinges and pulls. |
| | Minimum Charge | Job | | 232 | | 232 | |

# Cabinets and Countertops

| Base Cabinets | | Unit | Material | Labor | Equip. | Total | Specification |
|---|---|---|---|---|---|---|---|
| **36″ Wide** | | | | | | | |
| | Demolish | Ea. | | 30.50 | | 30.50 | Includes material and labor to install |
| | Install | Ea. | 900 | 58 | | 958 | custom modular unit with solid |
| | Demolish and Install | Ea. | 900 | 88.50 | | 988.50 | hardwood faced frames, hardwood |
| | Reinstall | Ea. | | 46.48 | | 46.48 | door frames and drawer fronts, |
| | Clean | Ea. | 1.23 | 32.50 | | 33.73 | hardwood veneer on raised door |
| | Paint | Ea. | 3.36 | 77.50 | | 80.86 | panels, hinges and pulls. |
| | Minimum Charge | Job | | 232 | | 232 | |
| **48″ Wide** | | | | | | | |
| | Demolish | Ea. | | 39.50 | | 39.50 | Includes material and labor to install |
| | Install | Ea. | 1400 | 77.50 | | 1477.50 | custom modular unit with solid |
| | Demolish and Install | Ea. | 1400 | 117 | | 1517 | hardwood faced frames, hardwood |
| | Reinstall | Ea. | | 61.97 | | 61.97 | door frames and drawer fronts, |
| | Clean | Ea. | 1.47 | 39.50 | | 40.97 | hardwood veneer on raised door |
| | Paint | Ea. | 3.85 | 96.50 | | 100.35 | panels, hinges and pulls. |
| | Minimum Charge | Job | | 232 | | 232 | |
| **60″ Wide** | | | | | | | |
| | Demolish | Ea. | | 44 | | 44 | Includes material and labor to install |
| | Install | Ea. | 1575 | 97 | | 1672 | custom modular unit with solid |
| | Demolish and Install | Ea. | 1575 | 141 | | 1716 | hardwood faced frames, hardwood |
| | Reinstall | Ea. | | 77.47 | | 77.47 | door frames and drawer fronts, |
| | Clean | Ea. | 1.84 | 50.50 | | 52.34 | hardwood veneer on raised door |
| | Paint | Ea. | 3.85 | 129 | | 132.85 | panels, hinges and pulls. |
| | Minimum Charge | Job | | 232 | | 232 | |
| **Clean Interior** | | | | | | | |
| | Clean | L.F. | .30 | 5.40 | | 5.70 | Includes labor and materials to clean |
| | Minimum Charge | Job | | 178 | | 178 | the interior of a 27″ cabinet. |
| **Liquor Bar** | | | | | | | |
| | Demolish | L.F. | | 9.85 | | 9.85 | Includes material and labor to install |
| | Install | L.F. | 215 | 46.50 | | 261.50 | custom modular unit with solid |
| | Demolish and Install | L.F. | 215 | 56.35 | | 271.35 | hardwood faced frames, hardwood |
| | Reinstall | L.F. | | 37.18 | | 37.18 | door frames and drawer fronts, |
| | Clean | L.F. | .30 | 7.40 | | 7.70 | hardwood veneer on raised door |
| | Paint | L.F. | 5.90 | 17.55 | | 23.45 | panels, hinges and pulls. |
| | Minimum Charge | Job | | 232 | | 232 | |
| **Clean Interior** | | | | | | | |
| | Clean | L.F. | .30 | 5.40 | | 5.70 | Includes labor and materials to clean |
| | Minimum Charge | Job | | 178 | | 178 | the interior of a 27″ cabinet. |
| **Seal Interior** | | | | | | | |
| | Install | L.F. | .57 | 6.35 | | 6.92 | Includes labor and materials to seal a |
| | Minimum Charge | Job | | 193 | | 193 | cabinet interior. |
| **Back Bar** | | | | | | | |
| | Demolish | L.F. | | 9.85 | | 9.85 | Includes material and labor to install |
| | Install | L.F. | 215 | 46.50 | | 261.50 | custom modular unit with solid |
| | Demolish and Install | L.F. | 215 | 56.35 | | 271.35 | hardwood faced frames, hardwood |
| | Reinstall | L.F. | | 37.18 | | 37.18 | door frames and drawer fronts, |
| | Clean | L.F. | .30 | 7.40 | | 7.70 | hardwood veneer on raised door |
| | Paint | L.F. | 5.90 | 17.55 | | 23.45 | panels, hinges and pulls. |
| | Minimum Charge | Job | | 232 | | 232 | |
| **Clean Interior** | | | | | | | |
| | Clean | L.F. | .30 | 5.40 | | 5.70 | Includes labor and materials to clean |
| | Minimum Charge | Job | | 178 | | 178 | the interior of a 27″ cabinet. |

# Cabinets and Countertops

## Base Cabinets

| Base Cabinets | | Unit | Material | Labor | Equip. | Total | Specification |
|---|---|---|---|---|---|---|---|
| **Seal Interior** | | | | | | | |
| | Install | L.F. | .57 | 6.35 | | 6.92 | Includes labor and materials to seal a |
| | Minimum Charge | Job | | 193 | | 193 | cabinet interior. |
| **Storage (garage)** | | | | | | | |
| | Demolish | L.F. | | 9.85 | | 9.85 | Includes material and labor to install |
| | Install | L.F. | 161 | 46.50 | | 207.50 | good modular unit with melamine |
| | Demolish and Install | L.F. | 161 | 56.35 | | 217.35 | laminate, textured colors or wood |
| | Reinstall | L.F. | | 37.18 | | 37.18 | grain print finish including hinges and |
| | Clean | L.F. | .30 | 7.40 | | 7.70 | pulls. |
| | Paint | L.F. | 5.90 | 17.55 | | 23.45 | |
| | Minimum Charge | Job | | 232 | | 232 | |
| **Clean Interior** | | | | | | | |
| | Clean | L.F. | .30 | 5.40 | | 5.70 | Includes labor and materials to clean |
| | Minimum Charge | Job | | 178 | | 178 | the interior of a 27" cabinet. |
| **Seal Interior** | | | | | | | |
| | Install | L.F. | .57 | 6.35 | | 6.92 | Includes labor and materials to seal a |
| | Minimum Charge | Job | | 193 | | 193 | cabinet interior. |
| **Built-in Desk** | | | | | | | |
| | Demolish | L.F. | | 9.85 | | 9.85 | Includes material and labor to install |
| | Install | L.F. | 215 | 46.50 | | 261.50 | custom modular unit with solid |
| | Demolish and Install | L.F. | 215 | 56.35 | | 271.35 | hardwood faced frames, hardwood |
| | Reinstall | L.F. | | 37.18 | | 37.18 | door frames and drawer fronts, |
| | Clean | L.F. | .30 | 7.40 | | 7.70 | hardwood veneer on raised door |
| | Paint | L.F. | 5.90 | 17.55 | | 23.45 | panels, hinges and pulls. |
| | Minimum Charge | Job | | 232 | | 232 | |
| **Built-in Bookcase** | | | | | | | |
| | Demolish | L.F. | | 13.15 | | 13.15 | Includes material and labor to install |
| | Install | L.F. | 44.50 | 46.50 | | 91 | custom modular bookcase unit with |
| | Demolish and Install | L.F. | 44.50 | 59.65 | | 104.15 | clear pine faced frames, shelves 12" |
| | Reinstall | L.F. | | 37.18 | | 37.18 | O.C., 7' high, 8" deep. |
| | Clean | L.F. | .59 | 12.70 | | 13.29 | |
| | Paint | L.F. | 1.73 | 29.50 | | 31.23 | |
| | Minimum Charge | Job | | 232 | | 232 | |
| **Strip and Refinish** | | | | | | | |
| | Install | L.F. | 5.35 | 48.50 | | 53.85 | Includes labor and material to strip, |
| | Minimum Charge | Job | | 193 | | 193 | prep and refinish exterior of cabinets. |
| **Stain** | | | | | | | |
| | Install | L.F. | .75 | 1.80 | | 2.55 | Includes labor and material to stain |
| | Minimum Charge | Job | | 232 | | 232 | door and exterior of cabinets. |

## Countertop

| Countertop | | Unit | Material | Labor | Equip. | Total | Specification |
|---|---|---|---|---|---|---|---|
| **Laminated** | | | | | | | |
| **With Splash** | | | | | | | |
| | Demolish | L.F. | | 6.60 | | 6.60 | Includes material and labor to install |
| | Install | L.F. | 48 | 16.60 | | 64.60 | one piece laminated top with 4" |
| | Demolish and Install | L.F. | 48 | 23.20 | | 71.20 | backsplash. |
| | Reinstall | L.F. | | 13.28 | | 13.28 | |
| | Clean | L.F. | .03 | .89 | | .92 | |
| | Minimum Charge | Job | | 232 | | 232 | |

229

# Cabinets and Countertops

| Countertop | | Unit | Material | Labor | Equip. | Total | Specification |
|---|---|---|---|---|---|---|---|
| **Roll Top** | | | | | | | |
| | Demolish | L.F. | | 6.60 | | 6.60 | Includes material and labor to install |
| | Install | L.F. | 13.85 | 15.50 | | 29.35 | one piece laminated top with rolled |
| | Demolish and Install | L.F. | 13.85 | 22.10 | | 35.95 | drip edge (post formed) and |
| | Reinstall | L.F. | | 12.39 | | 12.39 | backsplash. |
| | Clean | L.F. | .06 | 1.35 | | 1.41 | |
| | Minimum Charge | Job | | 232 | | 232 | |
| **Ceramic Tile** | | | | | | | |
| | Demolish | S.F. | | 3.29 | | 3.29 | Cost includes material and labor to |
| | Install | S.F. | 21.50 | 9.30 | | 30.80 | install 4-1/4" x 4-1/4" to 6" x 6" |
| | Demolish and Install | S.F. | 21.50 | 12.59 | | 34.09 | glazed tile set in mortar bed and grout |
| | Clean | S.F. | .02 | .68 | | .70 | on particle board substrate. |
| | Minimum Charge | Job | | 219 | | 219 | |
| **Cultured Marble** | | | | | | | |
| | Demolish | L.F. | | 6.60 | | 6.60 | Includes material and labor to install |
| | Install | L.F. | 59.50 | 46.50 | | 106 | marble countertop, 24" wide, no |
| | Demolish and Install | L.F. | 59.50 | 53.10 | | 112.60 | backsplash. |
| | Reinstall | L.F. | | 37.12 | | 37.12 | |
| | Clean | L.F. | .06 | 1.35 | | 1.41 | |
| | Minimum Charge | Job | | 219 | | 219 | |
| **With Splash** | | | | | | | |
| | Demolish | L.F. | | 6.60 | | 6.60 | Includes material and labor to install |
| | Install | L.F. | 68 | 46.50 | | 114.50 | marble countertop, 24" wide, with |
| | Demolish and Install | L.F. | 68 | 53.10 | | 121.10 | backsplash. |
| | Reinstall | L.F. | | 37.12 | | 37.12 | |
| | Clean | L.F. | .06 | 1.35 | | 1.41 | |
| | Minimum Charge | Job | | 219 | | 219 | |
| **Quarry Tile** | | | | | | | |
| | Demolish | S.F. | | 3.29 | | 3.29 | Includes material and labor to install |
| | Install | S.F. | 8.55 | 9.30 | | 17.85 | quarry tile countertop with no |
| | Demolish and Install | S.F. | 8.55 | 12.59 | | 21.14 | backsplash. |
| | Clean | S.F. | .02 | .68 | | .70 | |
| | Minimum Charge | Job | | 219 | | 219 | |
| **With Splash** | | | | | | | |
| | Demolish | S.F. | | 3.29 | | 3.29 | Includes material and labor to install |
| | Install | S.F. | 11.60 | 10.35 | | 21.95 | quarry tile countertop with backsplash. |
| | Demolish and Install | S.F. | 11.60 | 13.64 | | 25.24 | |
| | Clean | S.F. | .02 | .68 | | .70 | |
| | Minimum Charge | Job | | 219 | | 219 | |
| **Butcher Block** | | | | | | | |
| | Demolish | S.F. | | 3.29 | | 3.29 | Includes material and labor to install |
| | Install | S.F. | 48.50 | 8.30 | | 56.80 | solid laminated maple countertop with |
| | Demolish and Install | S.F. | 48.50 | 11.59 | | 60.09 | no backsplash. |
| | Reinstall | S.F. | | 6.64 | | 6.64 | |
| | Clean | S.F. | .04 | .28 | | .32 | |
| | Minimum Charge | Job | | 232 | | 232 | |
| **Solid Surface** | | | | | | | |
| | Demolish | L.F. | | 6.60 | | 6.60 | Includes material and labor to install |
| | Install | L.F. | 58 | 33 | | 91 | solid surface counter top 22" deep. |
| | Demolish and Install | L.F. | 58 | 39.60 | | 97.60 | |
| | Reinstall | L.F. | | 26.56 | | 26.56 | |
| | Clean | L.F. | .06 | 1.35 | | 1.41 | |
| | Minimum Charge | Job | | 232 | | 232 | |

# Cabinets and Countertops

| Vanity Cabinets | Unit | Material | Labor | Equip. | Total | Specification |
|---|---|---|---|---|---|---|
| **Good Grade Laminated** | | | | | | |
| **24" Wide, 2 Door** | | | | | | |
| Demolish | Ea. | | 19.75 | | 19.75 | Includes material and labor to install |
| Install | Ea. | 289 | 46.50 | | 335.50 | modular unit with melamine laminated |
| Demolish and Install | Ea. | 289 | 66.25 | | 355.25 | to particle board, textured colors or |
| Reinstall | Ea. | | 37.18 | | 37.18 | wood grain print finish, hinges and |
| Clean | Ea. | .56 | 14.20 | | 14.76 | pulls. |
| Paint | Ea. | 1.47 | 32 | | 33.47 | |
| Minimum Charge | Job | | 232 | | 232 | |
| **30" Wide, 2 Door** | | | | | | |
| Demolish | Ea. | | 24.50 | | 24.50 | Includes material and labor to install |
| Install | Ea. | 345 | 58 | | 403 | modular unit with melamine laminated |
| Demolish and Install | Ea. | 345 | 82.50 | | 427.50 | to particle board, textured colors or |
| Reinstall | Ea. | | 46.48 | | 46.48 | wood grain print finish, hinges and |
| Clean | Ea. | .74 | 17.75 | | 18.49 | pulls. |
| Paint | Ea. | 1.85 | 38.50 | | 40.35 | |
| Minimum Charge | Job | | 232 | | 232 | |
| **36" Wide, 2 Door** | | | | | | |
| Demolish | Ea. | | 30.50 | | 30.50 | Includes material and labor to install |
| Install | Ea. | 335 | 69.50 | | 404.50 | modular unit with melamine laminated |
| Demolish and Install | Ea. | 335 | 100 | | 435 | to particle board, textured colors or |
| Reinstall | Ea. | | 55.79 | | 55.79 | wood grain print finish, hinges and |
| Clean | Ea. | .81 | 21 | | 21.81 | pulls. |
| Paint | Ea. | 2.23 | 48.50 | | 50.73 | |
| Minimum Charge | Job | | 232 | | 232 | |
| **42" w, 3 Doors / Drawer** | | | | | | |
| Demolish | Ea. | | 36 | | 36 | Includes material and labor to install |
| Install | Ea. | 555 | 75 | | 630 | modular unit with melamine laminated |
| Demolish and Install | Ea. | 555 | 111 | | 666 | to particle board, textured colors or |
| Reinstall | Ea. | | 60.07 | | 60.07 | wood grain print finish, hinges and |
| Clean | Ea. | 1.06 | 25.50 | | 26.56 | pulls. |
| Paint | Ea. | 2.56 | 55 | | 57.56 | |
| Minimum Charge | Job | | 232 | | 232 | |
| **48" w, 3 Doors / Drawer** | | | | | | |
| Demolish | Ea. | | 39.50 | | 39.50 | Includes material and labor to install |
| Install | Ea. | 440 | 81.50 | | 521.50 | modular unit with melamine laminated |
| Demolish and Install | Ea. | 440 | 121 | | 561 | to particle board, textured colors or |
| Reinstall | Ea. | | 65.06 | | 65.06 | wood grain print finish, hinges and |
| Clean | Ea. | 1.47 | 39.50 | | 40.97 | pulls. |
| Paint | Ea. | 3.85 | 96.50 | | 100.35 | |
| Minimum Charge | Job | | 232 | | 232 | |
| **Clean Interior** | | | | | | |
| Install | L.F. | .30 | 7.40 | | 7.70 | Includes labor and materials to clean |
| Minimum Charge | Job | | 178 | | 178 | base cabinetry. |
| **Better Grade Wood** | | | | | | |
| **24" Wide, 2 Door** | | | | | | |
| Demolish | Ea. | | 19.75 | | 19.75 | Includes material and labor to install |
| Install | Ea. | 385 | 46.50 | | 431.50 | custom modular unit with solid |
| Demolish and Install | Ea. | 385 | 66.25 | | 451.25 | hardwood faced frames, hardwood |
| Reinstall | Ea. | | 37.18 | | 37.18 | door frames and drawer fronts, |
| Clean | Ea. | .56 | 14.20 | | 14.76 | hardwood veneer on raised door |
| Paint | Ea. | 1.47 | 35 | | 36.47 | panels, hinges and pulls. |
| Minimum Charge | Job | | 232 | | 232 | |

231

# Cabinets and Countertops

| Vanity Cabinets | Unit | Material | Labor | Equip. | Total | Specification |
|---|---|---|---|---|---|---|
| **30" Wide, 2 Door** | | | | | | |
| Demolish | Ea. | | 24.50 | | 24.50 | Includes material and labor to install |
| Install | Ea. | 460 | 58 | | 518 | custom modular unit with solid |
| Demolish and Install | Ea. | 460 | 82.50 | | 542.50 | hardwood faced frames, hardwood |
| Reinstall | Ea. | | 46.48 | | 46.48 | door frames and drawer fronts, |
| Clean | Ea. | .74 | 17.75 | | 18.49 | hardwood veneer on raised door |
| Paint | Ea. | 1.85 | 43 | | 44.85 | panels, hinges and pulls. |
| Minimum Charge | Job | | 232 | | 232 | |
| **36" Wide, 2 Door** | | | | | | |
| Demolish | Ea. | | 30.50 | | 30.50 | Includes material and labor to install |
| Install | Ea. | 445 | 69.50 | | 514.50 | custom modular unit with solid |
| Demolish and Install | Ea. | 445 | 100 | | 545 | hardwood faced frames, hardwood |
| Reinstall | Ea. | | 55.79 | | 55.79 | door frames and drawer fronts, |
| Clean | Ea. | .81 | 21 | | 21.81 | hardwood veneer on raised door |
| Paint | Ea. | 2.23 | 55 | | 57.23 | panels, hinges and pulls. |
| Minimum Charge | Job | | 232 | | 232 | |
| **42" w, 3 Doors / Drawer** | | | | | | |
| Demolish | Ea. | | 36 | | 36 | Includes material and labor to install |
| Install | Ea. | 555 | 75 | | 630 | custom modular unit with solid |
| Demolish and Install | Ea. | 555 | 111 | | 666 | hardwood faced frames, hardwood |
| Reinstall | Ea. | | 60.07 | | 60.07 | door frames and drawer fronts, |
| Clean | Ea. | 1.06 | 25.50 | | 26.56 | hardwood veneer on raised door |
| Paint | Ea. | 2.56 | 64.50 | | 67.06 | panels, hinges and pulls. |
| Minimum Charge | Job | | 232 | | 232 | |
| **48" w, 3 Doors / Drawer** | | | | | | |
| Demolish | Ea. | | 39.50 | | 39.50 | Includes material and labor to install |
| Install | Ea. | 585 | 81.50 | | 666.50 | custom modular unit with solid |
| Demolish and Install | Ea. | 585 | 121 | | 706 | hardwood faced frames, hardwood |
| Reinstall | Ea. | | 65.06 | | 65.06 | door frames and drawer fronts, |
| Clean | Ea. | 1.47 | 39.50 | | 40.97 | hardwood veneer on raised door |
| Paint | Ea. | 3.85 | 96.50 | | 100.35 | panels, hinges and pulls. |
| Minimum Charge | Job | | 232 | | 232 | |
| **Clean Interior** | | | | | | |
| Install | L.F. | .30 | 7.40 | | 7.70 | Includes labor and materials to clean |
| Minimum Charge | Job | | 178 | | 178 | base cabinetry. |
| **Premium Grade Wood** | | | | | | |
| **24" Wide, 2 Door** | | | | | | |
| Demolish | Ea. | | 19.75 | | 19.75 | Includes material and labor to install |
| Install | Ea. | 405 | 46.50 | | 451.50 | premium modular unit with solid |
| Demolish and Install | Ea. | 405 | 66.25 | | 471.25 | hardwood faced frames, hardwood |
| Reinstall | Ea. | | 37.18 | | 37.18 | door frames and drawer fronts, |
| Clean | Ea. | .56 | 14.20 | | 14.76 | hardwood veneer on raised door |
| Paint | Ea. | 1.47 | 38.50 | | 39.97 | panels, hinges and pulls. |
| Minimum Charge | Job | | 232 | | 232 | |
| **30" Wide, 2 Door** | | | | | | |
| Demolish | Ea. | | 24.50 | | 24.50 | Includes material and labor to install |
| Install | Ea. | 565 | 58 | | 623 | premium modular unit with solid |
| Demolish and Install | Ea. | 565 | 82.50 | | 647.50 | hardwood faced frames, hardwood |
| Reinstall | Ea. | | 46.48 | | 46.48 | door frames and drawer fronts, |
| Clean | Ea. | .74 | 17.75 | | 18.49 | hardwood veneer on raised door |
| Paint | Ea. | 1.85 | 48.50 | | 50.35 | panels, hinges and pulls. |
| Minimum Charge | Job | | 232 | | 232 | |

## Vanity Cabinets

| Vanity Cabinets | Unit | Material | Labor | Equip. | Total | Specification |
|---|---|---|---|---|---|---|
| **36″ Wide, 2 Door** | | | | | | |
| Demolish | Ea. | | 30.50 | | 30.50 | Includes material and labor to install |
| Install | Ea. | 550 | 69.50 | | 619.50 | premium modular unit with solid |
| Demolish and Install | Ea. | 550 | 100 | | 650 | hardwood faced frames, hardwood |
| Reinstall | Ea. | | 55.79 | | 55.79 | door frames and drawer fronts, |
| Clean | Ea. | .81 | 21 | | 21.81 | hardwood veneer on raised door |
| Paint | Ea. | 2.23 | 55 | | 57.23 | panels, hinges and pulls. |
| Minimum Charge | Job | | 232 | | 232 | |
| **42″ w, 3 Doors / Drawer** | | | | | | |
| Demolish | Ea. | | 36 | | 36 | Includes material and labor to install |
| Install | Ea. | 645 | 75 | | 720 | premium modular unit with solid |
| Demolish and Install | Ea. | 645 | 111 | | 756 | hardwood faced frames, hardwood |
| Reinstall | Ea. | | 60.07 | | 60.07 | door frames and drawer fronts, |
| Clean | Ea. | 1.06 | 25.50 | | 26.56 | hardwood veneer on raised door |
| Paint | Ea. | 2.56 | 64.50 | | 67.06 | panels, hinges and pulls. |
| Minimum Charge | Job | | 232 | | 232 | |
| **48″ w, 3 Doors / Drawer** | | | | | | |
| Demolish | Ea. | | 39.50 | | 39.50 | Includes material and labor to install |
| Install | Ea. | 670 | 81.50 | | 751.50 | premium modular unit with solid |
| Demolish and Install | Ea. | 670 | 121 | | 791 | hardwood faced frames, hardwood |
| Reinstall | Ea. | | 65.06 | | 65.06 | door frames and drawer fronts, |
| Clean | Ea. | 1.47 | 39.50 | | 40.97 | hardwood veneer on raised door |
| Paint | Ea. | 3.85 | 96.50 | | 100.35 | panels, hinges and pulls. |
| Minimum Charge | Job | | 232 | | 232 | |
| **Clean Interior** | | | | | | |
| Install | L.F. | .30 | 7.40 | | 7.70 | Includes labor and materials to clean |
| Minimum Charge | Job | | 178 | | 178 | base cabinetry. |

## Premium Hardwood

| Premium Hardwood | Unit | Material | Labor | Equip. | Total | Specification |
|---|---|---|---|---|---|---|
| **24″ Wide, 2 Door** | | | | | | |
| Demolish | Ea. | | 19.75 | | 19.75 | Includes material and labor to install |
| Install | Ea. | 480 | 46.50 | | 526.50 | premium modular unit with solid |
| Demolish and Install | Ea. | 480 | 66.25 | | 546.25 | hardwood faced frames, hardwood |
| Reinstall | Ea. | | 37.18 | | 37.18 | door frames and drawer fronts, |
| Clean | Ea. | .56 | 14.20 | | 14.76 | hardwood veneer on raised door |
| Paint | Ea. | 1.47 | 43 | | 44.47 | panels, hinges and pulls. |
| Minimum Charge | Job | | 232 | | 232 | |
| **30″ Wide, 2 Door** | | | | | | |
| Demolish | Ea. | | 24.50 | | 24.50 | Includes material and labor to install |
| Install | Ea. | 720 | 58 | | 778 | premium modular unit with solid |
| Demolish and Install | Ea. | 720 | 82.50 | | 802.50 | hardwood faced frames, hardwood |
| Reinstall | Ea. | | 46.48 | | 46.48 | door frames and drawer fronts, |
| Clean | Ea. | .74 | 17.75 | | 18.49 | hardwood veneer on raised door |
| Paint | Ea. | 1.85 | 48.50 | | 50.35 | panels, hinges and pulls. |
| Minimum Charge | Job | | 232 | | 232 | |
| **36″ Wide, 2 Door** | | | | | | |
| Demolish | Ea. | | 30.50 | | 30.50 | Includes material and labor to install |
| Install | Ea. | 1000 | 69.50 | | 1069.50 | premium modular unit with solid |
| Demolish and Install | Ea. | 1000 | 100 | | 1100 | hardwood faced frames, hardwood |
| Reinstall | Ea. | | 55.79 | | 55.79 | door frames and drawer fronts, |
| Clean | Ea. | .81 | 21 | | 21.81 | hardwood veneer on raised door |
| Paint | Ea. | 2.23 | 64.50 | | 66.73 | panels, hinges and pulls. |
| Minimum Charge | Job | | 232 | | 232 | |

233

For customer support on your **Contractor's Pricing Guide: Residential Repair & Remodeling**, call 800.448.8182.

# Cabinets and Countertops

## Vanity Cabinets

| | Unit | Material | Labor | Equip. | Total | Specification |
|---|---|---|---|---|---|---|
| **42" w, 3 Doors / Drawer** | | | | | | |
| Demolish | Ea. | | 36 | | 36 | Includes material and labor to install |
| Install | Ea. | 650 | 75 | | 725 | premium modular unit with solid |
| Demolish and Install | Ea. | 650 | 111 | | 761 | hardwood faced frames, hardwood |
| Reinstall | Ea. | | 60.07 | | 60.07 | door frames and drawer fronts, |
| Clean | Ea. | 1.06 | 25.50 | | 26.56 | hardwood veneer on raised door |
| Paint | Ea. | 2.56 | 77.50 | | 80.06 | panels, hinges and pulls. |
| Minimum Charge | Job | | 232 | | 232 | |
| **48" w, 3 Doors / Drawer** | | | | | | |
| Demolish | Ea. | | 39.50 | | 39.50 | Includes material and labor to install |
| Install | Ea. | 720 | 81.50 | | 801.50 | premium modular unit with solid |
| Demolish and Install | Ea. | 720 | 121 | | 841 | hardwood faced frames, hardwood |
| Reinstall | Ea. | | 65.06 | | 65.06 | door frames and drawer fronts, |
| Clean | Ea. | 1.47 | 39.50 | | 40.97 | hardwood veneer on raised door |
| Paint | Ea. | 3.85 | 96.50 | | 100.35 | panels, hinges and pulls. |
| Minimum Charge | Job | | 232 | | 232 | |
| **Clean Interior** | | | | | | |
| Install | L.F. | .30 | 7.40 | | 7.70 | Includes labor and materials to clean |
| Minimum Charge | Job | | 178 | | 178 | base cabinetry. |
| **Strip and Refinish** | | | | | | |
| Install | L.F. | 5.35 | 48.50 | | 53.85 | Includes labor and material to strip, |
| Minimum Charge | Job | | 193 | | 193 | prep and refinish exterior of cabinets. |

## Vanity Cabinet Tops

| | Unit | Material | Labor | Equip. | Total | Specification |
|---|---|---|---|---|---|---|
| **Cultured Marble Top** | | | | | | |
| Demolish | L.F. | | 6.60 | | 6.60 | Includes material and labor to install |
| Install | L.F. | 54 | 27.50 | | 81.50 | cultured marble vanity top with |
| Demolish and Install | L.F. | 54 | 34.10 | | 88.10 | integral sink, 22" deep. |
| Reinstall | L.F. | | 21.84 | | 21.84 | |
| Clean | L.F. | .06 | 1.35 | | 1.41 | |
| Minimum Charge | Job | | 219 | | 219 | |
| **Laminated Top** | | | | | | |
| Demolish | L.F. | | 3.95 | | 3.95 | Includes material and labor to install |
| Install | L.F. | 48 | 16.60 | | 64.60 | one piece laminated top with 4" |
| Demolish and Install | L.F. | 48 | 20.55 | | 68.55 | backsplash. |
| Reinstall | L.F. | | 13.28 | | 13.28 | |
| Clean | L.F. | .03 | .89 | | .92 | |
| Minimum Charge | Job | | 232 | | 232 | |
| **Solid Surface** | | | | | | |
| Demolish | L.F. | | 6.60 | | 6.60 | Includes material and labor to install |
| Install | L.F. | 85.50 | 34.50 | | 120 | solid surface vanity top with integral |
| Demolish and Install | L.F. | 85.50 | 41.10 | | 126.60 | sink installed, 22" deep. |
| Reinstall | L.F. | | 27.54 | | 27.54 | |
| Clean | L.F. | .06 | 1.35 | | 1.41 | |
| Minimum Charge | Job | | 232 | | 232 | |

**Roller Handle**

**Paint Brush**

| Paint Preparation, Walls | | Unit | Material | Labor | Equip. | Total | Specification |
|---|---|---|---|---|---|---|---|
| **Cover / Protect Floors** | Install | S.F. | .03 | .15 | | .18 | Includes material and labor to install |
| | Minimum Charge | Job | | 193 | | 193 | plastic masking sheet on large uninterrupted areas. |
| **Cover / Protect Walls** | Install | S.F. | .03 | .15 | | .18 | Includes material and labor to install |
| | Minimum Charge | Job | | 193 | | 193 | plastic masking sheet on walls. |
| **Clean Walls** | | | | | | | |
| Light | Clean | S.F. | | .12 | | .12 | Includes labor to wash gypsum |
| | Minimum Charge | Job | | 193 | | 193 | drywall or plaster wall surfaces. |
| Heavy | Clean | S.F. | | .15 | | .15 | Includes labor and material for heavy |
| | Minimum Charge | Job | | 193 | | 193 | cleaning (multiple applications) with detergent and solvent. |
| **Prep Walls** | Paint | S.F. | .04 | .19 | | .23 | Includes labor and material to prepare |
| | Minimum Charge | Job | | 193 | | 193 | for painting including scraping, patching and puttying. |
| **Sand Walls** | Paint | S.F. | | .13 | | .13 | Includes labor to sand gypsum drywall |
| | Minimum Charge | Job | | 193 | | 193 | wall surfaces. |
| **Seal Walls** | | | | | | | |
| Large Area | Paint | S.F. | .07 | .29 | | .36 | Includes labor and material to paint |
| | Minimum Charge | Job | | 193 | | 193 | with primer / sealer by roller. |
| Spot | Paint | S.F. | .07 | .34 | | .41 | Includes labor and material to seal |
| | Minimum Charge | Job | | 193 | | 193 | existing drywall with shellac-based material. |

For customer support on your Contractor's Pricing Guide: Residential Repair & Remodeling, call 800.448.8182.

235

# Painting

## Paint Preparation, Ceilings

| Paint Preparation, Ceilings | | Unit | Material | Labor | Equip. | Total | Specification |
|---|---|---|---|---|---|---|---|
| **Cover / Protect Floors** | | | | | | | |
| | Install | S.F. | .03 | .15 | | .18 | Includes material and labor to install |
| | Minimum Charge | Job | | 193 | | 193 | plastic masking sheet on large uninterrupted areas. |
| **Cover / Protect Walls** | | | | | | | |
| | Install | S.F. | .03 | .15 | | .18 | Includes material and labor to install |
| | Minimum Charge | Job | | 193 | | 193 | plastic masking sheet on walls. |
| **Prep Ceiling** | | | | | | | |
| | Paint | S.F. | .07 | .81 | | .88 | Includes labor and material to fill in |
| | Minimum Charge | Job | | 193 | | 193 | hairline cracks up to 1/8" with filler, sand and prep for paint. |
| **Sand Ceiling** | | | | | | | |
| | Paint | S.F. | | .18 | | .18 | Includes labor to sand gypsum drywall |
| | Minimum Charge | Job | | 193 | | 193 | ceiling surfaces. |
| **Seal Ceiling** | | | | | | | |
| **Large Area** | | | | | | | |
| | Paint | S.F. | .08 | .30 | | .38 | Includes labor and material to paint |
| | Minimum Charge | Job | | 193 | | 193 | ceiling, one coat flat latex. |
| **Spot** | | | | | | | |
| | Paint | S.F. | .07 | .34 | | .41 | Includes labor and material to seal |
| | Minimum Charge | Job | | 193 | | 193 | existing drywall with shellac-based material. |

## Paint / Texture, Walls

| Paint / Texture, Walls | | Unit | Material | Labor | Equip. | Total | Specification |
|---|---|---|---|---|---|---|---|
| **Texture Walls** | | | | | | | |
| **Spray** | | | | | | | |
| | Install | S.F. | .04 | .56 | | .60 | Includes labor and material to install |
| | Minimum Charge | Job | | 193 | | 193 | by spray, texture finish. |
| **Trowel** | | | | | | | |
| | Install | S.F. | .43 | .36 | | .79 | Includes labor and material to apply |
| | Minimum Charge | Job | | 193 | | 193 | hand troweled texture. |
| **Paint Walls 1 Coat** | | | | | | | |
| | Paint | S.F. | .08 | .30 | | .38 | Includes labor and material to paint, |
| | Minimum Charge | Job | | 193 | | 193 | one coat flat latex. |
| **Paint Walls 2 Coats** | | | | | | | |
| | Paint | S.F. | .15 | .48 | | .63 | Includes labor and material to paint, |
| | Minimum Charge | Job | | 193 | | 193 | two coats flat latex. |
| **Paint Walls 3 Coats** | | | | | | | |
| | Paint | S.F. | .24 | .59 | | .83 | Includes labor and material to paint, |
| | Minimum Charge | Job | | 193 | | 193 | three coats flat latex. |
| **Clean and Seal** | | | | | | | |
| | Paint | S.F. | .41 | 1.03 | | 1.44 | Includes labor and material to do light |
| | Minimum Charge | Job | | 193 | | 193 | cleaning, sealing with 2 coats of latex paint. |

| Paint / Texture, Ceilings | | Unit | Material | Labor | Equip. | Total | Specification |
|---|---|---|---|---|---|---|---|
| **Texture Ceiling** | | | | | | | |
| Spray | | | | | | | |
| | Install | S.F. | .04 | .56 | | .60 | Includes labor and material to install |
| | Minimum Charge | Job | | 193 | | 193 | by spray, texture finish. |
| Trowel | | | | | | | |
| | Install | S.F. | .43 | .36 | | .79 | Includes labor and material to apply |
| | Minimum Charge | Job | | 193 | | 193 | hand troweled texture. |
| **Paint Ceiling, 1 Coat** | | | | | | | |
| | Paint | S.F. | .08 | .30 | | .38 | Includes labor and material to paint |
| | Minimum Charge | Job | | 193 | | 193 | ceiling, one coat flat latex. |
| **Paint Ceiling, 2 Coats** | | | | | | | |
| | Paint | S.F. | .25 | .48 | | .73 | Includes labor and material to paint |
| | Minimum Charge | Job | | 193 | | 193 | two coats flat latex paint. |
| **Paint Ceiling, 3 Coats** | | | | | | | |
| | Paint | S.F. | .37 | .57 | | .94 | Includes labor and material to paint |
| | Minimum Charge | Job | | 193 | | 193 | ceiling, three coats flat latex. |
| **Clean and Seal** | | | | | | | |
| | Paint | S.F. | .37 | .81 | | 1.18 | Includes labor and material for light |
| | Minimum Charge | Job | | 193 | | 193 | cleaning, sealing and paint 3 coats of flat latex paint. |

237

| Plank Flooring | Wood Strip Flooring | Tile Flooring |

| Wood Plank Flooring | | Unit | Material | Labor | Equip. | Total | Specification |
|---|---|---|---|---|---|---|---|
| **Maple** | | | | | | | |
| 2-1/4" | | | | | | | |
| | Demolish | S.F. | | 1.58 | | 1.58 | Includes material and labor to install |
| | Install | S.F. | 4.45 | 2.32 | | 6.77 | unfinished T&G maple flooring, |
| | Demolish and Install | S.F. | 4.45 | 3.90 | | 8.35 | 25/32" thick, 2-1/4" wide, random |
| | Reinstall | S.F. | | 1.85 | | 1.85 | 3' to 16' lengths including felt |
| | Clean | S.F. | | .44 | | .44 | underlayment, nailed in place over |
| | Paint | S.F. | .23 | 2.73 | | 2.96 | prepared subfloor. |
| | Minimum Charge | Job | | 232 | | 232 | |
| 3-1/4" | | | | | | | |
| | Demolish | S.F. | | 1.58 | | 1.58 | Includes material and labor to install |
| | Install | S.F. | 5.10 | 2.32 | | 7.42 | unfinished T&G maple flooring, |
| | Demolish and Install | S.F. | 5.10 | 3.90 | | 9 | 33/32" thick, 3-1/4" wide, random |
| | Reinstall | S.F. | | 1.85 | | 1.85 | 3' to 16' lengths including felt |
| | Clean | S.F. | | .44 | | .44 | underlayment, nailed in place over |
| | Paint | S.F. | .23 | 2.73 | | 2.96 | prepared subfloor. |
| | Minimum Charge | Job | | 232 | | 232 | |
| **Sand and Finish** | | | | | | | |
| | Install | S.F. | .23 | .89 | | 1.12 | Includes labor and material to refinish |
| | Minimum Charge | Job | | 232 | | 232 | wood floor. |
| **Oak** | | | | | | | |
| 2-1/4" | | | | | | | |
| | Demolish | S.F. | | 1.58 | | 1.58 | Includes material and labor to install |
| | Install | S.F. | 4.96 | 2.32 | | 7.28 | unfinished T&G oak flooring, 25/32" |
| | Demolish and Install | S.F. | 4.96 | 3.90 | | 8.86 | thick, 2-1/4" wide, random 3' to 16' |
| | Reinstall | S.F. | | 1.85 | | 1.85 | lengths including felt underlayment, |
| | Clean | S.F. | | .44 | | .44 | nailed in place over prepared |
| | Paint | S.F. | .23 | 2.73 | | 2.96 | subfloor. |
| | Minimum Charge | Job | | 232 | | 232 | |
| 3-1/4" | | | | | | | |
| | Demolish | S.F. | | 1.58 | | 1.58 | Includes material and labor to install |
| | Install | S.F. | 4.74 | 1.64 | | 6.38 | unfinished T&G oak flooring, 25/32" |
| | Demolish and Install | S.F. | 4.74 | 3.22 | | 7.96 | thick, 3-1/4" wide, random 3' to 16' |
| | Reinstall | S.F. | | 1.31 | | 1.31 | lengths including felt underlayment, |
| | Clean | S.F. | | .44 | | .44 | nailed in place over prepared |
| | Paint | S.F. | .23 | 2.73 | | 2.96 | subfloor. |
| | Minimum Charge | Job | | 232 | | 232 | |
| **Sand and Finish** | | | | | | | |
| | Install | S.F. | .23 | 2.73 | | 2.96 | Includes labor and material to refinish |
| | Minimum Charge | Job | | 232 | | 232 | wood floor. |

| Wood Plank Flooring | Unit | Material | Labor | Equip. | Total | Specification |
|---|---|---|---|---|---|---|
| **Walnut** | | | | | | |
| **3″ to 7″** | | | | | | |
| Demolish | S.F. | | 1.58 | | 1.58 | Includes material and labor to install |
| Install | S.F. | 14.30 | 1.64 | | 15.94 | unfinished T&G walnut flooring, |
| Demolish and Install | S.F. | 14.30 | 3.22 | | 17.52 | 25/32″ thick, random 3′ to 16′ |
| Reinstall | S.F. | | 1.31 | | 1.31 | lengths including felt underlayment, |
| Clean | S.F. | | .44 | | .44 | nailed in place over prepared |
| Paint | S.F. | .23 | 2.73 | | 2.96 | subfloor. |
| Minimum Charge | Job | | 232 | | 232 | |
| **Sand and Finish** | | | | | | |
| Install | S.F. | .23 | 2.73 | | 2.96 | Includes labor and material to refinish |
| Minimum Charge | Job | | 232 | | 232 | wood floor. |
| **Pine** | | | | | | |
| **2-1/4″** | | | | | | |
| Demolish | S.F. | | 1.58 | | 1.58 | Includes material and labor to install |
| Install | S.F. | 7.90 | 1.75 | | 9.65 | unfinished T&G pine flooring, 25/32″ |
| Demolish and Install | S.F. | 7.90 | 3.33 | | 11.23 | thick, 2-1/4″ wide, random 3′ to 16′ |
| Reinstall | S.F. | | 1.40 | | 1.40 | lengths including felt underlayment, |
| Clean | S.F. | | .44 | | .44 | nailed in place over prepared |
| Paint | S.F. | .23 | .89 | | 1.12 | subfloor. |
| Minimum Charge | Job | | 232 | | 232 | |
| **Sand and Finish** | | | | | | |
| Install | S.F. | .23 | 2.73 | | 2.96 | Includes labor and material to refinish |
| Minimum Charge | Job | | 232 | | 232 | wood floor. |
| **Prefinished Oak** | | | | | | |
| **2-1/4″** | | | | | | |
| Demolish | S.F. | | 1.58 | | 1.58 | Includes material and labor to install |
| Install | S.F. | 4.74 | 2.32 | | 7.06 | prefinished T&G oak flooring, 2-1/4″ |
| Demolish and Install | S.F. | 4.74 | 3.90 | | 8.64 | wide, random 3′ to 16′ lengths |
| Reinstall | S.F. | | 1.85 | | 1.85 | including felt underlayment, nailed in |
| Clean | S.F. | | .44 | | .44 | place over prepared subfloor. |
| Minimum Charge | Job | | 232 | | 232 | |
| **3-1/4″** | | | | | | |
| Demolish | S.F. | | 1.58 | | 1.58 | Includes material and labor to install |
| Install | S.F. | 5.15 | 2.13 | | 7.28 | prefinished T&G oak flooring, 25/32″ |
| Demolish and Install | S.F. | 5.15 | 3.71 | | 8.86 | thick, 3″ to 7″ wide, random 3′ to 16′ |
| Reinstall | S.F. | | 1.70 | | 1.70 | lengths including felt underlayment, |
| Clean | S.F. | | .44 | | .44 | nailed in place over prepared |
| Minimum Charge | Job | | 232 | | 232 | subfloor. |
| **Sand and Finish New** | | | | | | |
| Install | S.F. | .23 | 1.20 | | 1.43 | Includes labor and material to refinish |
| Minimum Charge | Job | | 232 | | 232 | wood floor. |
| **Refinish Existing** | | | | | | |
| Install | S.F. | .23 | .89 | | 1.12 | Includes labor and material to refinish |
| Minimum Charge | Job | | 232 | | 232 | wood floor. |

# Flooring

## Wood Plank Flooring

| Wood Plank Flooring | | Unit | Material | Labor | Equip. | Total | Specification |
|---|---|---|---|---|---|---|---|
| **Urethane Coat** | | | | | | | |
| | Install | S.F. | .17 | .20 | | .37 | Includes labor and material to install |
| | Minimum Charge | Job | | 232 | | 232 | two coats of urethane, on existing |
| | | | | | | | floor. |
| **Clean and Wax** | | | | | | | |
| | Clean | S.F. | .04 | .28 | | .32 | Includes labor and material to wash |
| | Minimum Charge | Job | | 232 | | 232 | and wax hardwood. |
| **Underlayment Per S.F.** | | | | | | | |
| | Demolish | S.F. | | .68 | | .68 | Cost includes material and labor to |
| | Install | S.F. | 1.03 | .62 | | 1.65 | install 1/4" lauan subfloor, standard |
| | Demolish and Install | S.F. | 1.03 | 1.30 | | 2.33 | interior grade, nailed every 6". |
| | Clean | S.F. | | .34 | | .34 | |
| | Minimum Charge | Job | | 232 | | 232 | |

## Wood Parquet Tile Floor

| Wood Parquet Tile Floor | | Unit | Material | Labor | Equip. | Total | Specification |
|---|---|---|---|---|---|---|---|
| **Oak** | | | | | | | |
| **9" x 9"** | | | | | | | |
| | Demolish | S.F. | | 1.15 | | 1.15 | Includes material and labor to install |
| | Install | S.F. | 7.25 | 3.72 | | 10.97 | prefinished oak 9" x 9" parquet block |
| | Demolish and Install | S.F. | 7.25 | 4.87 | | 12.12 | flooring, 5/16" thick, installed in |
| | Reinstall | S.F. | | 2.97 | | 2.97 | mastic. |
| | Clean | S.F. | | .44 | | .44 | |
| | Paint | S.F. | .23 | 2.73 | | 2.96 | |
| | Minimum Charge | Job | | 232 | | 232 | |
| **13" x 13"** | | | | | | | |
| | Demolish | S.F. | | 1.15 | | 1.15 | Includes material and labor to install |
| | Install | S.F. | 10.35 | 2.91 | | 13.26 | prefinished oak 13" x 13" parquet |
| | Demolish and Install | S.F. | 10.35 | 4.06 | | 14.41 | block flooring, 5/16" thick, installed in |
| | Reinstall | S.F. | | 2.32 | | 2.32 | mastic. |
| | Clean | S.F. | | .44 | | .44 | |
| | Paint | S.F. | .23 | 2.73 | | 2.96 | |
| | Minimum Charge | Job | | 232 | | 232 | |
| **Cherry** | | | | | | | |
| **9" x 9"** | | | | | | | |
| | Demolish | S.F. | | 1.15 | | 1.15 | Includes material and labor to install |
| | Install | S.F. | 19.40 | 3.72 | | 23.12 | prefinished cherry 9" x 9" parquet |
| | Demolish and Install | S.F. | 19.40 | 4.87 | | 24.27 | block flooring, 5/16" thick, installed in |
| | Reinstall | S.F. | | 3.72 | | 3.72 | mastic. |
| | Clean | S.F. | | .44 | | .44 | |
| | Paint | S.F. | .23 | 2.73 | | 2.96 | |
| | Minimum Charge | Job | | 232 | | 232 | |
| **13" x 13"** | | | | | | | |
| | Demolish | S.F. | | 1.15 | | 1.15 | Includes material and labor to install |
| | Install | S.F. | 19.40 | 2.91 | | 22.31 | prefinished cherry 13" x 13" parquet |
| | Demolish and Install | S.F. | 19.40 | 4.06 | | 23.46 | block flooring, 5/16" thick, installed in |
| | Reinstall | S.F. | | 2.91 | | 2.91 | mastic. |
| | Clean | S.F. | | .44 | | .44 | |
| | Paint | S.F. | .23 | 2.73 | | 2.96 | |
| | Minimum Charge | Job | | 232 | | 232 | |

| Wood Parquet Tile Floor | Unit | Material | Labor | Equip. | Total | Specification |
|---|---|---|---|---|---|---|
| **Walnut** | | | | | | |
| **9″ x 9″** | | | | | | |
| Demolish | S.F. | | 1.15 | | 1.15 | Includes material and labor to install |
| Install | S.F. | 6.95 | 3.72 | | 10.67 | prefinished walnut 9″ x 9″ parquet |
| Demolish and Install | S.F. | 6.95 | 4.87 | | 11.82 | flooring in mastic. |
| Reinstall | S.F. | | 2.97 | | 2.97 | |
| Clean | S.F. | | .44 | | .44 | |
| Paint | S.F. | .23 | 2.73 | | 2.96 | |
| Minimum Charge | Job | | 232 | | 232 | |
| **13″ x 13″** | | | | | | |
| Demolish | S.F. | | 1.15 | | 1.15 | Includes material and labor to install |
| Install | S.F. | 12.40 | 2.91 | | 15.31 | prefinished walnut 13″ x 13″ parquet |
| Demolish and Install | S.F. | 12.40 | 4.06 | | 16.46 | block flooring, 5/16″ thick, installed in |
| Reinstall | S.F. | | 2.91 | | 2.91 | mastic. |
| Clean | S.F. | | .44 | | .44 | |
| Paint | S.F. | .23 | 2.73 | | 2.96 | |
| Minimum Charge | Job | | 232 | | 232 | |
| **Acrylic Impregnated** | | | | | | |
| **Oak 12″ x 12″** | | | | | | |
| Demolish | S.F. | | 1.15 | | 1.15 | Includes material and labor to install |
| Install | S.F. | 11.50 | 3.32 | | 14.82 | red oak 12″ x 12″ acrylic |
| Demolish and Install | S.F. | 11.50 | 4.47 | | 15.97 | impregnated parquet block flooring, |
| Reinstall | S.F. | | 3.32 | | 3.32 | 5/16″ thick, installed in mastic. |
| Clean | S.F. | | .44 | | .44 | |
| Minimum Charge | Job | | 232 | | 232 | |
| **Cherry 12″ x 12″** | | | | | | |
| Demolish | S.F. | | 1.15 | | 1.15 | Includes material and labor to install |
| Install | S.F. | 22 | 3.32 | | 25.32 | prefinished cherry 12″ x 12″ acrylic |
| Demolish and Install | S.F. | 22 | 4.47 | | 26.47 | impregnated parquet block flooring, |
| Reinstall | S.F. | | 3.32 | | 3.32 | 5/16″ thick, installed in mastic. |
| Clean | S.F. | | .44 | | .44 | |
| Minimum Charge | Job | | 232 | | 232 | |
| **Ash 12″ x 12″** | | | | | | |
| Demolish | S.F. | | 1.15 | | 1.15 | Includes material and labor to install |
| Install | S.F. | 20.50 | 3.32 | | 23.82 | ash 12″ x 12″ acrylic impregnated |
| Demolish and Install | S.F. | 20.50 | 4.47 | | 24.97 | parquet block flooring, 5/16″ thick in |
| Reinstall | S.F. | | 3.32 | | 3.32 | mastic. |
| Clean | S.F. | | .44 | | .44 | |
| Minimum Charge | Job | | 232 | | 232 | |
| **Teak** | | | | | | |
| Demolish | S.F. | | 1.15 | | 1.15 | Includes material and labor to install |
| Install | S.F. | 9.55 | 2.91 | | 12.46 | prefinished teak parquet block |
| Demolish and Install | S.F. | 9.55 | 4.06 | | 13.61 | flooring, 5/16″ thick, and installed in |
| Reinstall | S.F. | | 2.91 | | 2.91 | mastic. |
| Clean | S.F. | | .44 | | .44 | |
| Paint | S.F. | .23 | 2.73 | | 2.96 | |
| Minimum Charge | Job | | 232 | | 232 | |
| **Clean and Wax** | | | | | | |
| Clean | S.F. | .04 | .28 | | .32 | Includes labor and material to wash |
| Minimum Charge | Job | | 232 | | 232 | and wax hardwood. |

# Flooring

## Ceramic Tile Flooring

| Ceramic Tile Flooring | Unit | Material | Labor | Equip. | Total | Specification |
|---|---|---|---|---|---|---|
| **Economy Grade** | | | | | | |
| Demolish | S.F. | | 1.17 | | 1.17 | Cost includes material and labor to |
| Install | S.F. | 8.95 | 8.60 | | 17.55 | install 4-1/4" x 4-1/4" ceramic tile in |
| Demolish and Install | S.F. | 8.95 | 9.77 | | 18.72 | mortar bed including grout. |
| Clean | S.F. | | .68 | | .68 | |
| Minimum Charge | Job | | 219 | | 219 | |
| **Average Grade** | | | | | | |
| Demolish | S.F. | | 1.17 | | 1.17 | Cost includes material and labor to |
| Install | S.F. | 11.35 | 8.75 | | 20.10 | install 4-1/4" x 4-1/4" ceramic tile in |
| Demolish and Install | S.F. | 11.35 | 9.92 | | 21.27 | mortar bed including grout. |
| Clean | S.F. | | .68 | | .68 | |
| Minimum Charge | Job | | 219 | | 219 | |
| **Premium Grade** | | | | | | |
| Demolish | S.F. | | 1.17 | | 1.17 | Cost includes material and labor to |
| Install | S.F. | 12.45 | 8.75 | | 21.20 | install 4-1/4" x 4-1/4" ceramic tile in |
| Demolish and Install | S.F. | 12.45 | 9.92 | | 22.37 | mortar bed including grout. |
| Clean | S.F. | | .68 | | .68 | |
| Minimum Charge | Job | | 219 | | 219 | |
| **Re-grout** | | | | | | |
| Install | S.F. | .07 | 3.50 | | 3.57 | Includes material and labor to regrout |
| Minimum Charge | Job | | 219 | | 219 | tile floors. |

## Hard Tile Flooring

| Hard Tile Flooring | Unit | Material | Labor | Equip. | Total | Specification |
|---|---|---|---|---|---|---|
| **Thick Set Paver** | | | | | | |
| **Brick** | | | | | | |
| Demolish | S.F. | | 2.47 | | 2.47 | Cost includes material and labor to |
| Install | S.F. | 1.58 | 20 | | 21.58 | install 6" x 12" adobe brick paver with |
| Demolish and Install | S.F. | 1.58 | 22.47 | | 24.05 | 1/2" mortar joints. |
| Clean | S.F. | | .68 | | .68 | |
| Minimum Charge | Job | | 219 | | 219 | |
| **Mexican Red** | | | | | | |
| Demolish | S.F. | | 2.47 | | 2.47 | Cost includes material and labor to |
| Install | S.F. | 2.04 | 9.10 | | 11.14 | install 12" x 12" red Mexican paver |
| Demolish and Install | S.F. | 2.04 | 11.57 | | 13.61 | tile in mortar bed with grout. |
| Clean | S.F. | | .68 | | .68 | |
| Minimum Charge | Job | | 219 | | 219 | |
| **Saltillo** | | | | | | |
| Demolish | S.F. | | 1.17 | | 1.17 | Cost includes material and labor to |
| Install | S.F. | 1.64 | 9.10 | | 10.74 | install 12" x 12" saltillo tile in mortar |
| Demolish and Install | S.F. | 1.64 | 10.27 | | 11.91 | bed with grout. |
| Clean | S.F. | | .68 | | .68 | |
| Minimum Charge | Job | | 219 | | 219 | |
| **Marble** | | | | | | |
| **Premium Grade** | | | | | | |
| Demolish | S.F. | | 1.17 | | 1.17 | Cost includes material and labor to |
| Install | S.F. | 21 | 19.05 | | 40.05 | install 3/8" x 12" x 12" marble, in |
| Demolish and Install | S.F. | 21 | 20.22 | | 41.22 | mortar bed with grout. |
| Clean | S.F. | | .68 | | .68 | |
| Minimum Charge | Job | | 219 | | 219 | |

243

# Flooring

| Hard Tile Flooring | Unit | Material | Labor | Equip. | Total | Specification |
|---|---|---|---|---|---|---|
| **Terrazzo** | | | | | | |
| **Gray Cement** | | | | | | |
| Demolish | S.F. | | 1.57 | | 1.57 | Cost includes material and labor to |
| Install | S.F. | 3.93 | 10.70 | 4.57 | 19.20 | install 1-3/4" terrazzo , #1 and #2 |
| Demolish and Install | S.F. | 3.93 | 12.27 | 4.57 | 20.77 | chips in gray Portland cement. |
| Clean | S.F. | | .68 | | .68 | |
| Minimum Charge | Job | | 400 | 171 | 571 | |
| **White Cement** | | | | | | |
| Demolish | S.F. | | 1.57 | | 1.57 | Cost includes material and labor to |
| Install | S.F. | 4.58 | 10.70 | 4.57 | 19.85 | install 1-3/4" terrazzo, #1 and #2 |
| Demolish and Install | S.F. | 4.58 | 12.27 | 4.57 | 21.42 | chips in white Portland cement. |
| Clean | S.F. | | .68 | | .68 | |
| Minimum Charge | Job | | 400 | 171 | 571 | |
| **Non-skid Gray** | | | | | | |
| Demolish | S.F. | | 1.57 | | 1.57 | Cost includes material and labor to |
| Install | S.F. | 7.25 | 13.40 | 5.70 | 26.35 | install 1-3/4" terrazzo, #1 and #2 |
| Demolish and Install | S.F. | 7.25 | 14.97 | 5.70 | 27.92 | chips in gray Portland cement with |
| Clean | S.F. | | .68 | | .68 | light non-skid abrasive. |
| Minimum Charge | Job | | 400 | 171 | 571 | |
| **Non-skid White** | | | | | | |
| Demolish | S.F. | | 1.57 | | 1.57 | Cost includes material and labor to |
| Install | S.F. | 8 | 13.40 | 5.70 | 27.10 | install 1-3/4" terrazzo, #1 and #2 |
| Demolish and Install | S.F. | 8 | 14.97 | 5.70 | 28.67 | chips in white Portland cement with |
| Clean | S.F. | | .68 | | .68 | light non-skid abrasive. |
| Minimum Charge | Job | | 400 | 171 | 571 | |
| **Gray w / Brass Divider** | | | | | | |
| Demolish | S.F. | | 1.58 | | 1.58 | Cost includes material and labor to |
| Install | S.F. | 11.20 | 14.60 | 6.25 | 32.05 | install 1-3/4" terrazzo, #1 and #2 |
| Demolish and Install | S.F. | 11.20 | 16.18 | 6.25 | 33.63 | chips in gray Portland cement with |
| Clean | S.F. | | .68 | | .68 | brass strips 2' O.C. each way. |
| Minimum Charge | Job | | 400 | 171 | 571 | |
| **Clean** | | | | | | |
| Clean | S.F. | .07 | .45 | | .52 | Includes labor and material to clean |
| Minimum Charge | Job | | 400 | 171 | 571 | terrazzo flooring. |
| **Slate** | | | | | | |
| Demolish | S.F. | | 1.41 | | 1.41 | Includes material and labor to install |
| Install | S.F. | 8.45 | 4.31 | | 12.76 | slate tile flooring in thin set with grout. |
| Demolish and Install | S.F. | 8.45 | 5.72 | | 14.17 | |
| Clean | S.F. | | .68 | | .68 | |
| Minimum Charge | Job | | 219 | | 219 | |
| **Granite** | | | | | | |
| Demolish | S.F. | | 1.16 | | 1.16 | Includes material and labor to install |
| Install | S.F. | 28.50 | 4.38 | | 32.88 | granite tile flooring in thin set with |
| Demolish and Install | S.F. | 28.50 | 5.54 | | 34.04 | grout. |
| Reinstall | S.F. | | 4.38 | | 4.38 | |
| Clean | S.F. | | .68 | | .68 | |
| Minimum Charge | Job | | 219 | | 219 | |
| **Re-grout** | | | | | | |
| Install | S.F. | .07 | 3.50 | | 3.57 | Includes material and labor to regrout |
| Minimum Charge | Job | | 219 | | 219 | tile floors. |

# Flooring

## Sheet Vinyl Per S.F.

| | Unit | Material | Labor | Equip. | Total | Specification |
|---|---|---|---|---|---|---|
| **Economy Grade** | | | | | | |
| Demolish | S.F. | | .56 | | .56 | Includes material and labor to install |
| Install | S.F. | 2.49 | 1.37 | | 3.86 | resilient sheet vinyl flooring. |
| Demolish and Install | S.F. | 2.49 | 1.93 | | 4.42 | |
| Clean | S.F. | | .44 | | .44 | |
| Minimum Charge | Job | | 219 | | 219 | |
| **Average Grade** | | | | | | |
| Demolish | S.F. | | .56 | | .56 | Includes material and labor to install |
| Install | S.F. | 3.73 | 1.37 | | 5.10 | resilient sheet vinyl flooring. |
| Demolish and Install | S.F. | 3.73 | 1.93 | | 5.66 | |
| Clean | S.F. | | .44 | | .44 | |
| Minimum Charge | Job | | 219 | | 219 | |
| **Premium Grade** | | | | | | |
| Demolish | S.F. | | .56 | | .56 | Includes material and labor to install |
| Install | S.F. | 3.46 | 1.37 | | 4.83 | resilient sheet vinyl flooring. |
| Demolish and Install | S.F. | 3.46 | 1.93 | | 5.39 | |
| Clean | S.F. | | .44 | | .44 | |
| Minimum Charge | Job | | 219 | | 219 | |
| **Clean and Wax** | | | | | | |
| Clean | S.F. | .04 | .49 | | .53 | Includes labor and material to wash |
| Minimum Charge | Job | | 219 | | 219 | and wax a vinyl tile floor. |

## Sheet Vinyl

| | Unit | Material | Labor | Equip. | Total | Specification |
|---|---|---|---|---|---|---|
| **Economy Grade** | | | | | | |
| Demolish | S.Y. | | 5.10 | | 5.10 | Includes material and labor to install |
| Install | S.Y. | 22.50 | 12.30 | | 34.80 | no wax sheet vinyl flooring. |
| Demolish and Install | S.Y. | 22.50 | 17.40 | | 39.90 | |
| Clean | S.Y. | | 3.99 | | 3.99 | |
| Minimum Charge | Job | | 219 | | 219 | |
| **Average Grade** | | | | | | |
| Demolish | S.Y. | | 5.10 | | 5.10 | Includes material and labor to install |
| Install | S.Y. | 33.50 | 12.30 | | 45.80 | no wax sheet vinyl flooring. |
| Demolish and Install | S.Y. | 33.50 | 17.40 | | 50.90 | |
| Clean | S.Y. | | 3.99 | | 3.99 | |
| Minimum Charge | Job | | 219 | | 219 | |
| **Premium Grade** | | | | | | |
| Demolish | S.Y. | | 5.10 | | 5.10 | Includes material and labor to install |
| Install | S.Y. | 31 | 12.30 | | 43.30 | no wax sheet vinyl flooring. |
| Demolish and Install | S.Y. | 31 | 17.40 | | 48.40 | |
| Clean | S.Y. | | 3.99 | | 3.99 | |
| Minimum Charge | Job | | 219 | | 219 | |
| **Clean and Wax** | | | | | | |
| Clean | S.F. | .04 | .49 | | .53 | Includes labor and material to wash |
| Minimum Charge | Job | | 219 | | 219 | and wax a vinyl tile floor. |

## Vinyl Tile

| | Unit | Material | Labor | Equip. | Total | Specification |
|---|---|---|---|---|---|---|
| **Economy Grade** | | | | | | |
| Demolish | S.F. | | .79 | | .79 | Includes material and labor to install |
| Install | S.F. | 4.08 | .88 | | 4.96 | no wax 12" x 12" vinyl tile. |
| Demolish and Install | S.F. | 4.08 | 1.67 | | 5.75 | |
| Clean | S.F. | | .44 | | .44 | |
| Minimum Charge | Job | | 219 | | 219 | |

For customer support on your Contractor's Pricing Guide: Residential Repair & Remodeling, call 800.448.8182.

245

# Flooring

## Vinyl Tile

| | Unit | Material | Labor | Equip. | Total | Specification |
|---|---|---|---|---|---|---|
| **Average Grade** | | | | | | |
| Demolish | S.F. | | .79 | | .79 | Includes material and labor to install |
| Install | S.F. | 3.55 | .88 | | 4.43 | no wax 12" x 12" vinyl tile. |
| Demolish and Install | S.F. | 3.55 | 1.67 | | 5.22 | |
| Clean | S.F. | | .44 | | .44 | |
| Minimum Charge | Job | | 219 | | 219 | |
| **Premium Grade** | | | | | | |
| Demolish | S.F. | | .79 | | .79 | Includes material and labor to install |
| Install | S.F. | 6.90 | .88 | | 7.78 | no wax 12" x 12" vinyl tile. |
| Demolish and Install | S.F. | 6.90 | 1.67 | | 8.57 | |
| Clean | S.F. | | .44 | | .44 | |
| Minimum Charge | Job | | 219 | | 219 | |
| **Clean and Wax** | | | | | | |
| Clean | S.F. | .04 | .49 | | .53 | Includes labor and material to wash |
| Minimum Charge | Job | | 219 | | 219 | and wax a vinyl tile floor. |

## Carpeting Per S.F.

| | Unit | Material | Labor | Equip. | Total | Specification |
|---|---|---|---|---|---|---|
| **Economy Grade** | | | | | | |
| Demolish | S.F. | | .09 | | .09 | Includes material and labor to install |
| Install | S.F. | 1.32 | .65 | | 1.97 | carpet including tack strips and hot |
| Demolish and Install | S.F. | 1.32 | .74 | | 2.06 | melt tape on seams. |
| Clean | S.F. | | .62 | | .62 | |
| Minimum Charge | Job | | 219 | | 219 | |
| **Average Grade** | | | | | | |
| Demolish | S.F. | | .09 | | .09 | Includes material and labor to install |
| Install | S.F. | 1.64 | .65 | | 2.29 | carpet including tack strips and hot |
| Demolish and Install | S.F. | 1.64 | .74 | | 2.38 | melt tape on seams. |
| Clean | S.F. | | .62 | | .62 | |
| Minimum Charge | Job | | 219 | | 219 | |
| **Premium Grade** | | | | | | |
| Demolish | S.F. | | .09 | | .09 | Includes material and labor to install |
| Install | S.F. | 2.39 | .65 | | 3.04 | carpet including tack strips and hot |
| Demolish and Install | S.F. | 2.39 | .74 | | 3.13 | melt tape on seams. |
| Clean | S.F. | | .62 | | .62 | |
| Minimum Charge | Job | | 219 | | 219 | |
| **Indoor/Outdoor** | | | | | | |
| Demolish | S.F. | | .09 | | .09 | Includes material and labor to install |
| Install | S.F. | 1.18 | .65 | | 1.83 | indoor-outdoor carpet. |
| Demolish and Install | S.F. | 1.18 | .74 | | 1.92 | |
| Clean | S.F. | | .62 | | .62 | |
| Minimum Charge | Job | | 219 | | 219 | |
| **100% Wool** | | | | | | |
| **Average Grade** | | | | | | |
| Demolish | S.F. | | .09 | | .09 | Includes material and labor to install |
| Install | S.F. | 8.15 | .87 | | 9.02 | carpet including tack strips and hot |
| Demolish and Install | S.F. | 8.15 | .96 | | 9.11 | melt tape on seams. |
| Clean | S.F. | | .62 | | .62 | |
| Minimum Charge | Job | | 219 | | 219 | |
| **Premium Grade** | | | | | | |
| Demolish | S.F. | | .09 | | .09 | Includes material and labor to install |
| Install | S.F. | 11.85 | .87 | | 12.72 | carpet including tack strips and hot |
| Demolish and Install | S.F. | 11.85 | .96 | | 12.81 | melt tape on seams. |
| Clean | S.F. | | .62 | | .62 | |
| Minimum Charge | Job | | 219 | | 219 | |

## Carpeting Per S.F.

| | Unit | Material | Labor | Equip. | Total | Specification |
|---|---|---|---|---|---|---|
| **Luxury Grade** | | | | | | |
| Demolish | S.F. | | .09 | | .09 | Includes material and labor to install |
| Install | S.F. | 12 | .87 | | 12.87 | carpet including tack strips and hot |
| Demolish and Install | S.F. | 12 | .96 | | 12.96 | melt tape on seams. |
| Clean | S.F. | | .62 | | .62 | |
| Minimum Charge | Job | | 219 | | 219 | |
| **Berber** | | | | | | |
| **Average Grade** | | | | | | |
| Demolish | S.F. | | .09 | | .09 | Includes material and labor to install |
| Install | S.F. | 6.05 | 1.03 | | 7.08 | carpet including tack strips and hot |
| Demolish and Install | S.F. | 6.05 | 1.12 | | 7.17 | melt tape on seams. |
| Clean | S.F. | | .62 | | .62 | |
| Minimum Charge | Job | | 219 | | 219 | |
| **Premium Grade** | | | | | | |
| Demolish | S.F. | | .09 | | .09 | Includes material and labor to install |
| Install | S.F. | 8.35 | 1.03 | | 9.38 | carpet including tack strips and hot |
| Demolish and Install | S.F. | 8.35 | 1.12 | | 9.47 | melt tape on seams. |
| Clean | S.F. | | .62 | | .62 | |
| Minimum Charge | Job | | 219 | | 219 | |

## Carpeting

| | Unit | Material | Labor | Equip. | Total | Specification |
|---|---|---|---|---|---|---|
| **Indoor/Outdoor** | | | | | | |
| Demolish | S.Y. | | 1.32 | | 1.32 | Includes material and labor to install |
| Install | S.Y. | 10.55 | 5.85 | | 16.40 | indoor-outdoor carpet. |
| Demolish and Install | S.Y. | 10.55 | 7.17 | | 17.72 | |
| Reinstall | S.Y. | | 4.67 | | 4.67 | |
| Clean | S.Y. | | 5.55 | | 5.55 | |
| Minimum Charge | Job | | 219 | | 219 | |
| **Economy Grade** | | | | | | |
| Demolish | S.Y. | | 1.32 | | 1.32 | Includes material and labor to install |
| Install | S.Y. | 11.90 | 5.85 | | 17.75 | carpet including tack strips and hot |
| Demolish and Install | S.Y. | 11.90 | 7.17 | | 19.07 | melt tape on seams. |
| Reinstall | S.Y. | | 4.67 | | 4.67 | |
| Clean | S.Y. | | 5.55 | | 5.55 | |
| Minimum Charge | Job | | 219 | | 219 | |
| **Average Grade** | | | | | | |
| Demolish | S.Y. | | 1.32 | | 1.32 | Includes material and labor to install |
| Install | S.Y. | 14.80 | 5.85 | | 20.65 | carpet including tack strips and hot |
| Demolish and Install | S.Y. | 14.80 | 7.17 | | 21.97 | melt tape on seams. |
| Reinstall | S.Y. | | 4.67 | | 4.67 | |
| Clean | S.Y. | | 5.55 | | 5.55 | |
| Minimum Charge | Job | | 219 | | 219 | |
| **Premium Grade** | | | | | | |
| Demolish | S.Y. | | 1.32 | | 1.32 | Includes material and labor to install |
| Install | S.Y. | 21.50 | 5.85 | | 27.35 | carpet including tack strips and hot |
| Demolish and Install | S.Y. | 21.50 | 7.17 | | 28.67 | melt tape on seams. |
| Reinstall | S.Y. | | 4.67 | | 4.67 | |
| Clean | S.Y. | | 5.55 | | 5.55 | |
| Minimum Charge | Job | | 219 | | 219 | |

# Flooring

| Carpeting | | Unit | Material | Labor | Equip. | Total | Specification |
|---|---|---|---|---|---|---|---|
| **100% Wool** | | | | | | | |
| **Average Grade** | | | | | | | |
| | Demolish | S.Y. | | 1.32 | | 1.32 | Includes material and labor to install |
| | Install | S.Y. | 73 | 7.80 | | 80.80 | wool carpet including tack strips and |
| | Demolish and Install | S.Y. | 73 | 9.12 | | 82.12 | hot melt tape on seams. |
| | Reinstall | S.Y. | | 6.25 | | 6.25 | |
| | Clean | S.Y. | | 5.55 | | 5.55 | |
| | Minimum Charge | Job | | 219 | | 219 | |
| **Premium Grade** | | | | | | | |
| | Demolish | S.Y. | | 1.32 | | 1.32 | Includes material and labor to install |
| | Install | S.Y. | 106 | 7.80 | | 113.80 | carpet including tack strips and hot |
| | Demolish and Install | S.Y. | 106 | 9.12 | | 115.12 | melt tape on seams. |
| | Reinstall | S.Y. | | 6.25 | | 6.25 | |
| | Clean | S.Y. | | 5.55 | | 5.55 | |
| | Minimum Charge | Job | | 219 | | 219 | |
| **Luxury Grade** | | | | | | | |
| | Demolish | S.Y. | | 1.32 | | 1.32 | Includes material and labor to install |
| | Install | S.Y. | 108 | 7.80 | | 115.80 | carpet including tack strips and hot |
| | Demolish and Install | S.Y. | 108 | 9.12 | | 117.12 | melt tape on seams. |
| | Reinstall | S.Y. | | 6.25 | | 6.25 | |
| | Clean | S.Y. | | 5.55 | | 5.55 | |
| | Minimum Charge | Job | | 219 | | 219 | |
| **Berber** | | | | | | | |
| **Average Grade** | | | | | | | |
| | Demolish | S.Y. | | 1.32 | | 1.32 | Includes material and labor to install |
| | Install | S.Y. | 54.50 | 9.30 | | 63.80 | berber carpet including tack strips and |
| | Demolish and Install | S.Y. | 54.50 | 10.62 | | 65.12 | hot melt tape on seams. |
| | Reinstall | S.Y. | | 7.45 | | 7.45 | |
| | Clean | S.Y. | | 5.55 | | 5.55 | |
| | Minimum Charge | Job | | 219 | | 219 | |
| **Premium Grade** | | | | | | | |
| | Demolish | S.Y. | | 1.32 | | 1.32 | Includes material and labor to install |
| | Install | S.Y. | 75 | 9.30 | | 84.30 | carpet including tack strips and hot |
| | Demolish and Install | S.Y. | 75 | 10.62 | | 85.62 | melt tape on seams. |
| | Reinstall | S.Y. | | 7.45 | | 7.45 | |
| | Clean | S.Y. | | 5.55 | | 5.55 | |
| | Minimum Charge | Job | | 219 | | 219 | |

| Carpet Pad Per S.F. | | Unit | Material | Labor | Equip. | Total | Specification |
|---|---|---|---|---|---|---|---|
| **Urethane** | | | | | | | |
| | Demolish | S.F. | | .02 | | .02 | Includes material and labor to install |
| | Install | S.F. | .75 | .32 | | 1.07 | urethane carpet pad. |
| | Demolish and Install | S.F. | .75 | .34 | | 1.09 | |
| | Minimum Charge | Job | | 219 | | 219 | |
| **Foam Rubber Slab** | | | | | | | |
| | Demolish | S.F. | | .02 | | .02 | Includes material and labor to install |
| | Install | S.F. | .64 | .32 | | .96 | carpet pad, 3/8" thick foam rubber. |
| | Demolish and Install | S.F. | .64 | .34 | | .98 | |
| | Minimum Charge | Job | | 219 | | 219 | |
| **Waffle** | | | | | | | |
| | Demolish | S.F. | | .02 | | .02 | Includes material and labor to install |
| | Install | S.F. | .87 | .04 | | .91 | rubber waffle carpet pad. |
| | Demolish and Install | S.F. | .87 | .06 | | .93 | |
| | Minimum Charge | Job | | 219 | | 219 | |

# Flooring

## Carpet Pad Per S.F.

| | Unit | Material | Labor | Equip. | Total | Specification |
|---|---|---|---|---|---|---|
| **Jute** | | | | | | |
| Demolish | S.F. | | .02 | | .02 | Includes material and labor to install |
| Install | S.F. | .84 | .29 | | 1.13 | jute hair carpet pad. |
| Demolish and Install | S.F. | .84 | .31 | | 1.15 | |
| Minimum Charge | Job | | 219 | | 219 | |
| **Rebound** | | | | | | |
| Demolish | S.F. | | .02 | | .02 | Includes material and labor to install |
| Install | S.F. | .50 | .24 | | .74 | rebound carpet pad. |
| Demolish and Install | S.F. | .50 | .26 | | .76 | |
| Minimum Charge | Job | | 219 | | 219 | |

## Carpet Pad

| | Unit | Material | Labor | Equip. | Total | Specification |
|---|---|---|---|---|---|---|
| **Urethane** | | | | | | |
| Demolish | S.Y. | | .23 | | .23 | Includes material and labor to install |
| Install | S.Y. | 6.70 | 2.92 | | 9.62 | urethane carpet pad. |
| Demolish and Install | S.Y. | 6.70 | 3.15 | | 9.85 | |
| Reinstall | S.Y. | | 2.33 | | 2.33 | |
| Minimum Charge | Job | | 219 | | 219 | |
| **Foam Rubber Slab** | | | | | | |
| Demolish | S.Y. | | .23 | | .23 | Includes material and labor to install |
| Install | S.Y. | 5.70 | 2.92 | | 8.62 | rubber slab carpet pad. |
| Demolish and Install | S.Y. | 5.70 | 3.15 | | 8.85 | |
| Reinstall | S.Y. | | 2.33 | | 2.33 | |
| Minimum Charge | Job | | 219 | | 219 | |
| **Waffle** | | | | | | |
| Demolish | S.Y. | | .23 | | .23 | Includes material and labor to install |
| Install | S.Y. | 7.80 | .32 | | 8.12 | rubber waffle carpet pad. |
| Demolish and Install | S.Y. | 7.80 | .55 | | 8.35 | |
| Reinstall | S.Y. | | .26 | | .26 | |
| Minimum Charge | Job | | 219 | | 219 | |
| **Jute** | | | | | | |
| Demolish | S.Y. | | .23 | | .23 | Includes material and labor to install |
| Install | S.Y. | 7.55 | 2.57 | | 10.12 | jute hair carpet pad. |
| Demolish and Install | S.Y. | 7.55 | 2.80 | | 10.35 | |
| Reinstall | S.Y. | | 2.06 | | 2.06 | |
| Minimum Charge | Job | | 219 | | 219 | |
| **Rebound** | | | | | | |
| Demolish | S.Y. | | .23 | | .23 | Includes material and labor to install |
| Install | S.Y. | 4.50 | 2.18 | | 6.68 | rebound carpet pad. |
| Demolish and Install | S.Y. | 4.50 | 2.41 | | 6.91 | |
| Reinstall | S.Y. | | 1.75 | | 1.75 | |
| Minimum Charge | Job | | 219 | | 219 | |

## Stair Components

| | Unit | Material | Labor | Equip. | Total | Specification |
|---|---|---|---|---|---|---|
| **Carpeting** | | | | | | |
| Demolish | Ea. | | 2.39 | | 2.39 | Includes material and labor to install |
| Install | Riser | 58.50 | 11.50 | | 70 | carpet and pad on stairs. |
| Demolish and Install | Riser | 58.50 | 13.89 | | 72.39 | |
| Reinstall | Riser | | 9.21 | | 9.21 | |
| Clean | Ea. | | 5.55 | | 5.55 | |
| Minimum Charge | Job | | 219 | | 219 | |

**For customer support on your Contractor's Pricing Guide: Residential Repair & Remodeling, call 800.448.8182.**

# Finish Mechanical

**Tub/Shower**

**Stainless Steel Double Sink**

**Water Closet**

| Commode | Unit | Material | Labor | Equip. | Total | Specification |
|---|---|---|---|---|---|---|
| **Floor Mounted w / Tank** | | | | | | |
| Demolish | Ea. | | 52.50 | | 52.50 | Includes material and labor to install a |
| Install | Ea. | 253 | 178 | | 431 | toilet with valve, seat and cover. |
| Demolish and Install | Ea. | 253 | 230.50 | | 483.50 | |
| Reinstall | Ea. | | 142.44 | | 142.44 | |
| Clean | Ea. | .14 | 17.75 | | 17.89 | |
| Minimum Charge | Job | | 262 | | 262 | |
| **Wall w / Flush Valve** | | | | | | |
| Demolish | Ea. | | 52.50 | | 52.50 | Includes material and labor to install a |
| Install | Ea. | 1075 | 163 | | 1238 | wall hung toilet with flush valve, seat |
| Demolish and Install | Ea. | 1075 | 215.50 | | 1290.50 | and cover. |
| Reinstall | Ea. | | 130.16 | | 130.16 | |
| Clean | Ea. | .14 | 17.75 | | 17.89 | |
| Minimum Charge | Job | | 262 | | 262 | |
| **Designer Floor Mount** | | | | | | |
| Demolish | Ea. | | 52.50 | | 52.50 | Includes material and labor to install a |
| Install | Ea. | 840 | 178 | | 1018 | toilet with valve, seat and cover. |
| Demolish and Install | Ea. | 840 | 230.50 | | 1070.50 | |
| Reinstall | Ea. | | 142.44 | | 142.44 | |
| Clean | Ea. | .14 | 17.75 | | 17.89 | |
| Minimum Charge | Job | | 262 | | 262 | |
| **Urinal** | | | | | | |
| Demolish | Ea. | | 52.50 | | 52.50 | Includes material and labor to install a |
| Install | Ea. | 310 | 315 | | 625 | wall hung urinal with flush valve. |
| Demolish and Install | Ea. | 310 | 367.50 | | 677.50 | |
| Reinstall | Ea. | | 251.65 | | 251.65 | |
| Clean | Ea. | .14 | 17.75 | | 17.89 | |
| Minimum Charge | Job | | 262 | | 262 | |
| **Bidet** | | | | | | |
| **Vitreous China** | | | | | | |
| Demolish | Ea. | | 52.50 | | 52.50 | Includes material and labor to install a |
| Install | Ea. | 705 | 189 | | 894 | vitreous china bidet complete with |
| Demolish and Install | Ea. | 705 | 241.50 | | 946.50 | trim. |
| Reinstall | Ea. | | 150.99 | | 150.99 | |
| Clean | Ea. | .14 | 17.75 | | 17.89 | |
| Minimum Charge | Job | | 262 | | 262 | |

# Finish Mechanical

| Commode | | Unit | Material | Labor | Equip. | Total | Specification |
|---|---|---|---|---|---|---|---|
| **Chrome Fittings** | | | | | | | |
| | Install | Ea. | 286 | 65.50 | | 351.50 | Includes labor and materials to install chrome fittings for a bidet. |
| **Brass Fittings** | | | | | | | |
| | Install | Ea. | 315 | 65.50 | | 380.50 | Includes labor and materials to install brass fittings for a bidet. |
| **Seat** | | | | | | | |
| | Demolish | Ea. | | 22 | | 22 | Includes material and labor to install a |
| | Install | Ea. | 32.50 | 22 | | 54.50 | toilet seat and cover. |
| | Demolish and Install | Ea. | 32.50 | 44 | | 76.50 | |
| | Reinstall | Ea. | | 17.47 | | 17.47 | |
| | Clean | Ea. | .08 | 7.40 | | 7.48 | |
| | Minimum Charge | Job | | 262 | | 262 | |
| **Rough-in** | | | | | | | |
| | Install | Ea. | 435 | 330 | | 765 | Includes labor and materials to install water supply pipe with valves and drain, waste and vent pipe with all couplings, hangers and fasteners necessary. |

| Sink (assembly) | | Unit | Material | Labor | Equip. | Total | Specification |
|---|---|---|---|---|---|---|---|
| **Single** | | | | | | | |
| **Porcelain** | | | | | | | |
| | Demolish | Ea. | | 33 | | 33 | Includes material and labor to install a |
| | Install | Ea. | 725 | 169 | | 894 | single bowl enamel finished cast iron |
| | Demolish and Install | Ea. | 725 | 202 | | 927 | kitchen sink with faucet and drain. |
| | Reinstall | Ea. | | 134.81 | | 134.81 | |
| | Clean | Ea. | .10 | 12.70 | | 12.80 | |
| | Minimum Charge | Job | | 262 | | 262 | |
| **Stainless Steel** | | | | | | | |
| | Demolish | Ea. | | 33 | | 33 | Includes material and labor to install a |
| | Install | Ea. | 750 | 169 | | 919 | single bowl stainless steel kitchen sink |
| | Demolish and Install | Ea. | 750 | 202 | | 952 | with faucet and drain. |
| | Reinstall | Ea. | | 134.81 | | 134.81 | |
| | Clean | Ea. | .10 | 12.70 | | 12.80 | |
| | Minimum Charge | Job | | 262 | | 262 | |
| **Double** | | | | | | | |
| **Porcelain** | | | | | | | |
| | Demolish | Ea. | | 37.50 | | 37.50 | Includes material and labor to install a |
| | Install | Ea. | 415 | 197 | | 612 | double bowl enamel finished cast iron |
| | Demolish and Install | Ea. | 415 | 234.50 | | 649.50 | kitchen sink with faucet and drain. |
| | Reinstall | Ea. | | 157.28 | | 157.28 | |
| | Clean | Ea. | .14 | 17.75 | | 17.89 | |
| | Minimum Charge | Job | | 262 | | 262 | |
| **Stainless Steel** | | | | | | | |
| | Demolish | Ea. | | 37.50 | | 37.50 | Includes material and labor to install a |
| | Install | Ea. | 475 | 197 | | 672 | double bowl stainless steel kitchen sink |
| | Demolish and Install | Ea. | 475 | 234.50 | | 709.50 | with faucet and drain. |
| | Reinstall | Ea. | | 157.28 | | 157.28 | |
| | Clean | Ea. | .14 | 17.75 | | 17.89 | |
| | Minimum Charge | Job | | 262 | | 262 | |

# Finish Mechanical

## Sink (assembly)

| | Unit | Material | Labor | Equip. | Total | Specification |
|---|---|---|---|---|---|---|
| **Bar** | | | | | | |
| Demolish | Ea. | | 33 | | 33 | Includes material and labor to install a |
| Install | Ea. | 194 | 131 | | 325 | small single bowl stainless steel bar / |
| Demolish and Install | Ea. | 194 | 164 | | 358 | vegetable / kitchen sink with faucet |
| Reinstall | Ea. | | 104.80 | | 104.80 | and drain. |
| Clean | Ea. | .10 | 12.70 | | 12.80 | |
| Minimum Charge | Job | | 262 | | 262 | |
| **Floor** | | | | | | |
| Demolish | Ea. | | 105 | | 105 | Includes material and labor to install |
| Install | Ea. | 1125 | 214 | | 1339 | an enameled cast iron floor mounted |
| Demolish and Install | Ea. | 1125 | 319 | | 1444 | corner service sink with faucet and |
| Reinstall | Ea. | | 171.58 | | 171.58 | drain. |
| Clean | Ea. | .14 | 8.90 | | 9.04 | |
| Minimum Charge | Job | | 262 | | 262 | |
| **Pedestal** | | | | | | |
| Demolish | Ea. | | 33 | | 33 | Includes material and labor to install a |
| Install | Ea. | 350 | 147 | | 497 | vitreous china pedestal lavatory with |
| Demolish and Install | Ea. | 350 | 180 | | 530 | faucet set and pop-up drain. |
| Reinstall | Ea. | | 117.96 | | 117.96 | |
| Clean | Ea. | .10 | 12.70 | | 12.80 | |
| Minimum Charge | Job | | 262 | | 262 | |
| **Vanity Lavatory** | | | | | | |
| Demolish | Ea. | | 33 | | 33 | Includes material and labor to install a |
| Install | Ea. | 242 | 175 | | 417 | vitreous china lavatory with faucet set |
| Demolish and Install | Ea. | 242 | 208 | | 450 | and pop-up drain. |
| Reinstall | Ea. | | 139.80 | | 139.80 | |
| Clean | Ea. | .10 | 12.70 | | 12.80 | |
| Minimum Charge | Job | | 262 | | 262 | |
| **Rough-in** | | | | | | |
| Install | Ea. | 305 | 440 | | 745 | Includes labor and materials to install copper supply pipe with valves and drain, waste and vent pipe with all couplings, hangers and fasteners necessary. |
| **Laundry** | | | | | | |
| Demolish | Ea. | | 35.50 | | 35.50 | Includes material and labor to install a |
| Install | Ea. | 160 | 145 | | 305 | laundry sink (wall mounted or with |
| Demolish and Install | Ea. | 160 | 180.50 | | 340.50 | legs) including faucet and pop up |
| Clean | Ea. | .10 | 12.70 | | 12.80 | drain. |
| Minimum Charge | Job | | 262 | | 262 | |

## Sink Only

| | Unit | Material | Labor | Equip. | Total | Specification |
|---|---|---|---|---|---|---|
| **Single** | | | | | | |
| **Porcelain** | | | | | | |
| Demolish | Ea. | | 33 | | 33 | Includes material and labor to install a |
| Install | Ea. | 615 | 187 | | 802 | single bowl enamel finished cast iron |
| Demolish and Install | Ea. | 615 | 220 | | 835 | kitchen sink. |
| Reinstall | Ea. | | 149.71 | | 149.71 | |
| Clean | Ea. | .10 | 12.70 | | 12.80 | |
| Minimum Charge | Job | | 262 | | 262 | |

## Sink Only

| | Unit | Material | Labor | Equip. | Total | Specification |
|---|---|---|---|---|---|---|
| **Stainless Steel** | | | | | | |
| Demolish | Ea. | | 33 | | 33 | Includes material and labor to install a |
| Install | Ea. | 640 | 131 | | 771 | single bowl stainless steel kitchen sink. |
| Demolish and Install | Ea. | 640 | 164 | | 804 | |
| Reinstall | Ea. | | 104.80 | | 104.80 | |
| Clean | Ea. | .10 | 12.70 | | 12.80 | |
| Minimum Charge | Job | | 262 | | 262 | |
| **Double** | | | | | | |
| **Porcelain** | | | | | | |
| Demolish | Ea. | | 37.50 | | 37.50 | Includes material and labor to install a |
| Install | Ea. | 289 | 218 | | 507 | double bowl enamel finished cast iron |
| Demolish and Install | Ea. | 289 | 255.50 | | 544.50 | kitchen sink. |
| Reinstall | Ea. | | 174.67 | | 174.67 | |
| Clean | Ea. | .14 | 17.75 | | 17.89 | |
| Minimum Charge | Job | | 262 | | 262 | |
| **Stainless Steel** | | | | | | |
| Demolish | Ea. | | 37.50 | | 37.50 | Includes material and labor to install a |
| Install | Ea. | 955 | 169 | | 1124 | double bowl stainless steel kitchen |
| Demolish and Install | Ea. | 955 | 206.50 | | 1161.50 | sink. |
| Reinstall | Ea. | | 135.23 | | 135.23 | |
| Clean | Ea. | .14 | 17.75 | | 17.89 | |
| Minimum Charge | Job | | 262 | | 262 | |
| **Bar** | | | | | | |
| Demolish | Ea. | | 33 | | 33 | Includes material and labor to install a |
| Install | Ea. | 85.50 | 131 | | 216.50 | small single bowl stainless steel bar / |
| Demolish and Install | Ea. | 85.50 | 164 | | 249.50 | vegetable / kitchen sink. |
| Reinstall | Ea. | | 104.80 | | 104.80 | |
| Clean | Ea. | .10 | 12.70 | | 12.80 | |
| Minimum Charge | Job | | 262 | | 262 | |
| **Floor** | | | | | | |
| Demolish | Ea. | | 105 | | 105 | Includes material and labor to install |
| Install | Ea. | 995 | 238 | | 1233 | an enameled cast iron floor mounted |
| Demolish and Install | Ea. | 995 | 343 | | 1338 | corner service sink. |
| Reinstall | Ea. | | 190.55 | | 190.55 | |
| Clean | Ea. | .14 | 8.90 | | 9.04 | |
| Minimum Charge | Job | | 262 | | 262 | |
| **Pedestal** | | | | | | |
| Demolish | Ea. | | 33 | | 33 | Includes material and labor to install a |
| Install | Ea. | 287 | 164 | | 451 | vitreous china pedestal lavatory. |
| Demolish and Install | Ea. | 287 | 197 | | 484 | |
| Reinstall | Ea. | | 131 | | 131 | |
| Clean | Ea. | .10 | 12.70 | | 12.80 | |
| Minimum Charge | Job | | 262 | | 262 | |
| **Vanity Lavatory** | | | | | | |
| Demolish | Ea. | | 33 | | 33 | Includes material and labor to install a |
| Install | Ea. | 171 | 164 | | 335 | vitreous china lavatory. |
| Demolish and Install | Ea. | 171 | 197 | | 368 | |
| Reinstall | Ea. | | 131 | | 131 | |
| Clean | Ea. | .10 | 12.70 | | 12.80 | |
| Minimum Charge | Job | | 262 | | 262 | |
| **Rough-in** | | | | | | |
| Install | Ea. | 262 | 410 | | 672 | Includes labor and materials to install copper supply pipe with valves and drain, waste and vent pipe with all couplings, hangers and fasteners necessary. |

# Finish Mechanical

## Sink Only

| | Unit | Material | Labor | Equip. | Total | Specification |
|---|---|---|---|---|---|---|
| **Faucet Sink** | | | | | | |
| **Good Grade** | | | | | | |
| Demolish | Ea. | | 26 | | 26 | Includes material and labor to install a |
| Install | Ea. | 92 | 36 | | 128 | sink faucet and fittings. |
| Demolish and Install | Ea. | 92 | 62 | | 154 | |
| Reinstall | Ea. | | 28.93 | | 28.93 | |
| Clean | Ea. | .06 | 11.10 | | 11.16 | |
| Minimum Charge | Job | | 262 | | 262 | |
| **Better Grade** | | | | | | |
| Demolish | Ea. | | 26 | | 26 | Includes material and labor to install a |
| Install | Ea. | 154 | 36 | | 190 | sink faucet and fittings. |
| Demolish and Install | Ea. | 154 | 62 | | 216 | |
| Reinstall | Ea. | | 28.93 | | 28.93 | |
| Clean | Ea. | .06 | 11.10 | | 11.16 | |
| Minimum Charge | Job | | 262 | | 262 | |
| **Premium Grade** | | | | | | |
| Demolish | Ea. | | 26 | | 26 | Includes material and labor to install a |
| Install | Ea. | 300 | 45 | | 345 | sink faucet and fittings. |
| Demolish and Install | Ea. | 300 | 71 | | 371 | |
| Reinstall | Ea. | | 36.08 | | 36.08 | |
| Clean | Ea. | .06 | 11.10 | | 11.16 | |
| Minimum Charge | Job | | 262 | | 262 | |
| **Drain & Basket** | | | | | | |
| Demolish | Ea. | | 16.40 | | 16.40 | Includes material and labor to install a |
| Install | Ea. | 16.50 | 33 | | 49.50 | sink drain and basket. |
| Demolish and Install | Ea. | 16.50 | 49.40 | | 65.90 | |
| Reinstall | Ea. | | 26.20 | | 26.20 | |
| Clean | Ea. | .06 | 11.10 | | 11.16 | |
| Minimum Charge | Job | | 262 | | 262 | |

## Bathtub

| | Unit | Material | Labor | Equip. | Total | Specification |
|---|---|---|---|---|---|---|
| **Enameled** | | | | | | |
| **Steel** | | | | | | |
| Demolish | Ea. | | 94.50 | | 94.50 | Includes labor and material to install a |
| Install | Ea. | 555 | 172 | | 727 | formed steel bathtub with spout, |
| Demolish and Install | Ea. | 555 | 266.50 | | 821.50 | mixing valve, shower head and |
| Reinstall | Ea. | | 137.26 | | 137.26 | pop-up drain. |
| Clean | Ea. | .30 | 29.50 | | 29.80 | |
| Minimum Charge | Job | | 262 | | 262 | |
| **Cast Iron** | | | | | | |
| Demolish | Ea. | | 94.50 | | 94.50 | Includes material and labor to install a |
| Install | Ea. | 1300 | 214 | | 1514 | cast iron bathtub with spout, mixing |
| Demolish and Install | Ea. | 1300 | 308.50 | | 1608.50 | valve, shower head and pop-up drain. |
| Reinstall | Ea. | | 171.58 | | 171.58 | |
| Clean | Ea. | .30 | 29.50 | | 29.80 | |
| Minimum Charge | Job | | 262 | | 262 | |
| **Fiberglass** | | | | | | |
| Demolish | Ea. | | 94.50 | | 94.50 | Includes material and labor to install |
| Install | Ea. | 1850 | 172 | | 2022 | an acrylic soaking tub with spout, |
| Demolish and Install | Ea. | 1850 | 266.50 | | 2116.50 | mixing valve, shower head and |
| Reinstall | Ea. | | 137.26 | | 137.26 | pop-up drain. |
| Clean | Ea. | .30 | 29.50 | | 29.80 | |
| Minimum Charge | Job | | 262 | | 262 | |

# Finish Mechanical

## Bathtub

| | Unit | Material | Labor | Equip. | Total | Specification |
|---|---|---|---|---|---|---|
| **Institutional** | | | | | | |
| Demolish | Ea. | | 94.50 | | 94.50 | Includes material and labor to install a |
| Install | Ea. | 1450 | 315 | | 1765 | hospital / institutional type bathtub |
| Demolish and Install | Ea. | 1450 | 409.50 | | 1859.50 | with spout, mixing valve, shower head |
| Reinstall | Ea. | | 251.65 | | 251.65 | and pop-up drain. |
| Clean | Ea. | .30 | 29.50 | | 29.80 | |
| Minimum Charge | Job | | 262 | | 262 | |
| **Whirlpool (Acrylic)** | | | | | | |
| Demolish | Ea. | | 291 | | 291 | Includes material and labor to install a |
| Install | Ea. | 1300 | 945 | | 2245 | molded fiberglass whirlpool tub. |
| Demolish and Install | Ea. | 1300 | 1236 | | 2536 | |
| Reinstall | Ea. | | 754.94 | | 754.94 | |
| Clean | Ea. | .30 | 29.50 | | 29.80 | |
| Minimum Charge | Job | | 262 | | 262 | |
| **Pump / Motor** | | | | | | |
| Demolish | Ea. | | 73 | | 73 | Includes material and labor to install a |
| Install | Ea. | 297 | 157 | | 454 | whirlpool tub pump / motor. |
| Demolish and Install | Ea. | 297 | 230 | | 527 | |
| Reinstall | Ea. | | 125.89 | | 125.89 | |
| Clean | Ea. | .08 | 22 | | 22.08 | |
| Minimum Charge | Job | | 262 | | 262 | |
| **Heater / Motor** | | | | | | |
| Demolish | Ea. | | 73 | | 73 | Includes material and labor to install a |
| Install | Ea. | 202 | 233 | | 435 | whirlpool tub heater. |
| Demolish and Install | Ea. | 202 | 306 | | 508 | |
| Reinstall | Ea. | | 186.31 | | 186.31 | |
| Clean | Ea. | .08 | 22 | | 22.08 | |
| Minimum Charge | Job | | 262 | | 262 | |
| **Thermal Cover** | | | | | | |
| Demolish | Ea. | | 11.65 | | 11.65 | Includes material and labor to install a |
| Install | Ea. | 420 | 33 | | 453 | whirlpool tub thermal cover. |
| Demolish and Install | Ea. | 420 | 44.65 | | 464.65 | |
| Reinstall | Ea. | | 26.20 | | 26.20 | |
| Clean | Ea. | .74 | 22 | | 22.74 | |
| Minimum Charge | Job | | 262 | | 262 | |
| **Tub / Shower Combination** | | | | | | |
| Demolish | Ea. | | 83 | | 83 | Includes material and labor to install a |
| Install | Ea. | 835 | 236 | | 1071 | fiberglass module tub with shower |
| Demolish and Install | Ea. | 835 | 319 | | 1154 | surround with spout, mixing valve, |
| Reinstall | Ea. | | 188.74 | | 188.74 | shower head and pop-up drain. |
| Clean | Ea. | .36 | 44.50 | | 44.86 | |
| Minimum Charge | Job | | 262 | | 262 | |
| **Accessories** | | | | | | |
| **Sliding Door** | | | | | | |
| Demolish | Ea. | | 24.50 | | 24.50 | Includes material and labor to install a |
| Install | Ea. | 665 | 107 | | 772 | 48" aluminum framed tempered glass |
| Demolish and Install | Ea. | 665 | 131.50 | | 796.50 | shower door. |
| Reinstall | Ea. | | 85.83 | | 85.83 | |
| Clean | Ea. | .62 | 14.80 | | 15.42 | |
| Minimum Charge | Job | | 262 | | 262 | |
| **Shower Head** | | | | | | |
| Demolish | Ea. | | 6.55 | | 6.55 | Includes material and labor to install a |
| Install | Ea. | 54.50 | 22 | | 76.50 | water saving shower head. |
| Demolish and Install | Ea. | 54.50 | 28.55 | | 83.05 | |
| Reinstall | Ea. | | 17.47 | | 17.47 | |
| Clean | Ea. | .01 | 4.44 | | 4.45 | |
| Minimum Charge | Job | | 262 | | 262 | |

# Finish Mechanical

## Bathtub

| Bathtub | | Unit | Material | Labor | Equip. | Total | Specification |
|---|---|---|---|---|---|---|---|
| **Faucet Set** | | | | | | | |
| | Demolish | Ea. | | 33 | | 33 | Includes material and labor to install a |
| | Install | Ea. | 94.50 | 65.50 | | 160 | combination spout / diverter for a |
| | Demolish and Install | Ea. | 94.50 | 98.50 | | 193 | bathtub. |
| | Reinstall | Ea. | | 52.40 | | 52.40 | |
| | Clean | Ea. | .06 | 11.10 | | 11.16 | |
| | Minimum Charge | Job | | 262 | | 262 | |
| **Shower Rod** | | | | | | | |
| | Demolish | Ea. | | 4.11 | | 4.11 | Includes material and labor to install a |
| | Install | Ea. | 31 | 36 | | 67 | chrome plated shower rod. |
| | Demolish and Install | Ea. | 31 | 40.11 | | 71.11 | |
| | Reinstall | Ea. | | 28.60 | | 28.60 | |
| | Clean | Ea. | | 4.44 | | 4.44 | |
| | Minimum Charge | Job | | 262 | | 262 | |
| **Rough-in** | | | | | | | |
| | Install | Ea. | 475 | 455 | | 930 | Includes labor and materials to install copper supply pipe with valves and drain, waste and vent pipe with all couplings, hangers and fasteners necessary. |

## Shower

| Shower | | Unit | Material | Labor | Equip. | Total | Specification |
|---|---|---|---|---|---|---|---|
| **Fiberglass** | | | | | | | |
| **32" x 32"** | | | | | | | |
| | Demolish | Ea. | | 197 | | 197 | Includes material and labor to install a |
| | Install | Ea. | 380 | 172 | | 552 | fiberglass shower stall with door, |
| | Demolish and Install | Ea. | 380 | 369 | | 749 | mixing valve and shower head and |
| | Reinstall | Ea. | | 137.26 | | 137.26 | drain fitting. |
| | Clean | Ea. | .30 | 29.50 | | 29.80 | |
| | Minimum Charge | Job | | 262 | | 262 | |
| **36" x 36"** | | | | | | | |
| | Demolish | Ea. | | 197 | | 197 | Includes material and labor to install a |
| | Install | Ea. | 440 | 172 | | 612 | fiberglass shower stall with door, |
| | Demolish and Install | Ea. | 440 | 369 | | 809 | mixing valve and shower head and |
| | Reinstall | Ea. | | 137.26 | | 137.26 | drain fitting. |
| | Clean | Ea. | .30 | 29.50 | | 29.80 | |
| | Minimum Charge | Job | | 262 | | 262 | |
| **35" x 60"** | | | | | | | |
| | Demolish | Ea. | | 197 | | 197 | Includes material and labor to install a |
| | Install | Ea. | 725 | 236 | | 961 | handicap fiberglass shower stall with |
| | Demolish and Install | Ea. | 725 | 433 | | 1158 | door & seat, mixing valve and shower |
| | Reinstall | Ea. | | 188.74 | | 188.74 | head and drain fitting. |
| | Clean | Ea. | .30 | 29.50 | | 29.80 | |
| | Minimum Charge | Job | | 262 | | 262 | |
| **Accessories** | | | | | | | |
| **Single Door** | | | | | | | |
| | Demolish | Ea. | | 24.50 | | 24.50 | Includes material and labor to install a |
| | Install | Ea. | 151 | 56.50 | | 207.50 | shower door. |
| | Demolish and Install | Ea. | 151 | 81 | | 232 | |
| | Reinstall | Ea. | | 45.30 | | 45.30 | |
| | Clean | Ea. | .62 | 14.80 | | 15.42 | |
| | Minimum Charge | Job | | 262 | | 262 | |

257

# Finish Mechanical

## Shower

| Shower | Unit | Material | Labor | Equip. | Total | Specification |
|---|---|---|---|---|---|---|
| **Shower Head** | | | | | | Includes material and labor to install a |
| Demolish | Ea. | | 6.55 | | 6.55 | water saving shower head. |
| Install | Ea. | 54.50 | 22 | | 76.50 | |
| Demolish and Install | Ea. | 54.50 | 28.55 | | 83.05 | |
| Reinstall | Ea. | | 17.47 | | 17.47 | |
| Clean | Ea. | .01 | 4.44 | | 4.45 | |
| Minimum Charge | Job | | 262 | | 262 | |
| **Faucet Set** | | | | | | Includes material and labor to install a |
| Demolish | Ea. | | 33 | | 33 | combination spout / diverter for a |
| Install | Ea. | 94.50 | 65.50 | | 160 | bathtub. |
| Demolish and Install | Ea. | 94.50 | 98.50 | | 193 | |
| Reinstall | Ea. | | 52.40 | | 52.40 | |
| Clean | Ea. | .06 | 11.10 | | 11.16 | |
| Minimum Charge | Job | | 262 | | 262 | |
| **Shower Pan** | | | | | | Includes material and labor to install |
| Demolish | Ea. | | 29 | | 29 | the base portion of a fiberglass |
| Install | Ea. | 112 | 65.50 | | 177.50 | shower unit. |
| Demolish and Install | Ea. | 112 | 94.50 | | 206.50 | |
| Reinstall | Ea. | | 52.40 | | 52.40 | |
| Clean | Ea. | .30 | 29.50 | | 29.80 | |
| Minimum Charge | Job | | 262 | | 262 | |
| **Rough-in** | | | | | | Includes labor and materials to install |
| Install | Ea. | 405 | 460 | | 865 | copper supply pipe with valves and drain, waste and vent pipe with all couplings, hangers and fasteners necessary. |

## Ductwork

| Ductwork | Unit | Material | Labor | Equip. | Total | Specification |
|---|---|---|---|---|---|---|
| **Return Grill** | | | | | | |
| **Under 10" Wide** | | | | | | |
| Demolish | Ea. | | 11.30 | | 11.30 | Includes material and labor to install a |
| Install | Ea. | 25 | 28.50 | | 53.50 | 10" x 10" air return register. |
| Demolish and Install | Ea. | 25 | 39.80 | | 64.80 | |
| Reinstall | Ea. | | 22.65 | | 22.65 | |
| Clean | Ea. | .08 | 9.85 | | 9.93 | |
| Minimum Charge | Job | | 255 | | 255 | |
| **12" to 20" Wide** | | | | | | |
| Demolish | Ea. | | 11.30 | | 11.30 | Includes material and labor to install a |
| Install | Ea. | 48 | 30 | | 78 | 16" x 16" air return register. |
| Demolish and Install | Ea. | 48 | 41.30 | | 89.30 | |
| Reinstall | Ea. | | 23.98 | | 23.98 | |
| Clean | Ea. | .08 | 9.85 | | 9.93 | |
| Minimum Charge | Job | | 255 | | 255 | |
| **Over 20" Wide** | | | | | | |
| Demolish | Ea. | | 17.15 | | 17.15 | Includes material and labor to install a |
| Install | Ea. | 98 | 46.50 | | 144.50 | 24" x 24" air return register. |
| Demolish and Install | Ea. | 98 | 63.65 | | 161.65 | |
| Reinstall | Ea. | | 37.06 | | 37.06 | |
| Clean | Ea. | .08 | 11.10 | | 11.18 | |
| Minimum Charge | Job | | 255 | | 255 | |

| Ductwork | Unit | Material | Labor | Equip. | Total | Specification |
|---|---|---|---|---|---|---|
| **Supply Grill** | | | | | | |
| **Baseboard Type** | | | | | | |
| Demolish | Ea. | | 8.10 | | 8.10 | Includes material and labor to install a |
| Install | Ea. | 9.70 | 25.50 | | 35.20 | 14" x 6" baseboard air supply |
| Demolish and Install | Ea. | 9.70 | 33.60 | | 43.30 | register. |
| Reinstall | Ea. | | 20.38 | | 20.38 | |
| Clean | Ea. | .08 | 9.85 | | 9.93 | |
| Minimum Charge | Job | | 255 | | 255 | |
| **10" Wide** | | | | | | |
| Demolish | Ea. | | 9.45 | | 9.45 | Includes material and labor to install a |
| Install | Ea. | 20.50 | 25.50 | | 46 | 10" x 6" air supply register. |
| Demolish and Install | Ea. | 20.50 | 34.95 | | 55.45 | |
| Reinstall | Ea. | | 20.38 | | 20.38 | |
| Clean | Ea. | .08 | 9.85 | | 9.93 | |
| Minimum Charge | Job | | 255 | | 255 | |
| **12" to 15" Wide** | | | | | | |
| Demolish | Ea. | | 9.45 | | 9.45 | Includes material and labor to install a |
| Install | Ea. | 21 | 30 | | 51 | 14" x 8" air supply register. |
| Demolish and Install | Ea. | 21 | 39.45 | | 60.45 | |
| Reinstall | Ea. | | 23.98 | | 23.98 | |
| Clean | Ea. | .08 | 9.85 | | 9.93 | |
| Minimum Charge | Job | | 255 | | 255 | |
| **18" to 24" Wide** | | | | | | |
| Demolish | Ea. | | 11.30 | | 11.30 | Includes material and labor to install a |
| Install | Ea. | 39 | 39 | | 78 | 24" x 8" air supply register. |
| Demolish and Install | Ea. | 39 | 50.30 | | 89.30 | |
| Reinstall | Ea. | | 31.36 | | 31.36 | |
| Clean | Ea. | .08 | 11.10 | | 11.18 | |
| Minimum Charge | Job | | 255 | | 255 | |
| **30" to 36" Wide** | | | | | | |
| Demolish | Ea. | | 14.15 | | 14.15 | Includes material and labor to install a |
| Install | Ea. | 50.50 | 36.50 | | 87 | 30" x 8" air supply register. |
| Demolish and Install | Ea. | 50.50 | 50.65 | | 101.15 | |
| Reinstall | Ea. | | 29.12 | | 29.12 | |
| Clean | Ea. | .08 | 22 | | 22.08 | |
| Minimum Charge | Job | | 255 | | 255 | |
| **Diffuser** | | | | | | |
| **12" x 12"** | | | | | | |
| Demolish | Ea. | | 11.30 | | 11.30 | Includes material and labor to install a |
| Install | Ea. | 77.50 | 42.50 | | 120 | 12" x 12" aluminum diffuser. |
| Demolish and Install | Ea. | 77.50 | 53.80 | | 131.30 | |
| Reinstall | Ea. | | 33.97 | | 33.97 | |
| Clean | Ea. | .08 | 9.85 | | 9.93 | |
| Minimum Charge | Job | | 255 | | 255 | |
| **14" x 14"** | | | | | | |
| Demolish | Ea. | | 11.30 | | 11.30 | Includes material and labor to install a |
| Install | Ea. | 142 | 42.50 | | 184.50 | 14" x 14" aluminum diffuser. |
| Demolish and Install | Ea. | 142 | 53.80 | | 195.80 | |
| Reinstall | Ea. | | 33.97 | | 33.97 | |
| Clean | Ea. | .08 | 11.10 | | 11.18 | |
| Minimum Charge | Job | | 255 | | 255 | |
| **18" x 18"** | | | | | | |
| Demolish | Ea. | | 17.15 | | 17.15 | Includes material and labor to install a |
| Install | Ea. | 125 | 56.50 | | 181.50 | 18" x 18" aluminum diffuser. |
| Demolish and Install | Ea. | 125 | 73.65 | | 198.65 | |
| Reinstall | Ea. | | 45.30 | | 45.30 | |
| Clean | Ea. | .08 | 14.80 | | 14.88 | |
| Minimum Charge | Job | | 255 | | 255 | |

259

For customer support on your **Contractor's Pricing Guide: Residential Repair & Remodeling**, call 800.448.8182.

# Finish Mechanical

## Ductwork

| | Unit | Material | Labor | Equip. | Total | Specification |
|---|---|---|---|---|---|---|
| **12″ Round** | | | | | | |
| Demolish | Ea. | | 15.75 | | 15.75 | Includes material and labor to install a |
| Install | Ea. | 21 | 42.50 | | 63.50 | 12″ diameter aluminum diffuser with a |
| Demolish and Install | Ea. | 21 | 58.25 | | 79.25 | butterfly damper. |
| Reinstall | Ea. | | 33.97 | | 33.97 | |
| Clean | Ea. | .08 | 9.85 | | 9.93 | |
| Minimum Charge | Job | | 255 | | 255 | |
| **14″ to 20″ Round** | | | | | | |
| Demolish | Ea. | | 21 | | 21 | Includes material and labor to install a |
| Install | Ea. | 167 | 56.50 | | 223.50 | 20″ diameter aluminum diffuser with a |
| Demolish and Install | Ea. | 167 | 77.50 | | 244.50 | butterfly damper. |
| Reinstall | Ea. | | 45.30 | | 45.30 | |
| Clean | Ea. | .08 | 11.10 | | 11.18 | |
| Minimum Charge | Job | | 255 | | 255 | |
| **Flush-out / Sanitize** | | | | | | |
| Install | L.F. | .30 | 4.44 | | 4.74 | Includes labor and material costs to |
| Minimum Charge | Job | | 255 | | 255 | clean ductwork or flue. |

| Lighting Fixture | | | Duplex Receptacle | | | |

| Electrical Per S.F. | | Unit | Material | Labor | Equip. | Total | Specification |
|---|---|---|---|---|---|---|---|
| **Residential** | | | | | | | |
| **Light Fixtures** | | | | | | | |
| | Install | S.F. | .24 | .26 | | .50 | Includes labor and material to replace house light fixtures. Apply S.F. cost to floor area including garages but not basements. |

| Circuits | | Unit | Material | Labor | Equip. | Total | Specification |
|---|---|---|---|---|---|---|---|
| **120V** | | | | | | | |
| **Outlet with Wiring** | | | | | | | |
| | Demolish | Ea. | | 29 | | 29 | Includes material and labor to install |
| | Install | Ea. | 9.25 | 36 | | 45.25 | general purpose outlet with cover, up |
| | Demolish and Install | Ea. | 9.25 | 65 | | 74.25 | to 20 feet of #12/2 wire. |
| | Reinstall | Ea. | | 28.88 | | 28.88 | |
| | Clean | Ea. | | 2.22 | | 2.22 | |
| | Minimum Charge | Job | | 263 | | 263 | |
| **GFCI Outlet w / Wiring** | | | | | | | |
| | Demolish | Ea. | | 29 | | 29 | Cost includes material and labor to |
| | Install | Ea. | 23 | 42.50 | | 65.50 | install 120V GFCI receptacle and wall |
| | Demolish and Install | Ea. | 23 | 71.50 | | 94.50 | plate, with up to 20 feet of #12/2 |
| | Reinstall | Ea. | | 34.13 | | 34.13 | wire. |
| | Clean | Ea. | | 2.22 | | 2.22 | |
| | Minimum Charge | Job | | 263 | | 263 | |
| **Exterior Outlet w / Wiring** | | | | | | | |
| | Demolish | Ea. | | 29 | | 29 | Cost includes material and labor to |
| | Install | Ea. | 44.50 | 44 | | 88.50 | install 110V waterproof external |
| | Demolish and Install | Ea. | 44.50 | 73 | | 117.50 | receptacle and cover plate with up to |
| | Reinstall | Ea. | | 35.01 | | 35.01 | 20' of #12/2 wire. |
| | Clean | Ea. | | 2.22 | | 2.22 | |
| | Minimum Charge | Job | | 263 | | 263 | |
| **240V Outlet w / Wiring** | | | | | | | |
| | Demolish | Ea. | | 29 | | 29 | Cost includes material and labor to |
| | Install | Ea. | 59 | 82 | | 141 | install 240V receptacle and cover |
| | Demolish and Install | Ea. | 59 | 111 | | 170 | plate with up to 20' of #12/2 wire. |
| | Reinstall | Ea. | | 65.55 | | 65.55 | |
| | Clean | Ea. | | 2.22 | | 2.22 | |
| | Minimum Charge | Job | | 263 | | 263 | |

261

For customer support on your **Contractor's Pricing Guide: Residential Repair & Remodeling,** call 800.448.8182.

# Finish Electrical

## Circuits

| Circuits | Unit | Material | Labor | Equip. | Total | Specification |
|---|---|---|---|---|---|---|
| **TV / Cable Jack w / Wiring** | | | | | | Includes material and labor to install |
| Demolish | Ea. | | 29 | | 29 | TV/cable outlet box and cover plate |
| Install | Ea. | 17.10 | 33 | | 50.10 | including 20' of wiring. |
| Demolish and Install | Ea. | 17.10 | 62 | | 79.10 | |
| Reinstall | Ea. | | 26.26 | | 26.26 | |
| Clean | Ea. | | 2.22 | | 2.22 | |
| Minimum Charge | Job | | 263 | | 263 | |
| **Phone Jack with Wiring** | | | | | | Includes material and labor to install |
| Demolish | Ea. | | 29 | | 29 | phone jack and cover plate including |
| Install | Ea. | 8.80 | 20 | | 28.80 | up to 20' of telephone cable. |
| Demolish and Install | Ea. | 8.80 | 49 | | 57.80 | |
| Reinstall | Ea. | | 16.16 | | 16.16 | |
| Clean | Ea. | | 1.39 | | 1.39 | |
| Minimum Charge | Job | | 263 | | 263 | |

## Light Switch

| Light Switch | Unit | Material | Labor | Equip. | Total | Specification |
|---|---|---|---|---|---|---|
| **Single** | | | | | | |
| **Switch** | | | | | | |
| Demolish | Ea. | | 5.85 | | 5.85 | Includes material and labor to install |
| Install | Ea. | 13 | 25 | | 38 | single switch and wall plate. Wiring is |
| Demolish and Install | Ea. | 13 | 30.85 | | 43.85 | not included. |
| Reinstall | Ea. | | 20.01 | | 20.01 | |
| Clean | Ea. | | 2.22 | | 2.22 | |
| Minimum Charge | Job | | 263 | | 263 | |
| **Plate** | | | | | | |
| Demolish | Ea. | | 2.92 | | 2.92 | Includes material and labor to install a |
| Install | Ea. | .42 | 6.55 | | 6.97 | one gang brown wall switch plate. |
| Demolish and Install | Ea. | .42 | 9.47 | | 9.89 | |
| Reinstall | Ea. | | 5.25 | | 5.25 | |
| Clean | Ea. | .03 | .44 | | .47 | |
| Minimum Charge | Job | | 263 | | 263 | |
| **Double** | | | | | | |
| **Switch** | | | | | | |
| Demolish | Ea. | | 11.65 | | 11.65 | Includes material and labor to install |
| Install | Ea. | 14.15 | 37.50 | | 51.65 | double switch and wall plate. Wiring |
| Demolish and Install | Ea. | 14.15 | 49.15 | | 63.30 | is not included. |
| Reinstall | Ea. | | 30.01 | | 30.01 | |
| Clean | Ea. | .03 | 7.10 | | 7.13 | |
| Minimum Charge | Job | | 263 | | 263 | |
| **Plate** | | | | | | |
| Demolish | Ea. | | 2.92 | | 2.92 | Includes material and labor to install a |
| Install | Ea. | .83 | 9.90 | | 10.73 | two gang brown wall switch plate. |
| Demolish and Install | Ea. | .83 | 12.82 | | 13.65 | |
| Reinstall | Ea. | | 7.93 | | 7.93 | |
| Clean | Ea. | .03 | .44 | | .47 | |
| Minimum Charge | Job | | 263 | | 263 | |
| **Triple** | | | | | | |
| **Switch** | | | | | | |
| Demolish | Ea. | | 17.70 | | 17.70 | Includes material and labor to install |
| Install | Ea. | 8.25 | 44 | | 52.25 | triple switch and wall plate. Wiring is |
| Demolish and Install | Ea. | 8.25 | 61.70 | | 69.95 | not included. |
| Reinstall | Ea. | | 35.01 | | 35.01 | |
| Clean | Ea. | .03 | 10.65 | | 10.68 | |
| Minimum Charge | Job | | 263 | | 263 | |

# Finish Electrical

| Light Switch | | Unit | Material | Labor | Equip. | Total | Specification |
|---|---|---|---|---|---|---|---|
| **Plate** | | | | | | | |
| | Demolish | Ea. | | 2.92 | | 2.92 | Includes material and labor to install a |
| | Install | Ea. | 1.22 | 16.40 | | 17.62 | three gang brown wall switch plate. |
| | Demolish and Install | Ea. | 1.22 | 19.32 | | 20.54 | |
| | Reinstall | Ea. | | 13.13 | | 13.13 | |
| | Clean | Ea. | .03 | .44 | | .47 | |
| | Minimum Charge | Job | | 263 | | 263 | |
| **Dimmer** | | | | | | | |
| **Switch** | | | | | | | |
| | Demolish | Ea. | | 5.85 | | 5.85 | Includes material and labor to install |
| | Install | Ea. | 33 | 31 | | 64 | dimmer switch. Wiring is not included. |
| | Demolish and Install | Ea. | 33 | 36.85 | | 69.85 | |
| | Reinstall | Ea. | | 24.72 | | 24.72 | |
| | Clean | Ea. | | 2.22 | | 2.22 | |
| | Minimum Charge | Job | | 263 | | 263 | |
| **Plate** | | | | | | | |
| | Demolish | Ea. | | 2.92 | | 2.92 | Includes material and labor to install a |
| | Install | Ea. | .42 | 6.55 | | 6.97 | one gang brown wall switch plate. |
| | Demolish and Install | Ea. | .42 | 9.47 | | 9.89 | |
| | Reinstall | Ea. | | 5.25 | | 5.25 | |
| | Clean | Ea. | .03 | .44 | | .47 | |
| | Minimum Charge | Job | | 263 | | 263 | |
| **With Wiring** | | | | | | | |
| **Single** | | | | | | | |
| | Demolish | Ea. | | 29 | | 29 | Includes material and labor to install |
| | Install | Ea. | 17.05 | 30.50 | | 47.55 | single switch and plate including up to |
| | Demolish and Install | Ea. | 17.05 | 59.50 | | 76.55 | 20' of romex cable. |
| | Reinstall | Ea. | | 24.57 | | 24.57 | |
| | Clean | Ea. | | 2.22 | | 2.22 | |
| | Minimum Charge | Job | | 263 | | 263 | |
| **Double** | | | | | | | |
| | Demolish | Ea. | | 29 | | 29 | Includes material and labor to install |
| | Install | Ea. | 24 | 52.50 | | 76.50 | double switch and plate including up |
| | Demolish and Install | Ea. | 24 | 81.50 | | 105.50 | to 20' of romex cable. |
| | Reinstall | Ea. | | 42.02 | | 42.02 | |
| | Clean | Ea. | .03 | 7.10 | | 7.13 | |
| | Minimum Charge | Job | | 263 | | 263 | |
| **Triple** | | | | | | | |
| | Demolish | Ea. | | 44 | | 44 | Includes material and labor to install |
| | Install | Ea. | 13.85 | 59 | | 72.85 | triple switch and plate including up to |
| | Demolish and Install | Ea. | 13.85 | 103 | | 116.85 | 20' of romex cable. |
| | Reinstall | Ea. | | 47.26 | | 47.26 | |
| | Clean | Ea. | .03 | 10.65 | | 10.68 | |
| | Minimum Charge | Job | | 263 | | 263 | |
| **Dimmer** | | | | | | | |
| | Demolish | Ea. | | 29 | | 29 | Includes material and labor to install |
| | Install | Ea. | 37 | 36 | | 73 | dimmer switch and cover plate |
| | Demolish and Install | Ea. | 37 | 65 | | 102 | including up to 20' of romex cable. |
| | Reinstall | Ea. | | 28.88 | | 28.88 | |
| | Clean | Ea. | | 2.22 | | 2.22 | |
| | Minimum Charge | Job | | 263 | | 263 | |

# Finish Electrical

## Cover for Outlet / Switch

| Cover for Outlet / Switch | Unit | Material | Labor | Equip. | Total | Specification |
|---|---|---|---|---|---|---|
| **High Grade** | | | | | | |
| Demolish | Ea. | | 2.92 | | 2.92 | Includes material and labor to install |
| Install | Ea. | 2.82 | 6.55 | | 9.37 | wall plate, stainless steel, 1 gang. |
| Demolish and Install | Ea. | 2.82 | 9.47 | | 12.29 | |
| Reinstall | Ea. | | 5.25 | | 5.25 | |
| Clean | Ea. | .03 | .44 | | .47 | |
| Minimum Charge | Job | | 263 | | 263 | |
| **Deluxe Grade** | | | | | | |
| Demolish | Ea. | | 2.92 | | 2.92 | Includes material and labor to install |
| Install | Ea. | 4.69 | 6.55 | | 11.24 | wall plate, brushed brass, 1 gang. |
| Demolish and Install | Ea. | 4.69 | 9.47 | | 14.16 | |
| Reinstall | Ea. | | 5.25 | | 5.25 | |
| Clean | Ea. | .03 | .44 | | .47 | |
| Minimum Charge | Job | | 263 | | 263 | |

## Residential Light Fixture

| Residential Light Fixture | Unit | Material | Labor | Equip. | Total | Specification |
|---|---|---|---|---|---|---|
| **Ceiling** | | | | | | |
| **Good Quality** | | | | | | |
| Demolish | Ea. | | 12.15 | | 12.15 | Includes material and labor to install |
| Install | Ea. | 25.50 | 13.15 | | 38.65 | standard quality incandescent ceiling |
| Demolish and Install | Ea. | 25.50 | 25.30 | | 50.80 | light fixture. Wiring is not included. |
| Reinstall | Ea. | | 10.50 | | 10.50 | |
| Clean | Ea. | .03 | 29.50 | | 29.53 | |
| Minimum Charge | Job | | 263 | | 263 | |
| **Better Quality** | | | | | | |
| Demolish | Ea. | | 12.15 | | 12.15 | Includes material and labor to install |
| Install | Ea. | 79 | 27.50 | | 106.50 | custom quality incandescent ceiling |
| Demolish and Install | Ea. | 79 | 39.65 | | 118.65 | light fixture. Wiring is not included. |
| Reinstall | Ea. | | 22.11 | | 22.11 | |
| Clean | Ea. | .03 | 29.50 | | 29.53 | |
| Minimum Charge | Job | | 263 | | 263 | |
| **Premium Quality** | | | | | | |
| Demolish | Ea. | | 12.15 | | 12.15 | Includes material and labor to install |
| Install | Ea. | 335 | 27.50 | | 362.50 | deluxe quality incandescent ceiling |
| Demolish and Install | Ea. | 335 | 39.65 | | 374.65 | light fixture. Wiring is not included. |
| Reinstall | Ea. | | 22.11 | | 22.11 | |
| Clean | Ea. | .03 | 29.50 | | 29.53 | |
| Minimum Charge | Job | | 263 | | 263 | |
| **Recessed Ceiling** | | | | | | |
| Demolish | Ea. | | 24.50 | | 24.50 | Includes material and labor to install |
| Install | Ea. | 69 | 17.50 | | 86.50 | standard quality recessed eyeball |
| Demolish and Install | Ea. | 69 | 42 | | 111 | spotlight with housing. Wiring is not |
| Reinstall | Ea. | | 14.01 | | 14.01 | included. |
| Clean | Ea. | .01 | 3.70 | | 3.71 | |
| Minimum Charge | Job | | 263 | | 263 | |
| **Recessed Eyeball** | | | | | | |
| Demolish | Ea. | | 24.50 | | 24.50 | Includes material and labor to install |
| Install | Ea. | 73 | 18.75 | | 91.75 | custom quality recessed eyeball |
| Demolish and Install | Ea. | 73 | 43.25 | | 116.25 | spotlight with housing. Wiring is not |
| Reinstall | Ea. | | 15.01 | | 15.01 | included. |
| Clean | Ea. | .01 | 3.70 | | 3.71 | |
| Minimum Charge | Job | | 263 | | 263 | |

| Residential Light Fixture | Unit | Material | Labor | Equip. | Total | Specification |
|---|---|---|---|---|---|---|
| **Recessed Shower Type** | | | | | | |
| Demolish | Ea. | | 24.50 | | 24.50 | Includes material and labor to install |
| Install | Ea. | 94.50 | 17.50 | | 112 | deluxe quality recessed eyeball |
| Demolish and Install | Ea. | 94.50 | 42 | | 136.50 | spotlight with housing. Wiring is not |
| Reinstall | Ea. | | 14.01 | | 14.01 | included. |
| Clean | Ea. | .01 | 3.70 | | 3.71 | |
| Minimum Charge | Job | | 263 | | 263 | |
| **Run Wiring** | | | | | | |
| Install | Ea. | 16 | 21 | | 37 | Includes labor and material to install |
| Minimum Charge | Job | | 263 | | 263 | an electric fixture utility box with |
| | | | | | | non-metallic cable. |
| **Fluorescent** | | | | | | |
| **Good Quality** | | | | | | |
| Demolish | Ea. | | 26.50 | | 26.50 | Includes material and labor to install |
| Install | Ea. | 43 | 75 | | 118 | standard quality 12" wide by 48" long |
| Demolish and Install | Ea. | 43 | 101.50 | | 144.50 | decorative surface mounted fluorescent |
| Reinstall | Ea. | | 60.02 | | 60.02 | fixture w/acrylic diffuser. |
| Clean | Ea. | .01 | 3.70 | | 3.71 | |
| Minimum Charge | Job | | 263 | | 263 | |
| **Better Quality** | | | | | | |
| Demolish | Ea. | | 26.50 | | 26.50 | Includes material and labor to install |
| Install | Ea. | 88 | 84.50 | | 172.50 | an interior fluorescent lighting fixture. |
| Demolish and Install | Ea. | 88 | 111 | | 199 | |
| Reinstall | Ea. | | 67.77 | | 67.77 | |
| Clean | Ea. | .01 | 3.70 | | 3.71 | |
| Minimum Charge | Job | | 263 | | 263 | |
| **Premium Quality** | | | | | | |
| Demolish | Ea. | | 26.50 | | 26.50 | Includes material and labor to install |
| Install | Ea. | 93.50 | 99 | | 192.50 | deluxe quality 24" wide by 48" long |
| Demolish and Install | Ea. | 93.50 | 125.50 | | 219 | decorative surface mounted fluorescent |
| Reinstall | Ea. | | 79.28 | | 79.28 | fixture w/acrylic diffuser. |
| Clean | Ea. | .01 | 3.70 | | 3.71 | |
| Minimum Charge | Job | | 263 | | 263 | |
| **Run Wiring** | | | | | | |
| Install | Ea. | 16 | 21 | | 37 | Includes labor and material to install |
| Minimum Charge | Job | | 263 | | 263 | an electric fixture utility box with |
| | | | | | | non-metallic cable. |
| **Exterior** | | | | | | |
| **Good Quality** | | | | | | |
| Demolish | Ea. | | 23.50 | | 23.50 | Includes material and labor to install |
| Install | Ea. | 66.50 | 33 | | 99.50 | standard quality incandescent outdoor |
| Demolish and Install | Ea. | 66.50 | 56.50 | | 123 | light fixture. Wiring is not included. |
| Reinstall | Ea. | | 26.26 | | 26.26 | |
| Clean | Ea. | .08 | 14.80 | | 14.88 | |
| Minimum Charge | Job | | 263 | | 263 | |
| **Better Quality** | | | | | | |
| Demolish | Ea. | | 23.50 | | 23.50 | Includes material and labor to install |
| Install | Ea. | 116 | 33 | | 149 | custom quality incandescent outdoor |
| Demolish and Install | Ea. | 116 | 56.50 | | 172.50 | light fixture. Wiring is not included. |
| Reinstall | Ea. | | 26.26 | | 26.26 | |
| Clean | Ea. | .08 | 14.80 | | 14.88 | |
| Minimum Charge | Job | | 263 | | 263 | |
| **Premium Quality** | | | | | | |
| Demolish | Ea. | | 23.50 | | 23.50 | Includes material and labor to install |
| Install | Ea. | 259 | 131 | | 390 | deluxe quality incandescent outdoor |
| Demolish and Install | Ea. | 259 | 154.50 | | 413.50 | light fixture. Wiring is not included. |
| Reinstall | Ea. | | 105.04 | | 105.04 | |
| Clean | Ea. | .08 | 14.80 | | 14.88 | |
| Minimum Charge | Job | | 263 | | 263 | |

# Finish Electrical

## Residential Light Fixture

| Residential Light Fixture | Unit | Material | Labor | Equip. | Total | Specification |
|---|---|---|---|---|---|---|
| **Outdoor w / Pole** | | | | | | Includes material and labor to install |
| Demolish | Ea. | | 23.50 | | 23.50 | post lantern type incandescent outdoor |
| Install | Ea. | 297 | 131 | | 428 | light fixture. Wiring is not included. |
| Demolish and Install | Ea. | 297 | 154.50 | | 451.50 | |
| Reinstall | Ea. | | 105.04 | | 105.04 | |
| Clean | Ea. | .08 | 14.80 | | 14.88 | |
| Minimum Charge | Job | | 263 | | 263 | |
| **Run Wiring** | | | | | | Includes labor and material to install |
| Install | Ea. | 16 | 21 | | 37 | an electric fixture utility box with |
| Minimum Charge | Job | | 263 | | 263 | non-metallic cable. |
| **Track Lighting** | | | | | | Includes material and labor to install |
| Demolish | L.F. | | 8.10 | | 8.10 | track lighting. |
| Install | L.F. | 22 | 22 | | 44 | |
| Demolish and Install | L.F. | 22 | 30.10 | | 52.10 | |
| Reinstall | L.F. | | 17.51 | | 17.51 | |
| Clean | L.F. | .08 | 7.10 | | 7.18 | |
| Minimum Charge | Job | | 263 | | 263 | |

## Commercial Light Fixture

| Commercial Light Fixture | Unit | Material | Labor | Equip. | Total | Specification |
|---|---|---|---|---|---|---|
| **Fluorescent** | | | | | | |
| **2 Bulb, 2'** | | | | | | Includes material and labor to install a |
| Demolish | Ea. | | 23.50 | | 23.50 | 2' long surface mounted fluorescent |
| Install | Ea. | 37 | 65.50 | | 102.50 | light fixture with two 40 watt bulbs. |
| Demolish and Install | Ea. | 37 | 89 | | 126 | |
| Reinstall | Ea. | | 52.52 | | 52.52 | |
| Clean | Ea. | .36 | 25.50 | | 25.86 | |
| Minimum Charge | Job | | 263 | | 263 | |
| **2 Bulb, 4'** | | | | | | Includes material and labor to install a |
| Demolish | Ea. | | 23.50 | | 23.50 | 4' long surface mounted fluorescent |
| Install | Ea. | 80 | 75 | | 155 | light fixture with two 40 watt bulbs. |
| Demolish and Install | Ea. | 80 | 98.50 | | 178.50 | |
| Reinstall | Ea. | | 60.02 | | 60.02 | |
| Clean | Ea. | .36 | 25.50 | | 25.86 | |
| Minimum Charge | Job | | 263 | | 263 | |
| **3 Bulb, 4'** | | | | | | Includes material and labor to install a |
| Demolish | Ea. | | 23.50 | | 23.50 | 4' long pendant mounted fluorescent |
| Install | Ea. | 106 | 75 | | 181 | light fixture with three 34 watt bulbs. |
| Demolish and Install | Ea. | 106 | 98.50 | | 204.50 | |
| Reinstall | Ea. | | 60.02 | | 60.02 | |
| Clean | Ea. | .36 | 25.50 | | 25.86 | |
| Minimum Charge | Job | | 263 | | 263 | |
| **4 Bulb, 4'** | | | | | | Includes material and labor to install a |
| Demolish | Ea. | | 23.50 | | 23.50 | 4' long pendant mounted fluorescent |
| Install | Ea. | 96.50 | 81 | | 177.50 | light fixture with four 34 watt bulbs. |
| Demolish and Install | Ea. | 96.50 | 104.50 | | 201 | |
| Reinstall | Ea. | | 64.64 | | 64.64 | |
| Clean | Ea. | .36 | 25.50 | | 25.86 | |
| Minimum Charge | Job | | 263 | | 263 | |
| **2 Bulb, 8'** | | | | | | Includes material and labor to install |
| Demolish | Ea. | | 29 | | 29 | an 8' long surface mounted |
| Install | Ea. | 119 | 99 | | 218 | fluorescent light fixture with two 110 |
| Demolish and Install | Ea. | 119 | 128 | | 247 | watt bulbs. |
| Reinstall | Ea. | | 79.28 | | 79.28 | |
| Clean | Ea. | .36 | 25.50 | | 25.86 | |
| Minimum Charge | Job | | 263 | | 263 | |

| Commercial Light Fixture | | Unit | Material | Labor | Equip. | Total | Specification |
|---|---|---|---|---|---|---|---|
| **Run Wiring** | | | | | | | |
| | Install | Ea. | 16 | 21 | | 37 | Includes labor and material to install |
| | Minimum Charge | Job | | 263 | | 263 | an electric fixture utility box with |
| | | | | | | | non-metallic cable. |
| **Ceiling, Recessed** | | | | | | | |
| **Eyeball Type** | | | | | | | |
| | Demolish | Ea. | | 24.50 | | 24.50 | Includes material and labor to install |
| | Install | Ea. | 73 | 18.75 | | 91.75 | custom quality recessed eyeball |
| | Demolish and Install | Ea. | 73 | 43.25 | | 116.25 | spotlight with housing. Wiring is not |
| | Reinstall | Ea. | | 15.01 | | 15.01 | included. |
| | Clean | Ea. | .01 | 3.70 | | 3.71 | |
| | Minimum Charge | Job | | 263 | | 263 | |
| **Shower Type** | | | | | | | |
| | Demolish | Ea. | | 24.50 | | 24.50 | Includes material and labor to install |
| | Install | Ea. | 94.50 | 17.50 | | 112 | deluxe quality recessed eyeball |
| | Demolish and Install | Ea. | 94.50 | 42 | | 136.50 | spotlight with housing. Wiring is not |
| | Reinstall | Ea. | | 14.01 | | 14.01 | included. |
| | Clean | Ea. | .01 | 3.70 | | 3.71 | |
| | Minimum Charge | Job | | 263 | | 263 | |
| **Run Wiring** | | | | | | | |
| | Install | Ea. | 16 | 21 | | 37 | Includes labor and material to install |
| | Minimum Charge | Job | | 263 | | 263 | an electric fixture utility box with |
| | | | | | | | non-metallic cable. |
| **Ceiling Grid Type** | | | | | | | |
| **2 Tube** | | | | | | | |
| | Demolish | Ea. | | 34.50 | | 34.50 | Cost includes material and labor to |
| | Install | Ea. | 59.50 | 99 | | 158.50 | install 2 lamp fluorescent light fixture |
| | Demolish and Install | Ea. | 59.50 | 133.50 | | 193 | in 2' x 4' ceiling grid system including |
| | Reinstall | Ea. | | 79.28 | | 79.28 | lens and tubes. |
| | Clean | Ea. | .36 | 25.50 | | 25.86 | |
| | Minimum Charge | Job | | 263 | | 263 | |
| **4 Tube** | | | | | | | |
| | Demolish | Ea. | | 39 | | 39 | Cost includes material and labor to |
| | Install | Ea. | 68 | 112 | | 180 | install 4 lamp fluorescent light fixture |
| | Demolish and Install | Ea. | 68 | 151 | | 219 | in 2' x 4' ceiling grid system including |
| | Reinstall | Ea. | | 89.40 | | 89.40 | lens and tubes. |
| | Clean | Ea. | .36 | 25.50 | | 25.86 | |
| | Minimum Charge | Job | | 263 | | 263 | |
| **Acrylic Diffuser** | | | | | | | |
| | Demolish | Ea. | | 6.10 | | 6.10 | Includes material and labor to install |
| | Install | Ea. | 97.50 | 15.90 | | 113.40 | fluorescent light fixture w/acrylic |
| | Demolish and Install | Ea. | 97.50 | 22 | | 119.50 | diffuser. |
| | Reinstall | Ea. | | 12.73 | | 12.73 | |
| | Clean | Ea. | .23 | 4.44 | | 4.67 | |
| | Minimum Charge | Job | | 263 | | 263 | |
| **Louver Diffuser** | | | | | | | |
| | Demolish | Ea. | | 7.30 | | 7.30 | Includes material and labor to install |
| | Install | Ea. | 109 | 15.90 | | 124.90 | fluorescent light fixture w/louver |
| | Demolish and Install | Ea. | 109 | 23.20 | | 132.20 | diffuser. |
| | Reinstall | Ea. | | 12.73 | | 12.73 | |
| | Clean | Ea. | .23 | 4.44 | | 4.67 | |
| | Minimum Charge | Job | | 263 | | 263 | |
| **Run Wiring** | | | | | | | |
| | Install | Ea. | 16 | 21 | | 37 | Includes labor and material to install |
| | Minimum Charge | Job | | 263 | | 263 | an electric fixture utility box with |
| | | | | | | | non-metallic cable. |

267

# Finish Electrical

| Commercial Light Fixture | Unit | Material | Labor | Equip. | Total | Specification |
|---|---|---|---|---|---|---|
| **Track Lighting** | | | | | | |
| Demolish | L.F. | | 8.10 | | 8.10 | Includes material and labor to install |
| Install | L.F. | 22 | 22 | | 44 | track lighting. |
| Demolish and Install | L.F. | 22 | 30.10 | | 52.10 | |
| Reinstall | L.F. | | 17.51 | | 17.51 | |
| Clean | L.F. | .08 | 7.10 | | 7.18 | |
| Minimum Charge | Job | | 263 | | 263 | |
| **Emergency** | | | | | | |
| Demolish | Ea. | | 41.50 | | 41.50 | Includes material and labor to install |
| Install | Ea. | 158 | 131 | | 289 | battery-powered emergency lighting |
| Demolish and Install | Ea. | 158 | 172.50 | | 330.50 | unit with two (2) lights, including |
| Reinstall | Ea. | | 105.04 | | 105.04 | wiring. |
| Clean | Ea. | .19 | 25.50 | | 25.69 | |
| Minimum Charge | Job | | 263 | | 263 | |
| **Run Wiring** | | | | | | |
| Install | Ea. | 16 | 21 | | 37 | Includes labor and material to install |
| Minimum Charge | Job | | 263 | | 263 | an electric fixture utility box with non-metallic cable. |

| Fan | Unit | Material | Labor | Equip. | Total | Specification |
|---|---|---|---|---|---|---|
| **Ceiling Paddle** | | | | | | |
| **Good Quality** | | | | | | |
| Demolish | Ea. | | 32.50 | | 32.50 | Includes material and labor to install |
| Install | Ea. | 91 | 52.50 | | 143.50 | standard ceiling fan. |
| Demolish and Install | Ea. | 91 | 85 | | 176 | |
| Reinstall | Ea. | | 42.02 | | 42.02 | |
| Clean | Ea. | .14 | 22 | | 22.14 | |
| Minimum Charge | Job | | 263 | | 263 | |
| **Better Quality** | | | | | | |
| Demolish | Ea. | | 32.50 | | 32.50 | Includes material and labor to install |
| Install | Ea. | 209 | 52.50 | | 261.50 | better ceiling fan. |
| Demolish and Install | Ea. | 209 | 85 | | 294 | |
| Reinstall | Ea. | | 42.02 | | 42.02 | |
| Clean | Ea. | .14 | 22 | | 22.14 | |
| Minimum Charge | Job | | 263 | | 263 | |
| **Premium Quality** | | | | | | |
| Demolish | Ea. | | 32.50 | | 32.50 | Includes material and labor to install |
| Install | Ea. | 298 | 65.50 | | 363.50 | deluxe ceiling fan. |
| Demolish and Install | Ea. | 298 | 98 | | 396 | |
| Reinstall | Ea. | | 52.52 | | 52.52 | |
| Clean | Ea. | .14 | 22 | | 22.14 | |
| Minimum Charge | Job | | 263 | | 263 | |
| **Light Kit** | | | | | | |
| Install | Ea. | 43.50 | 65.50 | | 109 | Includes labor and material to install a |
| Minimum Charge | Job | | 263 | | 263 | ceiling fan light kit. |
| **Bathroom Exhaust** | | | | | | |
| **Ceiling Mounted** | | | | | | |
| Demolish | Ea. | | 26 | | 26 | Cost includes material and labor to |
| Install | Ea. | 46.50 | 52.50 | | 99 | install 60 CFM ceiling mounted |
| Demolish and Install | Ea. | 46.50 | 78.50 | | 125 | bathroom exhaust fan, up to 10' of |
| Reinstall | Ea. | | 41.89 | | 41.89 | wiring with one wall switch. |
| Clean | Ea. | .08 | 8.90 | | 8.98 | |
| Minimum Charge | Job | | 263 | | 263 | |

## Fan

| | Unit | Material | Labor | Equip. | Total | Specification |
|---|---|---|---|---|---|---|
| **Wall Mounted** | | | | | | |
| Demolish | Ea. | | 26 | | 26 | Cost includes material and labor to |
| Install | Ea. | 69.50 | 52.50 | | 122 | install 110 CFM wall mounted |
| Demolish and Install | Ea. | 69.50 | 78.50 | | 148 | bathroom exhaust fan, up to 10' of |
| Reinstall | Ea. | | 41.89 | | 41.89 | wiring with one wall switch. |
| Clean | Ea. | .08 | 8.90 | | 8.98 | |
| Minimum Charge | Job | | 263 | | 263 | |
| **Lighted** | | | | | | |
| Demolish | Ea. | | 26 | | 26 | Cost includes material and labor to |
| Install | Ea. | 56 | 52.50 | | 108.50 | install 50 CFM ceiling mounted lighted |
| Demolish and Install | Ea. | 56 | 78.50 | | 134.50 | exhaust fan, up to 10' of wiring with |
| Reinstall | Ea. | | 41.89 | | 41.89 | one wall switch.. |
| Clean | Ea. | .08 | 8.90 | | 8.98 | |
| Minimum Charge | Job | | 263 | | 263 | |
| **Blower Heater w / Light** | | | | | | |
| Demolish | Ea. | | 26 | | 26 | Includes material and labor to install |
| Install | Ea. | 86 | 87.50 | | 173.50 | ceiling blower type ventilator with |
| Demolish and Install | Ea. | 86 | 113.50 | | 199.50 | switch, light and snap-on grill. Includes |
| Reinstall | Ea. | | 69.82 | | 69.82 | 1300 watt/120 volt heater. |
| Clean | Ea. | .08 | 8.90 | | 8.98 | |
| Minimum Charge | Job | | 263 | | 263 | |
| **Fan Heater w / Light** | | | | | | |
| Demolish | Ea. | | 26 | | 26 | Includes material and labor to install |
| Install | Ea. | 116 | 87.50 | | 203.50 | circular ceiling fan heater unit, 1500 |
| Demolish and Install | Ea. | 116 | 113.50 | | 229.50 | watt, surface mounted. |
| Reinstall | Ea. | | 69.82 | | 69.82 | |
| Clean | Ea. | .08 | 8.90 | | 8.98 | |
| Minimum Charge | Job | | 263 | | 263 | |
| **Whole House Exhaust** | | | | | | |
| **36" Ceiling Mounted** | | | | | | |
| Install | Ea. | 1000 | 131 | | 1131 | Includes material and labor to install a |
| Clean | Ea. | .02 | 6.65 | | 6.67 | whole house exhaust fan. |
| Minimum Charge | Job | | 263 | | 263 | |
| **Kitchen Exhaust** | | | | | | |
| **Standard Model** | | | | | | |
| Install | Ea. | 75 | 35 | | 110 | Includes labor and materials to install |
| Clean | Ea. | .02 | 6.65 | | 6.67 | a bathroom or kitchen vent fan. |
| Minimum Charge | Job | | 263 | | 263 | |
| **Low Noise Model** | | | | | | |
| Install | Ea. | 101 | 35 | | 136 | Includes labor and materials to install |
| Clean | Ea. | .02 | 6.65 | | 6.67 | a bathroom or kitchen vent fan, low |
| Minimum Charge | Job | | 263 | | 263 | noise model. |

## Low Voltage Systems

| | Unit | Material | Labor | Equip. | Total | Specification |
|---|---|---|---|---|---|---|
| **Smoke Detector** | | | | | | |
| **Pre-wired** | | | | | | |
| Demolish | Ea. | | 12.15 | | 12.15 | Includes material and labor to install |
| Install | Ea. | 22.50 | 24.50 | | 47 | AC type smoke detector including |
| Demolish and Install | Ea. | 22.50 | 36.65 | | 59.15 | wiring. |
| Reinstall | Ea. | | 19.70 | | 19.70 | |
| Clean | Ea. | .03 | 11.10 | | 11.13 | |
| Minimum Charge | Job | | 263 | | 263 | |

269

For customer support on your Contractor's Pricing Guide: Residential Repair & Remodeling, call 800.448.8182.

| Low Voltage Systems | Unit | Material | Labor | Equip. | Total | Specification |
|---|---|---|---|---|---|---|
| **Run Wiring** | | | | | | |
| Install | Ea. | 16 | 21 | | 37 | Includes labor and material to install |
| Minimum Charge | Job | | 263 | | 263 | an electric fixture utility box with |
| | | | | | | non-metallic cable. |
| **Door Bell / Chimes** | | | | | | |
| **Good Quality** | | | | | | |
| Demolish | Ea. | | 36.50 | | 36.50 | Includes material and labor to install |
| Install | Ea. | 50.50 | 131 | | 181.50 | surface mounted 2 note door bell. |
| Demolish and Install | Ea. | 50.50 | 167.50 | | 218 | Wiring is not included. |
| Reinstall | Ea. | | 105.04 | | 105.04 | |
| Clean | Ea. | .08 | 11.10 | | 11.18 | |
| Minimum Charge | Job | | 263 | | 263 | |
| **Better Quality** | | | | | | |
| Demolish | Ea. | | 36.50 | | 36.50 | Includes material and labor to install |
| Install | Ea. | 76 | 131 | | 207 | standard quality surface mounted 2 |
| Demolish and Install | Ea. | 76 | 167.50 | | 243.50 | note door bell. Wiring is not included. |
| Reinstall | Ea. | | 105.04 | | 105.04 | |
| Clean | Ea. | .08 | 11.10 | | 11.18 | |
| Minimum Charge | Job | | 263 | | 263 | |
| **Premium Quality** | | | | | | |
| Demolish | Ea. | | 36.50 | | 36.50 | Includes material and labor to install |
| Install | Ea. | 105 | 131 | | 236 | deluxe quality surface mounted 2 note |
| Demolish and Install | Ea. | 105 | 167.50 | | 272.50 | door bell. Wiring is not included. |
| Reinstall | Ea. | | 105.04 | | 105.04 | |
| Clean | Ea. | .08 | 11.10 | | 11.18 | |
| Minimum Charge | Job | | 263 | | 263 | |
| **Run Wiring** | | | | | | |
| Install | Ea. | 16 | 21 | | 37 | Includes labor and material to install |
| Minimum Charge | Job | | 263 | | 263 | an electric fixture utility box with |
| | | | | | | non-metallic cable. |
| **Intercom System** | | | | | | |
| **Intercom (master)** | | | | | | |
| Demolish | Ea. | | 97.50 | | 97.50 | Includes material and labor to install |
| Install | Ea. | 760 | 263 | | 1023 | master intercom station with AM/FM |
| Demolish and Install | Ea. | 760 | 360.50 | | 1120.50 | music and room monitoring. Wiring is |
| Reinstall | Ea. | | 210.08 | | 210.08 | not included. |
| Clean | Ea. | .08 | 29.50 | | 29.58 | |
| Minimum Charge | Job | | 263 | | 263 | |
| **Intercom (remote)** | | | | | | |
| Demolish | Ea. | | 58.50 | | 58.50 | Includes material and labor to install |
| Install | Ea. | 69.50 | 65.50 | | 135 | remote station for master intercom |
| Demolish and Install | Ea. | 69.50 | 124 | | 193.50 | system. Wiring is not included. |
| Reinstall | Ea. | | 52.52 | | 52.52 | |
| Clean | Ea. | .08 | 11.10 | | 11.18 | |
| Minimum Charge | Job | | 263 | | 263 | |
| **Security System** | | | | | | |
| **Alarm Panel** | | | | | | |
| Demolish | Ea. | | 117 | | 117 | Includes material and labor to install a |
| Install | Ea. | 145 | 263 | | 408 | 6 zone burglar alarm panel. |
| Demolish and Install | Ea. | 145 | 380 | | 525 | |
| Reinstall | Ea. | | 210.08 | | 210.08 | |
| Clean | Ea. | .08 | 11.10 | | 11.18 | |
| Minimum Charge | Job | | 263 | | 263 | |

# Finish Electrical

| Low Voltage Systems | | Unit | Material | Labor | Equip. | Total | Specification |
|---|---|---|---|---|---|---|---|
| **Motion Detector** | | | | | | | |
| | Demolish | Ea. | | 83.50 | | 83.50 | Includes material and labor to install |
| | Install | Ea. | 213 | 228 | | 441 | an ultrasonic motion detector for a |
| | Demolish and Install | Ea. | 213 | 311.50 | | 524.50 | security system. |
| | Reinstall | Ea. | | 182.68 | | 182.68 | |
| | Clean | Ea. | .08 | 11.10 | | 11.18 | |
| | Minimum Charge | Job | | 263 | | 263 | |
| **Door Switch** | | | | | | | |
| | Demolish | Ea. | | 36.50 | | 36.50 | Includes material and labor to install a |
| | Install | Ea. | 67.50 | 99 | | 166.50 | door / window contact for a security |
| | Demolish and Install | Ea. | 67.50 | 135.50 | | 203 | system. |
| | Reinstall | Ea. | | 79.28 | | 79.28 | |
| | Clean | Ea. | .08 | 3.55 | | 3.63 | |
| | Minimum Charge | Job | | 263 | | 263 | |
| **Window Switch** | | | | | | | |
| | Demolish | Ea. | | 36.50 | | 36.50 | Includes material and labor to install a |
| | Install | Ea. | 67.50 | 99 | | 166.50 | door / window contact for a security |
| | Demolish and Install | Ea. | 67.50 | 135.50 | | 203 | system. |
| | Reinstall | Ea. | | 79.28 | | 79.28 | |
| | Clean | Ea. | .08 | 3.55 | | 3.63 | |
| | Minimum Charge | Job | | 263 | | 263 | |
| **Satellite System** | | | | | | | |
| **Dish** | | | | | | | |
| | Demolish | Ea. | | 162 | | 162 | Includes material, labor and |
| | Install | Ea. | 1850 | 440 | | 2290 | equipment to install 10' mesh dish. |
| | Demolish and Install | Ea. | 1850 | 602 | | 2452 | |
| | Reinstall | Ea. | | 350.13 | | 350.13 | |
| | Clean | Ea. | 1.47 | 44.50 | | 45.97 | |
| | Minimum Charge | Job | | 263 | | 263 | |
| **Motor** | | | | | | | |
| | Demolish | Ea. | | 81 | | 81 | Includes material and labor to install a |
| | Install | Ea. | 285 | 219 | | 504 | television satellite dish motor unit. |
| | Demolish and Install | Ea. | 285 | 300 | | 585 | |
| | Reinstall | Ea. | | 175.07 | | 175.07 | |
| | Clean | Ea. | .08 | 11.10 | | 11.18 | |
| | Minimum Charge | Job | | 263 | | 263 | |
| **Television Antenna** | | | | | | | |
| **40' Pole** | | | | | | | |
| | Demolish | Ea. | | 91 | | 91 | Includes material and labor to install |
| | Install | Ea. | 22 | 164 | | 186 | stand-alone 1"-2" metal pole, 10' long. |
| | Demolish and Install | Ea. | 22 | 255 | | 277 | Cost does not include antenna. |
| | Reinstall | Ea. | | 131.30 | | 131.30 | |
| | Clean | Ea. | .30 | 44.50 | | 44.80 | |
| | Paint | Ea. | 8.35 | 48.50 | | 56.85 | |
| | Minimum Charge | Job | | 263 | | 263 | |
| **Rotor Unit** | | | | | | | |
| | Demolish | Ea. | | 73 | | 73 | Includes material and labor to install a |
| | Install | Ea. | 69.50 | 65.50 | | 135 | television antenna rotor. |
| | Demolish and Install | Ea. | 69.50 | 138.50 | | 208 | |
| | Reinstall | Ea. | | 52.52 | | 52.52 | |
| | Clean | Ea. | .30 | 44.50 | | 44.80 | |
| | Minimum Charge | Job | | 263 | | 263 | |
| **Single Booster** | | | | | | | |
| | Demolish | Ea. | | 73 | | 73 | Includes material and labor to install |
| | Install | Ea. | 34.50 | 65.50 | | 100 | standard grade VHF television signal |
| | Demolish and Install | Ea. | 34.50 | 138.50 | | 173 | booster unit and all hardware required |
| | Reinstall | Ea. | | 52.52 | | 52.52 | for connection to TV cable. |
| | Clean | Ea. | .01 | 5.55 | | 5.56 | |
| | Minimum Charge | Job | | 263 | | 263 | |

271

| Gazebo | Gutter | Shutter |
|---|---|---|

| Window Treatment | Unit | Material | Labor | Equip. | Total | Specification |
|---|---|---|---|---|---|---|
| **Drapery** | | | | | | |
| **Good Grade** | | | | | | |
| Install | L.F. | 46 | 7.10 | | 53.10 | Includes material and labor to install |
| Reinstall | L.F. | | 5.68 | | 5.68 | good quality drapery. |
| Clean | L.F. | 1.18 | .18 | | 1.36 | |
| Minimum Charge | Job | | 178 | | 178 | |
| **Premium Grade** | | | | | | |
| Install | L.F. | 75.50 | 7.10 | | 82.60 | Includes material and labor to install |
| Reinstall | L.F. | | 5.68 | | 5.68 | premium quality drapery. |
| Clean | L.F. | 1.18 | .20 | | 1.38 | |
| Minimum Charge | Job | | 178 | | 178 | |
| **Lining** | | | | | | |
| Install | L.F. | 25.50 | 7.10 | | 32.60 | Includes material and labor to install |
| Reinstall | L.F. | | 5.68 | | 5.68 | drapery lining. |
| Clean | L.F. | 1.18 | .10 | | 1.28 | |
| Minimum Charge | Job | | 178 | | 178 | |
| **Valance Board** | | | | | | |
| Demolish | L.F. | | 1.97 | | 1.97 | Includes material and labor to install a |
| Install | L.F. | 7.45 | 14.20 | | 21.65 | drapery valance board. |
| Demolish and Install | L.F. | 7.45 | 16.17 | | 23.62 | |
| Reinstall | L.F. | | 11.37 | | 11.37 | |
| Clean | L.F. | 1.18 | .36 | | 1.54 | |
| Minimum Charge | Job | | 178 | | 178 | |
| **Rod** | | | | | | |
| Demolish | L.F. | | 4.93 | | 4.93 | Includes material and labor to install |
| Install | L.F. | 6.75 | 7.90 | | 14.65 | standard traverse drapery hardware. |
| Demolish and Install | L.F. | 6.75 | 12.83 | | 19.58 | |
| Reinstall | L.F. | | 6.30 | | 6.30 | |
| Clean | L.F. | .03 | .33 | | .36 | |
| Paint | L.F. | .07 | .74 | | .81 | |
| Minimum Charge | Job | | 178 | | 178 | |
| **Blinds** | | | | | | |
| **Mini-type** | | | | | | |
| Demolish | S.F. | | 1.01 | | 1.01 | Includes material and labor to install |
| Install | S.F. | 6.50 | .79 | | 7.29 | horizontal 1" slat blinds. |
| Demolish and Install | S.F. | 6.50 | 1.80 | | 8.30 | |
| Reinstall | S.F. | | .63 | | .63 | |
| Clean | S.F. | 1.33 | .04 | | 1.37 | |
| Minimum Charge | Job | | 178 | | 178 | |

## Window Treatment

| Window Treatment | Unit | Material | Labor | Equip. | Total | Specification |
|---|---|---|---|---|---|---|
| **Vertical Type** | | | | | | |
| Demolish | S.F. | | 1.01 | | 1.01 | Includes material and labor to install |
| Install | S.F. | 9.50 | 1.01 | | 10.51 | vertical 3" - 5" cloth or PVC blinds. |
| Demolish and Install | S.F. | 9.50 | 2.02 | | 11.52 | |
| Reinstall | S.F. | | .81 | | .81 | |
| Clean | S.F. | 1.33 | .04 | | 1.37 | |
| Minimum Charge | Job | | 178 | | 178 | |
| **Vertical Type per L.F.** | | | | | | |
| Demolish | L.F. | | 1.64 | | 1.64 | Includes material and labor to install |
| Install | L.F. | 57 | 4.74 | | 61.74 | vertical 3" - 5" cloth or PVC blinds x |
| Demolish and Install | L.F. | 57 | 6.38 | | 63.38 | 72" long. |
| Reinstall | L.F. | | 3.79 | | 3.79 | |
| Clean | L.F. | 8.35 | .25 | | 8.60 | |
| Minimum Charge | Job | | 178 | | 178 | |
| **Venetian Type** | | | | | | |
| Demolish | S.F. | | 1.01 | | 1.01 | Includes material and labor to install |
| Install | S.F. | 5.35 | .79 | | 6.14 | horizontal 2" slat blinds. |
| Demolish and Install | S.F. | 5.35 | 1.80 | | 7.15 | |
| Reinstall | S.F. | | .63 | | .63 | |
| Clean | S.F. | 1.33 | .04 | | 1.37 | |
| Minimum Charge | Job | | 178 | | 178 | |

## Wood Shutter

### 27" x 36", 4 Panel

| | Unit | Material | Labor | Equip. | Total | Specification |
|---|---|---|---|---|---|---|
| Demolish | Ea. | | 32.50 | | 32.50 | Includes material and labor to install |
| Install | Pr. | 117 | 27.50 | | 144.50 | movable louver, up to 33" x 36", 4 |
| Demolish and Install | Pr. | 117 | 60 | | 177 | panel shutter. |
| Reinstall | Pr. | | 21.87 | | 21.87 | |
| Clean | Ea. | .03 | 11.10 | | 11.13 | |
| Paint | Ea. | 1.33 | 27.50 | | 28.83 | |
| Minimum Charge | Job | | 232 | | 232 | |

### 33" x 36", 4 Panel

| | Unit | Material | Labor | Equip. | Total | Specification |
|---|---|---|---|---|---|---|
| Demolish | Ea. | | 32.50 | | 32.50 | Includes material and labor to install |
| Install | Pr. | 117 | 27.50 | | 144.50 | movable louver, up to 33" x 36", 4 |
| Demolish and Install | Pr. | 117 | 60 | | 177 | panel shutter. |
| Reinstall | Pr. | | 21.87 | | 21.87 | |
| Clean | Ea. | .03 | 11.10 | | 11.13 | |
| Paint | Ea. | 1.39 | 29.50 | | 30.89 | |
| Minimum Charge | Job | | 232 | | 232 | |

### 39" x 36", 4 Panel

| | Unit | Material | Labor | Equip. | Total | Specification |
|---|---|---|---|---|---|---|
| Demolish | Ea. | | 32.50 | | 32.50 | Includes material and labor to install |
| Install | Pr. | 127 | 27.50 | | 154.50 | movable louver, up to 47" x 36", 4 |
| Demolish and Install | Pr. | 127 | 60 | | 187 | panel shutter. |
| Reinstall | Pr. | | 21.87 | | 21.87 | |
| Clean | Ea. | .03 | 11.10 | | 11.13 | |
| Paint | Ea. | 1.39 | 29.50 | | 30.89 | |
| Minimum Charge | Job | | 232 | | 232 | |

### 47" x 36", 4 Panel

| | Unit | Material | Labor | Equip. | Total | Specification |
|---|---|---|---|---|---|---|
| Demolish | Ea. | | 32.50 | | 32.50 | Includes material and labor to install |
| Install | Pr. | 127 | 27.50 | | 154.50 | movable louver, up to 47" x 36", 4 |
| Demolish and Install | Pr. | 127 | 60 | | 187 | panel shutter. |
| Reinstall | Pr. | | 21.87 | | 21.87 | |
| Clean | Ea. | .03 | 11.10 | | 11.13 | |
| Paint | Ea. | 1.45 | 32 | | 33.45 | |
| Minimum Charge | Job | | 232 | | 232 | |

## Stairs

| | Unit | Material | Labor | Equip. | Total | Specification |
|---|---|---|---|---|---|---|
| **Concrete Steps** | | | | | | |
| **Precast** | | | | | | |
| Demolish | Ea. | | 67 | 44.50 | 111.50 | Includes material and labor to install |
| Install | Flight | 605 | 161 | 32 | 798 | precast concrete front entrance stairs, |
| Demolish and Install | Flight | 605 | 228 | 76.50 | 909.50 | 4' wide, with 48" platform and 2 |
| Clean | Ea. | | 4.93 | | 4.93 | risers. |
| Paint | Ea. | 4.02 | 48.50 | | 52.52 | |
| Minimum Charge | Job | | 425 | | 425 | |
| **Metal** | | | | | | |
| **Concrete Filled** | | | | | | |
| Demolish | Ea. | | 19.75 | | 19.75 | Includes material and labor to install |
| Install | Riser | 690 | 74.50 | 3.61 | 768.11 | steel, cement filled metal pan stairs |
| Demolish and Install | Riser | 690 | 94.25 | 3.61 | 787.86 | with picket rail. |
| Clean | Ea. | .04 | 2.99 | | 3.03 | |
| Paint | Ea. | .70 | 2.86 | | 3.56 | |
| Minimum Charge | Job | | 275 | | 275 | |
| **Landing** | | | | | | |
| Demolish | S.F. | | 4.93 | | 4.93 | Includes material and labor to install |
| Install | S.F. | 145 | 8.75 | .42 | 154.17 | pre-erected conventional steel pan |
| Demolish and Install | S.F. | 145 | 13.68 | .42 | 159.10 | landing. |
| Clean | S.F. | .04 | .63 | | .67 | |
| Paint | S.F. | .24 | .82 | | 1.06 | |
| Minimum Charge | Job | | 275 | | 275 | |
| **Railing** | | | | | | |
| Demolish | L.F. | | 2.72 | | 2.72 | Includes material and labor to install |
| Install | L.F. | 19.70 | 12.70 | .61 | 33.01 | primed steel pipe. |
| Demolish and Install | L.F. | 19.70 | 15.42 | .61 | 35.73 | |
| Clean | L.F. | .30 | 2.22 | | 2.52 | |
| Paint | L.F. | .80 | 6.45 | | 7.25 | |
| Minimum Charge | Job | | 275 | | 275 | |

## Elevator

| | Unit | Material | Labor | Equip. | Total | Specification |
|---|---|---|---|---|---|---|
| **2 Stop Residential** | | | | | | |
| Install | Ea. | 11900 | 7075 | | 18975 | Includes labor, material and |
| Clean | Ea. | .47 | 22 | | 22.47 | equipment to install 2 stop elevator for |
| Minimum Charge | Job | | 710 | | 710 | residential use. |
| **2 Stop Commercial** | | | | | | |
| Install | Ea. | 48200 | 14200 | | 62400 | Includes labor, material and |
| Clean | Ea. | .47 | 22 | | 22.47 | equipment to install 2 stop elevator for |
| Minimum Charge | Job | | 710 | | 710 | apartment or commercial building with |
| | | | | | | 2,000 lb/13 passenger capacity. |
| **3 Stop Commercial** | | | | | | |
| Install | Ea. | 59000 | 21800 | | 80800 | Includes labor, material and |
| Clean | Ea. | .47 | 22 | | 22.47 | equipment to install 3 stop elevator for |
| Minimum Charge | Job | | 710 | | 710 | apartment or commercial building with |
| | | | | | | 2,000 lb/13 passenger capacity. |
| **4 Stop Commercial** | | | | | | |
| Install | Ea. | 70000 | 29500 | | 99500 | Includes labor, material and |
| Clean | Ea. | .47 | 22 | | 22.47 | equipment to install 4 stop elevator for |
| Minimum Charge | Job | | 710 | | 710 | apartment or commercial building with |
| | | | | | | 2,000 lb/13 passenger capacity. |

# Improvements / Appliances / Treatments

| Elevator | | Unit | Material | Labor | Equip. | Total | Specification |
|---|---|---|---|---|---|---|---|
| **5 Stop Commercial** | | | | | | | Includes labor, material and |
| | Install | Ea. | 81000 | 37200 | | 118200 | equipment to install 5 stop elevator for |
| | Clean | Ea. | .47 | 22 | | 22.47 | apartment or commercial building with |
| | Minimum Charge | Job | | 710 | | 710 | 2,000 lb/13 passenger capacity. |

| Rangetop | | Unit | Material | Labor | Equip. | Total | Specification |
|---|---|---|---|---|---|---|---|
| **4-burner** | | | | | | | Cost includes material and labor to |
| | Demolish | Ea. | | 36 | | 36 | install 4-burner electric surface unit. |
| | Install | Ea. | 360 | 87.50 | | 447.50 | |
| | Demolish and Install | Ea. | 360 | 123.50 | | 483.50 | |
| | Reinstall | Ea. | | 70.03 | | 70.03 | |
| | Clean | Ea. | .02 | 33.50 | | 33.52 | |
| | Minimum Charge | Job | | 89 | | 89 | |
| **6-burner** | | | | | | | Includes material and labor to install a |
| | Demolish | Ea. | | 36 | | 36 | 6-burner cooktop unit. |
| | Install | Ea. | 1900 | 105 | | 2005 | |
| | Demolish and Install | Ea. | 1900 | 141 | | 2041 | |
| | Reinstall | Ea. | | 84.03 | | 84.03 | |
| | Clean | Ea. | .02 | 33.50 | | 33.52 | |
| | Minimum Charge | Job | | 89 | | 89 | |
| **Premium Brand w / Grill** | | | | | | | Includes material and labor to install |
| | Demolish | Ea. | | 36 | | 36 | electric downdraft cooktop with grill. |
| | Install | Ea. | 1575 | 105 | | 1680 | |
| | Demolish and Install | Ea. | 1575 | 141 | | 1716 | |
| | Reinstall | Ea. | | 84.03 | | 84.03 | |
| | Clean | Ea. | .02 | 33.50 | | 33.52 | |
| | Minimum Charge | Job | | 89 | | 89 | |

| Rangehood | | Unit | Material | Labor | Equip. | Total | Specification |
|---|---|---|---|---|---|---|---|
| **Vented** | | | | | | | |
| **30"** | | | | | | | |
| | Demolish | Ea. | | 33 | | 33 | Includes material and labor to install a |
| | Install | Ea. | 102 | 105 | | 207 | vented range hood. |
| | Demolish and Install | Ea. | 102 | 138 | | 240 | |
| | Reinstall | Ea. | | 84.03 | | 84.03 | |
| | Clean | Ea. | .02 | 9.55 | | 9.57 | |
| | Minimum Charge | Job | | 89 | | 89 | |
| **36"** | | | | | | | |
| | Demolish | Ea. | | 33 | | 33 | Includes material and labor to install a |
| | Install | Ea. | 131 | 105 | | 236 | vented range hood. |
| | Demolish and Install | Ea. | 131 | 138 | | 269 | |
| | Reinstall | Ea. | | 84.03 | | 84.03 | |
| | Clean | Ea. | .02 | 9.55 | | 9.57 | |
| | Minimum Charge | Job | | 89 | | 89 | |
| **42"** | | | | | | | |
| | Demolish | Ea. | | 33 | | 33 | Includes material and labor to install a |
| | Install | Ea. | 187 | 119 | | 306 | vented range hood. |
| | Demolish and Install | Ea. | 187 | 152 | | 339 | |
| | Reinstall | Ea. | | 95.38 | | 95.38 | |
| | Clean | Ea. | .02 | 9.55 | | 9.57 | |
| | Minimum Charge | Job | | 89 | | 89 | |

| Rangehood | Unit | Material | Labor | Equip. | Total | Specification |
|---|---|---|---|---|---|---|
| **Ventless** | | | | | | |
| **30"** | | | | | | |
| Demolish | Ea. | | 33 | | 33 | Includes material and labor to install a |
| Install | Ea. | 119 | 65.50 | | 184.50 | ventless range hood. |
| Demolish and Install | Ea. | 119 | 98.50 | | 217.50 | |
| Reinstall | Ea. | | 52.52 | | 52.52 | |
| Clean | Ea. | .02 | 9.55 | | 9.57 | |
| Minimum Charge | Job | | 89 | | 89 | |
| **36"** | | | | | | |
| Demolish | Ea. | | 33 | | 33 | Includes material and labor to install a |
| Install | Ea. | 193 | 105 | | 298 | ventless range hood. |
| Demolish and Install | Ea. | 193 | 138 | | 331 | |
| Reinstall | Ea. | | 84.03 | | 84.03 | |
| Clean | Ea. | .02 | 9.55 | | 9.57 | |
| Minimum Charge | Job | | 89 | | 89 | |
| **42"** | | | | | | |
| Demolish | Ea. | | 33 | | 33 | Includes material and labor to install a |
| Install | Ea. | 185 | 87.50 | | 272.50 | ventless range hood. |
| Demolish and Install | Ea. | 185 | 120.50 | | 305.50 | |
| Reinstall | Ea. | | 70.03 | | 70.03 | |
| Clean | Ea. | .02 | 9.55 | | 9.57 | |
| Minimum Charge | Job | | 89 | | 89 | |

| Oven | Unit | Material | Labor | Equip. | Total | Specification |
|---|---|---|---|---|---|---|
| **Free Standing** | | | | | | |
| Demolish | Ea. | | 36 | | 36 | Cost includes material and labor to |
| Install | Ea. | 625 | 71 | | 696 | install 30" free standing electric range |
| Demolish and Install | Ea. | 625 | 107 | | 732 | with self-cleaning oven. |
| Reinstall | Ea. | | 56.83 | | 56.83 | |
| Clean | Ea. | .02 | 33.50 | | 33.52 | |
| Minimum Charge | Job | | 89 | | 89 | |
| **Single Wall Oven** | | | | | | |
| Demolish | Ea. | | 36 | | 36 | Includes material and labor to install a |
| Install | Ea. | 990 | 131 | | 1121 | built-in oven. |
| Demolish and Install | Ea. | 990 | 167 | | 1157 | |
| Reinstall | Ea. | | 105.04 | | 105.04 | |
| Clean | Ea. | .02 | 33.50 | | 33.52 | |
| Minimum Charge | Job | | 89 | | 89 | |
| **Double Wall Oven** | | | | | | |
| Demolish | Ea. | | 36 | | 36 | Includes material and labor to install a |
| Install | Ea. | 1275 | 232 | | 1507 | conventional double oven. |
| Demolish and Install | Ea. | 1275 | 268 | | 1543 | |
| Reinstall | Ea. | | 185.92 | | 185.92 | |
| Clean | Ea. | .02 | 33.50 | | 33.52 | |
| Minimum Charge | Job | | 89 | | 89 | |
| **Drop-in Type w / Range** | | | | | | |
| Demolish | Ea. | | 36 | | 36 | Includes material and labor to install a |
| Install | Ea. | 880 | 87.50 | | 967.50 | built-in cooking range with oven. |
| Demolish and Install | Ea. | 880 | 123.50 | | 1003.50 | |
| Reinstall | Ea. | | 70.03 | | 70.03 | |
| Clean | Ea. | .02 | 33.50 | | 33.52 | |
| Minimum Charge | Job | | 89 | | 89 | |

For customer support on your Contractor's Pricing Guide: Residential Repair & Remodeling, call 800.448.8182.

277

# Improvements / Appliances / Treatments

## Oven

| Oven | | Unit | Material | Labor | Equip. | Total | Specification |
|---|---|---|---|---|---|---|---|
| **Hi/Lo w / Range, Microwave** | | | | | | | Includes material and labor to install a |
| | Demolish | Ea. | | 36 | | 36 | double oven, one conventional, one |
| | Install | Ea. | 2500 | 465 | | 2965 | microwave. |
| | Demolish and Install | Ea. | 2500 | 501 | | 3001 | |
| | Reinstall | Ea. | | 371.84 | | 371.84 | |
| | Clean | Ea. | .02 | 33.50 | | 33.52 | |
| | Minimum Charge | Job | | 89 | | 89 | |
| **Hi/Lo w/ Range / Double Oven** | | | | | | | Includes material and labor to install a |
| | Demolish | Ea. | | 36 | | 36 | conventional double oven. |
| | Install | Ea. | 1275 | 232 | | 1507 | |
| | Demolish and Install | Ea. | 1275 | 268 | | 1543 | |
| | Reinstall | Ea. | | 185.92 | | 185.92 | |
| | Clean | Ea. | .02 | 33.50 | | 33.52 | |
| | Minimum Charge | Job | | 89 | | 89 | |
| **Countertop Microwave** | | | | | | | Includes material and labor to install a |
| | Demolish | Ea. | | 12.35 | | 12.35 | 1.0 cubic foot microwave oven. |
| | Install | Ea. | 174 | 44.50 | | 218.50 | |
| | Demolish and Install | Ea. | 174 | 56.85 | | 230.85 | |
| | Reinstall | Ea. | | 35.52 | | 35.52 | |
| | Clean | Ea. | .02 | 16.70 | | 16.72 | |
| | Minimum Charge | Job | | 89 | | 89 | |
| **Cabinet / Wall Mounted Microwave** | | | | | | | Includes material and labor to install a |
| | Demolish | Ea. | | 24.50 | | 24.50 | space saver microwave oven. |
| | Install | Ea. | 840 | 131 | | 971 | |
| | Demolish and Install | Ea. | 840 | 155.50 | | 995.50 | |
| | Reinstall | Ea. | | 105.04 | | 105.04 | |
| | Clean | Ea. | .02 | 16.70 | | 16.72 | |
| | Minimum Charge | Job | | 89 | | 89 | |

## Dryer

| Dryer | | Unit | Material | Labor | Equip. | Total | Specification |
|---|---|---|---|---|---|---|---|
| **Basic Grade** | | | | | | | Includes material and labor to install |
| | Demolish | Ea. | | 16.45 | | 16.45 | electric or gas dryer. Vent kit not |
| | Install | Ea. | 485 | 274 | | 759 | included. |
| | Demolish and Install | Ea. | 485 | 290.45 | | 775.45 | |
| | Reinstall | Ea. | | 219.31 | | 219.31 | |
| | Clean | Ea. | .02 | 33.50 | | 33.52 | |
| | Minimum Charge | Job | | 89 | | 89 | |
| **Good Grade** | | | | | | | Includes material and labor to install |
| | Demolish | Ea. | | 16.45 | | 16.45 | electric or gas dryer. Vent kit not |
| | Install | Ea. | 605 | 274 | | 879 | included. |
| | Demolish and Install | Ea. | 605 | 290.45 | | 895.45 | |
| | Reinstall | Ea. | | 219.31 | | 219.31 | |
| | Clean | Ea. | .02 | 33.50 | | 33.52 | |
| | Minimum Charge | Job | | 89 | | 89 | |
| **Premium Grade** | | | | | | | Includes material and labor to install |
| | Demolish | Ea. | | 16.45 | | 16.45 | electric or gas dryer. Vent kit not |
| | Install | Ea. | 1300 | 410 | | 1710 | included. |
| | Demolish and Install | Ea. | 1300 | 426.45 | | 1726.45 | |
| | Reinstall | Ea. | | 328.96 | | 328.96 | |
| | Clean | Ea. | .02 | 33.50 | | 33.52 | |
| | Minimum Charge | Job | | 89 | | 89 | |

# Improvements / Appliances / Treatments

## Dryer

| | Unit | Material | Labor | Equip. | Total | Specification |
|---|---|---|---|---|---|---|
| **Washer / Dryer Combination Unit** | | | | | | |
| Demolish | Ea. | | 16.45 | | 16.45 | Includes material and labor to install |
| Install | Ea. | 1675 | 71 | | 1746 | electric washer/dryer combination |
| Demolish and Install | Ea. | 1675 | 87.45 | | 1762.45 | unit, 27" wide, stackable or unitized. |
| Reinstall | Ea. | | 56.83 | | 56.83 | |
| Clean | Ea. | .02 | 33.50 | | 33.52 | |
| Minimum Charge | Job | | 89 | | 89 | |

## Washer

| | Unit | Material | Labor | Equip. | Total | Specification |
|---|---|---|---|---|---|---|
| **Basic Grade** | | | | | | |
| Demolish | Ea. | | 16.45 | | 16.45 | Includes material and labor to install |
| Install | Ea. | 710 | 175 | | 885 | washing machine. New hoses not |
| Demolish and Install | Ea. | 710 | 191.45 | | 901.45 | included. |
| Reinstall | Ea. | | 139.73 | | 139.73 | |
| Clean | Ea. | .02 | 33.50 | | 33.52 | |
| Minimum Charge | Job | | 89 | | 89 | |
| **Good Grade** | | | | | | |
| Demolish | Ea. | | 16.45 | | 16.45 | Includes material and labor to install |
| Install | Ea. | 1150 | 262 | | 1412 | washing machine. New hoses not |
| Demolish and Install | Ea. | 1150 | 278.45 | | 1428.45 | included. |
| Reinstall | Ea. | | 209.60 | | 209.60 | |
| Clean | Ea. | .02 | 33.50 | | 33.52 | |
| Minimum Charge | Job | | 89 | | 89 | |
| **Premium Grade** | | | | | | |
| Demolish | Ea. | | 16.45 | | 16.45 | Includes material and labor to install |
| Install | Ea. | 1475 | 525 | | 2000 | washing machine with digital |
| Demolish and Install | Ea. | 1475 | 541.45 | | 2016.45 | computer readouts and large capacity. |
| Reinstall | Ea. | | 419.20 | | 419.20 | New hoses not included. |
| Clean | Ea. | .02 | 33.50 | | 33.52 | |
| Minimum Charge | Job | | 89 | | 89 | |
| **Washer / Dryer Combination Unit** | | | | | | |
| Demolish | Ea. | | 16.45 | | 16.45 | Includes material and labor to install |
| Install | Ea. | 1675 | 71 | | 1746 | electric washer/dryer combination |
| Demolish and Install | Ea. | 1675 | 87.45 | | 1762.45 | unit, 27" wide, stackable or unitized. |
| Reinstall | Ea. | | 56.83 | | 56.83 | |
| Clean | Ea. | .02 | 33.50 | | 33.52 | |
| Minimum Charge | Job | | 89 | | 89 | |

## Refrigerator

| | Unit | Material | Labor | Equip. | Total | Specification |
|---|---|---|---|---|---|---|
| **12 Cubic Foot** | | | | | | |
| Demolish | Ea. | | 49.50 | | 49.50 | Includes material and labor to install a |
| Install | Ea. | 485 | 71 | | 556 | refrigerator. |
| Demolish and Install | Ea. | 485 | 120.50 | | 605.50 | |
| Reinstall | Ea. | | 56.83 | | 56.83 | |
| Clean | Ea. | .02 | 33.50 | | 33.52 | |
| Minimum Charge | Job | | 89 | | 89 | |
| **16 Cubic Foot** | | | | | | |
| Demolish | Ea. | | 49.50 | | 49.50 | Includes material and labor to install a |
| Install | Ea. | 610 | 79 | | 689 | refrigerator. |
| Demolish and Install | Ea. | 610 | 128.50 | | 738.50 | |
| Reinstall | Ea. | | 63.15 | | 63.15 | |
| Clean | Ea. | .02 | 33.50 | | 33.52 | |
| Minimum Charge | Job | | 89 | | 89 | |

**For customer support on your Contractor's Pricing Guide: Residential Repair & Remodeling, call 800.448.8182.**

# Improvements / Appliances / Treatments

| Refrigerator | Unit | Material | Labor | Equip. | Total | Specification |
|---|---|---|---|---|---|---|
| **18 Cubic Foot** | | | | | | |
| Demolish | Ea. | | 49.50 | | 49.50 | Includes material and labor to install a |
| Install | Ea. | 885 | 89 | | 974 | refrigerator. |
| Demolish and Install | Ea. | 885 | 138.50 | | 1023.50 | |
| Reinstall | Ea. | | 71.04 | | 71.04 | |
| Clean | Ea. | .02 | 33.50 | | 33.52 | |
| Minimum Charge | Job | | 89 | | 89 | |
| **21 Cubic Foot** | | | | | | |
| Demolish | Ea. | | 49.50 | | 49.50 | Includes material and labor to install a |
| Install | Ea. | 1175 | 101 | | 1276 | refrigerator. |
| Demolish and Install | Ea. | 1175 | 150.50 | | 1325.50 | |
| Reinstall | Ea. | | 81.19 | | 81.19 | |
| Clean | Ea. | .02 | 33.50 | | 33.52 | |
| Minimum Charge | Job | | 89 | | 89 | |
| **24 Cubic Foot** | | | | | | |
| Demolish | Ea. | | 49.50 | | 49.50 | Includes material and labor to install a |
| Install | Ea. | 1575 | 101 | | 1676 | refrigerator. |
| Demolish and Install | Ea. | 1575 | 150.50 | | 1725.50 | |
| Reinstall | Ea. | | 81.19 | | 81.19 | |
| Clean | Ea. | .02 | 33.50 | | 33.52 | |
| Minimum Charge | Job | | 89 | | 89 | |
| **27 Cubic Foot** | | | | | | |
| Demolish | Ea. | | 49.50 | | 49.50 | Includes material and labor to install a |
| Install | Ea. | 2850 | 71 | | 2921 | refrigerator. |
| Demolish and Install | Ea. | 2850 | 120.50 | | 2970.50 | |
| Reinstall | Ea. | | 56.83 | | 56.83 | |
| Clean | Ea. | .02 | 33.50 | | 33.52 | |
| Minimum Charge | Job | | 89 | | 89 | |
| **Ice Maker** | | | | | | |
| Install | Ea. | 88 | 65.50 | | 153.50 | Includes labor and material to install |
| Minimum Charge | Job | | 89 | | 89 | an automatic ice maker in a |
| | | | | | | refrigerator. |

| Freezer | Unit | Material | Labor | Equip. | Total | Specification |
|---|---|---|---|---|---|---|
| **15 Cubic Foot** | | | | | | |
| Demolish | Ea. | | 49.50 | | 49.50 | Cost includes material and labor to |
| Install | Ea. | 655 | 71 | | 726 | install 15 cubic foot freezer. |
| Demolish and Install | Ea. | 655 | 120.50 | | 775.50 | |
| Reinstall | Ea. | | 56.83 | | 56.83 | |
| Clean | Ea. | .02 | 33.50 | | 33.52 | |
| Minimum Charge | Job | | 89 | | 89 | |
| **18 Cubic Foot** | | | | | | |
| Demolish | Ea. | | 49.50 | | 49.50 | Cost includes material and labor to |
| Install | Ea. | 875 | 71 | | 946 | install 18 cubic foot freezer. |
| Demolish and Install | Ea. | 875 | 120.50 | | 995.50 | |
| Reinstall | Ea. | | 56.83 | | 56.83 | |
| Clean | Ea. | .02 | 33.50 | | 33.52 | |
| Minimum Charge | Job | | 89 | | 89 | |
| **21 Cubic Foot** | | | | | | |
| Demolish | Ea. | | 49.50 | | 49.50 | Includes material and labor to install a |
| Install | Ea. | 800 | 71 | | 871 | 15 to 23 cubic foot deep freezer. |
| Demolish and Install | Ea. | 800 | 120.50 | | 920.50 | |
| Reinstall | Ea. | | 56.83 | | 56.83 | |
| Clean | Ea. | .02 | 33.50 | | 33.52 | |
| Minimum Charge | Job | | 89 | | 89 | |

# Improvements / Appliances / Treatments

## Freezer

| | Unit | Material | Labor | Equip. | Total | Specification |
|---|---|---|---|---|---|---|
| **24 Cubic Foot** | | | | | | |
| Demolish | Ea. | | 49.50 | | 49.50 | Includes material and labor to install a |
| Install | Ea. | 885 | 142 | | 1027 | 24 cubic foot deep freezer. |
| Demolish and Install | Ea. | 885 | 191.50 | | 1076.50 | |
| Reinstall | Ea. | | 113.66 | | 113.66 | |
| Clean | Ea. | .02 | 33.50 | | 33.52 | |
| Minimum Charge | Job | | 89 | | 89 | |
| **Chest-type** | | | | | | |
| Demolish | Ea. | | 49.50 | | 49.50 | Cost includes material and labor to |
| Install | Ea. | 655 | 71 | | 726 | install 15 cubic foot freezer. |
| Demolish and Install | Ea. | 655 | 120.50 | | 775.50 | |
| Clean | Ea. | .02 | 33.50 | | 33.52 | |
| Minimum Charge | Job | | 89 | | 89 | |

## Dishwasher

| | Unit | Material | Labor | Equip. | Total | Specification |
|---|---|---|---|---|---|---|
| **4 Cycle** | | | | | | |
| Demolish | Ea. | | 39.50 | | 39.50 | Includes material and labor to install |
| Install | Ea. | 440 | 164 | | 604 | good grade automatic dishwasher. |
| Demolish and Install | Ea. | 440 | 203.50 | | 643.50 | |
| Reinstall | Ea. | | 131.06 | | 131.06 | |
| Clean | Ea. | .02 | 33.50 | | 33.52 | |
| Minimum Charge | Job | | 89 | | 89 | |

## Garbage Disposal

| | Unit | Material | Labor | Equip. | Total | Specification |
|---|---|---|---|---|---|---|
| **In Sink Unit** | | | | | | |
| Demolish | Ea. | | 36.50 | | 36.50 | Cost includes material and labor to |
| Install | Ea. | 216 | 65.50 | | 281.50 | install 1/2 HP custom disposal unit. |
| Demolish and Install | Ea. | 216 | 102 | | 318 | |
| Reinstall | Ea. | | 52.42 | | 52.42 | |
| Clean | Ea. | .01 | 9.55 | | 9.56 | |
| Minimum Charge | Job | | 89 | | 89 | |

## Trash Compactor

| | Unit | Material | Labor | Equip. | Total | Specification |
|---|---|---|---|---|---|---|
| **4 to 1 Compaction** | | | | | | |
| Demolish | Ea. | | 36.50 | | 36.50 | Includes material and labor to install a |
| Install | Ea. | 1000 | 93 | | 1093 | trash compactor. |
| Demolish and Install | Ea. | 1000 | 129.50 | | 1129.50 | |
| Reinstall | Ea. | | 74.37 | | 74.37 | |
| Clean | Ea. | .02 | 22 | | 22.02 | |
| Minimum Charge | Job | | 89 | | 89 | |

## Kitchenette Unit

| | Unit | Material | Labor | Equip. | Total | Specification |
|---|---|---|---|---|---|---|
| **Range, Refrigerator and Sink** | | | | | | |
| Demolish | Ea. | | 146 | | 146 | Includes material and labor to install |
| Install | Ea. | 1875 | 545 | | 2420 | range/refrigerator/sink unit, 72" |
| Demolish and Install | Ea. | 1875 | 691 | | 2566 | wide, base unit only. |
| Reinstall | Ea. | | 436.87 | | 436.87 | |
| Clean | Ea. | .36 | 16.70 | | 17.06 | |
| Minimum Charge | Job | | 89 | | 89 | |

For customer support on your Contractor's Pricing Guide: Residential Repair & Remodeling, call 800.448.8182.

# Improvements / Appliances / Treatments

| Bath Accessories | | Unit | Material | Labor | Equip. | Total | Specification |
|---|---|---|---|---|---|---|---|
| **Towel Bar** | | | | | | | |
| | Demolish | Ea. | | 8.20 | | 8.20 | Includes material and labor to install |
| | Install | Ea. | 51.50 | 19.35 | | 70.85 | towel bar up to 24" long. |
| | Demolish and Install | Ea. | 51.50 | 27.55 | | 79.05 | |
| | Reinstall | Ea. | | 15.49 | | 15.49 | |
| | Clean | Ea. | | 4.44 | | 4.44 | |
| | Minimum Charge | Job | | 232 | | 232 | |
| **Toothbrush Holder** | | | | | | | |
| | Demolish | Ea. | | 8.20 | | 8.20 | Includes material and labor to install a |
| | Install | Ea. | 14.80 | 23 | | 37.80 | surface mounted tumbler and |
| | Demolish and Install | Ea. | 14.80 | 31.20 | | 46 | toothbrush holder. |
| | Reinstall | Ea. | | 18.59 | | 18.59 | |
| | Clean | Ea. | | 4.44 | | 4.44 | |
| | Minimum Charge | Job | | 232 | | 232 | |
| **Grab Rail** | | | | | | | |
| | Demolish | Ea. | | 12.35 | | 12.35 | Includes material and labor to install a |
| | Install | Ea. | 34.50 | 20 | | 54.50 | 24" long grab bar. Blocking is not |
| | Demolish and Install | Ea. | 34.50 | 32.35 | | 66.85 | included. |
| | Reinstall | Ea. | | 16.17 | | 16.17 | |
| | Clean | Ea. | .14 | 7.40 | | 7.54 | |
| | Minimum Charge | Job | | 232 | | 232 | |
| **Soap Dish** | | | | | | | |
| | Demolish | Ea. | | 8.20 | | 8.20 | Includes material and labor to install a |
| | Install | Ea. | 5.95 | 23 | | 28.95 | surface mounted soap dish. |
| | Demolish and Install | Ea. | 5.95 | 31.20 | | 37.15 | |
| | Reinstall | Ea. | | 18.59 | | 18.59 | |
| | Clean | Ea. | | 4.44 | | 4.44 | |
| | Minimum Charge | Job | | 232 | | 232 | |
| **Toilet Paper Roller** | | | | | | | |
| | Demolish | Ea. | | 8.20 | | 8.20 | Includes material and labor to install a |
| | Install | Ea. | 19.85 | 15.50 | | 35.35 | surface mounted toilet tissue dispenser. |
| | Demolish and Install | Ea. | 19.85 | 23.70 | | 43.55 | |
| | Reinstall | Ea. | | 12.39 | | 12.39 | |
| | Clean | Ea. | .14 | 11.10 | | 11.24 | |
| | Minimum Charge | Job | | 232 | | 232 | |
| **Dispenser** | | | | | | | |
| **Soap** | | | | | | | |
| | Demolish | Ea. | | 10.95 | | 10.95 | Includes material and labor to install a |
| | Install | Ea. | 56 | 23 | | 79 | surface mounted soap dispenser. |
| | Demolish and Install | Ea. | 56 | 33.95 | | 89.95 | |
| | Reinstall | Ea. | | 18.59 | | 18.59 | |
| | Clean | Ea. | .14 | 11.10 | | 11.24 | |
| | Minimum Charge | Job | | 232 | | 232 | |
| **Towel** | | | | | | | |
| | Demolish | Ea. | | 12.35 | | 12.35 | Includes material and labor to install a |
| | Install | Ea. | 47.50 | 29 | | 76.50 | surface mounted towel dispenser. |
| | Demolish and Install | Ea. | 47.50 | 41.35 | | 88.85 | |
| | Reinstall | Ea. | | 23.24 | | 23.24 | |
| | Clean | Ea. | .14 | 14.80 | | 14.94 | |
| | Minimum Charge | Job | | 232 | | 232 | |
| **Seat Cover** | | | | | | | |
| | Demolish | Ea. | | 8.20 | | 8.20 | Includes material and labor to install a |
| | Install | Ea. | 33 | 31 | | 64 | surface mounted toilet seat cover |
| | Demolish and Install | Ea. | 33 | 39.20 | | 72.20 | dispenser. |
| | Reinstall | Ea. | | 24.79 | | 24.79 | |
| | Clean | Ea. | .14 | 11.10 | | 11.24 | |
| | Minimum Charge | Job | | 232 | | 232 | |

| Bath Accessories | Unit | Material | Labor | Equip. | Total | Specification |
|---|---|---|---|---|---|---|
| **Sanitary Napkin** | | | | | | |
| Demolish | Ea. | | 21 | | 21 | Includes material and labor to install a |
| Install | Ea. | 37 | 49 | | 86 | surface mounted sanitary napkin |
| Demolish and Install | Ea. | 37 | 70 | | 107 | dispenser. |
| Reinstall | Ea. | | 39.14 | | 39.14 | |
| Clean | Ea. | .08 | 14.80 | | 14.88 | |
| Minimum Charge | Job | | 232 | | 232 | |
| **Partition** | | | | | | |
| **Urinal** | | | | | | |
| Demolish | Ea. | | 24.50 | | 24.50 | Includes material and labor to install a |
| Install | Ea. | 253 | 116 | | 369 | urinal screen. |
| Demolish and Install | Ea. | 253 | 140.50 | | 393.50 | |
| Reinstall | Ea. | | 92.96 | | 92.96 | |
| Clean | Ea. | .22 | 9.85 | | 10.07 | |
| Minimum Charge | Job | | 232 | | 232 | |
| **Toilet** | | | | | | |
| Demolish | Ea. | | 73 | | 73 | Includes material and labor to install |
| Install | Ea. | 660 | 186 | | 846 | baked enamel standard size partition, |
| Demolish and Install | Ea. | 660 | 259 | | 919 | floor or ceiling mounted with one wall |
| Reinstall | Ea. | | 148.74 | | 148.74 | and one door. |
| Clean | Ea. | 1.47 | 22 | | 23.47 | |
| Minimum Charge | Job | | 232 | | 232 | |
| **Handicap** | | | | | | |
| Demolish | Ea. | | 73 | | 73 | Includes material and labor to install |
| Install | Ea. | 1025 | 186 | | 1211 | baked enamel handicap partition, |
| Demolish and Install | Ea. | 1025 | 259 | | 1284 | floor or ceiling mounted with one wall |
| Reinstall | Ea. | | 148.74 | | 148.74 | and one door. |
| Clean | Ea. | 1.47 | 22 | | 23.47 | |
| Minimum Charge | Job | | 232 | | 232 | |

| Fire Prevention / Protection | Unit | Material | Labor | Equip. | Total | Specification |
|---|---|---|---|---|---|---|
| **Fire Extinguisher** | | | | | | |
| **5# Carbon Dioxide** | | | | | | |
| Install | Ea. | 230 | | | 230 | Includes labor and material to install one wall mounted 5 lb. carbon dioxide factory charged unit, complete with hose, horn and wall mounting bracket. |
| **10# Carbon Dioxide** | | | | | | |
| Install | Ea. | 330 | | | 330 | Includes labor and material to install one wall mounted 10 lb. carbon dioxide factory charged unit, complete with hose, horn and wall mounting bracket. |
| **15# Carbon Dioxide** | | | | | | |
| Install | Ea. | 425 | | | 425 | Includes labor and material to install one wall mounted 15 lb. carbon dioxide factory charged unit, complete with hose, horn and wall mounting bracket. |

| Fire Prevention / Protection | Unit | Material | Labor | Equip. | Total | Specification |
|---|---|---|---|---|---|---|
| **5# Dry Chemical** | | | | | | |
| Install | Ea. | 61.50 | | | 61.50 | Includes labor and material to install one wall mounted 5 lb. dry chemical factory charged unit, complete with hose, horn and wall mounting bracket. |
| **20# Dry Chemical** | | | | | | |
| Install | Ea. | 152 | | | 152 | Includes labor and material to install one wall mounted 20 lb. dry chemical factory charged unit, complete with hose, horn and wall mounting bracket. |

| Gutters | Unit | Material | Labor | Equip. | Total | Specification |
|---|---|---|---|---|---|---|
| **Galvanized** | | | | | | |
| Demolish | L.F. | | 1.64 | | 1.64 | Cost includes material and labor to |
| Install | L.F. | 2.51 | 4.08 | | 6.59 | install 5" galvanized steel gutter with |
| Demolish and Install | L.F. | 2.51 | 5.72 | | 8.23 | enamel finish, fittings, hangers, and |
| Reinstall | L.F. | | 3.26 | | 3.26 | corners. |
| Clean | L.F. | | .47 | | .47 | |
| Paint | L.F. | .22 | 1.19 | | 1.41 | |
| Minimum Charge | Job | | 255 | | 255 | |
| **Aluminum** | | | | | | |
| Demolish | L.F. | | 1.64 | | 1.64 | Cost includes material and labor to |
| Install | L.F. | 3.08 | 4.08 | | 7.16 | install 5" aluminum gutter with enamel |
| Demolish and Install | L.F. | 3.08 | 5.72 | | 8.80 | finish, fittings, hangers, and corners. |
| Reinstall | L.F. | | 3.26 | | 3.26 | |
| Clean | L.F. | | .47 | | .47 | |
| Paint | L.F. | .22 | 1.19 | | 1.41 | |
| Minimum Charge | Job | | 255 | | 255 | |
| **Copper** | | | | | | |
| Demolish | L.F. | | 1.64 | | 1.64 | Cost includes material and labor to |
| Install | L.F. | 7.80 | 4.08 | | 11.88 | install 5" half-round copper gutter with |
| Demolish and Install | L.F. | 7.80 | 5.72 | | 13.52 | fittings, hangers, and corners. |
| Reinstall | L.F. | | 3.26 | | 3.26 | |
| Clean | L.F. | | .47 | | .47 | |
| Minimum Charge | Job | | 255 | | 255 | |
| **Plastic** | | | | | | |
| Demolish | L.F. | | 1.64 | | 1.64 | Cost includes material and labor to |
| Install | L.F. | 1.47 | 4.04 | | 5.51 | install 4" white vinyl gutter fittings, |
| Demolish and Install | L.F. | 1.47 | 5.68 | | 7.15 | hangers, and corners. |
| Reinstall | L.F. | | 3.23 | | 3.23 | |
| Clean | L.F. | | .47 | | .47 | |
| Paint | L.F. | .22 | 1.19 | | 1.41 | |
| Minimum Charge | Job | | 255 | | 255 | |

| Downspouts | Unit | Material | Labor | Equip. | Total | Specification |
|---|---|---|---|---|---|---|
| **Galvanized** | | | | | | |
| Demolish | L.F. | | 1.13 | | 1.13 | Cost includes material and labor to |
| Install | L.F. | 2.29 | 3.51 | | 5.80 | install 4" diameter downspout with |
| Demolish and Install | L.F. | 2.29 | 4.64 | | 6.93 | enamel finish. |
| Reinstall | L.F. | | 2.81 | | 2.81 | |
| Clean | L.F. | | .47 | | .47 | |
| Paint | L.F. | .22 | 1.19 | | 1.41 | |
| Minimum Charge | Job | | 255 | | 255 | |

| Downspouts | Unit | Material | Labor | Equip. | Total | Specification |
|---|---|---|---|---|---|---|
| **Aluminum** | | | | | | |
| Demolish | L.F. | | 1.13 | | 1.13 | Includes material and labor to install |
| Install | L.F. | 2.15 | 3.64 | | 5.79 | aluminum enameled downspout. |
| Demolish and Install | L.F. | 2.15 | 4.77 | | 6.92 | |
| Reinstall | L.F. | | 2.91 | | 2.91 | |
| Clean | L.F. | | .47 | | .47 | |
| Paint | L.F. | .22 | 1.19 | | 1.41 | |
| Minimum Charge | Job | | 255 | | 255 | |
| **Copper** | | | | | | |
| Demolish | L.F. | | 1.13 | | 1.13 | Includes material and labor to install |
| Install | L.F. | 9.45 | 2.68 | | 12.13 | round copper downspout. |
| Demolish and Install | L.F. | 9.45 | 3.81 | | 13.26 | |
| Reinstall | L.F. | | 2.15 | | 2.15 | |
| Clean | L.F. | | .47 | | .47 | |
| Paint | L.F. | .22 | 1.19 | | 1.41 | |
| Minimum Charge | Job | | 255 | | 255 | |
| **Plastic** | | | | | | |
| Demolish | L.F. | | 1.13 | | 1.13 | Cost includes material and labor to |
| Install | L.F. | 2.09 | 2.43 | | 4.52 | install 2″ x 3″ vinyl rectangular |
| Demolish and Install | L.F. | 2.09 | 3.56 | | 5.65 | downspouts. |
| Reinstall | L.F. | | 1.94 | | 1.94 | |
| Clean | L.F. | | .47 | | .47 | |
| Paint | L.F. | .22 | 1.19 | | 1.41 | |
| Minimum Charge | Job | | 255 | | 255 | |

| Shutters | Unit | Material | Labor | Equip. | Total | Specification |
|---|---|---|---|---|---|---|
| **Exterior** | | | | | | |
| **25″ High** | | | | | | |
| Demolish | Ea. | | 12.35 | | 12.35 | Includes material and labor to install |
| Install | Pr. | 188 | 46.50 | | 234.50 | pair of exterior shutters 25″ x 16″, |
| Demolish and Install | Pr. | 188 | 58.85 | | 246.85 | unpainted, installed as fixed shutters |
| Reinstall | Pr. | | 46.48 | | 46.48 | on concrete or wood frame. |
| Clean | Ea. | .03 | 11.10 | | 11.13 | |
| Paint | Ea. | 1.77 | 21 | | 22.77 | |
| Minimum Charge | Job | | 232 | | 232 | |
| **39″ High** | | | | | | |
| Demolish | Ea. | | 12.35 | | 12.35 | Includes material and labor to install |
| Install | Pr. | 252 | 46.50 | | 298.50 | pair of shutters 39″ x 16″, unpainted, |
| Demolish and Install | Pr. | 252 | 58.85 | | 310.85 | installed as fixed shutters on concrete |
| Reinstall | Pr. | | 46.48 | | 46.48 | or wood frame. |
| Clean | Ea. | .03 | 11.10 | | 11.13 | |
| Paint | Ea. | 2.75 | 28 | | 30.75 | |
| Minimum Charge | Job | | 232 | | 232 | |
| **51″ High** | | | | | | |
| Demolish | Ea. | | 12.35 | | 12.35 | Includes material and labor to install |
| Install | Pr. | 325 | 46.50 | | 371.50 | pair of shutters 51″ x 16″, unpainted, |
| Demolish and Install | Pr. | 325 | 58.85 | | 383.85 | installed as fixed shutters on concrete |
| Reinstall | Pr. | | 46.48 | | 46.48 | or wood frame. |
| Clean | Ea. | .03 | 11.10 | | 11.13 | |
| Paint | Ea. | 3.58 | 29 | | 32.58 | |
| Minimum Charge | Job | | 232 | | 232 | |

| Shutters | Unit | Material | Labor | Equip. | Total | Specification |
|---|---|---|---|---|---|---|
| **59" High** | | | | | | |
| Demolish | Ea. | | 12.35 | | 12.35 | Includes material and labor to install |
| Install | Pr. | 385 | 46.50 | | 431.50 | pair of shutters 59" x 16", unpainted, |
| Demolish and Install | Pr. | 385 | 58.85 | | 443.85 | installed as fixed shutters on concrete |
| Reinstall | Pr. | | 46.48 | | 46.48 | or wood frame. |
| Clean | Ea. | .03 | 11.10 | | 11.13 | |
| Paint | Ea. | 4.13 | 32 | | 36.13 | |
| Minimum Charge | Job | | 232 | | 232 | |
| **67" High** | | | | | | |
| Demolish | Ea. | | 12.35 | | 12.35 | Includes material and labor to install |
| Install | Pr. | 425 | 51.50 | | 476.50 | pair of shutters 67" x 16", unpainted, |
| Demolish and Install | Pr. | 425 | 63.85 | | 488.85 | installed as fixed shutters on concrete |
| Reinstall | Pr. | | 51.64 | | 51.64 | or wood frame. |
| Clean | Ea. | .03 | 11.10 | | 11.13 | |
| Paint | Ea. | 4.71 | 36 | | 40.71 | |
| Minimum Charge | Job | | 232 | | 232 | |
| **Interior Wood** | | | | | | |
| **23" x 24", 2 Panel** | | | | | | |
| Demolish | Ea. | | 12.35 | | 12.35 | Cost includes material and labor to |
| Install | Ea. | 140 | 27.50 | | 167.50 | install 23" x 24", 4 panel interior |
| Demolish and Install | Ea. | 140 | 39.85 | | 179.85 | movable pine shutter, 1-1/4" louver, |
| Reinstall | Ea. | | 21.87 | | 21.87 | knobs and hooks and hinges. |
| Clean | Ea. | .03 | 11.10 | | 11.13 | |
| Paint | Ea. | 1.33 | 27.50 | | 28.83 | |
| Minimum Charge | Job | | 232 | | 232 | |
| **27" x 24", 4 Panel** | | | | | | |
| Demolish | Ea. | | 12.35 | | 12.35 | Cost includes material and labor to |
| Install | Ea. | 154 | 27.50 | | 181.50 | install 27" x 24", 4 panel interior |
| Demolish and Install | Ea. | 154 | 39.85 | | 193.85 | movable pine shutter, 1-1/4" louver, |
| Reinstall | Ea. | | 21.87 | | 21.87 | knobs and hooks and hinges. |
| Clean | Ea. | .03 | 11.10 | | 11.13 | |
| Paint | Ea. | 1.33 | 27.50 | | 28.83 | |
| Minimum Charge | Job | | 232 | | 232 | |
| **31" x 24", 4 Panel** | | | | | | |
| Demolish | Ea. | | 12.35 | | 12.35 | Cost includes material and labor to |
| Install | Ea. | 169 | 27.50 | | 196.50 | install 31" x 24", 4 panel interior |
| Demolish and Install | Ea. | 169 | 39.85 | | 208.85 | movable pine shutter, 1-1/4" louver, |
| Reinstall | Ea. | | 21.87 | | 21.87 | knobs and hooks and hinges. |
| Clean | Ea. | .03 | 11.10 | | 11.13 | |
| Paint | Ea. | 1.33 | 27.50 | | 28.83 | |
| Minimum Charge | Job | | 232 | | 232 | |
| **35" x 24", 4 Panel** | | | | | | |
| Demolish | Ea. | | 12.35 | | 12.35 | Cost includes material and labor to |
| Install | Ea. | 182 | 27.50 | | 209.50 | install 35" x 24", 4 panel interior |
| Demolish and Install | Ea. | 182 | 39.85 | | 221.85 | movable pine shutter, 1-1/4" louver, |
| Reinstall | Ea. | | 21.87 | | 21.87 | knobs and hooks and hinges. |
| Clean | Ea. | .03 | 11.10 | | 11.13 | |
| Paint | Ea. | 1.33 | 27.50 | | 28.83 | |
| Minimum Charge | Job | | 232 | | 232 | |
| **39" x 24", 4 Panel** | | | | | | |
| Demolish | Ea. | | 12.35 | | 12.35 | Cost includes material and labor to |
| Install | Ea. | 202 | 27.50 | | 229.50 | install 39" x 24", 4 panel interior |
| Demolish and Install | Ea. | 202 | 39.85 | | 241.85 | movable pine shutter, 1-1/4" louver, |
| Reinstall | Ea. | | 21.87 | | 21.87 | knobs and hooks and hinges. |
| Clean | Ea. | .03 | 11.10 | | 11.13 | |
| Paint | Ea. | 1.33 | 27.50 | | 28.83 | |
| Minimum Charge | Job | | 232 | | 232 | |

## Shutters

| | Unit | Material | Labor | Equip. | Total | Specification |
|---|---|---|---|---|---|---|
| **47" x 24", 4 Panel** | | | | | | |
| Demolish | Ea. | | 12.35 | | 12.35 | Cost includes material and labor to |
| Install | Ea. | 230 | 27.50 | | 257.50 | install 47" x 24", 4 panel interior |
| Demolish and Install | Ea. | 230 | 39.85 | | 269.85 | movable pine shutter, 1-1/4" louver, |
| Reinstall | Ea. | | 21.87 | | 21.87 | knobs and hooks and hinges. |
| Clean | Ea. | .03 | 11.10 | | 11.13 | |
| Paint | Ea. | 1.39 | 29.50 | | 30.89 | |
| Minimum Charge | Job | | 232 | | 232 | |

## Awning

| | Unit | Material | Labor | Equip. | Total | Specification |
|---|---|---|---|---|---|---|
| **Standard Aluminum** | | | | | | |
| **36" Wide x 30" Deep** | | | | | | |
| Demolish | Ea. | | 12.35 | | 12.35 | Includes material and labor to install |
| Install | Ea. | 197 | 38.50 | | 235.50 | one aluminum awning with |
| Demolish and Install | Ea. | 197 | 50.85 | | 247.85 | weather-resistant finish. |
| Reinstall | Ea. | | 30.99 | | 30.99 | |
| Clean | Ea. | | 14.80 | | 14.80 | |
| Minimum Charge | Job | | 232 | | 232 | |
| **48" Wide x 30" Deep** | | | | | | |
| Demolish | Ea. | | 12.35 | | 12.35 | Includes material and labor to install |
| Install | Ea. | 234 | 46.50 | | 280.50 | one aluminum awning with |
| Demolish and Install | Ea. | 234 | 58.85 | | 292.85 | weather-resistant finish. |
| Reinstall | Ea. | | 37.18 | | 37.18 | |
| Clean | Ea. | | 14.80 | | 14.80 | |
| Minimum Charge | Job | | 232 | | 232 | |
| **60" Wide x 30" Deep** | | | | | | |
| Demolish | Ea. | | 24.50 | | 24.50 | Includes material and labor to install |
| Install | Ea. | 274 | 51.50 | | 325.50 | one aluminum awning with |
| Demolish and Install | Ea. | 274 | 76 | | 350 | weather-resistant finish. |
| Reinstall | Ea. | | 41.32 | | 41.32 | |
| Clean | Ea. | | 14.80 | | 14.80 | |
| Minimum Charge | Job | | 232 | | 232 | |
| **72" Wide x 30" Deep** | | | | | | |
| Demolish | Ea. | | 24.50 | | 24.50 | Includes material and labor to install |
| Install | Ea. | 305 | 58 | | 363 | one aluminum awning with |
| Demolish and Install | Ea. | 305 | 82.50 | | 387.50 | weather-resistant finish. |
| Reinstall | Ea. | | 46.48 | | 46.48 | |
| Clean | Ea. | | 14.80 | | 14.80 | |
| Minimum Charge | Job | | 232 | | 232 | |
| **Roll-up Type** | | | | | | |
| Demolish | S.F. | | .89 | | .89 | Includes material and labor to install |
| Install | S.F. | 16.60 | 4.65 | | 21.25 | an aluminum roll-up awning. |
| Demolish and Install | S.F. | 16.60 | 5.54 | | 22.14 | |
| Reinstall | S.F. | | 3.72 | | 3.72 | |
| Clean | S.F. | .01 | .74 | | .75 | |
| Minimum Charge | Job | | 232 | | 232 | |
| **Canvas** | | | | | | |
| **30" Wide** | | | | | | |
| Demolish | Ea. | | 7.40 | | 7.40 | Includes material and labor to install a |
| Install | Ea. | 215 | 38.50 | | 253.50 | canvas window awning. |
| Demolish and Install | Ea. | 215 | 45.90 | | 260.90 | |
| Reinstall | Ea. | | 30.99 | | 30.99 | |
| Clean | Ea. | | 4.44 | | 4.44 | |
| Minimum Charge | Job | | 232 | | 232 | |

287

| Awning | Unit | Material | Labor | Equip. | Total | Specification |
|---|---|---|---|---|---|---|
| **36″ Wide** | | | | | | |
| Demolish | Ea. | | 7.40 | | 7.40 | Includes material and labor to install a |
| Install | Ea. | 260 | 46.50 | | 306.50 | canvas window awning. |
| Demolish and Install | Ea. | 260 | 53.90 | | 313.90 | |
| Reinstall | Ea. | | 37.18 | | 37.18 | |
| Clean | Ea. | | 4.87 | | 4.87 | |
| Minimum Charge | Job | | 232 | | 232 | |
| **42″ Wide** | | | | | | |
| Demolish | Ea. | | 7.40 | | 7.40 | Includes material and labor to install a |
| Install | Ea. | 274 | 58 | | 332 | canvas window awning. |
| Demolish and Install | Ea. | 274 | 65.40 | | 339.40 | |
| Reinstall | Ea. | | 46.48 | | 46.48 | |
| Clean | Ea. | | 5.80 | | 5.80 | |
| Minimum Charge | Job | | 232 | | 232 | |
| **48″ Wide** | | | | | | |
| Demolish | Ea. | | 7.40 | | 7.40 | Includes material and labor to install a |
| Install | Ea. | 288 | 66.50 | | 354.50 | canvas window awning. |
| Demolish and Install | Ea. | 288 | 73.90 | | 361.90 | |
| Reinstall | Ea. | | 53.12 | | 53.12 | |
| Clean | Ea. | | 6.70 | | 6.70 | |
| Minimum Charge | Job | | 232 | | 232 | |
| **Cloth Canopy Cover** | | | | | | |
| **8′ x 10′** | | | | | | |
| Demolish | Ea. | | 400 | | 400 | Includes material and labor to install |
| Install | Ea. | 350 | 232 | | 582 | one 8′ x 10′ cloth canopy patio cover |
| Demolish and Install | Ea. | 350 | 632 | | 982 | with front bar and tension support |
| Reinstall | Ea. | | 185.92 | | 185.92 | rafters and 9″ valance. |
| Clean | Ea. | 1.18 | 59 | | 60.18 | |
| Minimum Charge | Job | | 465 | | 465 | |
| **8′ x 15′** | | | | | | |
| Demolish | Ea. | | 400 | | 400 | Includes material and labor to install |
| Install | Ea. | 530 | 232 | | 762 | one 8′ x 15′ cloth canopy patio cover |
| Demolish and Install | Ea. | 530 | 632 | | 1162 | with front bar and tension support |
| Reinstall | Ea. | | 185.92 | | 185.92 | rafters and 9″ valance. |
| Clean | Ea. | 1.18 | 89 | | 90.18 | |
| Minimum Charge | Job | | 465 | | 465 | |

| Vent | Unit | Material | Labor | Equip. | Total | Specification |
|---|---|---|---|---|---|---|
| **Small Roof Turbine** | | | | | | |
| Demolish | Ea. | | 12.35 | | 12.35 | Includes material and labor to install |
| Install | Ea. | 52.50 | 46 | | 98.50 | spinner ventilator, wind driven, |
| Demolish and Install | Ea. | 52.50 | 58.35 | | 110.85 | galvanized, 4″ neck diam, 180 CFM. |
| Reinstall | Ea. | | 36.70 | | 36.70 | |
| Clean | Ea. | | 7.90 | | 7.90 | |
| Minimum Charge | Job | | 262 | | 262 | |
| **Large Roof Turbine** | | | | | | |
| Demolish | Ea. | | 9.85 | | 9.85 | Includes material and labor to install |
| Install | Ea. | 61.50 | 65.50 | | 127 | spinner ventilator, wind driven, |
| Demolish and Install | Ea. | 61.50 | 75.35 | | 136.85 | galvanized, 8″ neck diam, 360 CFM. |
| Reinstall | Ea. | | 52.43 | | 52.43 | |
| Clean | Ea. | | 8.90 | | 8.90 | |
| Minimum Charge | Job | | 262 | | 262 | |

| Vent | Unit | Material | Labor | Equip. | Total | Specification |
|------|------|----------|-------|--------|-------|---------------|
| **Gable Louvered** | | | | | | |
| Demolish | Ea. | | 24.50 | | 24.50 | Includes material and labor to install a |
| Install | Ea. | 61 | 15.50 | | 76.50 | vinyl gable end vent. |
| Demolish and Install | Ea. | 61 | 40 | | 101 | |
| Reinstall | Ea. | | 12.39 | | 12.39 | |
| Clean | Ea. | | 4.74 | | 4.74 | |
| Paint | Ea. | 1.11 | 5.95 | | 7.06 | |
| Minimum Charge | Job | | 232 | | 232 | |
| **Ridge Vent** | | | | | | |
| Demolish | L.F. | | 1.27 | | 1.27 | Includes material and labor to install a |
| Install | L.F. | 4.32 | 3.29 | | 7.61 | mill finish aluminum ridge vent strip. |
| Demolish and Install | L.F. | 4.32 | 4.56 | | 8.88 | |
| Minimum Charge | Job | | 232 | | 232 | |
| **Foundation** | | | | | | |
| Demolish | Ea. | | 8.85 | | 8.85 | Includes material and labor to install |
| Install | Ea. | 17.80 | 15.45 | | 33.25 | galvanized foundation block vent. |
| Demolish and Install | Ea. | 17.80 | 24.30 | | 42.10 | |
| Reinstall | Ea. | | 12.37 | | 12.37 | |
| Clean | Ea. | .08 | 8.90 | | 8.98 | |
| Minimum Charge | Job | | 232 | | 232 | |
| **Frieze** | | | | | | |
| Demolish | L.F. | | .88 | | .88 | Includes material and labor to install |
| Install | L.F. | 1.46 | 1.86 | | 3.32 | pine frieze board. |
| Demolish and Install | L.F. | 1.46 | 2.74 | | 4.20 | |
| Reinstall | L.F. | | 1.49 | | 1.49 | |
| Clean | L.F. | .02 | .30 | | .32 | |
| Minimum Charge | Job | | 232 | | 232 | |
| **Water Heater Cap** | | | | | | |
| Demolish | Ea. | | 18.20 | | 18.20 | Includes material and labor to install a |
| Install | Ea. | 40.50 | 63.50 | | 104 | galvanized steel water heater cap. |
| Demolish and Install | Ea. | 40.50 | 81.70 | | 122.20 | |
| Reinstall | Ea. | | 50.96 | | 50.96 | |
| Clean | Ea. | .52 | 22 | | 22.52 | |

| Cupolas | Unit | Material | Labor | Equip. | Total | Specification |
|---------|------|----------|-------|--------|-------|---------------|
| **22" x 22"** | | | | | | |
| Demolish | Ea. | | 228 | | 228 | Includes material and labor to install a |
| Install | Ea. | 420 | 116 | | 536 | cedar cupola with aluminum roof |
| Demolish and Install | Ea. | 420 | 344 | | 764 | covering. |
| Reinstall | Ea. | | 92.96 | | 92.96 | |
| Minimum Charge | Job | | 232 | | 232 | |
| **29" x 29"** | | | | | | |
| Demolish | Ea. | | 228 | | 228 | Includes material and labor to install a |
| Install | Ea. | 590 | 116 | | 706 | cedar cupola with aluminum roof |
| Demolish and Install | Ea. | 590 | 344 | | 934 | covering. |
| Reinstall | Ea. | | 92.96 | | 92.96 | |
| Minimum Charge | Job | | 232 | | 232 | |
| **35" x 35"** | | | | | | |
| Demolish | Ea. | | 228 | | 228 | Includes material and labor to install a |
| Install | Ea. | 945 | 155 | | 1100 | cedar cupola with aluminum roof |
| Demolish and Install | Ea. | 945 | 383 | | 1328 | covering. |
| Reinstall | Ea. | | 123.95 | | 123.95 | |
| Minimum Charge | Job | | 232 | | 232 | |

## Cupolas

| | Unit | Material | Labor | Equip. | Total | Specification |
|---|---|---|---|---|---|---|
| **47" x 47"** | | | | | | |
| Demolish | Ea. | | 228 | | 228 | Includes material and labor to install a |
| Install | Ea. | 1850 | 232 | | 2082 | cedar cupola with aluminum roof |
| Demolish and Install | Ea. | 1850 | 460 | | 2310 | covering. |
| Reinstall | Ea. | | 185.92 | | 185.92 | |
| Minimum Charge | Job | | 232 | | 232 | |
| **Weather Vane** | | | | | | |
| Install | Ea. | 147 | 58 | | 205 | Includes minimum labor and |
| Minimum Charge | Job | | 232 | | 232 | equipment to install residential type weathervane. |

## Aluminum Carport

| | Unit | Material | Labor | Equip. | Total | Specification |
|---|---|---|---|---|---|---|
| **20' x 20' Complete** | | | | | | |
| Demolish | Ea. | | 247 | | 247 | Includes material, labor and |
| Install | Ea. | 9525 | 775 | 131 | 10431 | equipment to install an aluminum |
| Demolish and Install | Ea. | 9525 | 1022 | 131 | 10678 | carport. Foundations are not included. |
| Reinstall | Ea. | | 618.43 | 104.74 | 723.17 | |
| Clean | Ea. | .08 | 89 | | 89.08 | |
| Minimum Charge | Job | | 465 | | 465 | |
| **24' x 24' Complete** | | | | | | |
| Demolish | Ea. | | 247 | | 247 | Includes material, labor and |
| Install | Ea. | 12700 | 1025 | 175 | 13900 | equipment to install an aluminum |
| Demolish and Install | Ea. | 12700 | 1272 | 175 | 14147 | carport. Foundations are not included. |
| Reinstall | Ea. | | 824.58 | 139.65 | 964.23 | |
| Clean | Ea. | .08 | 178 | | 178.08 | |
| Minimum Charge | Job | | 465 | | 465 | |
| **Metal Support Posts** | | | | | | |
| Demolish | Ea. | | 3.29 | | 3.29 | Includes material and labor to install |
| Install | Ea. | 69.50 | 58 | | 127.50 | metal support posts for a carport. |
| Demolish and Install | Ea. | 69.50 | 61.29 | | 130.79 | |
| Reinstall | Ea. | | 46.48 | | 46.48 | |
| Clean | Ea. | .08 | 3.70 | | 3.78 | |
| Paint | Ea. | .20 | 7.75 | | 7.95 | |
| Minimum Charge | Job | | 465 | | 465 | |

## Storage Shed

| | Unit | Material | Labor | Equip. | Total | Specification |
|---|---|---|---|---|---|---|
| **Aluminum** | | | | | | |
| **Pre-fab (8' x 10')** | | | | | | |
| Demolish | Ea. | | 72 | | 72 | Includes material and labor to install |
| Install | Ea. | 2175 | 685 | | 2860 | an aluminum storage shed. |
| Demolish and Install | Ea. | 2175 | 757 | | 2932 | |
| Clean | Ea. | .50 | 44.50 | | 45 | |
| Minimum Charge | Ea. | | 550 | | 550 | |
| **Pre-fab (10' x 12')** | | | | | | |
| Demolish | Ea. | | 133 | | 133 | Includes material and labor to install |
| Install | Ea. | 1900 | 685 | | 2585 | an aluminum storage shed. |
| Demolish and Install | Ea. | 1900 | 818 | | 2718 | |
| Clean | Ea. | .50 | 49.50 | | 50 | |
| Minimum Charge | Ea. | | 550 | | 550 | |

# Improvements / Appliances / Treatments

## Storage Shed

| | Unit | Material | Labor | Equip. | Total | Specification |
|---|---|---|---|---|---|---|
| **Wood** | | | | | | |
| **Pre-fab (8' x 10')** | | | | | | |
| Demolish | Ea. | | 72 | | 72 | Includes material and labor to install a |
| Install | Ea. | 1925 | 685 | | 2610 | pre-fabricated wood storage shed. |
| Demolish and Install | Ea. | 1925 | 757 | | 2682 | |
| Clean | Ea. | .50 | 44.50 | | 45 | |
| Minimum Charge | Ea. | | 550 | | 550 | |
| **Pre-fab (10' x 12')** | | | | | | |
| Demolish | Ea. | | 133 | | 133 | Includes material and labor to install a |
| Install | Ea. | 2425 | 685 | | 3110 | pre-fabricated wood storage shed. |
| Demolish and Install | Ea. | 2425 | 818 | | 3243 | |
| Clean | Ea. | .50 | 49.50 | | 50 | |
| Minimum Charge | Ea. | | 550 | | 550 | |

## Gazebo

| | Unit | Material | Labor | Equip. | Total | Specification |
|---|---|---|---|---|---|---|
| **Good Grade** | | | | | | |
| Demolish | S.F. | | 3.95 | | 3.95 | Includes material and labor to install a |
| Install | S.F. | 76.50 | 35 | | 111.50 | gazebo. |
| Demolish and Install | S.F. | 76.50 | 38.95 | | 115.45 | |
| Reinstall | S.F. | | 27.89 | | 27.89 | |
| Clean | S.F. | .06 | 2.37 | | 2.43 | |
| Minimum Charge | Job | | 232 | | 232 | |
| **Custom Grade** | | | | | | |
| Demolish | S.F. | | 3.95 | | 3.95 | Includes material and labor to install a |
| Install | S.F. | 107 | 35 | | 142 | gazebo. |
| Demolish and Install | S.F. | 107 | 38.95 | | 145.95 | |
| Reinstall | S.F. | | 27.89 | | 27.89 | |
| Clean | S.F. | .06 | 2.37 | | 2.43 | |
| Minimum Charge | Job | | 232 | | 232 | |
| **Bench Seating** | | | | | | |
| Demolish | L.F. | | 6.60 | | 6.60 | Includes material and labor to install |
| Install | L.F. | 2.27 | 23 | | 25.27 | pressure treated bench seating. |
| Demolish and Install | L.F. | 2.27 | 29.60 | | 31.87 | |
| Reinstall | L.F. | | 18.59 | | 18.59 | |
| Clean | L.F. | .06 | 1.78 | | 1.84 | |
| Paint | L.F. | .69 | 3.09 | | 3.78 | |
| Minimum Charge | Job | | 232 | | 232 | |
| **Re-screen** | | | | | | |
| Install | S.F. | 1.83 | 1.86 | | 3.69 | Includes labor and material to |
| Minimum Charge | Job | | 232 | | 232 | re-screen wood frame. |

## Pool

| | Unit | Material | Labor | Equip. | Total | Specification |
|---|---|---|---|---|---|---|
| **Screen Enclosure** | | | | | | |
| **Re-screen** | | | | | | |
| Install | SF Wall | .03 | 5.50 | | 5.53 | Includes material and labor to install a |
| Minimum Charge | Job | | 232 | | 232 | screen enclosure for a pool. |
| **Screen Door** | | | | | | |
| Demolish | Ea. | | 9.20 | | 9.20 | Includes material and labor to install |
| Install | Ea. | 320 | 77.50 | | 397.50 | wood screen door with aluminum cloth |
| Demolish and Install | Ea. | 320 | 86.70 | | 406.70 | screen. Frame and hardware not |
| Reinstall | Ea. | | 61.97 | | 61.97 | included. |
| Clean | Ea. | .06 | 13.35 | | 13.41 | |
| Paint | Ea. | 4.81 | 15.45 | | 20.26 | |
| Minimum Charge | Job | | 232 | | 232 | |

For customer support on your Contractor's Pricing Guide: Residential Repair & Remodeling, call 800.448.8182.

| Pool | Unit | Material | Labor | Equip. | Total | Specification |
|---|---|---|---|---|---|---|
| **Heater / Motor** | | | | | | Includes material and labor to install |
| Demolish | Ea. | | 47 | | 47 | swimming pool heater, not incl. base |
| Install | Ea. | 2575 | 600 | | 3175 | or pad, elec., 12 kW, 4,800 gal |
| Demolish and Install | Ea. | 2575 | 647 | | 3222 | pool, incl. pump. |
| Reinstall | Ea. | | 478.62 | | 478.62 | |
| Clean | Ea. | 1.47 | 44.50 | | 45.97 | |
| Minimum Charge | Job | | 270 | | 270 | |
| **Filter System** | | | | | | Includes material, labor and |
| Demolish | Ea. | | 79 | | 79 | equipment to install sand filter system |
| Install | Total | 1375 | 420 | | 1795 | tank including hook-up to existing |
| Demolish and Install | Total | 1375 | 499 | | 1874 | equipment. Pump and motor not |
| Reinstall | Total | | 335.36 | | 335.36 | included. |
| Minimum Charge | Job | | 262 | | 262 | |
| **Pump / Motor** | | | | | | Includes material and labor to install a |
| Demolish | Ea. | | 47 | | 47 | 22 GPM pump. |
| Install | Ea. | 1000 | 105 | | 1105 | |
| Demolish and Install | Ea. | 1000 | 152 | | 1152 | |
| Reinstall | Ea. | | 83.84 | | 83.84 | |
| Clean | Ea. | 1.47 | 44.50 | | 45.97 | |
| Minimum Charge | Job | | 270 | | 270 | |
| **Pump Water** | | | | | | Includes labor and equipment to pump |
| Install | Day | | 480 | 93.50 | 573.50 | water from a swimming pool. |
| Minimum Charge | Job | | 225 | | 225 | |
| **Replaster** | | | | | | Includes labor and material to |
| Install | SF Surf | 1.03 | 11.85 | | 12.88 | replaster gunite pool with a surface |
| Minimum Charge | Job | | 225 | | 225 | area of 500 to 600 S.F. |
| **Solar Panels** | | | | | | Includes material and labor to install |
| Demolish | S.F. | | 1.94 | | 1.94 | solar panel. |
| Install | S.F. | 34.50 | 3.45 | | 37.95 | |
| Demolish and Install | S.F. | 34.50 | 5.39 | | 39.89 | |
| Reinstall | S.F. | | 2.76 | | 2.76 | |
| Minimum Charge | Ea. | | 550 | | 550 | |

| Construction Clean-up | Unit | Material | Labor | Equip. | Total | Specification |
|---|---|---|---|---|---|---|
| **Final** | | | | | | |
| Clean | S.F. | | .31 | | .31 | Includes labor for final site/job |
| Minimum Charge | Job | | 178 | | 178 | clean-up including detail cleaning. |

# Location Factors - Residential

Costs shown in *Residential Costs with RSMeans data* are based on national averages for materials and installation. To adjust these costs to a specific location, simply multiply the base cost by the factor for that city. The data is arranged alphabetically by state and postal zip code numbers. For a city not listed, use the factor for a nearby city with similar economic characteristics.

| STATE | CITY | Residential |
|---|---|---|
| **ALABAMA** | | |
| 350-352 | Birmingham | .83 |
| 354 | Tuscaloosa | .84 |
| 355 | Jasper | .83 |
| 356 | Decatur | .83 |
| 357-358 | Huntsville | .83 |
| 359 | Gadsden | .82 |
| 360-361 | Montgomery | .82 |
| 362 | Anniston | .81 |
| 363 | Dothan | .84 |
| 364 | Evergreen | .81 |
| 365-366 | Mobile | .83 |
| 367 | Selma | .82 |
| 368 | Phenix City | .83 |
| 369 | Butler | .81 |
| **ALASKA** | | |
| 995-996 | Anchorage | 1.22 |
| 997 | Fairbanks | 1.25 |
| 998 | Juneau | 1.23 |
| 999 | Ketchikan | 1.26 |
| **ARIZONA** | | |
| 850,853 | Phoenix | .87 |
| 851,852 | Mesa/Tempe | .86 |
| 855 | Globe | .85 |
| 856-857 | Tucson | .85 |
| 859 | Show Low | .86 |
| 860 | Flagstaff | .87 |
| 863 | Prescott | .86 |
| 864 | Kingman | .85 |
| 865 | Chambers | .87 |
| **ARKANSAS** | | |
| 716 | Pine Bluff | .78 |
| 717 | Camden | .76 |
| 718 | Texarkana | .78 |
| 719 | Hot Springs | .75 |
| 720-722 | Little Rock | .80 |
| 723 | West Memphis | .78 |
| 724 | Jonesboro | .77 |
| 725 | Batesville | .75 |
| 726 | Harrison | .76 |
| 727 | Fayetteville | .74 |
| 728 | Russellville | .75 |
| 729 | Fort Smith | .81 |
| **CALIFORNIA** | | |
| 900-902 | Los Angeles | 1.15 |
| 903-905 | Inglewood | 1.12 |
| 906-908 | Long Beach | 1.12 |
| 910-912 | Pasadena | 1.11 |
| 913-916 | Van Nuys | 1.14 |
| 917-918 | Alhambra | 1.15 |
| 919-921 | San Diego | 1.10 |
| 922 | Palm Springs | 1.11 |
| 923-924 | San Bernardino | 1.14 |
| 925 | Riverside | 1.14 |
| 926-927 | Santa Ana | 1.13 |
| 928 | Anaheim | 1.14 |
| 930 | Oxnard | 1.14 |
| 931 | Santa Barbara | 1.14 |
| 932-933 | Bakersfield | 1.11 |
| 934 | San Luis Obispo | 1.15 |
| 935 | Mojave | 1.12 |
| 936-938 | Fresno | 1.18 |
| 939 | Salinas | 1.21 |
| 940-941 | San Francisco | 1.33 |
| 942,956-958 | Sacramento | 1.19 |
| 943 | Palo Alto | 1.27 |
| 944 | San Mateo | 1.30 |
| 945 | Vallejo | 1.23 |
| 946 | Oakland | 1.29 |
| 947 | Berkeley | 1.32 |
| 948 | Richmond | 1.31 |
| 949 | San Rafael | 1.30 |
| 950 | Santa Cruz | 1.25 |
| 951 | San Jose | 1.30 |
| 952 | Stockton | 1.19 |
| 953 | Modesto | 1.18 |

| STATE | CITY | Residential |
|---|---|---|
| **CALIFORNIA (CONT'D)** | | |
| 954 | Santa Rosa | 1.27 |
| 955 | Eureka | 1.23 |
| 959 | Marysville | 1.20 |
| 960 | Redding | 1.24 |
| 961 | Susanville | 1.21 |
| **COLORADO** | | |
| 800-802 | Denver | .88 |
| 803 | Boulder | .90 |
| 804 | Golden | .85 |
| 805 | Fort Collins | .88 |
| 806 | Greeley | .88 |
| 807 | Fort Morgan | .85 |
| 808-809 | Colorado Springs | .84 |
| 810 | Pueblo | .84 |
| 811 | Alamosa | .84 |
| 812 | Salida | .82 |
| 813 | Durango | .86 |
| 814 | Montrose | .81 |
| 815 | Grand Junction | .92 |
| 816 | Glenwood Springs | .81 |
| **CONNECTICUT** | | |
| 060 | New Britain | 1.11 |
| 061 | Hartford | 1.09 |
| 062 | Willimantic | 1.11 |
| 063 | New London | 1.10 |
| 064 | Meriden | 1.10 |
| 065 | New Haven | 1.11 |
| 066 | Bridgeport | 1.12 |
| 067 | Waterbury | 1.11 |
| 068 | Norwalk | 1.11 |
| 069 | Stamford | 1.12 |
| **D.C.** | | |
| 200-205 | Washington | .92 |
| **DELAWARE** | | |
| 197 | Newark | 1.02 |
| 198 | Wilmington | 1.01 |
| 199 | Dover | 1.02 |
| **FLORIDA** | | |
| 320,322 | Jacksonville | .80 |
| 321 | Daytona Beach | .83 |
| 323 | Tallahassee | .80 |
| 324 | Panama City | .81 |
| 325 | Pensacola | .84 |
| 326,344 | Gainesville | .80 |
| 327-328,347 | Orlando | .81 |
| 329 | Melbourne | .83 |
| 330-332,340 | Miami | .79 |
| 333 | Fort Lauderdale | .80 |
| 334,349 | West Palm Beach | .80 |
| 335-336,346 | Tampa | .82 |
| 337 | St. Petersburg | .81 |
| 338 | Lakeland | .79 |
| 339,341 | Fort Myers | .79 |
| 342 | Sarasota | .83 |
| **GEORGIA** | | |
| 300-303,399 | Atlanta | .88 |
| 304 | Statesboro | .79 |
| 305 | Gainesville | .82 |
| 306 | Athens | .81 |
| 307 | Dalton | .84 |
| 308-309 | Augusta | .87 |
| 310-312 | Macon | .84 |
| 313-314 | Savannah | .85 |
| 315 | Waycross | .81 |
| 316 | Valdosta | .75 |
| 317,398 | Albany | .84 |
| 318-319 | Columbus | .84 |
| **HAWAII** | | |
| 967 | Hilo | 1.19 |
| 968 | Honolulu | 1.21 |

# Location Factors - Residential

| STATE | CITY | Residential | | STATE | CITY | Residential |
|---|---|---|---|---|---|---|
| **STATES & POSS.** | | | | **KENTUCKY (CONT'D)** | | |
| 969 | Guam | .95 | | 406 | Frankfort | .86 |
| | | | | 407-409 | Corbin | .82 |
| **IDAHO** | | | | 410 | Covington | .85 |
| 832 | Pocatello | .89 | | 411-412 | Ashland | .90 |
| 833 | Twin Falls | .89 | | 413-414 | Campton | .87 |
| 834 | Idaho Falls | .87 | | 415-416 | Pikeville | .86 |
| 835 | Lewiston | .98 | | 417-418 | Hazard | .86 |
| 836-837 | Boise | .90 | | 420 | Paducah | .88 |
| 838 | Coeur d'Alene | .98 | | 421-422 | Bowling Green | .87 |
| | | | | 423 | Owensboro | .88 |
| **ILLINOIS** | | | | 424 | Henderson | .88 |
| 600-603 | North Suburban | 1.22 | | 425-426 | Somerset | .84 |
| 604 | Joliet | 1.24 | | 427 | Elizabethtown | .84 |
| 605 | South Suburban | 1.22 | | | | |
| 606-608 | Chicago | 1.26 | | **LOUISIANA** | | |
| 609 | Kankakee | 1.16 | | 700-701 | New Orleans | .86 |
| 610-611 | Rockford | 1.13 | | 703 | Thibodaux | .82 |
| 612 | Rock Island | .98 | | 704 | Hammond | .78 |
| 613 | La Salle | 1.12 | | 705 | Lafayette | .86 |
| 614 | Galesburg | 1.04 | | 706 | Lake Charles | .87 |
| 615-616 | Peoria | 1.07 | | 707-708 | Baton Rouge | .86 |
| 617 | Bloomington | 1.04 | | 710-711 | Shreveport | .83 |
| 618-619 | Champaign | 1.05 | | 712 | Monroe | .81 |
| 620-622 | East St. Louis | 1.01 | | 713-714 | Alexandria | .82 |
| 623 | Quincy | 1.03 | | | | |
| 624 | Effingham | 1.04 | | **MAINE** | | |
| 625 | Decatur | 1.03 | | 039 | Kittery | .88 |
| 626-627 | Springfield | 1.04 | | 040-041 | Portland | .92 |
| 628 | Centralia | 1.01 | | 042 | Lewiston | .92 |
| 629 | Carbondale | 1.00 | | 043 | Augusta | .88 |
| | | | | 044 | Bangor | .91 |
| **INDIANA** | | | | 045 | Bath | .88 |
| 460 | Anderson | .90 | | 046 | Machias | .87 |
| 461-462 | Indianapolis | .92 | | 047 | Houlton | .88 |
| 463-464 | Gary | 1.04 | | 048 | Rockland | .87 |
| 465-466 | South Bend | .90 | | 049 | Waterville | .87 |
| 467-468 | Fort Wayne | .88 | | | | |
| 469 | Kokomo | .90 | | **MARYLAND** | | |
| 470 | Lawrenceburg | .87 | | 206 | Waldorf | .90 |
| 471 | New Albany | .87 | | 207-208 | College Park | .89 |
| 472 | Columbus | .91 | | 209 | Silver Spring | .90 |
| 473 | Muncie | .91 | | 210-212 | Baltimore | .92 |
| 474 | Bloomington | .93 | | 214 | Annapolis | .90 |
| 475 | Washington | .91 | | 215 | Cumberland | .91 |
| 476-477 | Evansville | .90 | | 216 | Easton | .87 |
| 478 | Terre Haute | .90 | | 217 | Hagerstown | .91 |
| 479 | Lafayette | .92 | | 218 | Salisbury | .85 |
| | | | | 219 | Elkton | .93 |
| **IOWA** | | | | | | |
| 500-503,509 | Des Moines | .92 | | **MASSACHUSETTS** | | |
| 504 | Mason City | .85 | | 010-011 | Springfield | 1.05 |
| 505 | Fort Dodge | .84 | | 012 | Pittsfield | 1.04 |
| 506-507 | Waterloo | .88 | | 013 | Greenfield | 1.03 |
| 508 | Creston | .88 | | 014 | Fitchburg | 1.11 |
| 510-511 | Sioux City | .89 | | 015-016 | Worcester | 1.12 |
| 512 | Sibley | .75 | | 017 | Framingham | 1.13 |
| 513 | Spencer | .76 | | 018 | Lowell | 1.14 |
| 514 | Carroll | .87 | | 019 | Lawrence | 1.14 |
| 515 | Council Bluffs | .87 | | 020-022, 024 | Boston | 1.19 |
| 516 | Shenandoah | .81 | | 023 | Brockton | 1.12 |
| 520 | Dubuque | .89 | | 025 | Buzzards Bay | 1.10 |
| 521 | Decorah | .87 | | 026 | Hyannis | 1.09 |
| 522-524 | Cedar Rapids | .94 | | 027 | New Bedford | 1.11 |
| 525 | Ottumwa | .87 | | | | |
| 526 | Burlington | .90 | | **MICHIGAN** | | |
| 527-528 | Davenport | .99 | | 480,483 | Royal Oak | .99 |
| | | | | 481 | Ann Arbor | 1.00 |
| **KANSAS** | | | | 482 | Detroit | 1.02 |
| 660-662 | Kansas City | 1.00 | | 484-485 | Flint | .93 |
| 664-666 | Topeka | .87 | | 486 | Saginaw | .90 |
| 667 | Fort Scott | .91 | | 487 | Bay City | .90 |
| 668 | Emporia | .84 | | 488-489 | Lansing | .90 |
| 669 | Belleville | .82 | | 490 | Battle Creek | .88 |
| 670-672 | Wichita | .82 | | 491 | Kalamazoo | .88 |
| 673 | Independence | .90 | | 492 | Jackson | .90 |
| 674 | Salina | .81 | | 493,495 | Grand Rapids | .88 |
| 675 | Hutchinson | .80 | | 494 | Muskegon | .86 |
| 676 | Hays | .83 | | 496 | Traverse City | .82 |
| 677 | Colby | .86 | | 497 | Gaylord | .86 |
| 678 | Dodge City | .83 | | 498-499 | Iron Mountain | .86 |
| 679 | Liberal | .82 | | | | |
| | | | | **MINNESOTA** | | |
| **KENTUCKY** | | | | 550-551 | Saint Paul | 1.06 |
| 400-402 | Louisville | .87 | | 553-555 | Minneapolis | 1.07 |
| 403-405 | Lexington | .87 | | 556-558 | Duluth | .99 |

| STATE | CITY | Residential |
|---|---|---|
| **MINNESOTA (CONT'D)** | | |
| 559 | Rochester | .99 |
| 560 | Mankato | .96 |
| 561 | Windom | .91 |
| 562 | Willmar | .93 |
| 563 | St. Cloud | 1.01 |
| 564 | Brainerd | .92 |
| 565 | Detroit Lakes | .90 |
| 566 | Bemidji | .91 |
| 567 | Thief River Falls | .90 |
| | | |
| **MISSISSIPPI** | | |
| 386 | Clarksdale | .73 |
| 387 | Greenville | .81 |
| 388 | Tupelo | .74 |
| 389 | Greenwood | .77 |
| 390-392 | Jackson | .82 |
| 393 | Meridian | .80 |
| 394 | Laurel | .78 |
| 395 | Biloxi | .84 |
| 396 | McComb | .74 |
| 397 | Columbus | .75 |
| | | |
| **MISSOURI** | | |
| 630-631 | St. Louis | 1.03 |
| 633 | Bowling Green | .97 |
| 634 | Hannibal | .95 |
| 635 | Kirksville | .91 |
| 636 | Flat River | .97 |
| 637 | Cape Girardeau | .91 |
| 638 | Sikeston | .89 |
| 639 | Poplar Bluff | .89 |
| 640-641 | Kansas City | 1.02 |
| 644-645 | St. Joseph | .97 |
| 646 | Chillicothe | .96 |
| 647 | Harrisonville | .99 |
| 648 | Joplin | .90 |
| 650-651 | Jefferson City | .93 |
| 652 | Columbia | .91 |
| 653 | Sedalia | .90 |
| 654-655 | Rolla | .96 |
| 656-658 | Springfield | .88 |
| | | |
| **MONTANA** | | |
| 590-591 | Billings | .87 |
| 592 | Wolf Point | .86 |
| 593 | Miles City | .87 |
| 594 | Great Falls | .87 |
| 595 | Havre | .83 |
| 596 | Helena | .87 |
| 597 | Butte | .84 |
| 598 | Missoula | .84 |
| 599 | Kalispell | .84 |
| | | |
| **NEBRASKA** | | |
| 680-681 | Omaha | .90 |
| 683-685 | Lincoln | .90 |
| 686 | Columbus | .89 |
| 687 | Norfolk | .89 |
| 688 | Grand Island | .87 |
| 689 | Hastings | .89 |
| 690 | McCook | .83 |
| 691 | North Platte | .86 |
| 692 | Valentine | .83 |
| 693 | Alliance | .82 |
| | | |
| **NEVADA** | | |
| 889-891 | Las Vegas | 1.02 |
| 893 | Ely | 1.00 |
| 894-895 | Reno | .90 |
| 897 | Carson City | .91 |
| 898 | Elko | .93 |
| | | |
| **NEW HAMPSHIRE** | | |
| 030 | Nashua | .98 |
| 031 | Manchester | .97 |
| 032-033 | Concord | .97 |
| 034 | Keene | .91 |
| 035 | Littleton | .94 |
| 036 | Charleston | .89 |
| 037 | Claremont | .89 |
| 038 | Portsmouth | .95 |

| STATE | CITY | Residential |
|---|---|---|
| **NEW JERSEY** | | |
| 070-071 | Newark | 1.20 |
| 072 | Elizabeth | 1.22 |
| 073 | Jersey City | 1.18 |
| 074-075 | Paterson | 1.20 |
| 076 | Hackensack | 1.19 |
| 077 | Long Branch | 1.19 |
| 078 | Dover | 1.19 |
| 079 | Summit | 1.20 |
| 080,083 | Vineland | 1.18 |
| 081 | Camden | 1.19 |
| 082,084 | Atlantic City | 1.22 |
| 085-086 | Trenton | 1.19 |
| 087 | Point Pleasant | 1.19 |
| 088-089 | New Brunswick | 1.22 |
| | | |
| **NEW MEXICO** | | |
| 870-872 | Albuquerque | .83 |
| 873 | Gallup | .82 |
| 874 | Farmington | .83 |
| 875 | Santa Fe | .84 |
| 877 | Las Vegas | .82 |
| 878 | Socorro | .82 |
| 879 | Truth/Consequences | .81 |
| 880 | Las Cruces | .82 |
| 881 | Clovis | .83 |
| 882 | Roswell | .83 |
| 883 | Carrizozo | .83 |
| 884 | Tucumcari | .84 |
| | | |
| **NEW YORK** | | |
| 100-102 | New York | 1.40 |
| 103 | Staten Island | 1.32 |
| 104 | Bronx | 1.36 |
| 105 | Mount Vernon | 1.17 |
| 106 | White Plains | 1.20 |
| 107 | Yonkers | 1.24 |
| 108 | New Rochelle | 1.21 |
| 109 | Suffern | 1.19 |
| 110 | Queens | 1.37 |
| 111 | Long Island City | 1.40 |
| 112 | Brooklyn | 1.38 |
| 113 | Flushing | 1.39 |
| 114 | Jamaica | 1.38 |
| 115,117,118 | Hicksville | 1.26 |
| 116 | Far Rockaway | 1.38 |
| 119 | Riverhead | 1.28 |
| 120-122 | Albany | 1.02 |
| 123 | Schenectady | 1.04 |
| 124 | Kingston | 1.13 |
| 125-126 | Poughkeepsie | 1.25 |
| 127 | Monticello | 1.14 |
| 128 | Glens Falls | .96 |
| 129 | Plattsburgh | 1.01 |
| 130-132 | Syracuse | .99 |
| 133-135 | Utica | .97 |
| 136 | Watertown | .95 |
| 137-139 | Binghamton | .99 |
| 140-142 | Buffalo | 1.09 |
| 143 | Niagara Falls | 1.05 |
| 144-146 | Rochester | 1.01 |
| 147 | Jamestown | .93 |
| 148-149 | Elmira | .96 |
| | | |
| **NORTH CAROLINA** | | |
| 270,272-274 | Greensboro | .94 |
| 271 | Winston-Salem | .94 |
| 275-276 | Raleigh | .93 |
| 277 | Durham | .94 |
| 278 | Rocky Mount | .92 |
| 279 | Elizabeth City | .91 |
| 280 | Gastonia | .96 |
| 281-282 | Charlotte | .94 |
| 283 | Fayetteville | .94 |
| 284 | Wilmington | .93 |
| 285 | Kinston | .92 |
| 286 | Hickory | .93 |
| 287-288 | Asheville | .94 |
| 289 | Murphy | .95 |
| | | |
| **NORTH DAKOTA** | | |
| 580-581 | Fargo | .90 |
| 582 | Grand Forks | .88 |
| 583 | Devils Lake | .90 |
| 584 | Jamestown | .88 |
| 585 | Bismarck | .89 |

297

# Location Factors - Residential

| STATE | CITY | Residential |
|---|---|---|
| **NORTH DAKOTA (CONT'D)** | | |
| 586 | Dickinson | .88 |
| 587 | Minot | .87 |
| 588 | Williston | .88 |
| | | |
| **OHIO** | | |
| 430-432 | Columbus | .91 |
| 433 | Marion | .90 |
| 434-436 | Toledo | .96 |
| 437-438 | Zanesville | .88 |
| 439 | Steubenville | .93 |
| 440 | Lorain | .94 |
| 441 | Cleveland | .97 |
| 442-443 | Akron | .96 |
| 444-445 | Youngstown | .93 |
| 446-447 | Canton | .92 |
| 448-449 | Mansfield | .90 |
| 450 | Hamilton | .89 |
| 451-452 | Cincinnati | .91 |
| 453-454 | Dayton | .89 |
| 455 | Springfield | .90 |
| 456 | Chillicothe | .93 |
| 457 | Athens | .91 |
| 458 | Lima | .89 |
| | | |
| **OKLAHOMA** | | |
| 730-731 | Oklahoma City | .81 |
| 734 | Ardmore | .82 |
| 735 | Lawton | .83 |
| 736 | Clinton | .82 |
| 737 | Enid | .83 |
| 738 | Woodward | .78 |
| 739 | Guymon | .82 |
| 740-741 | Tulsa | .82 |
| 743 | Miami | .80 |
| 744 | Muskogee | .80 |
| 745 | McAlester | .74 |
| 746 | Ponca City | .80 |
| 747 | Durant | .79 |
| 748 | Shawnee | .79 |
| 749 | Poteau | .79 |
| | | |
| **OREGON** | | |
| 970-972 | Portland | 1.00 |
| 973 | Salem | .99 |
| 974 | Eugene | 1.00 |
| 975 | Medford | .98 |
| 976 | Klamath Falls | .99 |
| 977 | Bend | 1.01 |
| 978 | Pendleton | .99 |
| 979 | Vale | .97 |
| | | |
| **PENNSYLVANIA** | | |
| 150-152 | Pittsburgh | 1.01 |
| 153 | Washington | .96 |
| 154 | Uniontown | .93 |
| 155 | Bedford | .90 |
| 156 | Greensburg | .96 |
| 157 | Indiana | .94 |
| 158 | Dubois | .91 |
| 159 | Johnstown | .91 |
| 160 | Butler | .92 |
| 161 | New Castle | .93 |
| 162 | Kittanning | .93 |
| 163 | Oil City | .93 |
| 164-165 | Erie | .94 |
| 166 | Altoona | .90 |
| 167 | Bradford | .96 |
| 168 | State College | .92 |
| 169 | Wellsboro | .92 |
| 170-171 | Harrisburg | .95 |
| 172 | Chambersburg | .89 |
| 173-174 | York | .91 |
| 175-176 | Lancaster | .94 |
| 177 | Williamsport | .90 |
| 178 | Sunbury | .92 |
| 179 | Pottsville | .92 |
| 180 | Lehigh Valley | 1.06 |
| 181 | Allentown | 1.03 |
| 182 | Hazleton | .91 |
| 183 | Stroudsburg | .96 |
| 184-185 | Scranton | .97 |
| 186-187 | Wilkes-Barre | .94 |
| 188 | Montrose | .93 |
| 189 | Doylestown | 1.07 |

| STATE | CITY | Residential |
|---|---|---|
| **PENNSYLVANIA (CONT'D)** | | |
| 190-191 | Philadelphia | 1.17 |
| 193 | Westchester | 1.09 |
| 194 | Norristown | 1.09 |
| 195-196 | Reading | .98 |
| | | |
| **PUERTO RICO** | | |
| 009 | San Juan | .77 |
| | | |
| **RHODE ISLAND** | | |
| 028 | Newport | 1.07 |
| 029 | Providence | 1.08 |
| | | |
| **SOUTH CAROLINA** | | |
| 290-292 | Columbia | .95 |
| 293 | Spartanburg | .95 |
| 294 | Charleston | .96 |
| 295 | Florence | .92 |
| 296 | Greenville | .94 |
| 297 | Rock Hill | .94 |
| 298 | Aiken | .95 |
| 299 | Beaufort | .85 |
| | | |
| **SOUTH DAKOTA** | | |
| 570-571 | Sioux Falls | .87 |
| 572 | Watertown | .76 |
| 573 | Mitchell | .78 |
| 574 | Aberdeen | .79 |
| 575 | Pierre | .83 |
| 576 | Mobridge | .78 |
| 577 | Rapid City | .86 |
| | | |
| **TENNESSEE** | | |
| 370-372 | Nashville | .83 |
| 373-374 | Chattanooga | .83 |
| 375,380-381 | Memphis | .84 |
| 376 | Johnson City | .76 |
| 377-379 | Knoxville | .80 |
| 382 | McKenzie | .72 |
| 383 | Jackson | .75 |
| 384 | Columbia | .79 |
| 385 | Cookeville | .70 |
| | | |
| **TEXAS** | | |
| 750 | McKinney | .80 |
| 751 | Waxahachie | .80 |
| 752-753 | Dallas | .83 |
| 754 | Greenville | .80 |
| 755 | Texarkana | .82 |
| 756 | Longview | .80 |
| 757 | Tyler | .82 |
| 758 | Palestine | .78 |
| 759 | Lufkin | .79 |
| 760-761 | Fort Worth | .82 |
| 762 | Denton | .84 |
| 763 | Wichita Falls | .84 |
| 764 | Eastland | .82 |
| 765 | Temple | .80 |
| 766-767 | Waco | .84 |
| 768 | Brownwood | .80 |
| 769 | San Angelo | .78 |
| 770-772 | Houston | .83 |
| 773 | Huntsville | .80 |
| 774 | Wharton | .80 |
| 775 | Galveston | .82 |
| 776-777 | Beaumont | .85 |
| 778 | Bryan | .78 |
| 779 | Victoria | .81 |
| 780 | Laredo | .80 |
| 781-782 | San Antonio | .80 |
| 783-784 | Corpus Christi | .83 |
| 785 | McAllen | .84 |
| 786-787 | Austin | .79 |
| 788 | Del Rio | .81 |
| 789 | Giddings | .80 |
| 790-791 | Amarillo | .79 |
| 792 | Childress | .81 |
| 793-794 | Lubbock | .80 |
| 795-796 | Abilene | .82 |
| 797 | Midland | .84 |
| 798-799,885 | El Paso | .79 |
| | | |
| **UTAH** | | |
| 840-841 | Salt Lake City | .84 |
| 842,844 | Ogden | .82 |
| 843 | Logan | .83 |

# Location Factors - Residential

| STATE | CITY | Residential |
|---|---|---|
| **UTAH (CONT'D)** | | |
| 845 | Price | .84 |
| 846-847 | Provo | .83 |
| | | |
| **VERMONT** | | |
| 050 | White River Jct. | .96 |
| 051 | Bellows Falls | 1.01 |
| 052 | Bennington | 1.00 |
| 053 | Brattleboro | 1.01 |
| 054 | Burlington | .93 |
| 056 | Montpelier | .99 |
| 057 | Rutland | .92 |
| 058 | St. Johnsbury | .97 |
| 059 | Guildhall | .97 |
| | | |
| **VIRGINIA** | | |
| 220-221 | Fairfax | 1.01 |
| 222 | Arlington | 1.03 |
| 223 | Alexandria | 1.04 |
| 224-225 | Fredericksburg | 1.00 |
| 226 | Winchester | 1.01 |
| 227 | Culpeper | 1.00 |
| 228 | Harrisonburg | .86 |
| 229 | Charlottesville | .88 |
| 230-232 | Richmond | 1.03 |
| 233-235 | Norfolk | .95 |
| 236 | Newport News | .95 |
| 237 | Portsmouth | .87 |
| 238 | Petersburg | .95 |
| 239 | Farmville | .85 |
| 240-241 | Roanoke | 1.00 |
| 242 | Bristol | .91 |
| 243 | Pulaski | .83 |
| 244 | Staunton | .86 |
| 245 | Lynchburg | .98 |
| 246 | Grundy | .81 |
| | | |
| **WASHINGTON** | | |
| 980-981,987 | Seattle | 1.05 |
| 982 | Everett | 1.06 |
| 983-984 | Tacoma | 1.03 |
| 985 | Olympia | 1.01 |
| 986 | Vancouver | .98 |
| 988 | Wenatchee | .95 |
| 989 | Yakima | 1.00 |
| 990-992 | Spokane | .97 |
| 993 | Richland | .97 |
| 994 | Clarkston | .94 |
| | | |
| **WEST VIRGINIA** | | |
| 247-248 | Bluefield | .93 |
| 249 | Lewisburg | .92 |
| 250-253 | Charleston | .95 |
| 254 | Martinsburg | .89 |
| 255-257 | Huntington | .97 |
| 258-259 | Beckley | .93 |
| 260 | Wheeling | .93 |
| 261 | Parkersburg | .91 |
| 262 | Buckhannon | .92 |
| 263-264 | Clarksburg | .92 |
| 265 | Morgantown | .92 |
| 266 | Gassaway | .92 |
| 267 | Romney | .89 |
| 268 | Petersburg | .89 |
| | | |
| **WISCONSIN** | | |
| 530,532 | Milwaukee | 1.05 |
| 531 | Kenosha | 1.05 |
| 534 | Racine | 1.04 |
| 535 | Beloit | .98 |
| 537 | Madison | .98 |
| 538 | Lancaster | .97 |
| 539 | Portage | .94 |
| 540 | New Richmond | .95 |
| 541-543 | Green Bay | 1.03 |
| 544 | Wausau | .96 |
| 545 | Rhinelander | .93 |
| 546 | La Crosse | .97 |
| 547 | Eau Claire | .99 |
| 548 | Superior | .95 |
| 549 | Oshkosh | .93 |
| | | |
| **WYOMING** | | |
| 820 | Cheyenne | .86 |
| 821 | Yellowstone Nat. Pk. | .83 |
| 822 | Wheatland | .79 |

| STATE | CITY | Residential |
|---|---|---|
| **WYOMING (CONT'D)** | | |
| 823 | Rawlins | .86 |
| 824 | Worland | .83 |
| 825 | Riverton | .82 |
| 826 | Casper | .83 |
| 827 | Newcastle | .85 |
| 828 | Sheridan | .86 |
| 829-831 | Rock Springs | .88 |
| | | |
| **CANADIAN FACTORS (reflect Canadian currency)** | | |
| **ALBERTA** | | |
| | Calgary | 1.07 |
| | Edmonton | 1.07 |
| | Fort McMurray | 1.10 |
| | Lethbridge | 1.07 |
| | Lloydminster | 1.02 |
| | Medicine Hat | 1.03 |
| | Red Deer | 1.03 |
| | | |
| **BRITISH COLUMBIA** | | |
| | Kamloops | 1.00 |
| | Prince George | .99 |
| | Vancouver | 1.01 |
| | Victoria | 1.01 |
| | | |
| **MANITOBA** | | |
| | Brandon | 1.07 |
| | Portage la Prairie | .97 |
| | Winnipeg | .95 |
| | | |
| **NEW BRUNSWICK** | | |
| | Bathurst | .89 |
| | Dalhousie | .90 |
| | Fredericton | .94 |
| | Moncton | .91 |
| | Newcastle | .90 |
| | St. John | .99 |
| | | |
| **NEWFOUNDLAND** | | |
| | Corner Brook | 1.06 |
| | St. Johns | 1.04 |
| | | |
| **NORTHWEST TERRITORIES** | | |
| | Yellowknife | 1.12 |
| | | |
| **NOVA SCOTIA** | | |
| | Bridgewater | .93 |
| | Dartmouth | 1.02 |
| | Halifax | 1.01 |
| | New Glasgow | 1.01 |
| | Sydney | 1.00 |
| | Truro | .93 |
| | Yarmouth | 1.01 |
| | | |
| **ONTARIO** | | |
| | Barrie | 1.10 |
| | Brantford | 1.09 |
| | Cornwall | 1.08 |
| | Hamilton | 1.06 |
| | Kingston | 1.08 |
| | Kitchener | 1.03 |
| | London | 1.05 |
| | North Bay | 1.16 |
| | Oshawa | 1.06 |
| | Ottawa | 1.06 |
| | Owen Sound | 1.09 |
| | Peterborough | 1.07 |
| | Sarnia | 1.10 |
| | Sault Ste. Marie | 1.04 |
| | St. Catharines | 1.04 |
| | Sudbury | 1.02 |
| | Thunder Bay | 1.07 |
| | Timmins | 1.06 |
| | Toronto | 1.08 |
| | Windsor | 1.06 |
| | | |
| **PRINCE EDWARD ISLAND** | | |
| | Charlottetown | .89 |
| | Summerside | .95 |
| | | |
| **QUEBEC** | | |
| | Cap-de-la-Madeleine | 1.07 |
| | Charlesbourg | 1.07 |
| | Chicoutimi | 1.10 |
| | Gatineau | 1.06 |
| | Granby | 1.06 |

**For customer support on your Contractor's Pricing Guide: Residential Repair & Remodeling, call 800.448.8182.**

# Location Factors - Residential

| STATE | CITY | Residential |
|---|---|---|
| QUEBEC (CONT'D) | | |
| | Hull | 1.06 |
| | Joliette | 1.07 |
| | Laval | 1.06 |
| | Montreal | 1.05 |
| | Quebec City | 1.06 |
| | Rimouski | 1.10 |
| | Rouyn-Noranda | 1.06 |
| | Saint-Hyacinthe | 1.06 |
| | Sherbrooke | 1.06 |
| | Sorel | 1.07 |
| | Saint-Jerome | 1.06 |
| | Trois-Rivieres | 1.17 |
| SASKATCHEWAN | | |
| | Moose Jaw | .89 |
| | Prince Albert | .88 |
| | Regina | 1.08 |
| | Saskatoon | 1.04 |
| YUKON | | |
| | Whitehorse | 1.03 |

**For customer support on your Contractor's Pricing Guide: Residential Repair & Remodeling, call 800.448.8182.**

# Abbreviations

| | |
|---|---|
| | Area |
| M | American Society for Testing and Materials |
| | Board feet |
| | Carpenter |
| | Cubic feet |
| | Composite wood joist |
| | Cubic yard |
| | Each |
| p. | Equipment |
| | Exposure |
| | Exterior |
| | Fahrenheit |
| | Foot, feet |
| | Gallon |
| | Hour |
| | Inch, inches |
| | Installation |
| | Interior |
| | Pound |
| | Linear feet |
| | Laminated veneer lumber |
| | Material |
| | Maximum |
| | Thousand board feet |
| | Thousand feet board measure |
| | Thousand square feet |
| | Minimum |
| | On center |
| | Overhead and profit |
| | Open web wood joist |
| | Ounce |
| | Pair |
| | Quantity |
| | Square foot |
| | Square, 100 square feet |
| | Square yard |
| | Vertical linear feet |
| | Foot, feet |
| | Inch, inches |
| | Degrees |

**For customer support on your Contractor's Pricing Guide: Residential Repair & Remodeling, call 800.448.8182.**

For more information visit our website at www.RSMeans.com

# Our Online Estimating Solution
## Competitive Cost Estimates Made Easy

Our online estimating solution is a web-based service that provides accurate and up-to-date cost information to help you build competitive estimates or budgets in less time.

Quick, intuitive, easy to use and automatically updated, you'll gain instant access to hundreds of thousands of material, labor, and equipment costs from RSMeans' comprehensive database, delivering the information you need to build budgets and competitive estimates every time.

With our online estimating solutions, you can perform quick searches to locate specific costs and adjust costs to reflect prices in your geographic area. Tag and store your favorites for fast access to your frequently used line item and assemblies and clone estimates to save time. System notifications will alert you as updated data becomes available. This data is automatically updated throughout the year.

Our visual, interactive estimating features help you create, manage, save and share estimates with ease! You'll enjoy increased flexibility with customizable advanced reports. Easily edit custom report templates and import you company logo onto your estimates.

| | Core | Advanced | Complete |
|---|:---:|:---:|:---:|
| Unit Prices | ✔ | ✔ | ✔ |
| Assemblies | ⊘ | ✔ | ✔ |
| Sq Foot Models | ⊘ | ⊘ | ✔ |
| Editable Sq Foot Models | ⊘ | ⊘ | ✔ |
| Editable Assembly Components | ⊘ | ✔ | ✔ |
| Custom Cost Data | ⊘ | ✔ | ✔ |
| User Defined Components | ⊘ | ✔ | ✔ |
| Advanced Reporting & Customization | ⊘ | ✔ | ✔ |
| Union Labor Type | ✔ | ✔ | ✔ |

Continue to check our website at www.RSMeans.com for more product offerings.

## Estimate with Precision
Find everything you need to develop complete, accurate estimates.
- Verified costs for construction materials
- Equipment rental costs
- Crew sizing, labor hours and labor rates
- Localized costs for U.S. and Canada

## Save Time & Increase Efficiency
Make cost estimating and calculating faster and easier than ever with secure, online estimating tools.
- Quickly locate costs in the searchable database
- Create estimates in minutes with RSMeans cost lines
- Tag and store favorites for fast access to frequently used items

## Improve Planning & Decision-Making
Back your estimates with complete, accurate and up-to-date cost data for informed business decisions.
- Verify construction costs from third parties
- Check validity of subcontractor proposals
- Evaluate material and assembly alternatives

## Increase Profits
Use our online estimating solution to estimate project quickly and accurately, so you can gain an edge over your competition.
- Create accurate and competitive bids
- Minimize the risk of cost overruns
- Reduce variability